Crystallography of Supramolecular Compounds

NATO ASI Series

Advanced Science Institutes Series

A Series presenting the results of activities sponsored by the NATO Science Committee, which aims at the dissemination of advanced scientific and technological knowledge, with a view to strengthening links between scientific communities.

The Series is published by an international board of publishers in conjunction with the NATO Scientific Affairs Division

A Life Sciences **B Physics**	Plenum Publishing Corporation London and New York
C Mathematical and Physical Sciences **D Behavioural and Social Sciences** **E Applied Sciences**	Kluwer Academic Publishers Dordrecht, Boston and London
F Computer and Systems Sciences **G Ecological Sciences** **H Cell Biology** **I Global Environmental Change**	Springer-Verlag Berlin, Heidelberg, New York, London, Paris and Tokyo

PARTNERSHIP SUB-SERIES

1. Disarmament Technologies	Kluwer Academic Publishers
2. Environment	Springer-Verlag / Kluwer Academic Publishers
3. High Technology	Kluwer Academic Publishers
4. Science and Technology Policy	Kluwer Academic Publishers
5. Computer Networking	Kluwer Academic Publishers

The Partnership Sub-Series incorporates activities undertaken in collaboration with NATO's Cooperation Partners, the countries of the CIS and Central and Eastern Europe, in Priority Areas of concern to those countries.

NATO-PCO-DATA BASE

The electronic index to the NATO ASI Series provides full bibliographical references (with keywords and/or abstracts) to more than 50000 contributions from international scientists published in all sections of the NATO ASI Series.
Access to the NATO-PCO-DATA BASE is possible in two ways:

– via online FILE 128 (NATO-PCO-DATA BASE) hosted by ESRIN,
Via Galileo Galilei, I-00044 Frascati, Italy.

– via CD-ROM "NATO-PCO-DATA BASE" with user-friendly retrieval software in English, French and German (© WTV GmbH and DATAWARE Technologies Inc. 1989).

The CD-ROM can be ordered through any member of the Board of Publishers or through NATO-PCO, Overijse, Belgium.

Series C: Mathematical and Physical Sciences – Vol. 480

Crystallography of Supramolecular Compounds

edited by

Georges Tsoucaris
CNRS UPR 180,
Université Paris-Sud,
Centre Pharmaceutique,
Châtenay-Malabry, France

Jerry L. Atwood
Department of Chemistry,
University of Missouri-Columbia,
Columbia, MO, U.S.A.

and

Janusz Lipkowski
Institute of Physical Chemistry,
Polish Academy of Sciences,
Warszawa, Poland

Kluwer Academic Publishers

Dordrecht / Boston / London

Published in cooperation with NATO Scientific Affairs Division

Proceedings of the NATO Advanced Study Institute on
Crystallography of Supramolecular Compounds
Erice, Italy
1–11 June 1995

A C.I.P. Catalogue record for this book is available from the Library of Congress

ISBN 0-7923-4051-5

Published by Kluwer Academic Publishers,
P.O. Box 17, 3300 AA Dordrecht, The Netherlands.

Kluwer Academic Publishers incorporates the publishing programmes of
D. Reidel, Martinus Nijhoff, Dr W. Junk and MTP Press.

Sold and distributed in the U.S.A. and Canada
by Kluwer Academic Publishers,
101 Philip Drive, Norwell, MA 02061, U.S.A.

In all other countries, sold and distributed
by Kluwer Academic Publishers Group,
P.O. Box 322, 3300 AH Dordrecht, The Netherlands.

Printed on acid-free paper

Printed in the Netherlands

This book contains the Proceedings of a NATO Advanced Study Institute held within the Programme of activities of the NATO Science Committee, Scientific and Environmental Affairs Division.

TABLE OF CONTENTS

APPENDIX

TUTORIALS ON MOLECULAR MODELLING

PREFACE

The Supramolecular Compounds refer to organised multi-molecular assemblies and associated phenomena. The stability and the properties of these compounds strongly involve structural three-dimensional (3D) information. The crystal itself can be considered as a giant supermolecule. Thus, a thorough understanding of crystal structures and crystal growth provides a unique information on the intermolecular interactions. Indeed, each crystal reflects in a particular way the recognition properties of molecules. More so, modern crystallography allows to study in detail two- or three-component crystalline solids in which the recognition processes can be seen from the structural standpoint. Crystallography of smaller and smaller single crystals, faster and faster experiments, time-resolved x-ray crystallography, are extremely potent source of physico-chemical information.

The present Advanced Study Institute (A.S.I.) - which was planned five years ago as the 22nd Course of the International School of Crystallography (director: T. L. BLUNDELL), 1-11 June 1995, E. Majorana Centre, Erice, Italy - is probably the first international meeting specifically devoted to the Crystallography of Supramolecular Compounds. The presence of crystallographers, chemists and physicists enhanced the coherence of the typical sequence:
Conception and Design - Synthesis - Structure and Visualisation - Properties.
The interactive and interdisciplinary character of this research is central to the development of general structural models for a large spectrum of compounds: ionophores, cryptates, fullerenes, calixarenes, cyclodextrins, cyclotriveratrylenes, pillar type compounds, zeolites, hydrates, solvates and others.

A major objective of the field is molecular recognition based on pre-organisation of molecular receptors leading to a large range of devices including switches and sensors. Further progress arises from imprinting the necessary structural information into molecular components so that they spontaneously proceed to self-organisation. Conversely, progress in molecular recognition, as well as transferring information from 2D monolayers to 3D structures, has proved to be a very useful step in further understanding of the crystalline assemblies.

Chemical reactivity is dependent on the 3D structure of the supramolecular compounds. An emerging paradigm arises with photochemical reactions in zeolites where the architecture of the host (space scale) is correlated with the dynamic behavior of the guest (time scale) and governs the reaction pathways.

Analogies with assembly principles in biological structures, even though they are not the only guidelines in the supramolecular field, still provide a strong motivation for a fruitful comparison between natural and synthetic systems, and lead to structural models of biological significance. Conversely, chemical catalysis by specifically raised antibodies, presented in the light of recent crystal structure determinations, provides a novel, supramolecular pathway towards chemical specificity by "enslaving" biological substances.

Despite the breadth of the disciplines involved in this meeting and the variety in the nature and size of the systems - from atoms to single 5000 MW molecules to high supramolecular polymers such as helicates - the importance of general characteristics has been emphasised: the supramolecular systems are governed by a set of non-covalent interactions (hydrogen bonds, electrostatic interactions, donor-acceptor bonding, solvation effects and others) with a precise, restricting geometry. Their study has been enhanced by combining the crystallographic results with information from different methodologies, including Molecular Dynamics, Thermodynamics, NMR and various Spectroscopies, Atomic Force Microscopy.

Summarising, this interdisciplinary A.S.I. has focused on general structural concepts as the basis for chemical synthesis and assembly of preconceived supramolecular compounds with increasing ability to ensure multiple functions. It has allowed further examination of fundamental questions in this rapidly evolving area, such as:
What is a supermolecule? What is considered as weak interaction? How the combination of several weak interactions leads to a definite supramolecular compound? What is the role of rigidity versus mobility, of hydrophobicity versus polarity?
An extended discussion session on the last afternoon, led by J. M. LEHN, endeavoured to provide answers as well to open new perspectives to the ever developing supramolecular domain.

The structural studies in the School have been enhanced by extensive use by the participants of modern information technologies including Data Banks and Molecular Graphic under the supervision of F. ALLEN (Cambridge), E. ELIOPOULOS (Athens) and G. WIPFF (Strasbourg), and their collaborators, O. JOHNSON, I. MAVRIDIS and A. VARNEK. We thank the Silicon Graphics Company for providing equipment and C. HENN for a scientific presentation.

It should be noted that the organisation of a variety of activities - Lectures, Discussion Sessions, Poster Sessions, Classroom Tutorials, Hands-on Tutorials - has greatly benefited from the experience and the considerable contribution of L. RIVA DI SANSEVERINO and P. SPADON, who also organised numerous events related to local archaeological, cultural and economic aspects.

The documents distributed to the Participants included a summary of each lecture, abstracts of the posters and documents for the hands-on tutorials. A chapter written by our late colleague Mario FARINA was included, with a memorandum highlighting his important contribution in our field and in our scientific community.
The Erice VACIAGO Award 1995, established to honour the memory of a widely appreciated Italian crystallographer, was delivered by J. M. LEHN to L. TOLEDO, MIT, USA, as the most dynamical young participant during scientific sessions.

The organisers express their gratitude to the NATO Science Committee for support.
In this A.S.I. the complementarity between crystallography and other methods, between chemists and physicists, between the organic and the inorganic area, appeared to be more than ever a reality and a necessity.
The organisers wish to warmly thank all participants for having reached throughout this meeting balance between competence and communication.

G. Tsoucaris
J. L. Atwood
J. Lipkowski

AN INTRODUCTION TO THE CRYSTALLOGRAPHY OF SUPRAMOLECULAR COMPOUNDS

Jerry L. Atwood
Department of Chemistry
University of Missouri-Columbia
Columbia, MO 65211

Before discussing the features of the crystallography of supramolecular compounds which set this emerging field apart from small molecule crystallography and from protein crystallography, it is appropriate to define the scope of supramolecular chemistry. This term, which was first used by Lehn [1], relates to work under the following headings:

.Chemistry Beyond the Molecule
.Chemistry of the Non-Covalent Bond
.Host-Guest Chemistry
.Molecular Recognition
.Inclusion Compounds
.Crystal Engineering
.Macrocycles
.Crown Ethers
.Cyclodextrins
.Calixarenes
.Layered Compounds
.Zeolites
.Clathrates

Of course, this list is not meant to be inclusive.

In 1996, Comprehensive Supramolecular Chemistry will be published. Insight into this field can be gained by a consideration of the titles of the ten volumes which comprise Comprehensive Supramolecular Chemistry:

.Molecular Recognition: Receptors for Cationic Guests
.Molecular Recognition: Receptors for Molecular Guests
.Cyclodextrins
.Supramolecular Reactivity and Transport: Bioorganic Systems
.Supramolecular Reactivity and Transport: Bioinorganic Systems
.Solid State Supramolecular Chemistry: Crystal Engineering
.Solid State Supramolecular Chemistry: Two- and Three-Dimensional
Inorganic Networks
.Physical Methods in Supramolecular Chemistry

G. Tsoucaris et al. (eds.), Crystallography of Supramolecular Compounds, 1–6.

.Templating, Self-Assembly and Self-Organization
.Molecular Devices and Applications of Supramolecular Technology

Having thus derived a definition based on common use of the term 'supramolecular compounds,' it is necessary to address that which is special about the crystallography of supramolecular compounds. It will be seen below that this area of crystallography does not present the same problems or expectations as does that of small molecules or proteins.

The major difficulties associated with the crystallography of supramolecular compounds may be divided into five categories.

Crystal Growth. One strives for the best possible crystals, but with large assemblies crystals are often small and/or poorly diffracting. The molecular unit, or assembly, is often organized by weak interactions, interactions without strong directionality. Thermal motion may be quite high. The clear voice of experience tells one to spend the time, within reason, to get the best possible crystalline sample. A wide range of crystallization techniques, including those utilized in the protein field, are available. However, often one is forced to proceed with data collection on a crystal of poorer quality than small molecule crystallographers normally accept.

Data Collection. If one begins with a poorly diffracting crystal, one will certainly suffer with a lack of observed data. Using conventional moving crystal, moving detector technology, it may be attractive to count each reflection for a long time in order to obtain as many 'observed' reflections as possible. This has two major disadvantages: an often overloaded diffractometer is tied up for an unacceptably long period of time and it is very difficult to gain a significant increase in the observed data. There is also a cost factor, since one pays either in funding or in access for data collection.

Structure Solution. A paucity of observed diffraction data quite often means that conventional direct methods fail to provide a solution to the problem. Many supramolecular compounds of interest do not possess heavy atoms, or, if they do, there may be too many of them (with regard to this point, making structure solution by heavy atom methods difficult). With many structures which are solved by conventional methods, there is still a time factor: the solution is not normally straightforward. Note that even for crystals which diffract well, some of the problems are beyond the reach of conventional techniques, problems with > 300 non-hydrogen atoms in the asymmetric unit.

Structure Refinement. The lack of observed data has already been recognized, as has the problem of high thermal motion. To these can often be added disorder. The non-directionality of bonding can lead to both static and dynamic disorder. (This also makes both the solution of the structure and the interpretation of the data more difficult.) The above-mentioned problems generally lead to high R values. When this author did a study of published calixarene structures some time ago, the average conventional R value was > 0.11. High R values may lead to difficulties in publication: many referees, editors, and journals now hold crystallographic structure determinations to the same standards, regardless of the compound, supramolecular or small-molecule. R values are not often discussed for protein structure determinations.

Structure Interpretation. In the view of this author, the standard for the R value should not be so rigidly enforced for supramolecular compounds. On the other hand, the standard for the interpretation of the structure should be more tightly enforced. Problems associated with a complex, complicated structure, held together by weak interactions, and displaying an array of types of disorder, lead to great difficulty in structure interpretation. To this one must add the reality that an interaction of little or no importance to one investigator may be of great interest to another.

All crystallographers realize that the problems discussed above originate in large part from initial crystal quality. If the supramolecular chemist does the best possible job in obtaining crystals, there is now new hope in the form of very sensitive area detectors, particularly those based on the CCD technology. Instruments based on this technology are capable of greatly improving the quality and quantity of diffraction data, and at a vast reduction in time for a large problem. This is illustrated in Example 2 below.

All crystallographers have files of troublesome compounds. Herein are offered two examples which are illustrative of the problems discussed above.

Example 1. Host-Guest Complexes Based on Fullerenes

Over the past four years our group, in conjunction with that of Prof. Colin L. Raston, has endeavored to prepare and study crystalline complexes of fullerenes with calixarenes and cyclotriveratrylenes (CTV, **1**). We have collected data on more than ten compounds. The longest low temperature data collection was six weeks on a CAD4. During this four-year period we have used approximately one year of single crystal diffractometer time. The only publication [2] to arise from this work is described briefly below.

The compound has the formula $(C_{60})_{1.5}(CTV)(C_7H_8)_{0.5}$, and presents the following crystal data: (298K, Enraf-Nonius CAD4 diffractometer): M = 1577.6, monoclinic, C2/m, a = 30.131(9), b = 17.436(5), c = 14.638(5) Å, β = 116.33(2)^o, U = 6892 Å^3, Z = 4, D_c = 1.53 g cm^{-3}, μ(Mo-K$_\alpha$) = 0.87 cm^{-1}, 5707 unique reflections with $2\theta_{max}$ = 50°, 1813 with $I > 3\sigma(I)$ used in the refinement. The structure was solved by direct methods (SHELX-86) and refined using alternating full matrix least squares and difference fourier synthesis (SHELX-76). The well-defined (C_{60})(CTV) complex resides across a mirror plane while the highly disordered second fullerene and toluene of crystallization are located on a 2/m site. Final R = 0.15.

Even though much time was devoted to the collection of the data (full data was collected on three different crystals), the best effort yielded only 32% observed data. Fortunately, this was sufficient for structure solution. However, much time and effort were required for such solution.

With regard to the interpretation of the structure, first note that the structure confirms the stoichiometry established by analysis: the asymmetric unit comprises 1/2 of a 1:1 (C_{60})(CTV) complex (the other half generated by a mirror plane), 1/4 of a highly disordered C_{60} residing over a *2/m* site symmetry, and possibly a 1/4 of a highly disordered toluene molecule of the same site symmetry. In the published account of the structure, the following observation was made: "While the structure is of low precision in consequence of this disorder, the (C_{60}.CTV) complex gave a meaningful refinement, clearly showing docking of the fullerene with the CTV, Figure 1." It was further noted: "The lack of disorder of this C_{60} relates to its contact with the rigid CTV molecule, and also to the matching of symmetry elements; the three-fold axis of the CTV coincides with a three-fold axis of C_{60} such that the C-9 ring of CTV lines up with a C-6 ring of the fullerene with the three immediate C-5 rings residing over the three aromatic rings of the CTV." Even at an R value of 0.15, important information was revealed by the structure. Much of the problem with this supramolecular compound doubtless results from the highly disordered C_{60} in the structure which appears to act as space filler in the packing of the ordered (C_{60}.CTV) units.

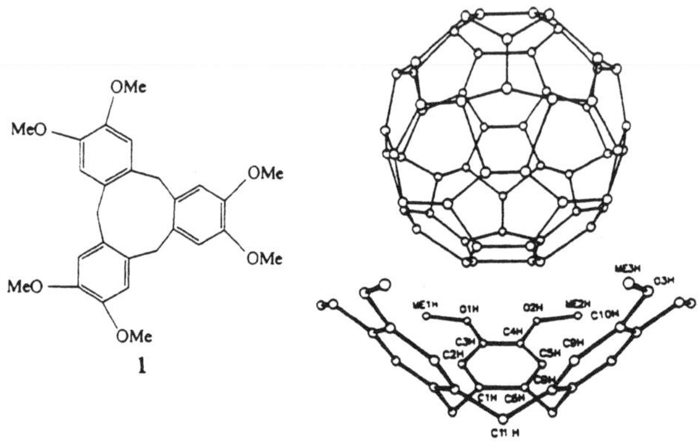

Figure 1. Ball and socket C_{60}.CTV structure (**1** = CTV).

The only structure solution of a complex of a calixarene with C_{60} is for the p-phenylcalix[4]arene. Here also, a small data set eventually afforded the structure. The C_{60} is not in the calixarene cavity, but rather is packed on the exterior, among the π-surfaces. Unfortunately, the best efforts, including data collection on a CCD-based diffractometer, has led only to R > 0.20.

Example 2. Anion Complexation by Metallated Calixarenes

Our group has experienced great success in producing a new class of anion complexation agents based on calixarenes and cyclotriveratrylenes. We have shown that the binding of transition metal ions to the carbocyclic rings of the outer face of calixarenes can markedly modify their inclusion properties, resulting in the inclusion of polar and anionic guests within the macrocyclic cavity [3,4]. Similar metallation of CTV results in the formation of new inclusion species in which the guest is held within the bowl of the CTV molecule in close proximity to a redox active transition metal center. The formulations of the compounds were confirmed by FAB mass spectrometry. The structure of a key supramolecular compound was revealed by a single crystal X-ray structure determination, Figure 2. The compound has the formula

$C_{57}H_{72}O_6Ru_3.6BF_4.9H_2O$, and the crystal data is as follows: M = 1839.36, yellow plate, cubic, Pa3, a = 25.173(2) Å, U = 15953(3) $Å^3$, Z = 8, D_c = 1.54 g cm^{-3}, μ(Mo-K_α) = 6.61 cm^{-1}. Of 5357 data collected on an Enraf-Nonius CAD4 diffractometer (2θ = 4 - 50° at 20°C) 3400 were independent and 1775 were judged observed [I > 3σ(I)]. The structure was solved and refined as for the fullerene complex in Example 1. The asymmetric unit was found to contain one third of one cationic molecule and a total of two tetrafluoroborate anions. The final difference electron density map also revealed the presence of three significant peaks corresponding to disordered solvent of crystallization, probably water. The remaining tetrafluoroborate anion was found to be disordered over three sites and, as a consequence, was extremely poorly defined. The final refinement converged with R = 0.079.

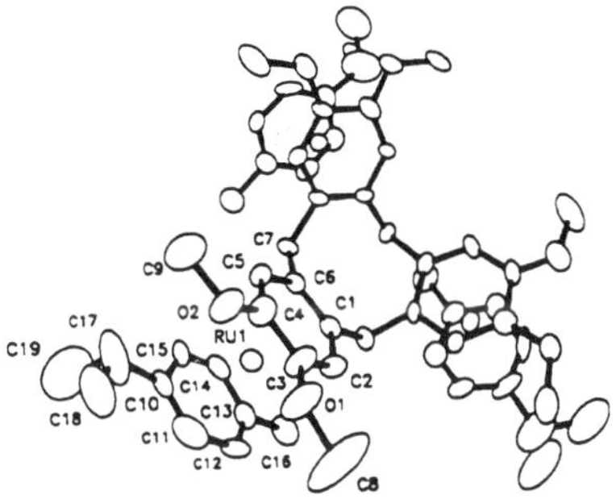

Figure 2. Cation in $[\{Ru(\eta^6\text{-}MeC_6H_4\text{-}4\text{-}CHMe_2)\}_3(\eta^6{:}\eta^6{:}\eta^6\text{-}CTV)][BF_4]_6$. The guest anion is badly disordered and is not displayed.

Within the asymmetric unit one of the two tetrafluoroborate anions is disordered over three sites; two of these sites are situated upon three-fold rotation axes while, most importantly, the final anion is distributed in a highly disordered fashion within the cavity of the CTV host. This disorder probably arises from a size mismatch between that small BF_4^- anion and the large, shallow cavity of the CTV. The structure does, however, demonstrate the principles (i) that the characteristic packing of free CTV may be disrupted such as to allow guest molecules to enter the cavity, and (ii) the electronic properties of the electron rich CTV cavity may be modified by appending of transition metal centers in order to allow the inclusion of anionic guest species, both interpretations which do not depend on precise metrical parameters.

This area of research has recently produced a success notable under this discussion. A compound similar to the CTV compound just discussed, but with a calixarene, provided only very small crystals. Indeed, no reflection data could be obtained on a CAD4 diffractometer, not even one reflection. Using a CCD-based Siemens system, 60% observed data were obtained in a 12-hour data collection. The structure was apparent at the 4 hour mark.

Supramolecular chemistry is presenting great challenges to the crystallographer. The field of supramolecular crystallography is now seen to lie somewhere between that of small molecule and protein crystallography. The supramolecular compounds are more difficult to solve than normal small molecules, but generally not so difficult as proteins. The data obtained from supramolecular compounds are not as metrically precise as for small molecules, but are much more so than for proteins.

References

1. Lehn, J.-M. (1995) Supramolecular Chemistry, Concepts and Perspectives, A Personal Account, VCH, Weinheim.

2. Steed, J. W., Junk, P. C., Atwood, J. L., Barnes, M. J., Raston, C. L. and Burkhalter, R. S. (1994) *J. Am. Chem. Soc.* **116**, 10346.

3. Steed, J. W., Juneja, R. K., Burkhalter, R. S. and Atwood, J. L. (1994) *J. Chem. Soc., Chem. Commun.*, 2205.

4. Steed, J. W., Juneja, R. K. and Atwood, J. L. (1994) *Angew. Chem. Int. Ed. Engl.* **33**, 2456.

ASSOCIATION OF HELICAL PEPTIDES AND ION CHANNELS

ISABELLA L. KARLE
Laboratory for the Structure of Matter
Naval Research Laboratory
Washington, D. C. 20375-5341

A number of naturally occurring peptides with ~9 to ~20 amino acid residues, $[\text{-NHCH(R)CO-}]_n$, where R varies from one residue to another, have been found to be physiologically active against various toxins at the surface of cells, or to function as an antibiotic by ion transport across cell membranes. Ion transport of a charged ion across a hydrophobic bilayer membrane can be accomplished by: (a) encapsulating the ion with a cyclic peptide that wraps around the ion to form a globular complex, with M^+—O ligands in the interior and a nonpolar surface consisting of the peptide side chains; or (b) by the formation of ion channels spanning the membrane by linear helical peptides, Figure 1. An example of encapsulation is provided by the cyclic dodecadepsipeptide valinomycin, cyclo[(-L-Val-D-Hyi-D-Val-L-Lac)$_3$-], that selectively transports K^+ ions across bilayer cellular membranes. Crystal structures of uncomplexed valinomycin show two quite different conformations, a collapsed bracelet conformation[2,3,4] that occurs in crystals grown from solvents such as ethanol, n-octane and acetone and a bowl shaped conformation that has nearly three-fold symmetry grown from a DMSO solution, Figure 2. The K^+ complex of valinomycin[6,7], Figure 3, has the K^+ ion surrounded by the peptide. Six K^+—O ligands are buried in the central portion of the complex while the exterior surface consists of a hydrophobic environment (isopropyl and methyl side chains) that is compatible with the hydrophobic character of the interior of a membrane. It is tempting to speculate that facile complexation can occur with the trigonal "petal" form of uncomplexed valinomycin, since all that is required mechanically is a simple closure of the three petals over the K^+ ion. Oxygens O(3), O(7) and O(11) near the bottom of the bowl, as shown in Figure 2(b) are already in position to make three K^+—O ligands with very minor movement.

Ion channel formation by helical peptides presents a much more complicated problem. Many of the linear peptides that form ion transport

G. Tsoucaris et al. (eds.), Crystallography of Supramolecular Compounds, 7–29.

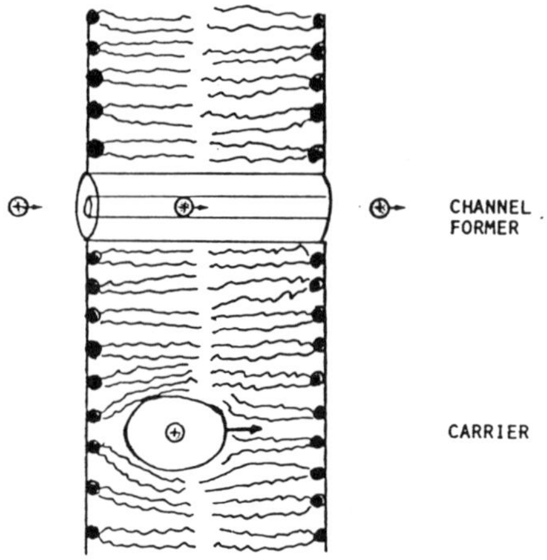

Figure 1. Schematic diagram of two modes for transport of K^+ ion through a bilayer membrane by encapsulation in a cyclic peptide and by ion channel formation.

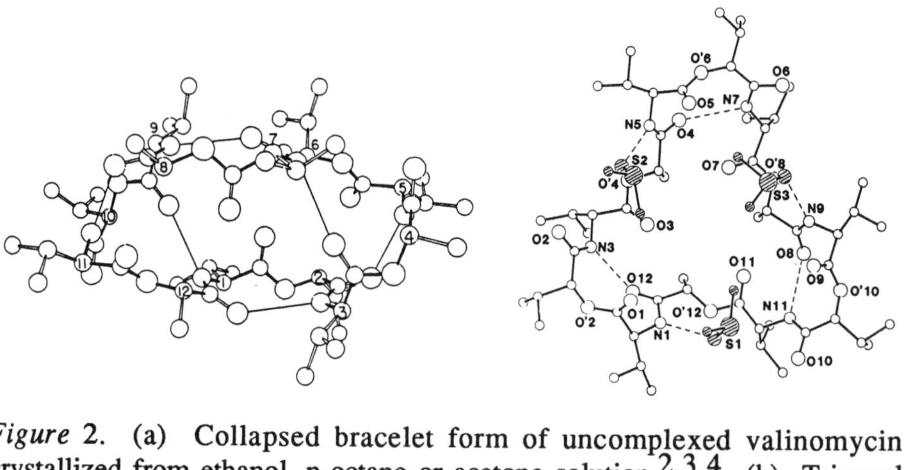

Figure 2. (a) Collapsed bracelet form of uncomplexed valinomycin crystallized from ethanol, n-octane or acetone solution.[2,3,4] (b) Trigonal "petal" form of valinomycin grown from DMSO solution.[5] The three shaded molecules are DMSO that form hydrogen bonds with the desipeptide.

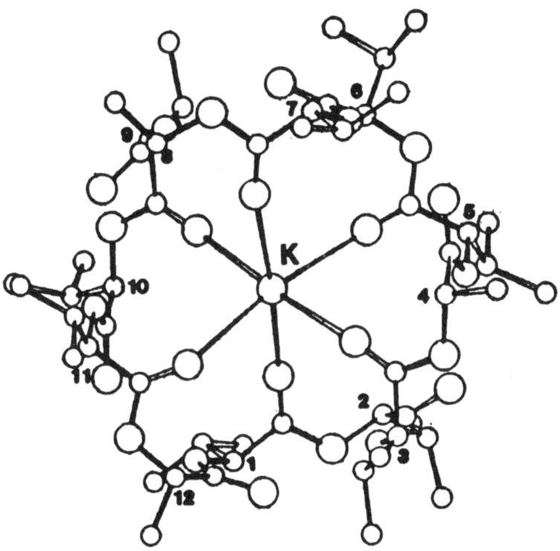

Figure 3. The K^+-complex of valinomycin.[6,7] The K^+ ion lies in the center of the peptide folded into a bracelet form and has a 6-fold coordination with the six carbonyl oxygens of the six amino acid residues.

channels are of the type that have alpha- or 3_{10}-helical conformations. An individual α- or 3_{10}-helix is tightly wound and has no interior space for a channel, Figure 4. Consequently, a number of helices are required to assemble into a bundle that forms a channel with a polar interior containing carbonyl and/or hydroxyl groups capable of forming transient M^+—O ligands as the M^+ ion passes through the channel. Furthermore the channel must be hydrophobic on the exterior to be compatible with membrane material. There have been serious questions about the form of the channel, the number if helices required to make an ion channel, the easy transition between a channel (passage of one M^+ ion at a time) and a pore (simultaneous passage of several or many M^+ ions), and other factors.[8] For these reasons, an extensive investigation of the structures and assembly of natural and synthetic crystalline peptides has been pursued. Of particular interest have been peptides with many hydrophobic residues and those containing the Aib residue (alpha aminoisobutyric acid) that promotes helix formation[9]

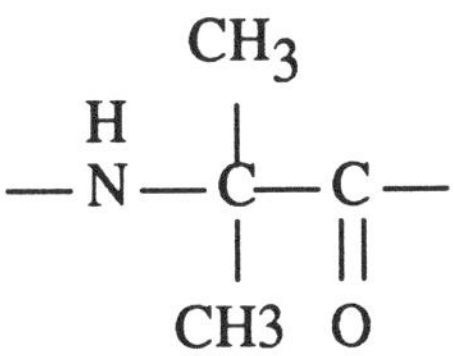

(Aib)

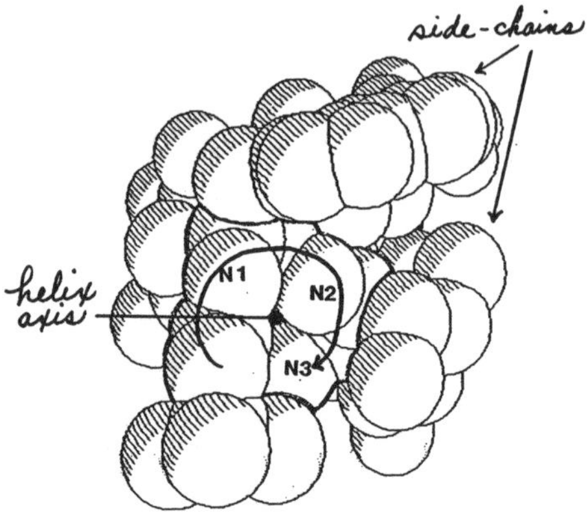

Figure 4. View into an α-helix from the top of the helix. A space filling diagram using the radii and coordinates of atoms as determined from crystal structure analyses shows that there is no space for a channel in the interior of the alpha-helix.

1. **Assembly of Apolar Helices**

1.1 HELIX FORMATION

Moderate-size peptides (up to ~20 residues) very rarely fold into helices if they contain only the usual amino acid residues. Naturally occurring peptides found in microbial/fungal sources that contain the Aib residue and synthetic peptides that contain the Aib residue almost invariably fold into an alpha- or 3_{10}-helix, Figure 5, or a mixed 3_{10}-/α-helix. The type of helix formed depends both on the total number of residues and the number of Aib residues. The alpha-helix is favored by increasing the number of residues while the 3_{10}-helix is favored by multiple inclusions of Aib residues.[9]

Recently the helix-folding properties of residues that have two side-chains of longer length have been established.

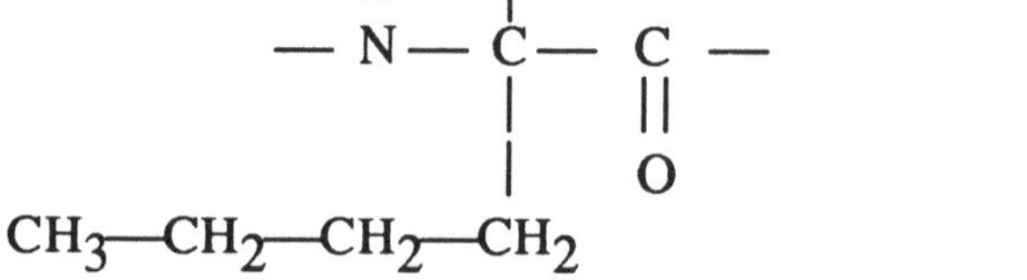

(Dbg)

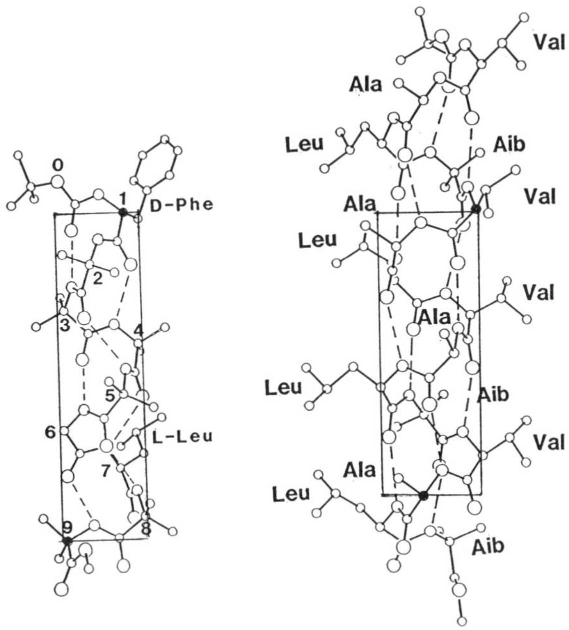

Figure 5. 3_{10}-helix on left and α-helix on right. The same number of residues are enclosed in the boxes.

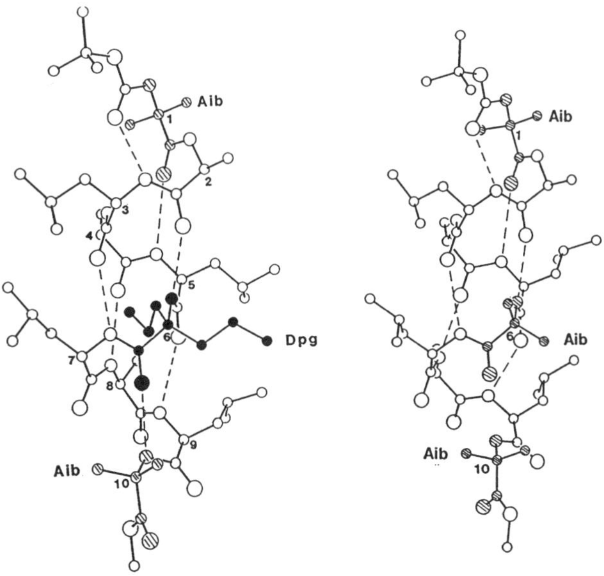

Figure 6. The replacement of Aib(6) with Dpg(6) (darkened).[10]

Earlier it had been assumed from modelling and calculations that helix formation was improbable. Synthetic peptides containing Dpg and Dbg all have folded into helices.[10,11] In fact, Dpg (dipropylglycine) can be substituted for Aib, Figure 6, without disturbing the helix or crystal packing, and having very nearly the same cell parameters. Dpg and Dbg appear to have helix forming propensities very similar to Aib. Helix forming propensities are illustrated in Figure 7 showing the conformation of Boc-Gly-Dpg-Pro-Val-Ala-Leu-Aib-Val-Ala-Leu-OMe that contain both helix breakers, Gly and Pro, and helix inducers, Aib and Dpg.

1.2 VERTICAL STACKING OF HELICES

In crystals, helical peptides form infinite columns by hydrogen bonding. A peptide backbone in a helix has free NH groups pointing upward in the first turn (the orientation used in this discussion) and free C=O groups in the last turn pointing downward; see Figure 8. Head-to-tail NH···O=C hydrogen bonds are formed when the helices are stacked over each other, either by translation or 2-fold screw rotation. The columns may be straight or somewhat zigzag. When the register between the head and the tail of two molecules is good, then three NH···O=C intermolecular hydrogen bonds are formed, and it is difficult to discern where the helix of one molecule ends and the other starts. More often there are only two, or even one, head-to-tail NH...O=C hydrogen bonds because of poor register. Usually (but by no means always) solvent molecules (water, MeOH, IprOH, etc.) span the gap by forming bridging type hydrogen bonds between the free NH and some of the free C=O moieties. In a number of structures, an NH group does not participate in any hydrogen bonding. Examples of various head-to-tail hydrogen bonding motifs are shown in Figure 8 with three direct NH···O=C bonds, Figure 9 with only one direct NH···O=C bond and numerous water molecules bridging the head NH with tail C=O moieties, and Figure 10, a rare example of no direct NH···O=C bonds but only NH···W···O=C bonds or NH···W···W···O=C bonds in the head-to-tail region.

1.3 LATERAL NEIGHBORS OF HELICES

The sides of helices composed of apolar residues are covered with hydrocarbon side chains. Except for a few cases where water (or solvent molecules) can approach the backbone atoms, the surface is composed of methyl groups or aromatic groups that preclude lateral hydrogen bond formation between helices. The various apolar side chains are not particularly selective with respect to packing except for spatial considerations. Bulges try

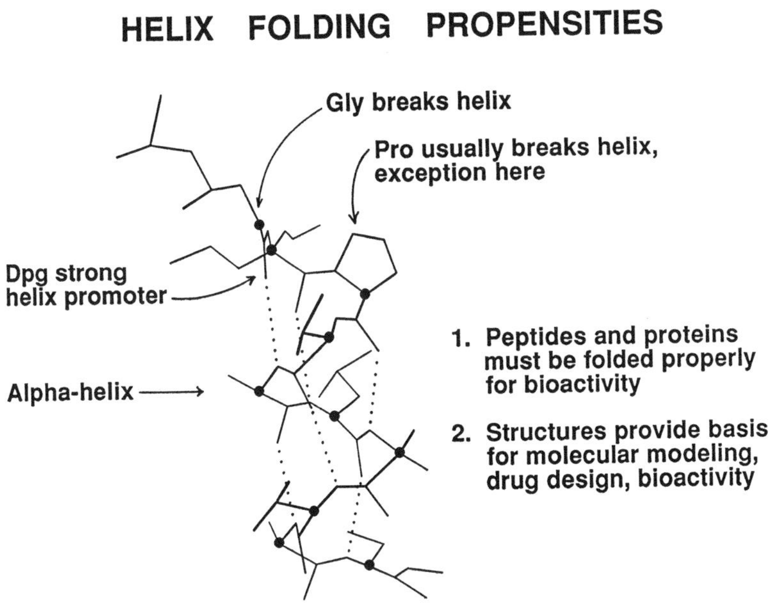

Figure 7. The conformation of Boc-Gly-Dpg-Pro-Val-Ala-Leu-Aib-Val-Ala-Leu-OMe. The dotted lines indiate NH···O=C hydrogen bonds in the alpha-helix portion.[11]

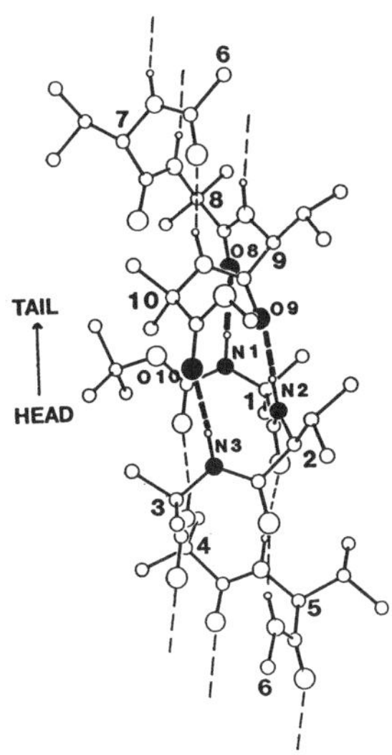

Figure 8. Three direct NH···O=C bonds (heavy dashed lines) in the head-to-tail region of Boc-Aib-Val-Aib-Val-Val-Val-Aib-Val-Aib-OMe.[12]

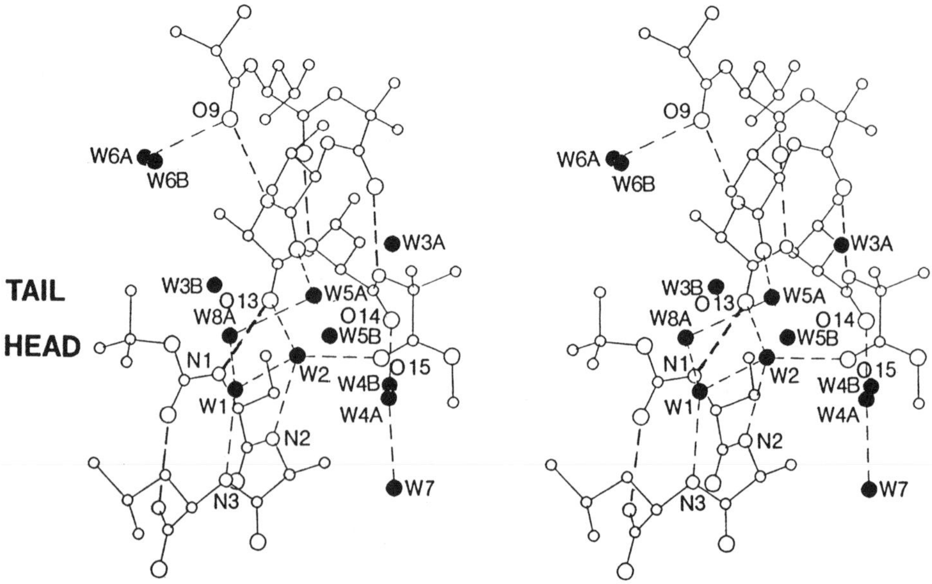

Figure 9. Stereodiagram of head-to-tail region of Boc-Val-Ala-Leu-Aib-Val-Ala-Leu-(Val-Ala-Leu-Aib)$_2$-OMe[13] with only one direct hydrogen bond, N(1)H···O(13), and many water molecules. (Those labelled with A and B are disordered between two sites.)

to fit into grooves. Figure 11 illustrates four different packing motifs found in different crystal forms of the same decapeptide. The closest intermolecular C···C approaches are between 3.8 and 4.0 Å. For most of the peptides, helical columns of peptides repeat in one lateral direction by translation and form a layer with all helix axes pointing in the same direction. Adjacent layers in a crystal assemble with axes pointed in the same direction, or with axes pointed in the opposite direction, or, in rare instances, the layers rotate so that the helix axes are skewed with respect to each other. The same or very similar peptides have been found to crystallize both in an all parallel fashion and in an antiparallel fashion in different polymorphs. Although antiparallel packing of helices prevails, packing of helices in which all helix axes point in the same direction occurs in one-third of the crystals studied.[3]

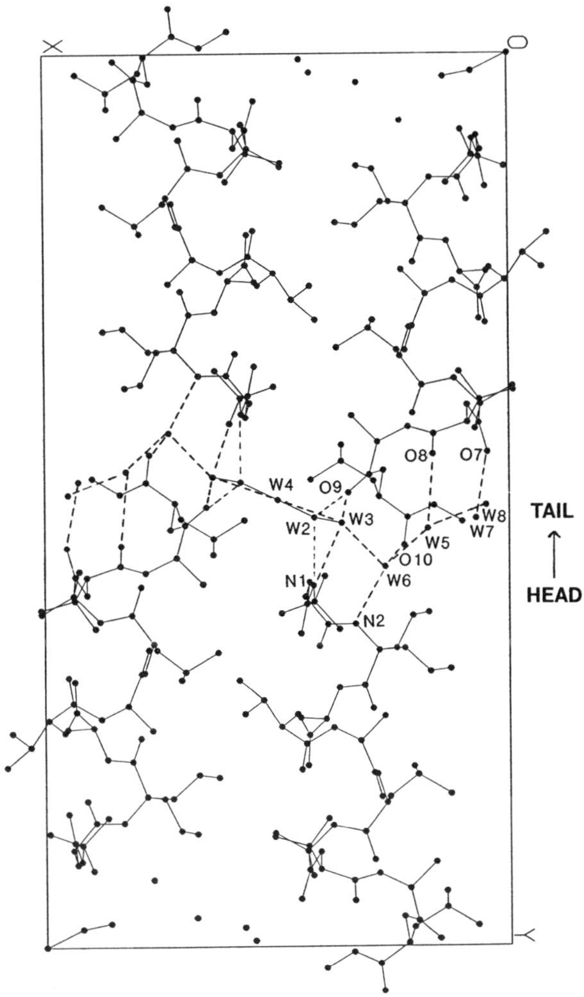

Figure 10. There are no direct NH···O=C hydrogen bonds in the head-to-tail region of Boc-Gly-Dpg-Ala-Val-Ala-Leu-Aib-Val-Ala-Leu-OMe.[11] Many water molecules bridge the head NH and tail C=O groups.

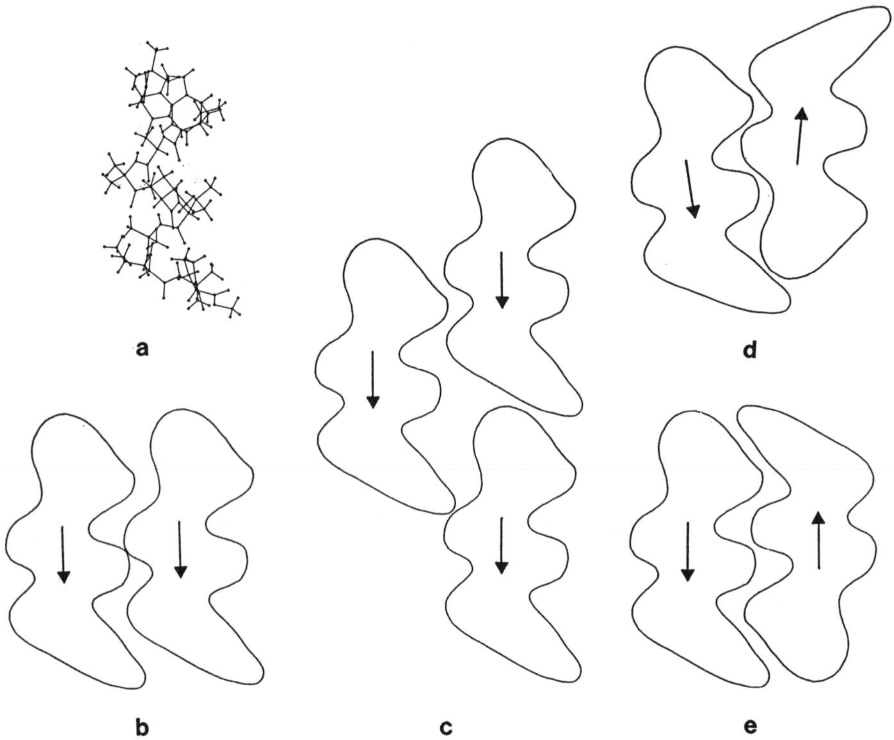

Figure 11. Antiparallel and completely parallel packing motifs found for apolar helices. Shape selection, that is, bulge fitting into groove, and van der Waals' attractions appear to be the dominating factors. (a) X-Trp-Ile-Ala-Aib-Ile-Val-Aib-Leu-Aib-Pro-OMe drawn with H atoms. (b) Envelope used to show all parallel packing in $P2_1$ with two-fold screw axis nearly coincident with helix axes. (c) All parallel packing in P1. (d) Antiparallel packing in $P2_1$ with two-fold screw horizontal. (e) Antiparallel packing with two-fold screw axis directed into plane of paper.

A special packing motif, in the form of a zipper, has been observed in a decapeptide with two very long side chains obtained by leaving blocking groups on the side chains of Glu^2 and Lys^9, separated by seven residues. The molecule assembles into a column by head-to-tail hydrogen bonds between the helical backbones and between the Glu^2 and Lys^9 side chains, resulting in a helix column with double teeth extending on one side. A pair of such columns interdigitate the double teeth in a zipper fashion, Figures 12 and 13.

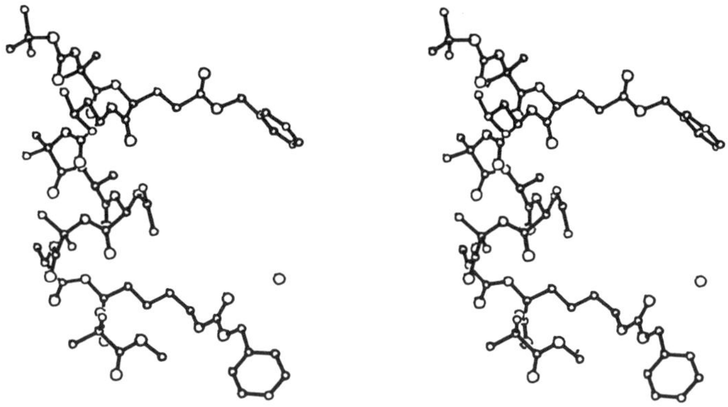

Figure 12. Stereodiagram of Boc-Aib-Glu(OBzl)-Leu-Aib-Ala-Leu-Aib-Ala-Lys(Z)-Aib-OMe; a helical backbone with two extra long side chains on the same side of the helix separated by seven residues.[14]

2.1 Polar Channel Formation

Zervamicin and related peptides[15], antibiotics isolated from spores in the soil, transport K+ ions through bilayer membranes.[8] They were presumed to form polar channels resembling the type represented schematically in Figure 14 by the aggregation of several helical molecules. The nonpolar side of each helix, in contact with the membrane, should be on the outside wall of the channel, and the polar side of the helix should be on the inside wall of the channel so that a K^+ ion can form transient K^+—O ligands with oxygens in carbonyl and hydroxyl moieties during the passage of the K^+ ion through the channel. A construction for a transmembrane ion channel is strongly suggested by the assembly of highly curved helices in the crystal structures of polymorphs of Leu-zervamicin (16-residues).[16-18] Alamethicin, a related ionophore with a curved helical backbone did not crystallize in a manner in which polar channels are formed.[19] The sequence of Leu^1-zervamicin is Ac-Leu-Ile-Gln-Iva-Ile^5-Thr-Aib-Leu-Aib-Hyp^{10}-Gln-Aib-Hyp-Aib-Pro^{15}-Phol where Phol=phenylalaninol; Iva=isovaline (α-ethylalanine); Hyp=4-hydroxyproline. The four Aib and one Iva residues are strong helix formers, whereas the two Hyp and one Pro residue tend to be helix breakers. The crystal structure shows that the molecule is helical but that the helix is bent severely at the Hyp^{10} residue, Figure 15. Further, four of the five polar residues, Gln^3, Thr^6, Hyp^{10}, Gln^{11} and Hyp^{13} are separated by 3 or 4 residues (3,6,10,13)

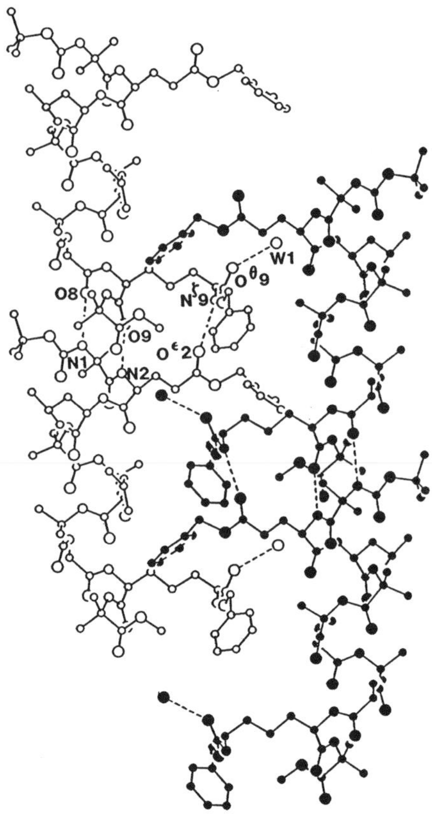

Figure 13. Vertical columns of helices (see *Figure* 12) (white molecules on the left and black molecules on the right), are formed by head-to-tail hydrogen bonding. A zipper motif is formed by interdigitating pairs of long, extended side chains in a double-tooth fashion. There are no hydrogen bonds between the white stack and the black stack.[14]

which is an appropriate separation to place them all on one side of a helix, giving the convex side a polar character while the concave side is hydrophobic due to the side chains of Leu, Ile, Iva and Pro. The one apparent inconsistency is Gln^{11} for which the C^{α} and C^{β} atoms occur on the hydrophobic side. However, the side chain of Gln^{11} is not extended away from the helical backbone, like the Leu and Ile side-chains, but is folded back around the backbone so that its $-CONH_2$ terminus is directed to the polar side of the helix.

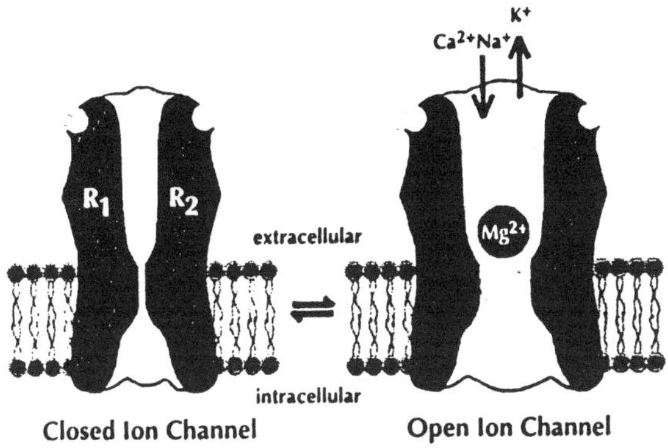

Figure 14. A schematic representation of an ion channel through a membrane often used in the literature. A circular cross-section is assumed and the unspecified number of peptides forming the channel is larger in the open state than in the closed state.

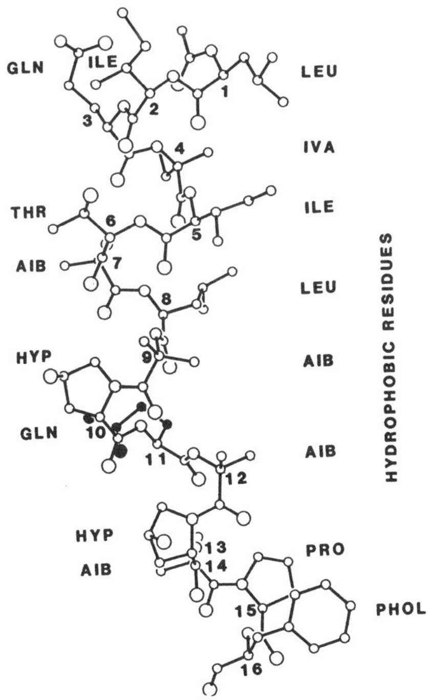

Figure 15. The bent, helical conformation of Leu^1-zervamicin. The polar face is on the left. The C^{α} atoms are labelled 1-16. The side chain of Gln^{11} (darkened) is not extended but folded back toward the polar face.

Three crystalline polymorphs were grown from methanol/water and ethanol/water solutions. Their cell parameters are shown Table 1. It is interesting to note that crystals A and B were found side-by-side in the same crystallizing vessel. In each of the crystal forms, the molecules had very

TABLE 1. Crystal parameters for three polymorphs

Parameter	Crystal A	Crystal B	Crystal C
space group	$P2_1$	$P2_1$	$P2_12_12_1$
Cell dimensions			
a, Å	23.068(6)	21.857(4)	10.337(2)
b, Å	9.162(3)	9.381(3)	28.389(7)
c, Å	26.727(9)	26.744(6)	39.864(11)
β, degrees	108.69(2)	105.22(2)	90
Growth solvent	CH_3OH/H_2O	CH_3OH/H_2O	CH_3CH_2OH

similar conformations except for the degree of bending of the helix axis.[16,18] Figure 16 shows a least-squares comparison of molecules A/B and C/B. Molecule C has the greatest bend with a value ~47°. There are no abrupt changes in torsional angles at any particular residue but small changes occur in a number of adjacent residues. In each crystal, very similar hour-glass shape channels are formed by the polar faces of the peptides that associate in an antiparallel manner. The channels differ in size according to the degree of bend in the helices and consequently the amount of water in each channel varies depending upon the size of the channel. A comparison of the water channels is shown in Figure 17 for the smallest, in crystal B, and the largest in crystal C. The water molecules in B are at full occupancy and form hydrogen bonds with carbonyl and hydroxyl oxygen atoms of the peptide as well as a few water-water hydrogen bonds. In C, there are many more water sites, most of them at partial occupancy, and many more water-water hydrogen bonds.

The hour-glass shaped channels are closed in the middle by the hydrogen bonds from $N^{\varepsilon}H_2$ of Gln^{11} of one molecule to O of Thr^6 and O^{δ} of Hyp^{10} of another molecule in crystals A and B. Crystal C has two conformations for the side chain of Gln^{11}. The major conformation is the same as in A and B. It is called the “closed” position in Figure 18. The minor conformation formed by a simple rotation about the C^{β}-C^{γ} bond is called the “open” position. In the “open” position, the $N^{\varepsilon}H_2$ moiety forms alternate hydrogen bonds in directions away from the channel.

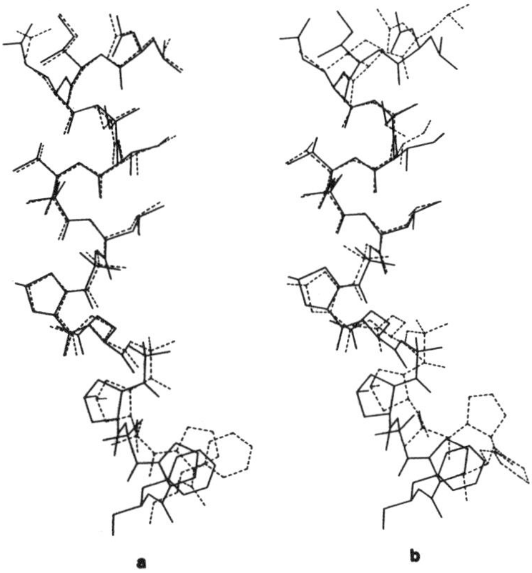

Figure 16. Comparison of Leu[1]-zervamicin in three crystal forms by least-squares fit of residues 3-9. (a) Comparison of crystal B conformation[17] (solid line) with crystal A conformation.[16] The helix is bent ~26° in B, ~36° in A. (b) Comparison of crystal B conformation (solid line) with crystal C conformation.[18] The helix is bent ~47° in C.

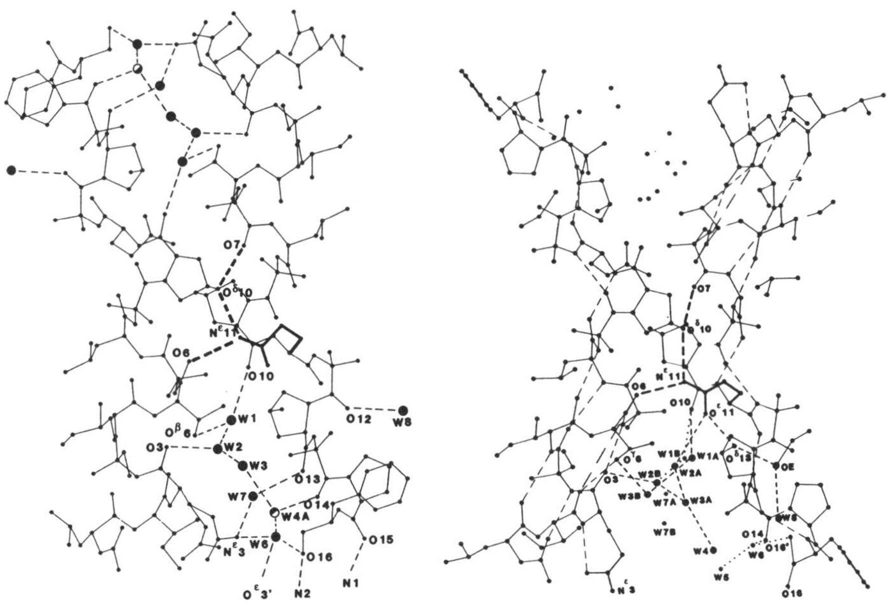

Figure 17. Interrupted water channels formed by antiparallel Leu-zervamicin molecules. (a) Narrow water channel in crystal B. (b) Wide water channel in crystal C.

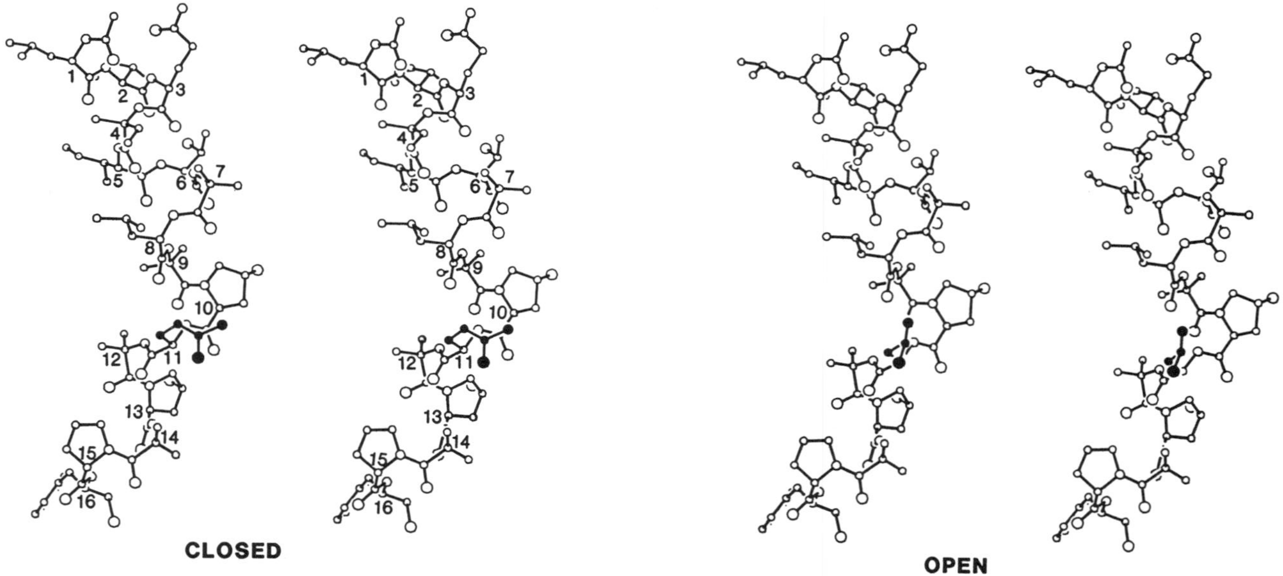

Figure 18. Stereodiagram of Leu^{1}-zervamicin (crystal C). Left: Gln^{11} side chain (darkened) is in the "closed" position (major conformation). Note that the Gln^{11} side chain has reversed direction in order to augment the polar convex face. The concave face is totally hydrophobic. Right: Gln^{11} side chain is in the "open" position (minor conformation).

The formation of a polar channel in crystal A, B and C and the possibility of opening and closing the channel in its narrow part led to an exercise that was performed in which a K^+ ion was guided through the static polar channel as it exists in the crystal. The channel is formed by three neighboring peptide molecules, two adjacent parallel peptides related simply by translation, and a third peptide antiparallel to the first two. It was assumed that K^+—O ligands are near 2.8 Å, that the K^+ ion could easily interchange any water of hydration for water of crystallization or for carbonyl O or OH groups. It was assumed that the K^+ ion could arrive at any water site by replacing the water or pushing it along the water channel. The approximately "best" x and z coordinates for the K^+ ion at various fixed y coordinates (corresponding to height along the channel) were found in a trial and error procedure by calculating the distances from the K^+ ion to all the nearest atoms lining the channel. A number of false paths were started that come to an end because the channel became too narrow at some point.

A successful path was found by placing the K^+ ion in the site for W(2A), Figure 17. In this location the K^+ ion could make ligands with $O^{\varepsilon}(11)$, W(1A), and W(3A). At this point the K^+ ion cannot travel any further along the channel unless the Gln^{11} side chain rotates to the open position, presumably with an applied potential. The "gate" must be in the open position as the K^+ ion travels from position A to E, Figure 19. Near position E the gate must close to allow the K^+ ion to travel from position E to J, after which point it will presumably stay closed, although the open or closed position is immaterial to the passage of the ion at these levels.[18]

The path described above required one opening of the Gln^{11} gate per passage of K^+ from one side to the other of the channel. The path is not through a straight pore, but is circuitous. All along the path, there are carbonyl O atoms or OH groups at the proper distance to make transient K^+—O ligands, but usually no more than two at any level. Atoms $O^{\varepsilon}(10)$, $O^{\gamma}(6)$, O(6), and O(7) appear most often at ligand distances for K^+. The ion path is offset to one side from an imaginary central vertical line. An exhaustive search of the available space in the channel at the constriction was not made, only a sampling. Other paths may be possible. Space that appears too narrow in a static system may increase in a dynamic system, as when a channel is formed in a membrane.

The analysis of the voltage and concentration dependence of macroscopic conductances for zervamicin IIB[20] suggested that, on average,

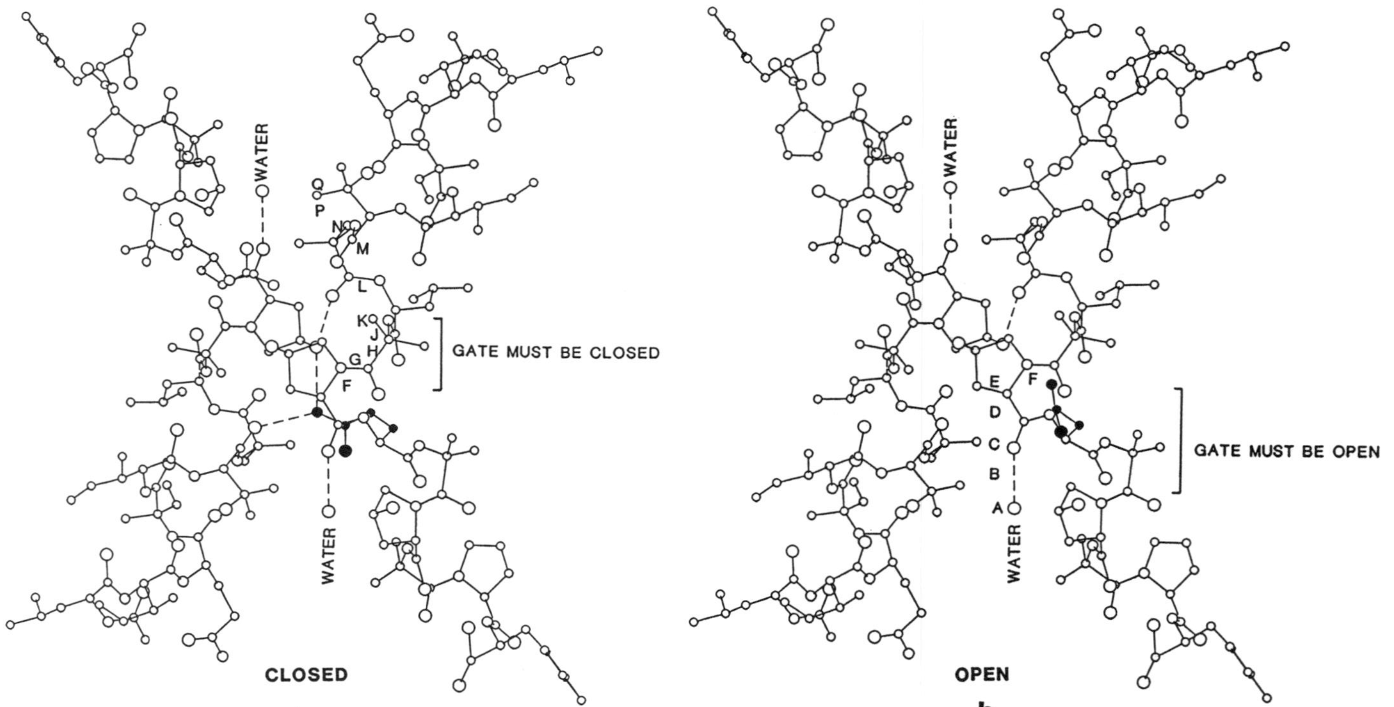

Figure 19. Interrupted channel formed by zervamicin in a crystal, possible mechanism for gating, and possible path for a K^+ ion. (Calculated positions A to Q for hypothetical passage of ion.)

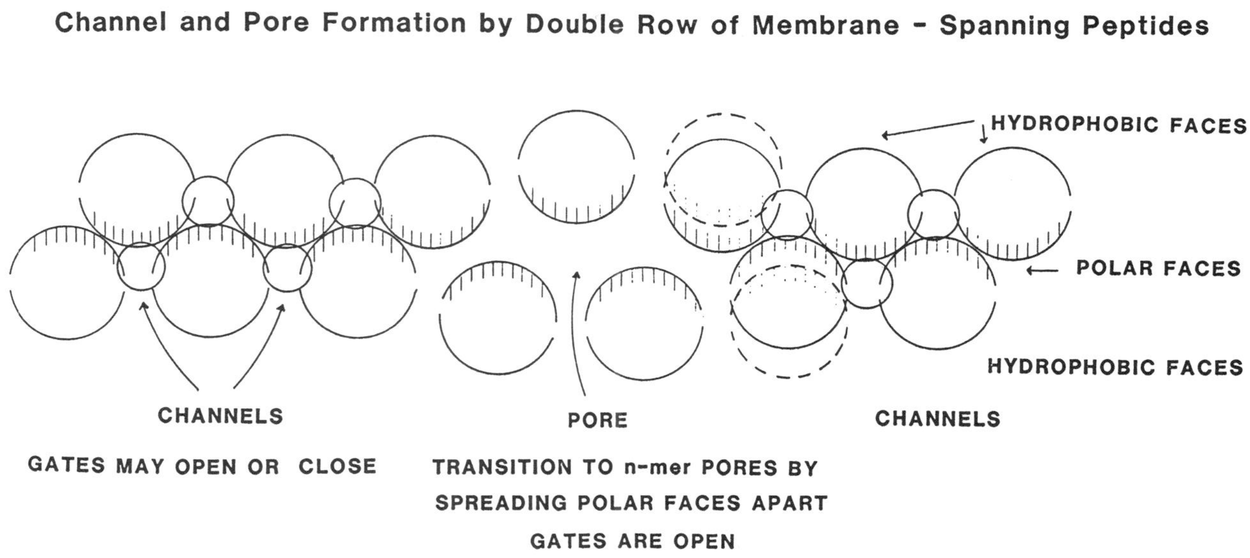

Figure 20. Schematic diagram of a finite portion of an extended double row of membrane-spanning peptide molecules based on the arrangement of Leu[1]-enkephalin molecules in crystals.[16-18] The large circles represent the middle sections of the sharply bent helical peptides with the polar sides of the helices facing each other and the hydrophobic sides of the helices on the exterior of the double row. The top and bottom of each bent helix (not shown) is curved into the hydrophobic direction . The small circles represent the circuitous polar channels. The relative size of the circles and their placement, at the extreme right and extreme left of this figure, is derived from the crystal structure. The middle of the diagram illustrates how small movements of adjacent peptides that would separate the upper and lower rows could create elongated polar pores under sufficient conditions of voltage and ion concentration. It is proposed that the n-mer pores could readily change the number of peptides forming them by the kind of zipper movement indicated by the dashed circles, probably involving reversible hydrogen-bond formation between the upper and lower rows.[18]

channels contain about 13 peptide monomers and that the pattern of successive conductance levels is consistent with a modified helix bundle in which the central pore in the higher order bundles is distorted toward a "torpedo"-shaped cross section.[20]

Sansom, in his review article,[8] had stated that alternative microscopic models for channels and pores may fit macroscopic data equally well. The packing motif for Leu-zervamicin helices in the three different crystal forms may provide an alternative for n-mer bundles. In the crystal, the bent helices associate with their polar sides facing each other, forming double rows in the close-packed arrangement shown in Figure 20. The circuitous polar channels (small circles in Figure 20) are formed by three peptide molecules and each peptide provides walls for three channels. The outside surfaces of the double row are hydrophobic. It can be envisioned that a finite portion of a double row of aggregated helices with single channels, similar to the arrangement in the crystal, is inserted into and spans the thickness of a membrane. As a consequence of a sufficient increase of applied voltage and concentration of K^+ ions, the double rows separate in the middle of the assembly, as shown in Figure 20, making approximately elliptical polar pores that could fill with electrolyte. This process would require the breaking of hydrogen bonds to Gln^{11} in the closed position and forming new hydrogen bonds in the open position. An elongated pore formed in this manner could contain a variable number of helices (n-mers) and it could switch readily from one n-mer to another by small motions in the double row aggregate (see dashed lines in Figure 20). The cross-sectional areas of elliptical or torpedo-shaped openings in n-mer pores (as compared to circular pores) can be adjusted to fit data for conductance levels. the macroscopic measurement of conductance may indicate the total number of peptide helices in an aggregated portion such as shown in Figure 20.[18]

3. REFERENCES

1. Tosteson, D. C., Cook, P., Andreoli, T. and Tieffenberg, M. J. (1967) The effect of valinomycin on potassium and sodium permeability of HK and LK sheep red cells, *J. Gen. Physiol.*. **50**, 2513-2519.

2. Karle, I. L. (1975) Conformation of Valinomycin in a Triclinic Crystal Form, *J. Am. Chem. Soc.* **97**, 4379-4386.

3. Duax, W. L., Hauptman, H., Weeks, C. M. and Norton, D. A. (1972) Crystal structure determination of valinomycin by direct methods, *Science* **176**, 911-914.

4. Smith, G. D., Duax, W. L., Langs, D. A., De Titta, G. T., Edmonds, J. W., Rohrer, D. C. and Weeks, C. M. (1975) Crystal and molecular structure of the triclinic and monoclinic forms of valinomycin, $C_{54}H_{90}N_6O_{18}$, *J. Am. Chem.Soc.* **97**, 7242-7247.

5. Karle, I. L. and Flippen-Anderson, J. L. (1988) A New Conformation Exhibiting Near-Threefold Symmetry for Uncomplexed Valinomycin in Crystals from Dimethyl Sulfoxide,*J. Am. Chem. Soc.* **110**, 3253-3257.

6. Pinkerton, M., Steinrauf, L. K. and Dawkins, P. (1969) Molecular structure and some transport properties of valinomycin, *Biochem. Biophys. Res. Commun.* **35**, 512-518.

7. Neupert-Laves, K. and Dobler, M. (1975) The Crystal Structure of a K^+ Complex of Valinomycin,*Helv. Chim. Acta* **58**, 432-442.

8. Samson, M. S. P. (1991) The Biophysics of Peptide Models of Ion Channels, *Prog. Biophys. Mol. Biol.* **55**, 139-235.

9. Karle, I. L. and Balaram, P. (1990) Structural Characteristics of α-Helical Peptide Molecules Containining Aib Residues, *Biochemistry* **29**, 6747-6756.

10. Karle, I. L., Rao, R. B., Prasad, S., Kaul, R. and Balaram, P. (1994) Nonstandard Amino Acids in Conformational Design of Peptides. Helical Structures in Crytals of 5-10 Residue Peptides Containing Dipropylglycine and Dibutylglycine,*J. Am. Chem. Soc.* **116**, 10355-10361.

11. Karle, I. L., Gurunath, R., Prasad, P., Kaul, R., Rao, R. B. and Balaram, P. (1995) Peptide Design. Structural Evaluation of Potential Non-Helical Segments Attached to Helical Modules. In publication.

12. Karle, I. L., Flippen-Anderson, J. L., Uma, K., Balaram, H. and Balaram, P. (1989) α-Helix and mixed 3_{10}/α-helix in cocrystallized conformers of Boc-Aib-Val-Aib-Aib-Val-Val-Val-Aib-Val-Aib-OMe, *Proc. Natl. Acad. Sci. USA* **86**, 765-769.

13. Karle, I. L., Flippen-Anderson, J. L., Uma, K., Sukumar, M. and Balaram, P. (1990) Modular Design of Synthetic Protein Mimics. Crystal Structures, Assembly, and Hydration of Two 15- and 16-Residue Apolar, Leucyl-Rich Helical Peptides, *J. Amer. Chem. Soc.* **112**, 9350-9356.

14. Karle, I. L., Flippen-Anderson, J. L., Uma, K. and Balaram, P. (1990) Parallel zippers formed by α-helical peptide columns in crystals of Boc-Aib-Glu(OBzl)-Leu-Aib-Ala-Leu-Aib-Ala-Lys(Z)-Aib-OMe, *Proc. Natl. Acad. Sci. USA* **87**, 7921-7925.

15. Rinehart, K. L., Jr., Pandey, R. C., Moore, M. L., Tarbox, S. R., Snelling, C. R., Cook, J. C., Jr. and Milberg, R. H. (1979) Mass Spectrometric Studies of Peptide Structures, in E. Gross and J. Meienhofer, (eds.), *Peptides, Structure and Biological Function,* Pierce Chemical Co., Rockford, Ill. pp. 59-71.

16. Karle, I. L., Flippen-Anderson, J. L., Agarwalla, S. and Balaram, P. (1991) Crystal Structure of [Leu^1]zervamicin, a membrane ion-channel peptide: Implications for gating mechanisms, *Proc. Natl. Acad. Sci. USA* **88**, 5307-5311.

17. Karle, I. L., Flippen-Anderson, J. L., Agarwalla, S. and Balaram, P. (1992) Implications for an Ion Channel in Leu-zervamicin: Crystal Structure of Polymorph B in R. H. Sarma and M. H. Sarma, (eds.) *Structure and Function: Vol. 2: Proteins. Proc. of the 7th Conversation in Biomolecular Stereodynamics*, Adenine Press, pp. 97-111.

18. Karle, I. L., Flippen-Anderson, J. L., Agarwalla, S. and Balaram, P. (1994) Conformation of the Flexible Bent Helix of Leu^1-Zervamicin in Crystal C and a Possible Gating Action for Ion Passage, *Biopolymers* **34**, 721-735.

19. Fox, R. O., and Richards, F. M. (1982) A voltage-gated ion channel model inferred from the crystal structure of alamethicin at 1.5 Å resolution, *Nature (London)* **300**, 325-330.

20. Balaram, P., Krishna, K., Sukumar, M., Mellor, I. R. and Sansom, M. S. P. (1992) The properties of ion channels formed by zervamicin, *Eur. Biophysc. J.* **21**, 117-128.

ON MOLECULAR RECOGNITION IN FULLERENE CHEMISTRY

H.B. BUERGI
Labor für Kristallographie
Freiestr. 3, Universität
CH-3012 BERN (Switzerland

1. Introduction

The synthesis, purification, as well as the study of analysis, structure, reactivity and practical applications of fullerenes has reached a high degree of sophistication in a time span of only 5-10 years. This is documented in „The Almost (but never quite) Complete Buckminsterfullerene Bibliography". The edition of Aug. 1994 lists more than 1900 papers distributed over more than 30 journals [1]. Given this abundance of information, a review of the field has to be selective. The approach chosen here may be summarized by two buzz words in current chemical research: Molecular Recognition. For the purpose of this review, we distinguish between a more analytical and a more constructive attitude towards Molecular Recognition. The former concentrates on analysing the interactions between molecules in whatever compounds happen to be available. The latter makes use of this knowledge for the deliberate construction of recognition sites which are able to accept or reject molecules with high selectivity and in a predictable manner. Molecular Recognition in Fullerene Chemistry seems still largely analytical, although there are a few convincing examples of successful design.

2. What is there to be recognized? Molecular structures.

The family tree of fullerene molecules has many branches, e.g. balls, tubes, onions and spiral shell particles. The ball shaped molecules have probably been characterized best - either in pristine or in derivative form - among them C_{60}, C_{70}, C_{76}, and C_{84} in particular . Their basic building blocks are threecoordinated carbon atoms which form five and sixmembered rings distributed over a spherical or ellipsoidal surface. To obtain a stable constitution each fivemembered ring has to be anellated to five sixmembered rings, but never to another fivemembered ring (isolated pentagon rule). Even with this restriction the higher fullerenes show constitutional isomers, some of them chiral [2].

G. Tsoucaris et al. (eds.), Crystallography of Supramolecular Compounds, 31–44.

The overall shape of C_{60} is that of a truncated icosahedron with icosahedral symmetry (Pointgroup I_h). All its carbon atoms must lie on a sphere. There are two types of bonds, a set of 30 6,6-bonds between hexagons and a set of 60 5,6-bonds between pentagons and hexagons. The structure has only two degrees of freedom: e.g. the distance R of the C-atoms from the molecular center and the difference between the 5,6- and the 6,6- bond distance, or the two different bond distances d_{56} and d_{66}, or other combinations of these. A geometric relationship between some of these quantities is

$$R^2 = 4(d_{56}^{\ 2} + d_{56}\, d_{66} + d_{66}^{\ 2} \sin^2 36^0) / (\sqrt{5}-1)^2$$

With increasing number of carbon atoms the geometrical aspects become rapidly more complicated (Table). However, the number of symmetry inequivalent bonds is always smaller than the number of degrees of freedom. This implies that the molecules can assume not only energetically optimal bond distances but can also distribute angular strain over the molecular surface.

<u>Table</u> Some geometric characteristics of C_{60}, C_{70}, C_{76}, C_{78}, C_{84}

	Symm	Degrees of freedom	Number of symmetry-inequivalent atoms	5,6 bonds	6,6 bonds	Diameter(s) (Å)
C_{60}	I_h	2	1	1	1	7.06
C_{70}	D_{5h}	12	5	4	4	7.10, 7.10, 7.91
C_{76}	D_2	57	19	15	15	6.8, 7.7, 8.8
C_{78}	C_{2v}	60	21	16	18	
	D_3	39	13	10	12	
C_{84}	D_{2d}	32	11	8	11	8.3, 8.3, 8.6
	D_2	63	21	15	17	similar

The detailed bond distances and angles of C_{60} have first been determined by *ab initio* calculation [3]; this was done before Krätschmer and Huffmann's synthetic method provided sufficient quantities of C_{60} for structure determination experiments [4] and was thus the only option at the time. It is probably fair to say that even today, high-level *ab initio* calculations are the best method to obtain reliable geometric parameters of pristine fullerene molecules. This is because fullerene crystal structures are notoriously disordered and show many other crystallographic difficulties as will be discussed in the next section. To our knowledge the best experimental structure determination of a fullerene molecule has been obtained for C_{60} from gas phase electron diffraction [5]. However, this is likely to remain a singular case due to the geometrical simplicity of C_{60} described above. It is unlikely that the spectra of interatomic distances obtainable from gas phase diffraction for higher fullerenes show sufficient resolution to distinguish between their many slightly different, symmetry inequivalent distances and angles (for an example see the Figure).

The structural elements available for molecular recognition may be regarded from two points of view. One, called „high resolution", is to look at atomic detail: atoms, bonds, rings and larger fragments. The other, called „low resolution", considers only the general shape of the molecules. A simple measure of shape is the curvature of the molecular envelope. For a sphere the curvature is the same everywhere. As seen from the outside, it is proportional to the inverse sphere radius 1/R; seen from the inside it is -1/R. Optimal recognition of C_{60} with outer curvature 1/R would be by another sphere with inner curvature $-1/(R+R_{vdw})$, where R_{vdw} is the van der Waals distance between C and the element(s) making up the outer sphere. A hypothetical example is C_{60} (R = 3.5 Å) enclosed by $B_{60}O_{90}$ (R~6,9 Å) in which the B-atoms form a truncated icosahedron and the O-atom bridge its edges. For a rotation ellipsoid with axes R_1 and R_2 (parallel and perpendicular to the rotation axis) the curvature at the poles is R_1/R_2^2; at the equator there are two different curvatures, the one along the equator is $1/R_2$, the one perpendicular to it is R_2/R_1^2. For the prolate C_{70} molecule, the curvature is largest at the poles, intermediate in the equator and smallest perpendicular to it.

A different, much simpler, but also less general measure of local curvature is the deviation from 360° of the sum of the three bond angles at a given carbon atom, it is 12° for C_{60} and 12.5, 12.5, 11.8, 9.2 and 6.6° for the five symmetry inequivalent carbon atoms of the ellipsoidal C_{70}, decreasing from the poles to the equator (Figure).

3. Selfrecognition of C_{60}, and C_{70}. Their crystal structures.

Crystallization of C_{60} and C_{70} from solution leads, more often than not, to cocrystals containing solvent and other molecules present in solution. These molecules are often flat (benzene, toluene) or flexible (alkanes, alkylethers) and may therefore cover larger parts of the surface of a fullerene molecule than another fullerene molecule may cover. The geometric shape and conformational properties of the solvating molecules would seem to be as important for explaining the propensity of fullerenes to form cocrystals as their donor/acceptor properties. Single crystals of pristine C_{60} suitable for diffraction studies are obtained by sublimation at ~500-600° C, or by He-gas transport at similar temperatures.

3.1. CRYSTAL STRUCTURE AND REORIENTATIONAL MOTION OF C_{60}.

As expected from its near spherical shape C_{60} is close packed [6]; most reports mention cubic close packing (ccp). There are three phases with transition temperatures at ~85K and ~250K [7]. All three phases show ccp; they differ in the orientational (dis-)order of C_{60}. In the intermediate temperature phase the four molecules in the cubic unit cell align one of their molecular threefold axes with one of the four non-intersecting crystallographic threefold axes (space group Pa3). However, each molecule instead of assuming a single, well defined orientation with respect to its axis, statistically adopts one of two orientations (a, b) differing by a rotation of ca. 60^0 about this axis [7,8]. The probabilities of finding orientation a or b are temperature dependent and their ratio follows a Boltzmann distribution ($p_a+p_b = 1$)

$$\Delta G = RT \ln p_a/p_b$$
$$= \Delta H - T\Delta S$$

From diffraction measurements on twinned crystals at 100, 110, 153 and 200 K, ΔH is found to be 1.06(5) kJ mol^{-1} and ΔS is 0.0(4) J mol^{-1} K^{-1} [8], i.e. the energetic differences between the two orientations are very small. The Boltzmann dependence also indicates little or no cooperativity during reorientation.

Due to the orientational disorder, four possible motifs of selfrecognition (intermolecular contacts) are found. Two of these are between a 6,6-bond of one molecule and a fivemembered ring of its neighbour, the other two are between a 6,6-bond and a sixmembered ring. An analysis of the intermolecular contact distances shows little difference between the four types of contact [8]. Shortest nonbonded distances are between 3.13 and 3.18 Å, ~ 0.1 Å shorter than in graphite. It is worth noting that packing calculations with atom-atom potentials lead to the wrong sign of ΔH. As one might expect, they favor the bump due to a 6,6-bond to face the larger hollow of the sixmembered ring rather than the smaller one of the fivemembered ring. If an ad-hoc shift of negative charge into the center of the 6,6-bond is postulated, the calculation can be made to agree with experiment [9]. The short observed distances (relative to graphite) are indicative of a specific attractive contribution in the intermolecular interactions but its nature seems as yet unknown.

There are two ways a C_{60} molecule can reorient in this phase, either by a rotation of 60^0 about the crystallographic threefold axes or by a rotation of ~ 42^0 about a molecular twofold axis (perpendicular to the crystallographic threefold) [7,8]. If the reorientation potential is assumed to be sinusoidal its barrier B can be estimated from the mean square molecular libration amplitudes $<\varphi^2>$ (in rad^2) which are obtained along with the atomic coordinates:

$$B = 2RT/(n^2<\varphi^2_n>)$$

where n = 360°/Rotation angle. The barrier for the 60^0-rotation is estimated to be 34, that of the 42^0 rotation to be 18.5(5) kJ mol^{-1}. The solid state NMR value is 17.6 kJ mol^{-1}. This is strong evidence that the preferred reorientation mechanism is a 42^0-rotational jump.

Below 85K reorientation seems to be very slow, no changes in p_a and p_b could be measured and the structure has therefore been described as an orientational glass phase [7]. Above 250K, reorientation becomes very facile, although not completely free; there are still preferred orientations of C_{60} [11]. However, the barriers to reorientation are on the order of RT. This implies that all possible orientations of the molecule can be seen and their probability determined. Using a classical Boltzmann distribution, the entire potential surface for reorientation can be determined without resorting to extrapolations based on mean square displacement parameters as described above for the intermediate temperature phase. The high mobility of C_{60} has been noted before [10]; the characteristic time of rotation (rotational correlation time) in crystalline C_{60} (9 ps, 283K) is only somewhat longer than that of free C_{60} molecules (3 ps, 283K) but shorter than that in tetrachloroethane solution (16 ps, 283K).

3.2. CRYSTAL STRUCTURE, ORIENTATIONAL AND STACKING DISORDER IN C_{70}.

The ellipsoidal shape of C_{70} complicates matters relative to C_{60}. Two families of polymorphs have been reported, one related to ccp and one to hcp [12,13]. The latter seems to be predominant and has been investigated in some detail. There are again three phases with transition temperatures at ~ 320 and ~ 280K.

At room temperature the long axes of all molecules are parallel, the molecules arrange in hexagonal close packed layers which are stacked on top of each other. This implies threefold symmetry axes perpendicular to the layers and therefore parallel to the molecular fivefold axis of the D_{5h}-symmetric C_{70} (Figure). The molecular fivefold axes are incompatible with the threefold axes required by the packing. The molecules therefore occupy one of three orientations, rotated 120° relative to each other, in a statistical way. The molecular electron density is found to consist essentially of nine parallel toruses, one in the molecular equator, four above and four below it [14]. This is consistent with virtually free rotation of C_{70} about its five fold axis, i.e. the molecule shows no well defined intermolecular contacts. The center-to-center distance between molecules within a layer is 10.11 Å, slightly larger than in the room temperature phase of C_{60} where it is 10.01 Å. Corresponding molecular diameters are 7.10 Å for C_{70} and 7.06 Å for C_{60}. The distance between the planes at the poles of C_{70} is 7.91 Å. From the point of view of molecular recognition all that can be said is that within a layer the C_{70} molecules match their directions of smallest curvature.

At higher temperatures the long molecular axis of C_{70} starts tumbling away from the direction normal to the molecular layers towards or even into directions parallel to these layers [12,13]. As a consequence the intra-layer intermolecular distance increases from 10.11 to 10.57 Å, the interlayer distances decreases from 18.58 to 17.56 Å. The ratio of the two distances at high temperature is 1.63 which corresponds to the ratio obtained for the close packing of spheres and suggests an essentially spherically symmetric orientational disorder. No details are known about the average potential governing the underlying tumbling motion.

1,2		1.453(3) Å
1,9	1.387(4)	
6,7		1.445(3)
7,8	1.378(3)	
7,22		1.447(2)
21,22		1.426(3)
22,23	1.414(2)	
23,24	1.462(4)	6.6 deg
		9.2
		11.8
		12.5
		12.5

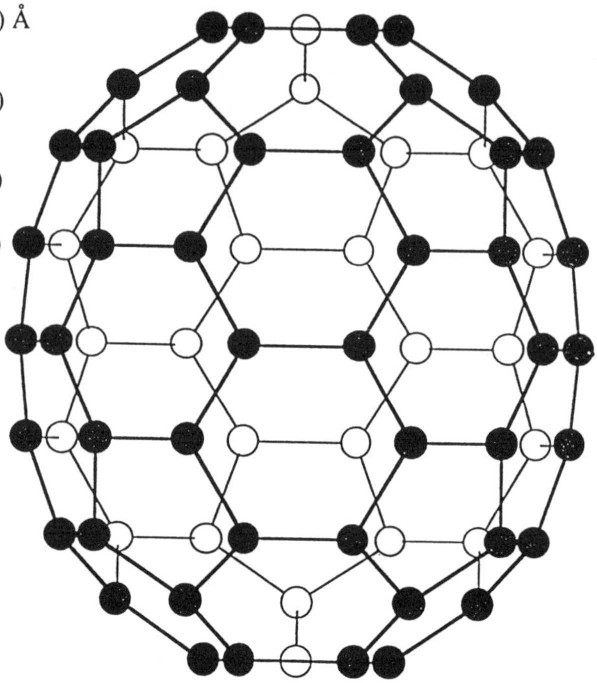

Figure C_{70}. Top left: bond types (1. column), 5,6-distances (2. column) 6,6-distances (3. column). Bottom left: 360 deg - sum of bond angles for the five divverent types of carbon atoms.

At low temperature the orientation with respect to the fivefold molecular axis locks into one of three positions [15]. The crystals are now triply twinned. X-ray studies at 100 and 220K have revealed details of the molecular geometry and the intermolecular contacts, but these results have to be taken with a grain of salt due to the stacking disorder mentioned above and discussed below.

In all three phases the stacking of layers shows disorder which layer has been investigated for the room temperature phase [16]. The phenomenon is related to a subtle recognition problem. Consider a close packed crystal of C_{70}. Most of the time the layers alternate between A and B, leading to a hexagonal stacking sequence ...ABAB... . Occasionally the stacking sequence is ...ABCABC... for a few layers. The top three layers will be either of type ABA, corresponding to hcp, or CBA, corresponding to ccp. If a new layer began to grow and „recognized" the three layers below perfectly, either an ordered hcp or an ordered ccp would result. Because the recognition is imperfect, stacking disorder results, i.e. a new layer growing on the surface has the choice of continuing the pattern of the top three layers or to make a mistake and switch from one pattern to the other. The probability to continue hcp is 0.83, that of switching to ccp is correspondingly 0.17. The probability to continue with ccp is 0.75, that of switching back to hcp is then 0.25. Since the probability to continue in hcp is somewhat larger than that to continue in ccp, the crystal is predominantly hcp. Although, the simple model with two probabilities describes the diffuse scattering very well, the interplay of forces responsible for this behaviour is very subtle and not understood.

3.3. SUBLIMATION ENTHALPIES AND SOME CONCLUSIONS.

The sublimation energy ΔH_{sub}(707K) of C_{60} has been estimated to be ~40kcal mol^{-1} (0.67kcal mol^{-1} per C-atom at 707K) [17]. If it is assumed that ΔH_{sub} is due primarily to the direct, contact zones between the C_{60}-spheres, one such contact contributes about 6.7kcal mol^{-1} to the sublimation energy. (The interaction energy associated with each of 12 ccp contact zones must be split equalliy between the two neighbouring molecules). The sublimation energy of C_{70} has been estimated at ~65kcal mol^{-1} [18]. The average contribution of 0.93 kcal mol^{-1} per C-atom is larger than for C_{60}. Since the interactions between C_{70} layers, i.e. between the poles of C_{70}, are geometrically similar to those in C_{60}, it may be assumed that their energy increments are approximately similar to those of C_{60} ($6 \cdot 6.7/2$ kcal mol^{-1}). The increment for an equatorial contact within a C_{70} layer would then be ~15kcal mol^{-1}. The energetic difference between the two types of contacts correlates with the difference in average curvature which is bigger at the poles than in the equator. The low barriers to fullerene reorientation may seem to contradict the high values of estimated sublimation energies. This is not so, however, since during molecular rotation the breaking of intermolecular contacts is constantly compensated by the formation of new ones, i.e. C_{60} and C_{70} behave like a kind of „spherical graphite“.

The high mobility of C_{60} and C_{70} in their crystalline phases seems to indicate that selfrecognition is weak and that attempts to construct specific non-covalent recognition sites for these molecules should focus primarily on their general, „low resolution“ shape rather than on „high resolution“ atomic detail. The most important „low resolution“ parameter available for recognition is the average radius of the fullerenes. The problem of constructing specific „low resolution“ recognition sites for fullerenes is thus comparable to that of differentiating between alkali or alkaline earth cations of different sizes. The basic principles to achieve this objective have been outlined by Lehn in the seventies [19]. Lehn recognized that one important parameter is the ionic radius and solved the problem by synthesizing cryptands with different ring sizes providing cavities of different diameters. The problem is likely to be more challenging in the case of fullerenes. For one thing, the fullerenes are neutral and one cannot make use of charge-dipole interactions as one can in the case of the alkali cryptates. Instead, one has to rely exclusively on weak van der Waals interactions between similar carbon atoms, similar bonds and rings of similar sizes. Second, fullerene molecules show similar radii for molecules with similar numbers of carbon atoms. Some fullerenes show subtle deviations from spherical shape which may differ for different isomers and which are probably difficult to distinguish.

4. Separation and analysis of fullerene mixtures.

In view of their spherical structures and lack of functional groups, separating fullerene mixtures into stochiometrically pure compounds, pure isomers or even pure enantiomers is an extremely difficult problem. Its solution requires not only techniques for separation but also analytical methods to monitor purity. Both are based, at least in part, on molecular recognition processes.

4.1. SEPARATION.

Some of the separation methods which are actively developed are chromatography, extraction and crystallization.

Chromatographic methods are based on the experience with separating polycyclic aromatic hydrocarbons. For these compounds π-acidic molecules covalently attached to a stationary silica phases have proven useful. They are also efficient for separating fullerenes as has been shown recently with a tetrachlorophthalimidopropyl-silica phase [20]. In another experiment, a porphyrin-silica stationary phase was used which allows separation of the endohedral $La_2@C_{84}$ from C_{60}, C_{70}, C_{76}, C_{78}, C_{82} and C_{84} in a single stage within ~ 30 minutes [21]. (p-Carboxyphenyl)triphenylporphyrin was attached covalently to a silica support and used either as the free base or as a zinc complex. The authors speculate that fullerene-porphyrin face-to-face, π-π interactions and fullerene-phenyl face-to-edge interactions „probably play a key role in the unique selectivity and strong retention observed“. Based on similar ideas, a bisconcave, tetrakis(9,10-dihydro-9,10-anthraceno)porphyrin has recently been designed and synthesized [22]. Its Co(II)-complex forms a cocrystal with C_{60} and 1,2-dichlorobenzene in a ratio of 1:2:2 showing the expected π, π-interactions [23]. C_{60} is docked nicely into both of the potential recognition sites, namely in the cups above and below the porphyrin plane and in the outer clefts formed by the dihydroanthracenyl fragments. At 146K every C_{60} molecule assumes four different orientations with approximately equal probability. The orientational disorder indicates that this porphyrin derivative recognizes the general shape of C_{60} rather than its molecular details.

Extraction from the liquid phase in the presence of suitable complexing agents is another way to separate fullerenes. Thus C_{60} - but not C_{70} - is solubilized in water in the presence of the octaglycoside γ-cyclodextrin. Precipitation, freeze-drying and recrystallization gave purple crystals of a C_{60}-(γ-cyclodextrin)$_2$ complex. The corresponding hexa- and hepta-glycosides α- and β-cyclodextrin do not form complexes [24]. In a similar experiment, a water soluble calix[8]arene was used to selectively extract C_{60} from toluene solution of C_{60} and C_{70} into the aqueous phase [25]. So far, none of the complexes has been characterized by crystal structure analysis.

Crystallization is often an efficient way of purification. An early possibility in this direction was stimulated by the observation that the crystal structure of pure hydroquinone shows two interpenetrating, hydrogenbonded networks of hydroquinone molecules (HQ) which are not connected to each other. In addition, the structure can accommodate small guest molecules like H_2S, CH_3OH or CH_3CN. It was hypothesized that removal of one of the networks and of the small molecules from the structure, would result in cavities that are of the proper size to accommodate one C_{60} molecule each. The expected composition $C_{60}(HQ)_3$ and the crystal architecture were confirmed by single crystal structure analysis [26]. Although the symmetry of the crystal would allow an ordered arrangement, the C_{60} molecule assumes a threefold orientational disorder in the entire temperature range between 100 and 400K [27]. Crystallization experiments with a 5:1 C_{60}/C_{70} mixture produced highly crystalline samples consisting largely of the $C_{60}(HQ)_3$ complex, thus suggesting „a simple method of separating C_{60} from C_{70} (and possibly from higher fullerenes)“ [26]. Somewhat later the structure of a cocrystal $(C_{70})_2(HQ)_9(\text{benzene})_2$ was described [28]. It showed a different, completely new and beautiful HQ host architecture with disordered C_{70} molecules. As far as the idea of fullerene separation by crystallization is concerned, the authors remark: „...C_{70} enclathration would appear to complicate C_{60} separation from higher fullerenes by the HQ inclusion crystallization...“. Nevertheless, the two beautiful host-guest architectures which resulted from this work were well worth the effort.

A practically useful method to purify C_{60} and C_{70} by crystallization is based on complexation with calixarenes [29]. On mixing toluene solutions of p-tert-butyl-calix[8]arene and crude fullerene mixtures a precipitate forms which contains mainly a C_{60}/calixarene 1:1 complex and some C_{70}. Two subsequent recrystallizations from toluene enrich the C_{60} content to 96 and then to >99.5%. A structure of the C_{60}/calixarene complex has not been reported so far.

4.2. MONITORING PURITY.

The performance of separation procedures is usually followed by NMR. There are two methods available, both with advantages and disadvantages. One is ^{13}C-NMR, the other ^{3}He-NMR. The former is experimentally straightforward, but in a mixture of unknown composition assignment of the signals to individual compounds can be a problem, especially for molecules of low symmetry with many different ^{13}C-peaks and present in low concentration..

An alternative is ^{3}He-NMR [30]. The raw fullerene extract is heated for several hours to 600°C in ^{3}He at a pressure of ~2500 atm, then it is worked up by extraction with toluene or carbon disulfide. The ^{3}He-labeled extract is then separated into its components as usual and each of them is analysed by ^{3}He-NMR. Present experience indicates that each labeled isomer gives rise to a single, usually resolved NMR signal; the number of signals and their relative intensity indicate the quality of the separation. The same method may be applied to detect the presence of isomers in chemical reactions leading to mixtures of fullerene derivatives.

5. Recognition in chemical reactions: Regioselectivty and enantioselectivity.

Among the chemical reactions leading to fullerene derivatives, osmylation, addition of Vaska-type complexes and Diels-Alder addition have been investigated with respect to their selectivity. The reactions incorporate a fullerene C-C-bond into a five-, three- or sixmembered ring, respectively. It is always a 6,6-bond but not at 5,6-bond that reacts. For C_{60}, addition of one reactant molecule leads to a single product whereas for the higher fullerenes isomers may be expected.

Reactions of fullerenes with inorganic and organometallic compounds including single and multiple adducts have been reviewed recently [31]. Here we discuss some monoadducts, selected either because of the observed regioselectivity in their formation or because their crystal structures show interesting recognition motifs.

5.1. VASKA-TYPE COMPLEXES.

$Ir(CO)Cl(PR_3)_2$ adds to C_{60} in benzene leading to a trigonal bipyramidal η^2-complex of iridium [32]. The reaction is reversible. The complex with two triphenylphosphines shows one phenylring of each phosphine in an intramolecular face-to-face, π-π contact with the fullerene-surface. The crystal structure also includes five benzene molecules which make several different types of contacts to C_{60}. This observation led to the idea of using $Ph_2P\text{-}C_6H_4\text{-}OCH_2\text{-}C_6H_5$ as a phosphine since it has additional aromatic surfaces linked flexibly to the triphenyl phosphine by the $\text{-}OCH_2$-connector. The terminal benzyl-groups were expected to replace the benzene solvent molecules in the crystal lattice. Indeed, $\eta^2\text{-}C_{60}Ir\ (CO)Cl(PPh_2C_6H_4OCH_2Ph)_2$ was found to crystallize without solvent. However, contrary to the PPh_3-complex, recognition between the long aromatic side chains and C_{60} is intermolecular [33].

The formation of the adduct $\eta^2\text{-}C_{70}\ Ir\ (CO)Cl\ (PPh_3)_2$ is also reversible. It crystallizes with 2.5 molecules of benzene. Its crystal structure provided the first detailed geometric parameters of a fullerene larger than C_{60} [34]. Only one isomer was found in which Ir coordinates to the 6,6-bond closest to the poles (1,9-bond, Figure). This bond connects the carbon atoms which are the most pyramidalized. Thus the skeletal reorganisation energy of C_{70} required for complexation to Ir is particularly small for this isomer. It is not clear, however, why the corresponding 7,8-isomer is not observed, since the pyramidal distortions in the 7,8-bond are of comparable magnitude to those of the 1,9-bond (Figure).

A benzene solution of C_{84} also forms an adduct with $Ir(CO)Cl(PPh_3)_2$. An -ray diffraction study of $C_{84}Ir(CO)Cl(PPh_3)_2 \cdot 4C_6H_6$ shows that C_{84} is present predominantly as the D_{2d}-isomer [35]. Iridium is coordinated on the S_4 (4) symmetry axes of C_{84}. The 6,6-bond bisected by this axis appears to connect the atoms with the largest angular distortions. The deviation of the bond angles from 360 deg is 10.6 deg, slightly larger than for the other two types of 6,6-bonds connecting two fivemembered rings (9.6, 9.8 deg). Residual electron density was found adjacent to three of the six 6,6-bonds bisected by the three molecular twofold symmetry axes. This density indicated the presence in the complex of isomers of C_{84}, other than the D_{2d} one.

5.5. DIELS-ALDER ADDITION.

Diene-addition to C_{60} occurs at 6,6-bonds. A particularly nice derivative is a Paddon-Row-type donor-acceptor system whose crystal structure shows aesthetically pleasing dimers [36]. In these, the curved, polycyclic hydrocarbon part of one molecule embraces the C_{60}-part of the other and vice-versa. The curvatures of the two parts are nicely complementary.

There are four different 6,6-bonds of C_{70} and correspondingly four possible Diels-Alder monoadducts. The two predominant isomers result from addition to the 1,9- and 7,8-bonds in a ratio of 2.4 to 1. The remaining two isomers are formed in very small quantities only (0.1-0.2 relative to the 7,8-adduct) [37]. The two main products which have been characterized by X-Ray crystallography are those expected from the pyramidal distortion, as discussed above [38].

5.3. OSMYLATION.

Reaction of C_{60} with an excess of OsO_4 in toluene, followed by addition of an excess of pyridine and exchange by 4-tert-butylpyridine (py) gives the $C_{60} \cdot OsO_4py_2$ adduct in 75% yield. The crystal structure of this compound is of historical interest, because it represents the first confirmation by X-ray diffraction of the soccer ball connectivity of C_{60} [39].

Osmylation of C_{70} with 0.75 equiv. of OsO_4 in pyridine/toluene solution at 0°C yielded two isomers of $C_{70}OsO_4py_2$ which were separated by HPLC chromatography [40]. Oxidation occurs predominantly at the 1,9- and 7,8- bonds with relative yields 2.1 and 1.0. The yields of the remaining two isomers are reported as <0.1. Overall the selectivity is very similar to that found for Diels-Alder addition and correlates with the curvature of different parts of the C_{70}-surface. Since the osmylation reaction is kinetically controlled, the reorganization energy implied by the additional pyramidalization necessary to form the C-O-bonds must be an important contribution to the activation energy of the osmylation reaction.

C_{76} and C_{78} show chiral isomers of symmetry D_2 and D_3, respectively. The main isomers of C_{84} are a chiral D_2 and a non-chiral D_{2d} form [2]. Osmylation has been applied to resolve C_{76}-, C_{78}- and C_{84}-racemates into their enantiomers [41]. In this reaction pyridine was replaced by an optically active Cinchona alkaloid. After partial osmylation optically active, unreacted starting material could be recovered and diastereomerically enriched osmate esters could be isolated. The latter were reduced with $SnCl_2$ and optically active fullerenes were obtained. In the case of C_{84}, the reaction was run to ~32% and ~73% conversion. In the first case C_{84} was regenerated from the osmate ester. In the second case unreacted C_{84} was recovered. The CD-spectra of the two samples are virtually mirror symmetric, thus proving that the two samples are enriched in enantiomers of opposite handedness. A number of questions remain unanswered: What is the degree of enrichment? Where on the C_{84}-molecule does osmylation take place? Are the C_{84}-mixtures consisting of isomers other than the D_2 and D_{2d}-forms (see section 4.2)? Whatever the answers to these questions will be, the demonstration of diastereomeric selectivity in fullerene chemistry is an impressive achievement.

6. Concluding remarks

The intention of this report was to summarize some basic observations pertaining to (1) the selfrecognition of fullerene molecules in their crystal structures, (2) the recognition of fullerene molecules by other molecules, especially those found in structures of cocrystals with a variety of other molecules (3) the lack of recognition of specific portions of fullerene molecules as evidenced by orientational disorder which is frequently found in fullerene cocrystals. (4) the selective reactivity of fullerenes in relation to their bond lengths and bond angles. A short survey of these topics is, by necessity, incomplete. Further important work is found in the references and in the „Bibliography“ mentioned in the Introduction [1].

As pointed out in the beginning, the approach adopted here is mainly descriptive. It is hoped, however, that it will provide stimulation and some of the information necessary to embark on the design and synthesis of molecules that specifically recognize and differentiate between different fullerene molecules.

References

1. To obtain a copy send the message BIBLIO to BUCKY@SOL1.LRSM.UPENN.EDU
2. A systematic tabulation is given by Fowler P.W., Manolopoulos, D.E. (1995) *An Atlas of Fullerenes,* Clarendon Press, Oxford.

3. For references see the report on a high-level calculation published at about the same time as the first experimental structure determinations: Häser, M., Almlöf, J., and Scuseria, G.E. (1991) *Chem. Phys. Lett.* . **181**, 497-500.
4. Krätschmer, W., Lamb, L.D., Fostiropoulos, D., and Huffman, D.R. (1990) *Nature*, **347**, 354-358.
5. Hedberg, K., Hedberg, L., Bethune, D.S., Brown, C.A., Dorn, H.C., Johnson, R.D., and de Vries, M. (1991) *Science*, **254**, 410-412.
6. Heiney, P.A., Fischer, J.E., McGhie, A.R., Romanow, W.J., Denenstein, A.M., McCauley, J.P. Jr., Smith A.B. III, and Cox, D.E. (1991) *Phys. Rev. Lett.* **66**, 2911-2914.
7. David, W.I.F., Ibberson, R.M., Dennis, T.J.S., Hare, J.P., and Prassides, K. (1992). *Europhys. Lett.* **18**, 219-225.
8. Bürgi, H.B., Blanc, E., Schwarzenbach, D., Lu, S., Lu, Y., Kappes, M.M. and Ibers, J.A. (1992) *Angew. Chem. Int. Ed. Engl.* **31**, 640-643.
9. Sprik, M., Cheng, A., and Klein, M.L. (1992) *J. Phys. Chem.* **96**, 2027-2029.
10. Johnson, R.D., Yannoni, C.S., Dorn, H.C., Salem, J.R., and Bethune, D.S. (1992) *Science* **255**, 1235-1238.
11. Bürgi, H.B., Restori, R., and Schwarzenbach, D. (1993) *Acta Cryst.* **B49**, 832-838.
12. Verheijen, M.A., Meekes, H., Meijer, G., Bennema, P., de Boer, J.L., van Smaalen, S., van Tendeloo, G., Amelinckx, S., Muto, S., and van Landuyt, J. (1992) *Chem. Phys.* **166**, 287-297.
13. Vaughan, G.B.M., Heiney, P.A., Cox, D.E., Fischer, J.E., McGhie, A.R., Smith, A.L., Strongin, R.M., Cichy, M.A., and Smith, A.B.III (1993) *Chem. Phys.* **178**, 599-613.
14. Blanc, E., Bürgi, H.B., Restori, R., Schwarzenbach, D., Stellberg, P., and Venugopalan, P. (1994) *Europhys. Lett.* **27**, 359.
15. van Smaalen, S., Petricek, V., de Boer, J.L., Dusek, M., Verheijen, M.A., and Meijer, G. (1994) *Chem. Phys. Lett.* **223**, 323-328.
16. Blanc, E., Bürgi, H.B., Restori, R., Schwarzenbach, D., and Ochsenbein, P. (1995) *submitted.*
17. Beckhaus, H.D., Rüchardt, C., Kao, M., Diederich, F., and Foote, C.S. (1992) *Angew. Chem. Int. Ed. Engl.* **31**, 63.
18. Beckhaus, H.D., Verevkin, S., Rüchardt, C., Diederich, F., Thilgen, C., ter Meer, H.U., Mohn, H., and Müller, W. (1994) .*Angew. Chem. Int. Ed. Engl.* **33**, 996.
19. Lehn, J.M. (1973) *Structure and Bonding* **16**, 1.
20. Herren, D., Thilgen, C., Calzaferri, G., and Diederich, F. (1993) *J. Chromatrogr.* **644**, 188.
21. Xiao, J., Savina, M.R., Martin, G.B., Francis, A.H., and Meyerhoff, M.E. (1994) *J. Am. Chem. Soc.* **116**, 9341-9342.
22. Ramondenc, Y., Schwenninger, R., Phau, T., Gruber, K., Kratky, C., and Kräutler, B. (1994) *Angew. Chem. Int. Ed. Engl.* **33**, 889.
23. Ochsenbein, P., Schwenninger, R., Gruber, K., Kratky, C., Kräutler, B., and Bürgi, H.B. (1995) in prearation.

24. Andersson, T., Nilsson, K., Sundahl, M., Westman, G., and Wennerström, O. (1992) *J. Chem. Soc. Chem. Commun.* 604-606. Yoshida, Z., Takekuma, H., Takekuma, S., and Matsubara, Y. (1994) *Angew. Chem. Int. Ed. Engl.* **33**, 1597.
25. Williams, R.M., and Verhoeven, J.W. (1992) *Rec. Trav. Chim. Pays-Bas* **111**, 531.
26. Ermer, O. (1991) *Helv. Chim. Acta* **74**, 1339-1351.
27. Bürgi, H.B., Förtsch, M., Venugopalan, P., Blanc, E., Restori, R., Schwarzenbach, D., and Ermer, O. (1995) in preparation.
28. Ermer, O., and Röbke, C. (119193) *J. Am. Chem. Soc.* **115**, 10077-10082.
29. Atwood, J.L., Kontsantonis, G.A., and Raston, C.L. (1994) *Nature* **368**, 229-231.
30. Saunders, M., Jiménez-Vázquez, H.A., Cross, R.J., Mroczkowski, S. Gross, M.L., Giblin, D.E., and Poreda, R.J. (1994) *J. Am. Chem. Soc.* **116**, 2193-2194. Saunders, M., Jiménez-Vázquez, H.A., Cross, R.J., Mroczkowski, S., Freedberg, D.I., and Anet, F.A.L. (1994) *Nature* **367**, 256-258.
31. Balch, A.L. in „*The Chemistry of Fullerenes*", Taylor, R., ed., World Scientific Publishing Company, in press.
32. Balch, A.L., Catalano, V.J., and Lee, J.W. (1991) *Inorg. Chem.* **30**, 3980-3981.
33. Balch, A.L., Catalano, V.J. , Lee, J.W., and Olmstead, M.M. (1992) *J. Am. Chem. Soc.* **114**, 5455-5457.
34. Balch, A.L., Catalano, V.J., Lee, J.W., Olmstead, M.M., and Parkin, S.R. (1991) *J. Am. Chem. Soc.* **113,** 8953-8955.
35. Balch, A.L., Ginwalla, A.S., Lee, J.W., Noll, B.C., and Olmstead, M.M. (1994) *J. Am. Chem. Soc.* **116**, 2227-2228.
36. Khan, S.I., Oliver, A.M., Paddon-Row, M.N., and Rubin, Y. (1993) *J. Am. Chem. Soc.* **115**, 4919-4920.
37. Herrmann, A., Diederich, F., Thilgen, C., ter Meer, H.U., and Müller, W.H. (1994) *Helv. Chim. Acta* **77**, 1689-1706.
38. Seiler, P., Herrmann, A., and Diederich, F. (1995) *Helv. Chim. Acta* **78**, 344-354.
39. Hawkins, J.M., Meyer, A., Lewis, T.A., Loren, S., and Hollander, F.J. (1991). *Science* **252**, 312-313.
40. Hawkins, J.M., Meyer, A., and Solow, M.A. (1993) *J. Am. Chem. Soc.* **115**, 7499-7500.
41. Hawkins, J.M., Nambu, M., and Meyer, A. (1994) *J. Am. Chem. Soc.* **116**, 7642-7645.

LONG RANGE ORDER IN ORGANIZED MONOLAYERS

D. MÖBIUS
Max-Planck-Institut für biophysikalische Chemie
Postfach 2841, D-37018 Göttingen, Germany

Long range order of molecules in insoluble monolayers due to particular packing of long hydrocarbon chains is easily observed by Brewster angle microscopy. In this technique that does not require the addititon of a fluorescent probe, plane polarized light is incident on the interface covered with the monolayer under the Brewster angle of the clean interface. Different types of long range order can be discriminated using an analyzer in the reflected beam. Long range order of the chains may be induced by addition of an non-amphiphilic component to the host monolayer filling free space between the chains of the host. In monolayers of particular dyes long range order of the chromophores is observed when large extended arrays of J-aggregates are formed.

1. Introduction

At interfaces between two media the environment of a molecule is anisotropic since many properties of the bulk phases change abruptly. Therefore, anisotropic molecules with functional groups may orient themselves in a particular way determined by the intermolecular interactions with either phase and with neighbor molecules. Adsorption of molecules at gas-solid and liquid-solid interfaces has been studied extensively in the view of catalysis and passivation of solids. The "self-assembly" technique of monolayer formation by chemisorption of reactive molecules at the solid-liquid interface [1,2] has gained much attention [3], e.g. in controlling wetting of solid surfaces [4,5].

The gas-water interface provides the unique possibility of organizing water insoluble amphiphilic molecules consisting of a hydrophilic head group and a sufficiently large hydrophobic part by reducing the available area with a mobile barrier, a technique invented more than 100 years ago [6]. The monomolecular layers (monolayers) formed at the water surface may show a surprising richness of phases [7,8] differing in molecular packing and long range order as characterized by X-ray studies [9] and optical investigations [10,11]. A monolayer may be regarded as the smallest subunit of a smectic liquid crystal. Therefore, similar types of long range order and textures are expected in both systems. Whereas the long hydrocarbon chains are most important for long range order in monolayers of typical amphiphilic molecules like fatty acids or phospholipids, the interactions between the chromophores of cyanine or merocyanine dyes determine long range order in two-dimensional aggregates. Dense packing of the long chains, however, may assist the formation of these aggregates as has been shown by spectroscopic investigations.

G. Tsoucaris et al. (eds.), Crystallography of Supramolecular Compounds, 45–54.

2. Formation of organized monolayers and monolayer phases

Insoluble monolayers at the gas-water interface are usually formed by spreading. A solution of the amphiphile in a volatile organic solvent is applied to the water surface either by carefully dropping or by gentle flow from a microsyringe. Ideally, the solution spreads, and after evaporation of the solvent the water insoluble molecules remain at the interface evenly distributed or associated to domains depending on the available area and temperature. From the concentration of the solution, the volume spread and the water surface area the average area per molecule is calculated. The process of spreading as well as compression and the resulting monomolecular layer can easily be observed if the amphiphiles are fluorescent [12] or when a fluorescent probe is present [13].

At large area, typically 100 nm^2/molecule, the amphiphilic molecules may not interact with each other, and the monolayer is in the gas-analog phase [14]. When the area is reduced continuously or in discrete steps with a mobile barrier in order to measure a surface pressure/area isotherm the monolayer undergoes a phase transition to the liquid expanded state followed by further transitions to liquid crystalline and solid phases depending on temperature and the nature of the amphiphile. Such phase transitions have been deduced mainly from abrupt changes of the slope or the occurrence of horizontal parts of the surface pressure-area isotherms. Phase transitions have been directly observed and analyzed using fluorescent probes [15-18]. In particular, the phases have been characterized at the air-water interface by synchrotron X-ray investigations[19,20], and unambiguous information on details of molecular packing have been obtained in this way. It is possible to visualize phase transitions directly using Brewster angle microscopy [21,22]. This method does not require fluorescent probes which may influence phase transitions like other impurities.

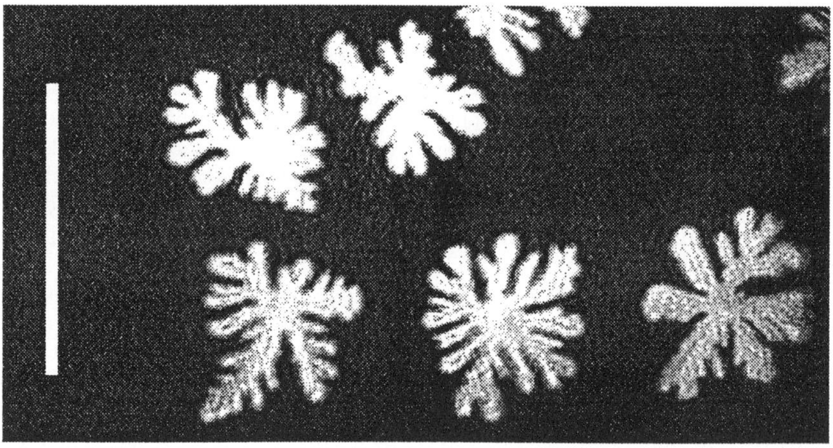

Figure 1. Brewster angle microscopy of a monolayer of dimyristoyl-phosphatidyl-ethanolamine on water; dendritic domains of solid phase (bright) surrounded by coexisting liquid phase; bar: 100 μm, surface pressure: 5 mN/m, temperature: 21°C.

Depending on the speed of compression and the material dendritic growth of domains formed at a phase transition is observed. An example of dendritic domains of solid phase coexisting with the surrounding liquid phase is shown in Figure 1 for a monolayer of dimyristoyl-phosphatidyl-ethanolamine. Even more pronounced dendritic structures have been observed in dioctadecyl-dimethylammonium bromide monolayers [23].

The contrast between the domains and the liquid phase monolayer arises from a different thickness and/or different refractive index of the two phases. A detailed analysis of the reflected light including the polarization provides insight in the structure of monolayer domains.

3. Types of long range order

With synchrotron radiation as intense X-ray source the packing of the molecules in a monolayer at the air-water interface has been determined for various amphiphiles and monolayer phases [9,24,25]. The highest degree of long range order is the crystalline arrangement of the molecules on a well defined lattice, i.e. the positional order. This type is observed in monolayers in the solid state.

In liquid crystals, most molecules do not reside on lattice sites. However, different types of long range order exist [26]. The molecules may be arranged such that the neighbors are in well defined directions over long distances. This bond orientational order (BO), typical e.g. for the smectic phase S_B, is also observed in particular monolayer phases which have been assigned recently the equivalent smectic phases [8]. A particular type of long range bond orientational order of vertically oriented hydrocarbon chains is the herringbone arrangement (smectic S_E) observed with molecules having an elliptical cross section. Very often the long hydrocarbon chains are tilted with respect to the surface normal. Such a tilt may be due to the presence of bulky head groups inhibiting close-packing of the chains in a vertical orientation. If the molecules have the same polar tilt angle and the same azimuthal angle over long range (without residing on lattice sites) the monolayer shows long range tilt orientational order (TO).

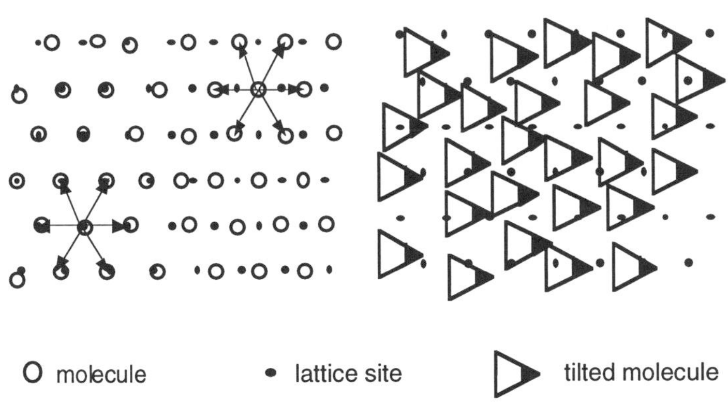

Figure 2. Long range order in monolayers: bond orientational order (left) and tilt orientational order (right); the molecules are not necessarily located on lattice sites.

Long range tilt orientational order gives rise to considerable optical anisotropy in the monolayer since the polarizability of a hydrocarbon chain is larger along the chain axis than perpendicular to the chain axis. Consequently, the refractive index is larger in the direction of the tilted chain than perpendicular to this direction.

4. Brewster angle microscopy: textures

The fundamental idea of Brewster angle microscopy (BAM) is surprisingly simple: p-polarized light incident on the interface between two media of different refractive indices under the Brewster angle is not reflected. With the light beam adjusted in this way, some light is reflected upon formation of a monolayer of a material with a refractive index different of those of the two bulk phases. The homogeneous monolayer or monolayer domains are seen against a zero background, and the method resembles in this sense fluorescence microscopy. The theory of the optical phenomena has been developed on the basis of the 4*4 matrix method of Berreman [27] using reasonable approximations [28]. A different theoretical approach has been published recently [29].

Images are taken by a CCD camera, and changes of the monolayer morphology upon compression/relaxation may be recorded in real time on a video recorder [30]. With an analyzer in the reflected beam the polarization of the reflected light provides information on long range order and molecular packing in individual domains and textures that are typical for liquid crystals and monolayers, like shell, stripe and star textures. An example of a shell texture observed in a monolayer of pentadecanoic acid on water is shown in Figure 3a. Such textures result from gradual or abrupt changes of the orientation of tilted molecules, taking into account that the polarizability of the hydrocarbon chains is larger parallel to the axis of the chain than perpendicular to the axis.

c

Figure 3. Shell texture in a monolayer of pentadecanoic acid at the air-water interface; (a) BAM image; (b) simulated pattern of reflected light intensity; (c) arrangement of tilted molecules, tail ends pointing to the interior of the domain (schematic).

Simulations of shell, star and stripe textures have been done using the simplified theory based on the 4*4 matrix method [10]. The simulation adequately decribing the shell texture of Figure 3a is shown in Figure 3b along with the schematic arrangement (Figure 3c) of the tilted molecules (arrows) in the domain. Here, the molecules are fixed to the water surface by the head groups, and the tails are pointing in a particular way to the interior of the domain. Other arrangements of the molecules lead to quite different patterns of the reflected intensity.

5. Tilt orientational and bond orientational order

So far, qualitative structural information on the type of long range order in particular monolayer phases as well as structural parameters like the tilt angle have been determined by X-ray scattering and X-ray reflection in time consuming experiments (mostly with synchrotron radiation). With the encouraging results of the simulations of textures, the theory has been applied to a quantitative analysis. By measuring the reflected intensity of selected domains under Brewster angle of the pure air-water interface the three parameters (two refractive indices and the tilt angle) of the uniaxial-tilted case adequate for the description of a phase with chains tilted to the nearest neighbour (L_2 or smectic L_I) have been evaluated [28] in excellent agreement with literature data based on X-ray studies. However, it should be stressed that the X-ray studies provide very valuable additional information as compared to Brewster angle microscopy.

Long range bond orientational order may cause a weak in plane anisotropy of the monolayer which is described by two different refractive indices in the plane (and a third normal to the plane). BO may be discriminated from TO in a simple qualitative way by turning the sample under the BAM, i.e. by changing the azimuth angle through 360° [11]. Under selected conditions, the reflectivity shows 2 maxima and 2 minima in the case of BO, whereas TO leads to 1 maximum and 1 minimum. This is experimentally quite simple with monolayers on solid substrates like glass. As exemplified in Figure 4, long range TO is found in a single monolayer of arachidic acid on glass (left), whereas BO is observed in systems of 5 monolayers of cadmium arachidate on glass (right) as well as in a monolayer of docosanoic acid at the air-water interface in the S (smectic S_E) phase (data not shown here).

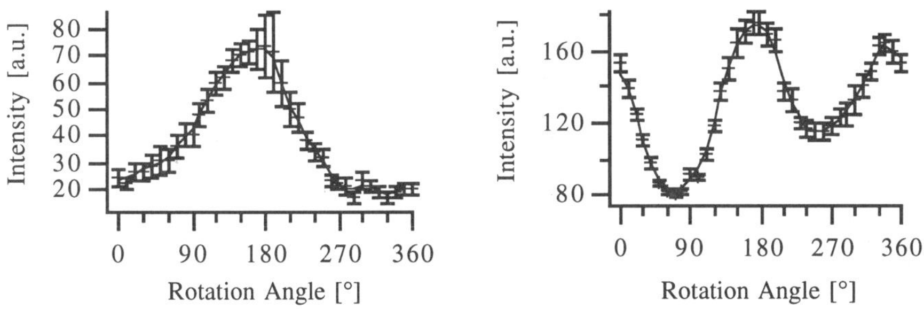

Figure 4. In-plane optical anisotropy of monolayers caused by long range tilt orientational (TO) order or long range bond orientational (BO) order. Discrimination of the two types of order by rotation of the sample around the surface normal in the Brewster angle microscope; reflected light intensity plotted vs. rotation angle; left: monolayer of arachidic acid on glass (1 maximum, TO); right: 5 monolayers of cadmium arachidate on glass (2 maxima, BO).

6. Induction of long range order

When the head group of an amphiphile is very bulky the long hydrocarbon chains are unable to pack, and therefore, form a quasi-liquid layer. This is observed in the case of monolayers of octadecyl-malonic acid (OMA) [31]. The two carboxyl groups prevent the long chains from ordering, and no optical anisotropy is seen by BAM upon compression of the monolayer [32]. If the cavities in the hydrophobic moieties could be filled by hydrocarbon chains, ordering of the chains should occur and a phase with optical anisotropy should be formed under certain conditions.

Indeed, with octadecane (OD) or chloro-octadecane (ClOD), respectively, added to the OMA in a molar ratio of 1:1, the surface pressure/area (π/A) isotherm changes dramatically, indicating dense-packing of the chains [33]. The π/A isotherm of a stearic acid monolayer and of the 1:1 mixed monolayer of OMA and OD resemble so closely, that these two systems have been used for the determination of the dipole moment contribution of the methyl end groups in dense-packed monolayers to the electrical surface potential (dipole compensation method, see [33]). A necessary requirement for this method is that the contribution of the head group region be the same for both systems. However, this has not been unambiguously proven so far.

In the Brewster angle microscope, a monolayer of ClOD : OMA = 1:1 shows formation of anisotropic domains immediately after spreading clearly indicating the induction of long range order in the hydrophobic moiety of the monolayer. The anisotropic domains are observed during compression up to π = 46 mN/m. At π = 47 mN/m a transition to a phase with vertically oriented chains and hexagonal arrangement takes place which appears homogeneous in the Brewster angle microscope [32]. The additive which does not form a stable monolayer itself, is incorporated in the host monolayer and induces the formation of a phase with long range. Therefore, organized monolayers with cavities may be excellent hosts for small or non-amphiphilic molecules that can not be oriented in a well defined way by other means. This is a truely supramolecular approach to ordering this type of molecules.

7. Extended dye aggregates

The concept of chain packing has been used to assist the formation of particular dye aggregates, the Scheibe- or J-aggregates [34], in monolayers of amphiphilic cyanine or merocyanine dyes. These aggregates which also form in solution of water soluble analogs at high concentrations are characterized by a strong and narrow absorption band shifted to longer waves with respect to the absorption band of the monomeric dye and unusual photophysical properties. In monolayers at the air-water interface, the extended two-dimensional crystals of sizes up to 1 mm length can be seen directly in an optical microscope since the light absorption is strongly polarized.

This behavior is due to the particular geometrical arrangement of the chromophores. According to the structural model of Hans Kuhn, the flat chromophores are packed like brickstones in a wall [35,36]. The spectral features of the aggregates have been described in excellent agreement with the experimental results by the extended dipole model. Different to the usual point dipole approximation of excitonic interactions according to Kasha, the dimensions of the molecular dipoles (transition moments) are taken into account since the molecules are in close contact. The unusual photophysical properties like the dependence of the excited state lifetime on temperature and an enhanced radiative rate as compared to the monomer have been explained by the model of coherent exciton in which a large number of dipoles (in the classical picture of an excited molecule) are coupled [37]. Consequently, the radiative rate of such an exciton is increased proportionally to the number of coupled dipoles as compared to the radiative rate of the monomer. First evidence for this effect was found in studies of the temperature dependence of energy transfer from an excited J-aggregate to an energy acceptor in monolayer systems [37].

The structural model of the J-aggregates has been confirmed by various techniques like X-ray diffraction, electron diffraction and second harmonic generation. Quite recently, the chromophore packing in transferred monolayers of an amphiphilic merocyanine dye on mica has been directly observed with scanning force microscopy [38], see Figure 5. The Fourier transform of the area shown (see inset Figure 5) exhibits discrete spots indicating a rectangular centered unit cell with the lattice constants of 0.85±0.06 nm and 1.28±0.06 nm. Consequently, the unit cell comprising two chromophores covers an area of 1.09±0.12 nm^2 in excellent agreement with the area per dye molecule, 0.55 nm^2 , taken from surface pressure/area isotherms.

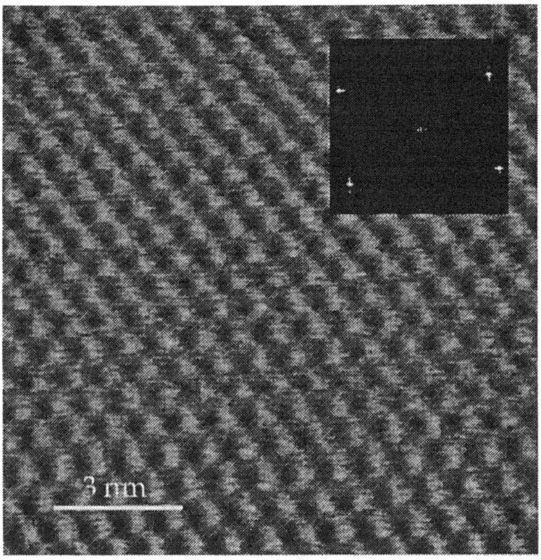

Figure 5. Scanning force microscope (SFM) image of transferred mixed monolayers of an amphiphilic merocyanine dye and octadecane, molar ratio 1:1, on mica; bar: 3 nm; inset: the Fourier transform shows discrete spots indicating a 2D crystalline arrangement of the chromophores in a rectangular centered unit cell.

In addition to the J-aggregates a variety of associates of cyanine dyes like dimers and H-aggregates with characteristic spectroscopic behavior are known. A particular type of cyanine dye aggregate characterized by two absorption bands, one shifted to shorter waves and the second to longer waves with respect to the monomer transition, has been found in monolayers of a water soluble cyanine dye adsorbed from the aqueous subphase to a matrix monolayer [39]. The observations have been rationalized in this case by a herringbone arrangement of the chromophores [40] giving rise to the two transitions oriented perpendicularly to each other.

J-aggregates in monolayers are unique entities with unusual optical and photophysical properties caused by strong chromophore interactions due to the particular crystalline order of the chromophores in large two-dimensional arrays.

8. Conclusion

The air-water interface provides exceptional possibilities to organize molecules to supramolecular units. In particular, long range order either of the hydrocarbon chains or of the hydrophilic chromophores is detected. The new technique of Brewster angle microscopy allows one to observe the textures typical for smectic liquid crystals, to discriminate the different types of long range order, and to determine structural parameters via numerical simulations. Organized monolayers are excellent systems for the systematic investigation of intermolecular interactions as a basis for assembling different molecules to functional supramolecular units.

9. References

1. Bigelow, W.C., Pickett, D.L., and Zisman, W.A. (1946) Oleophobic Monolayers. I. Films Adsorbed from Solution in Non-Polar Liquids, *J. Colloid Sci.* **1**, 513-538.
2. Sagiv, J. (1980) Organized Monolayers by Adsorption. 1. Formation and Structure of Oleophobic Mixed Monolayers on Solid Surfaces, *J. Am. Chem. Soc.* **102**, 92-98.
3. Ulman, A. (1991) *An introduction to ultrathin organic films*, Academic Press, San Diego.
4. Whitesides, G.M. and Laibinis, P.E. (1990) Wet Chemical Approaches to the Characterization of Organic Surfaces: Self-Assembled Monolayers, Wetting, and the Physical-Organic Chemistry of the Solid-Liquid Interface, *Langmuir* **6**, 87-96.
5. Abbott, N.L., Gorman, C.B., and Whitesides, G.M. (1995) Active control of wetting using applied electrical potentials and self-assembled monolayers, *Langmuir* **11**, 16-18.
6. Pockels, A. (1891) Surface Tension, *Nature (London)* **43**, 437-439.
7. Ställberg-Stenhagen, S. and Stenhagen, E. (1945) Phase Transitions in Condensed Monolayers of Normal Chain Carboxylic Acids, *Nature* **156**, 239-240.
8. Peterson, I.R. (1991) The Phase Behaviour of Amphiphilic Monolayers, *Ber. Bunsenges. Phys. Chem.* **95**, 1417-1420.
9. Kenn, R.M., Böhm, C., Bibo, A.M., Peterson, I.R., Möhwald, H., Als-Nielsen, J., and Kjaer, K. (1991) Mesophases and Crystalline Phases in Fatty Acid Monolayers, *J. Phys. Chem.* **95**, 2092-2097.
10. Overbeck, G., Hönig, D., and Möbius, D. (1994) Stars, Stripes and Shells in Monolayers: Simulation of the Molecular Arrangement in Schlieren Structures, *Thin Solid Films* **242**, 213-219.
11. Overbeck, G.A., Hönig, D., Wolthaus, L., Gnade, M., and Möbius, D. (1994) Observation of Bond Orientational Order in Floating and Transferred Monolayers with Brewster Angle Microscopy, *Thin Solid Films* **242**, 26-32.
12. Bücher, H., Drexhage, K.H., Fleck, M., Kuhn, H., Möbius, D., Schäfer, F.P., Sondermann, J., Sperling, W., Tillmann, P., and Wiegand, J. (1967) Controlled transfer of excitation energy through thin layers, *Mol. Cryst.* **2**, 199-230.
13. Fromherz, P. (1973) A new method for investigation of lipid assemblies with a lipoid pH indicator in monomolecular films, *Biochim. Biophys. Acta* **323**, 326-334.
14. Gershfeld, N.L. and Pagano, R.E. (1972) Physical chemistry of lipid films at the air-water interface. I. Intermolecular energies in single-component lipid films, *J. Phys. Chem.* **76**, 1231-1237.
15. Weis, R.M. and McConnell, H.M. (1984) Two-dimensional chiral crystals of phospholipid, *Nature* **310**, 47-49.

16. Knobler, C. (1990) Seeing phenomena in flatland: studies of monolayers by fluorescence microscopy, *Science* **249**, 870-874.
17. Möhwald, H. (1988) Direct characterization of monolayers at the air-water interface, *Thin Solid Films* **159**, 1-15.
18. Möhwald, H. (1990) Phospholipid and phospholipid-protein monolayers at the air/water interface, *Annu. Rev. Phys. Chem.* **41**, 441-476.
19. Kjaer, K., Als-Nielsen, J., Helm, C.A., Laxhuber, L.A., and Möhwald, H. (1987) Ordering in Lipid Monolayers Studied by Synchrotron X-Ray Diffraction and Fluorescence Microscopy, *Phys. Rev. Lett.* **58**, 2224-2227.
20. Dutta, P., Peng, J.B., Lin, B., Ketterson, J.B., Prakash, M., Georgopoulos, P., and Ehrlich, S. (1987) X-Ray Diffraction Studies of Organic Monolayers on the Surface of Water, *Phys. Rev. Lett.* **58**, 2228-2231.
21. Hénon, S. and Meunier, J. (1991) Microscope at the Brewster angle. Direct observation of first-order phase transitions in monolayers., *Rev. Sci. Instrum.* **62**, 936-939.
22. Hönig, D. and Möbius, D. (1991) Direct visualisation of monolayers at the air-water interface by Brewster angle microscopy, *J. Phys. Chem.* **95**, 4590-4592.
23. Ahuja, R.C., Caruso, P.-L., Hönig, D., Maack, J., Möbius, D., and Overbeck, G.A. (1994) Morphology of Organized Monolayers by Brewster Angle Microscopy, in H. Masuhara, F. C. De Schryver, N. Kitamura and N. Tamai (eds.), *Microchemistry, Spectroscopy and Chemistry in Small Domains*, Elsevier Science Publishers B.V., Amsterdam, pp. 211-223.
24. Shih, M.C., Bohanon, T.M., M., M.J., Zschack, P., and Dutta, P. (1992) X-ray-diffraction study of the superliquid region of the phase diagram of a Langmuir monolayer, *Phys. Rev. A* **45**, 5734-5737.
25. Als-Nielsen, J., Jacquemain, D., Kjaer, K., Leveiller, F., Lahav, M., and Leiserowitz, L. (1994) Principles and applications of grazing-incidence x-ray and neutron-scattering from ordered molecular monolayers at the air-water-interface, *Phys. Reports* **246(N5)**, 252-313.
26. Demus, D. and Richter, L. (1978) *Textures of Liquid Crystals*, Verlag Chemie, Weinheim.
27. Berreman, D.W. (1972) Optics in Stratified and Anisotropic Media: 4x4-Matrix Formulation, *J. Opt. Soc. Am.* **62**, 502-510.
28. Hönig, D. (1994) *Molekular organisierte Filme: laterale Struktur und optische Eigenschaften,* PhD Thesis, Georg-August-Universität Göttingen.
29. Tabe, Y. and Yokoyama, H. (1995) Fresnel Formula for optically anisotropic Langmuir monolayers: an application to Brewster angle microscopy, *Langmuir* **11**, 699-704.
30. Hönig, D., Overbeck, G.A., and Möbius, D. (1992) Morphology of Pentadecanoic Acid Monolayers at the Air/Water Interface Studied by BAM, *Adv. Mater.* **4**, 419-424.
31. Cordroch, W. and Möbius, D. (1992) Incorporation of non-amphiphilic compounds into host monolayers, *Thin Solid Films* **210/211**, 135-137.
32. Cordroch, W., Spohn, D.B., Hönig, D., and Möbius, D. Guest-host interactions in monolayers, *in preparation.*
33. Vogel, V. and Möbius, D. (1988) Local Surface Potentials and Electric Dipole Moments of Lipid Monolayers: Contributions of the Water/Lipid and the Lipid/Air Interfaces, *J. Colloid Interface Sci.* **126**, 408-420.
34. Möbius, D. (1995) Scheibe-Aggregates: highly ordered systems of strongly interacting chromophores, *Adv. Mater.* **7**, 437-444.
35. Czikkely, V., Försterling, H.D., and Kuhn, H. (1970) Extended dipole model for aggregates of dye molecules, *Chem. Phys. Lett.* **6**, 207-210.
36. Czikkely, V., Försterling, H.D., and Kuhn, H. (1970) Light absorption and structure of aggregates of dye molecules, *Chem. Phys. Lett.* **6**, 11-14.

37. Möbius, D. and Kuhn, H. (1979) Monolayer assemblies of dyes to study the role of thermal collisions in energy transfer, *Israel J. Chem.* **18**, 375-384.
38. Wolthaus, L., Schaper, A., and Möbius, D. (1994) Brickstone Arrangement in J-Aggregates of an Amphiphilic Merocyanine Dye, *Chem. Phys. Lett.* **225**, 322-326.
39. Kirstein, S., Möhwald, H., and Shimomura, M. (1989) Crystalline two-dimensional domains of cyanine dyes at interfaces, *Chem. Phys. Lett.* **154**, 303-308.
40. Kirstein, S. and Möhwald, H. (1992) Structure and optical properties of a monolayer single crystal of a cyanine dye, *Chem. Phys. Lett.* **189**, 408-413.

Abbreviations:

BAM	Brewster angle microscopy
BO	long range bond orientational order
ClOD	chloro-octadecane
DMPE	dimyristoyl-phosphatidyl-ethanolamine
OD	octadecane
OMA	octadecyl-malonic acid
TO	long range tilt orientational order

APPLICATIONS OF THE CAMBRIDGE STRUCTURAL DATABASE IN THE STUDY OF NON-COVALENT INTERACTIONS

FRANK H. ALLEN
Cambridge Crystallographic Data Centre
12 Union Road, Cambridge CB2 1EZ, UK

1. Introduction

In his Nobel Lecture, J.-M.Lehn [1] defined supramolecular chemistry as "the chemistry of the intermolecular bond" : a supermolecule is an assembly of covalently bonded units or ions that is organised according to the diverse weak forces that govern non-covalent interactions. Such a description, of course, is equally applicable to the extended structure of molecular crystals, where the crystalline supermolecule is assembled by molecular self-recognition or through the mediacy of additional components of the system, particularly solvent molecules or counter ions. If we are to make advances in the broad area of supramolecular chemistry, which must encompass such topics as the study of protein - ligand interactions in rational drug design and the design of molecular solids through crystal engineering, then we must acquire systematic information about non-covalent interactions. In particular, we need to know:

- What are the most important types of non-covalent interactions?
- What are their geometrical and directional characteristics?
- How strong are the interactions?

Modern molecular orbital calculations, for example the intermolecular perturbation theory of Hayes & Stone [2], now overcome basis set superposition errors and provide realistic models and good energy estimates for non-covalent interactions involving small molecular systems *in vacuo*. However, these techniques are cpu-intensive, and require both technical expertise and some knowledge of likely starting geometries. For these reasons, they are not well suited to systematic surveys, but are valuable for application to specific systems, particularly to provide answers to the third question posed above. Systematic answers to the first two questions can, however, be provided by in-depth analyses of crystalline supermolecules: the only direct experimental observations of non-covalent interactions that are available in sufficient quantity and quality. Quantitative estimates of the strengths of interactions cannot, of course, be obtained from crystal structure data, although qualitative relationships can sometimes be inferred.

In this chapter, then, we examine how the Cambridge Structural Database (CSD) System [3], a database of small-molecule structures together with appropriate search and analysis software, can make significant contributions to our knowledge of

G. Tsoucaris et al. (eds.), Crystallography of Supramolecular Compounds, 55–66.

non-covalent interactions. Indeed, its high structural diversity, coupled with the minimal cpu requirements for the analyses and the condensed-phase nature of the data, make the CSD a highly appropriate basis for these studies.

2. The Cambridge Structural Database (CSD) System

The CSD currently (October 1995) contains details of nearly 150,000 structures and the information content of each CSD structural entry is best summarised according to its 'dimensionality' :

- Bibliographical and chemical text (1D)

- Chemical structural formula (2D): a compact and fully searchable representation of the chemical connectivity in terms of element types, bond types, hydrogen counts, formal charges, etc.

- Crystallographic structural data (3D): cell parameters, space group, and atomic coordinates that are mapped directly to the relevant atoms of the 2D chemical formula.

The CSD software system comprises three main programs, each having a fully interactive graphical interface for user applications:

- QUEST3D is the main search program, permitting searches of all available information fields. Most important among these are (a) the location of 2D substructures, which may then be further constrained using a wide variety of 3D geometrical criteria, and (b) the location of non-covalent intermolecular contacts in the extended crystal structure, again using geometrical criteria. Searches of this type are fundamental to the studies summarised in this Chapter and relevant examples are illustrated in subsequent sections. The program will generate output files of user-defined geometrical parameters for any substructure (intramolecular or intermolecular) located in the search process.

- VISTA reads the geometrical table(s) generated by QUEST3D and provides extensive facilities for the graphical representation (histograms, scattergrams, polar plots, etc) and statistical analysis (summary statistics, regression, principal component analysis, etc) of the numerical data. Many of the illustrations for this Chapter were generated with VISTA.

- PLUTO is used to visualise crystal and molecular structures in a variety of styles.

Other databases within the total CSD System now comprise:

- A compacted version of the Protein Data Bank, in which all non-coordinate information of the PDB, including sequence information, can be searched through the QUEST3D program.

- A database of literature references (currently over 500) to scientific applications of the CSD in a research environment.

We now illustrate applications of the CSD to the study of a variety of non-bonded interactions.

3. Studies of Medium to Strong Hydrogen Bonds

The most ubiquitous and most important of non-covalent interactions is the hydrogen bond. Over the past forty years, most of our knowledge of hydrogen-bond geometry and directionality has been accumulated from crystal structure data, as evidenced by Jeffrey & Saenger's recent monograph [4]. In this Section, we exemplify the use of the CSD System through a study of N-H and O-H donation to thiocarbonyl (>C=S) acceptors [5].

The non-bonded contact search fragment for >C=S...H-(N or O) systems is shown in Figure 1. Formal chemical bond types are defined in the 2D search menus of the QUEST3D program. The hydrogen bond is defined according to distance constraints, either on the S....(O or N) contact, or on the S...H hydrogen bond itself. Since X-ray methods yield poor hydrogen geometry (X-H bonds lengths are typically foreshortened, although their vector directions are usually acceptable), the CSD software allows the H-atoms to be 're-positioned' along the X-ray vector direction, but at a distance that is typical of neutron-diffraction bond lengths. These 'normalised' H-positions were used throughout this Chapter when appropriate. For the S...H system of Figure 1, S...H distances out to the van der Waals radii limit (v(H) = 1.1, v(S) = 1.8, limit = 2.9A) were classified as hydrogen bonds.

For each fragment that fulfilled this basic distance criterion, additional geometrical parameters were calculated as shown in Figure 1. A further geometrical criterion, that $90 < \rho_H < 180^o$, was then imposed. Finally, to study the directionality of H-approach to S, we calculated the angles of elevation and rotation (θ and ϕ, Figure 1) of the S....H vector with respect to the >C=S plane. Thus, if H approaches S along an (assumed) sp^3 lone-pair direction, then θ and ϕ will be close to 0 and 120° respectively. In all CSD analyses of this type, it is also wise to impose some global secondary search constraints, particularly concerning the precision of crystal structures from which the hit fragments are retrieved.

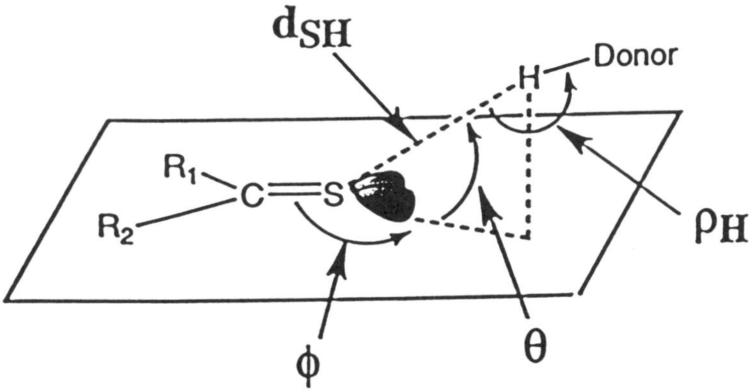

Figure 1. Substructure search and geometrical descriptors for >C=S...H-(N or O) systems

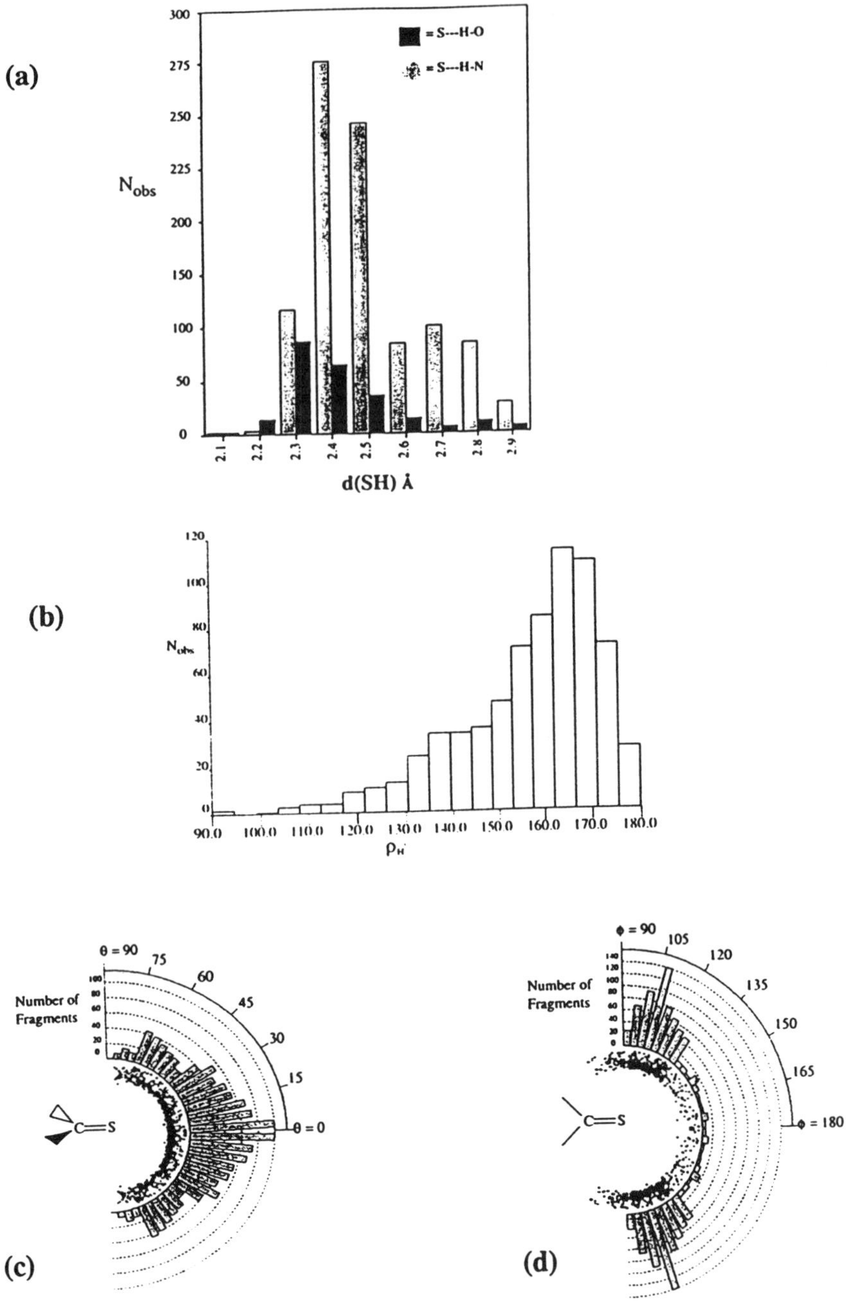

Figure 2 Normal and polar histograms of (a) d(SH), (b) ρ_H, (c) θ and (d) ϕ for >C=S...H-N,O systems

Figure 2 illustrates histograms of d(SH), ρ_H, θ and ϕ for >C=S...H-N,O systems. Figure 2a confirms that the choice of the 2.9A H-bond limit was a reasonable one. Further, mean d(SH) values derived from Figure 2a for O-H and N-H donors are some 0.25A longer than mean values obtained from analogous >C=O...H-N,O systems, even after allowance is made for the larger size of S. We may conclude that =S...H bonds are considerably weaker than =O...H bonds. Figure 2b shows the typical tendency of H-bonds towards linearity. The ϕ–histogram does indicate that there are preferred directions of H-approach to S, although the ϕ-peak occurs some 10° lower than would be expected for an $S(sp^3)$ lone pair - an indication, perhaps, of d-orbital participation. The θ-histogram, however, shows some wide variations from 0°, i.e. H approach to S is not uniquely close to the lone pair plane. Further examination of the dataset shows that many of the high-θ approaches involve multiple H-donation from water, a small molecule that can approach diffuse lone-pair density with minimal steric hindrance.

Figure 2 begs the obvious question: given that C and S have a minimal electronegativity difference, then why does >C=S accept hydrogen bonds at all? Detailed examination of the dataset [5] shows (Figure 3a) that only the longer C=S bonds accept H-bonds: only 5% of bonds having d(CS) = 1.63A form H-bonds, a figure that rises to more than 75% for d(CS) > 1.70A. Examination of the R1,R2-C=S substructures retrieved from the CSD indicates that systems having R1 and/or R2 = N are dominant. Here, electron donation from the N lone pair(s) (Figure 3b) induces a partial negative charge on S and lowers the C=S double bond character. Thus, H-bonding to C=S may be termed 'resonance-induced', an extension of the concept of resonance-assisted H-bonding developed by Gilli and co-workers [6].

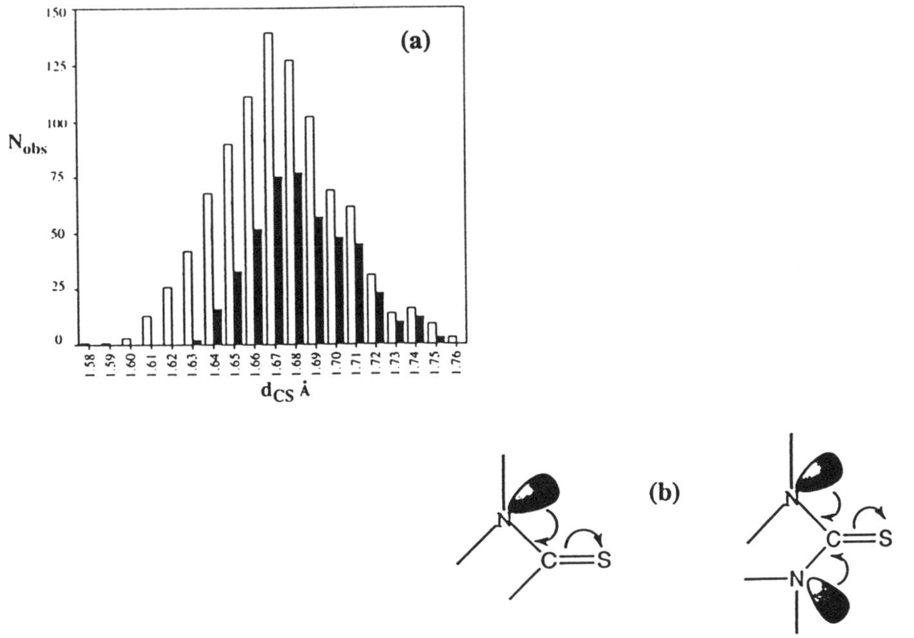

Figure 3. (a) Histogram of all d(CS) values (pale bars) and d(CS) values in H-bonded systems (black bars), (b) Resonance in R1,N-C=S and N2-C=S systems.

4. C-H...O Hydrogen Bonding

Despite early controversies that first enhanced and then contradicted a growing body of spectroscopic data, crystallographic evidence for the existence of C-H...X hydrogen bonds is now both well established [7] and increasingly well documented [4,8]. Although weak (4-8kJ.mol.$^{-1}$) these bonds, and particularly the C-H...O bond, are now seen as important secondary interactions in biological structures and can become primary, i.e. structure determining, in some small molecule structures. A continuing area of interest has been to establish the relative donor abilities of C-H from different chemical environments. Over the years, spectrosccopic data had indicated that donor ability decreased in the order Csp^1-H > Csp^2-H > Csp^3-H. This general H-acidity requirement was also noted in crystallographic studies [7] and has recently been systematically analysed using information retrieved from the CSD [9,10], resulting in a novel scale of carbon acidity based upon C...O separations from a wide variety of systems [10]. Acetylenic protons are particularly acidic and are the most potent C-H donors to carbonyl-O, as exemplified by the histogram of H...O distances (Figure 4a) and the scatterplot of d(H...O) vs the C-H...O angle (Figure 4b) that can rapidly be generated from the CSD. Figure 4a shows that a large proportion of the H...O contacts are significantly shorter than the sum of van der Waals radii (shaded area of the plot), while Figure 4b shows typical H-bond behaviour with the two parameters being highly correlated (r = -0.688).

5. X-H...π Hydrogen Bonding

Spectroscopic evidence for the existence of N,O-H..π hydrogen bonding to acetylenic, olefinic and aromatic acceptors is well documented [11]. A survey of crystallographic evidence for this type of bonding [12] has shown that intermolecular examples are clearly observed and that these bonds, although very weak, can be both structurally and energetically significant. Recently, Steiner [13,14] has presented novel crystal structures, database evidence and quantum chemical calculations on C≡C-H...π(C≡C) and π(phenyl) bonding. He cites H...C≡C (midpoint) distances as low as 2.51A, and observes H-bond cooperativity in extended systems with H-bond energies in the range 4.2-9.2 kJ.mol.$^{-1}$. Spectroscopic data [11] also indicate that cyclopropane rings can act as π(acceptors) in two modes: (a) via edge-on interactions with the C-C bonds in the ring plane, or (b) by face-on interactions with the ring centroid. Both bonding modes were observed and characterised in a recent CSD study [15]. The plot (Figure 5) maps the movement of donor-H about the cyclopropane ring and shows three clusters of data points: (a) edge-on contacts: points with long d(HY) values and the angle H-X1-Y tending towards 180°, (b) face-on contacts: points with the shortest d(HY) and H-X1-Y less than 90°, and (c) a few intermediate d(HY) values having H-X1-Y in the range 100-120°.

6. Atom...Group Interactions: Halogen...Nitro

The QUEST3D program also permits searches for Atom...Group interactions, i.e. the target atom can be constrained to lie within a preset distance of any atom of a functional group. An early CSD analysis of interactions between halogen atoms and the O-atoms of nitro groups [16] had indicated that the halogen prefers to approach in the plane of the

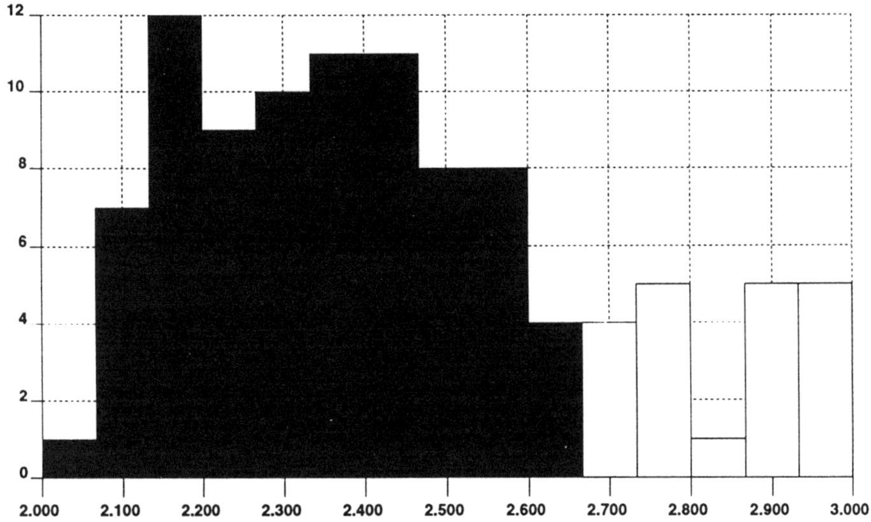

(a)

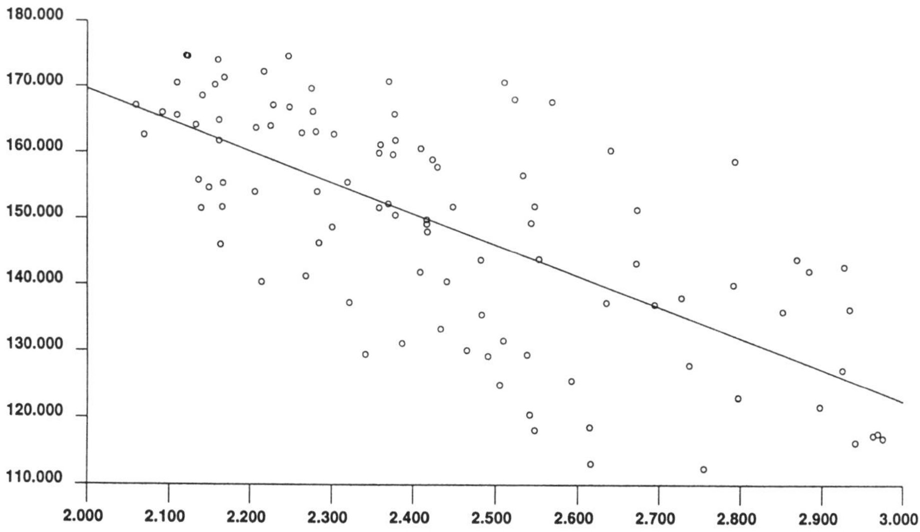

(b)

Figure 4. Csp^1-H...O(carbonyl) hydrogen bonds: (a) histogram of the H...O distance, and (b) scatterplot of the H...O distance versus the C-H...O hydrogen bond angle

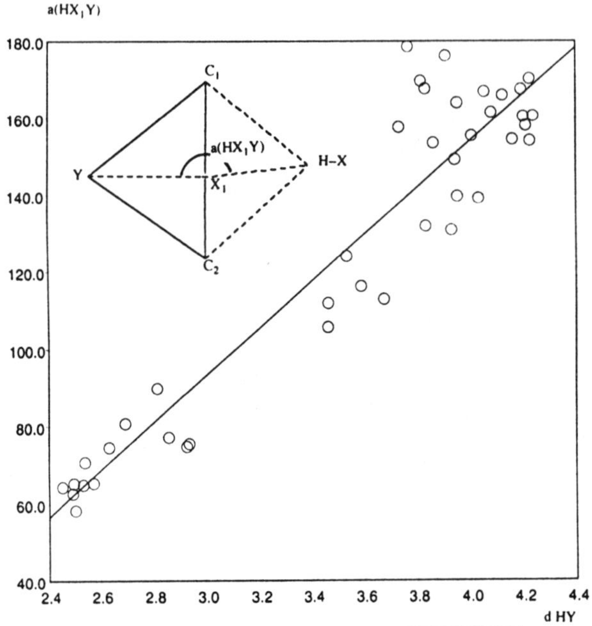

Figure 5. N,O-H....π(cyclopropane) hydrogen bonds: mapping of the H approach to the cyclopropane ring (see text).

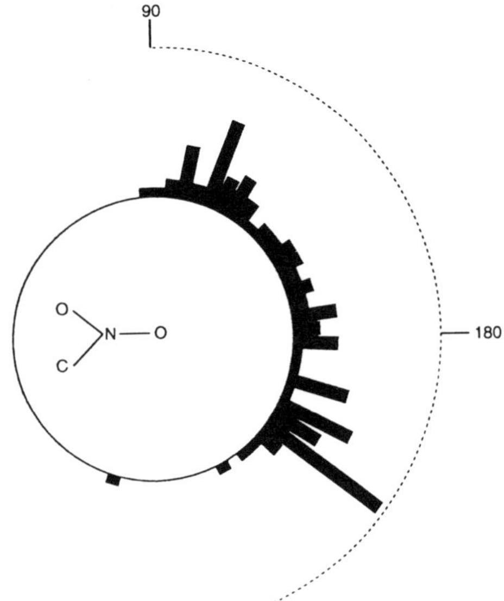

Figure 6. Circular histogram mapping the approach of Cl to the O atoms of nitro groups

nitro group, and either between the two oxygen atoms, or between the C and O atoms. This study has been repeated here for short chloro....nitro interactions using the new software and a constraint that Cl must be within 3.27A of either nitro O-atom. The circular histogram of Figure 6 maps the directionality of approach and confirms the general features of the earlier study. The importance of this interaction has recently been confirmed through the engineering of crystal structures in which iodo...nitro interactions are a key feature of the extended crystal structure [17].

7. Group....Group Interactions: Carbonyl...Carbonyl

The CSD software also permits the location of Group....Group contacts, i.e. a search constraint may be imposed that requires any atom of Group 1 to be less than some preset distance from any atom of Group 2. In Figure 7a, this distance constraint was set at 3.6A and the geometry of the interacting carbonyls was described by the C=O....C=O torsion angle (denoted as Tor) and by the angle of elevation of the carbonyl carbon of Group 2 with respect to the plane through Group 1. Earlier work [18] had indicated that carbonyl groups in crystal structures tend to form mutually antiparallel dipolar interaction motifs, with the C(δ+)=O(δ-) dipole of Group 2 overlying the O(δ-)=C(δ+) dipole of Group 1. The histograms of Figure 7b,c show this arrangement to be highly preferred: over 90% of the 574 observations of Tor are within a few degrees of zero, and the angle of elevation clusters around 90°. Molecular orbital calculations using the IMPT method [2] indicate an interaction energy of about $20kJ.mol.^{-1}$, equivalent to that of a medium strength hydrogen bond.

8. The Answer "No".

The previous sections have concentrated on illustrating the identification and characterisation of some of the the important non-covalent interactions that are observed in crystal structures. Equally important to supramolecular chemistry and molecular modelling applications is a knowledge of when such interactions do *not* exist, even though chemical sensibility might indicate otherwise. One example is the thioether fragment, C-S-C, where H-bond formation to the S-acceptor might be inferred by chemical analogy with the oxy-ethers C-O-C. A CSD study reveals that only 4.8% of more than 1000 thioether-S atoms form contacts to H-donors that are within van der Waals radii sums, and even when they do occur, these contacts are close to this distance limit. Simple MO calculations show the electroneutrality of S in dialkyl-thioethers, and an increasing δ-positivity of S as the alkyl groups are replaced by unsaturated centres [19], due to S-lone pair delocalisation in rings such as thiophene, etc. A similar mechanism would appear to operate in furan: out of 118 instances in which furan co-exists with N-H or O-H donors in crystal structures, the hetero-O atom forms hydrogen bonds on only 12 occasions.

9. Conclusion

The crystallographic databases, particularly the CSD and the Protein Data Bank, already play an important role in molecular modelling projects. Allied to search and analysis software of increasing sophistication, these mature databases now provide a sound basis for the acquisition of knowledge concerning those non-covalent interactions that are of fundamental importance in the supramolecular arena. Indeed, it would appear that a

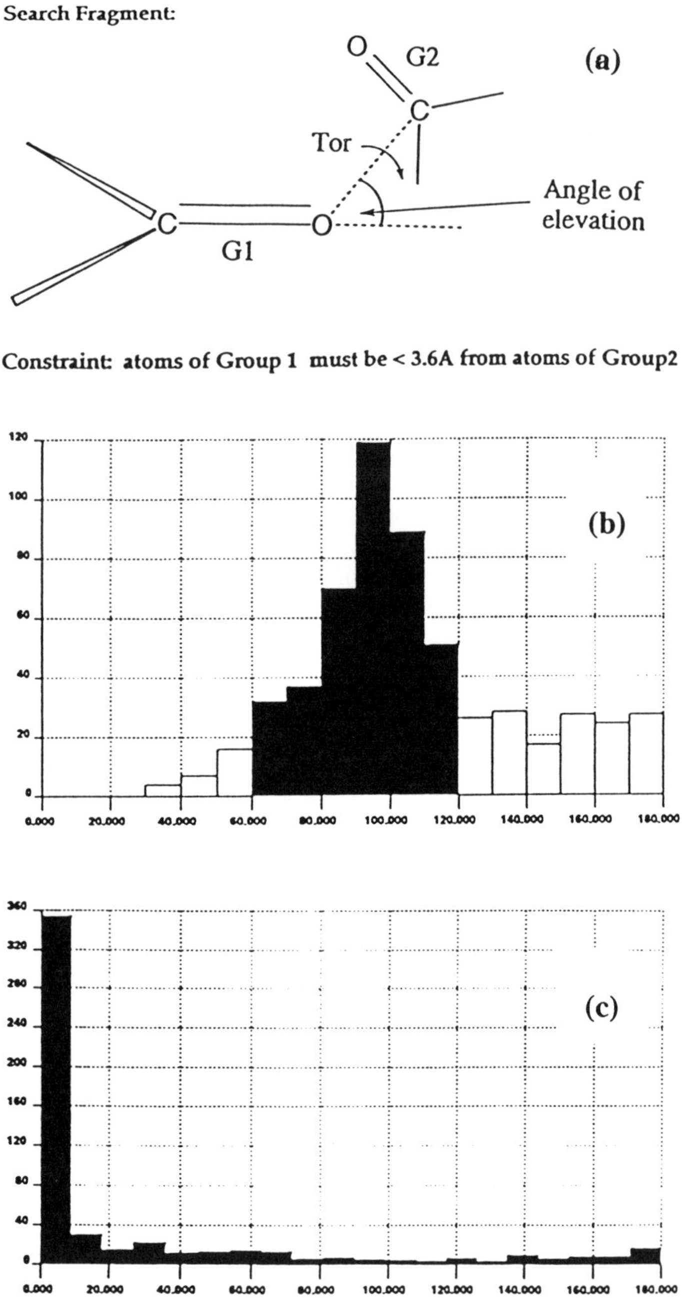

Figure 7. Carbonyl....carbonyl interactions: (a) CSD search fragment, search constraints and geometrical descriptors, (b) histogram of the angle of elevation, and (c) histogram of the torsion angle (Tor).

joint approach, involving database analyses coupled with high level molecular orbital calculations now seem to offer a chance to answer many important questions in this area.

However, we would note that the databases are just that: reservoirs of raw structural data from which knowledge must be acquired through intensive use of software for search, retrieval and data analysis. It has been the purpose of this Chapter to illustrate these procedures. The next development, obviously, is to capture the results of these endeavours in an organised and structured form in computerised systems. This development of computerised knowledge bases from the underlying databases, and incorporating relevant information from other sources, will undoubtedly be the key development of the next decade. Such knowledge bases form the central memory requirement for intelligent software systems that may then be designed to assist in problem solving in the domain of supramolecular chemistry [20].

10. References

1. Lehn, J.-M. (1988) Supramolecular Chemistry - Scope and Perspectives: Molecules, Supermolecules and Molecular Devices, *Angew.Chem. (Int.Ed.Engl.)*, **27**, 90-112.
2. Hayes, I.C. and Stone, A.J. (1984) Intermolecular Perturbation Theory, *Mol.Phys.*, **53**, 84-98.
3. Allen, F.H., Davies, J.E., Galloy, J.J., Johnson, O., Kennard, O., Macrae, C.F., Mitchell, E.M., Mitchell, G.F., Smith, J.M., and Watson, D.G. (1991) The Development of Versions 3 and 4 of the Cambridge Structural Database System, *J.Chem.Inf.Comput.Sci.*, **31**, 187-204.
4. Jeffrey, G.A. and Saenger, W. (1991) *Hydrogen Bonding in Biological Structures*, Springer Verlag, Berlin.
5. Allen, F.H., Bird, C.M. and Rowland, R.S. (1995) The Hydrogen-Bond Acceptor Properties of 'Univalent' Sulphur (X=S), *Acta Crystallogr.*, Submitted.
6. Gilli, G., Bellucci, F., Ferretti, V. and Bertolasi, V. (1989) Evidence for Resonance-Assisted Hydrogen Bonding from Crystal Structure Correlations on the Enol Form of the β-Diketone Fragment, *J.Am.Chem.Soc.*, **111**, 1023-1028.
7. Taylor, R. and Kennard, O. (1982) Crystallographic Evidence for the Existence of C-H...O, C-H...N and C-H...Cl Hydrogen Bonds, *J.Am.Chem.Soc.*, **104**, 5063-5070.
8. Desiraju, G.R. (1991) The C-H...O Hydrogen Bond in Crystals: What Is It?, *Acc.Chem.Res.*, **24,** 290-296.
9. Desiraju, G.R. and Murty, B.N. (1987) Correlations between Crystallographic and Spectroscopic Properties for C-H...O Bonds in Terminal Acetylenes, *Chem.Phys.Lett.*, **139**, 360-361.
10. Pedireddi, V.R. and Desiraju, G.R. (1992) A Crystallographic Scale of Carbon Acidity, *J.Chem.Soc., Chem.Commun.*, pp 988-990.
11. Joris, L., Schleyer, P. von R. and Gleiter, R. (1968) Cyclopropane Rings as Proton-Acceptor Groups in Hydrogen Bonding, *J.Am.Chem.Soc.*, **90**, 327-336.
12. Viswamitra, M.A., Radhakrishnan, R., Bandekar, J. and Desiraju, G.R. (1993) Evidence for O-H...C and N-H C Hydrogen Bonding in Crystalline Alkynes, Alkenes and Aromatics, *J.Am.Chem.Soc.*, **115,** 4868-4869.
13. Steiner, T. (1995) Cooperative C≡C-H...C≡C-H Interactions: Crystal Structure of DL-Prop-2-ynylglycine and Database Study of Terminal Alkynes, *J.Chem.Soc., Chem.Commun.*, pp 95-96.
14. Steiner, T., Starikov, E.B., Amado, A.M. and Teixeira-Dias, J.J.C. (1995) Weak Hydrogen Bonding. Part 2. The Hydrogen Bonding Nature of Short C-H...π

contacts: Crystallographic, Spectroscopic and Quantum Mechanical Studies of Some Terminal Alkynes, *J.Chem.Soc., Perkin Trans. 2,* pp 1321-1326.

15. Allen, F.H., Lommerse, J.P.M., Hoy, V.J., Howard, J.A.K. and Desiraju, G.R. (1995) The Hydrogen-Bond C-H Donor and π-Acceptor Characteristics of Three-Membered Rings, *Acta Crystallogr.*, Submitted.
16. Desiraju, G.R., Pedireddi, V.R., Sarma, J.A.R.P. and Zacharias, D.E. (1993) Determination of the Crystal Structure of 4-Iodo-β–Nitrostyrene with the aid of Packing Considerations: A Study of Halogen...Nitro Interactions, *Acta Chim. Hungarica*, **130,** 451-465.
17. Allen, F.H., Goud, S.B., Hoy, V.J., Howard, J.A.K. and Desiraju, G.R. (1994) The Crystal Structures of 4-Iodonitrobenzene and of the Complex Between 1,4-Diiodobenzene and 1,4-Dinitrobenzene, *J.Chem.Soc., Chem. Commun.*, pp 2729-2731.
18. Taylor, R., Mullaley, A. and Mullier, G.W. (1990) Use of Crystallographic Data in Searching for Isosteric Replacements: Composite Crystal-Field Environments of Nitro and Carbonyl Groups, *Pestic. Sci.*, **29**, 197-213.
19. Burling, F.T. and Goldstein, B.M. (1992) Computational Studies of Nonbonded Sulfur...Oxygen and Sulfur...Selenium Interactions in the Thiazole and Selenazole Nucleosides, *J.Am.Chem.Soc.*, **114,** 2313-2320.
20. Allen, F.H., Rowland, R.S., Fortier, S. and Glasgow, J.I. (1990) The Design of Intelligent Systems for Molecular Scene Analysis, *Tetrahedron Computer Methodology*, **3**, 757-771.

Solvent and Dynamic Effects on the Structure of Alkali Cation Complexes of the *t-Butyl*-calix[4]arene Anion: MD and FEP Computer Investigations on the Na^+ / Cs^+ Binding Affinity

A. Varnek, C. Sirlin and G. Wipff*
Laboratoire MSM, URA 422 CNRS,
Institut de Chimie, 4, rue B. Pascal, 67 000 Strasbourg France.

Keywords: Ionophores; calixarenes; molecular dynamics; preorganization; binding selectivity, solvent effect, molecular recognition.

Abstract. The *t-butyl*-calix[4]arene monooxyanion **L**$^-$, unsubstituted at the lower rim displays ionophoric behaviour and transports alkali cations through an organic phase, with the largest efficiency for Cs^+. The nature of such complexes *in solution* and the basis of the binding selectivity are unknown. In the solid state, the Cs^+ complex is *endo* (M^+ inside the cone of the host) while the Na^+ complex is *exo*. We report here a theoretical study based on molecular dynamics and free energy simulations *in vacuo*, in pure water, acetonitrile and chloroform, on both *endo* and *exo* complexes. For a given cation, we find that Na^+ to Cs^+ complexes display similar structures in the gas phase, in acetonitrile and in chloroform. Intrinsically, the *endo* forms are more stable than *exo*, whereas solvation favours *exo* complexes. In water, all M^+ complexes dissociate, except that of Li^+. The role of the *t-butyl* "gate" and pH on the stability of *endo* complexes were investigated. Based on free energy perturbation calculations, we predict in the gas phase or in chloroform that the binding selectivity is largest for Na^+, the order being $Na^+ > K^+ > Rb^+ > Cs^+$ for *exo* or *endo* complexes and for complexes in acetonitrile. The polarization energy is shown to favour *endo* complexation *in vacuo* and in chloroform. In acetonitrile, as in polar solvents, *exo* forms are

G. Tsoucaris et al. (eds.), Crystallography of Supramolecular Compounds, 67–99.

preferred because of the better solvation of the complex. Concerning the cation extraction from water to an organic phase, we calculated that Cs^+ is best extracted ($Cs^+ > Rb^+ > K^+ > Na^+$) in qualitative agreement with experimental data. We therefore ascribe the high Cs^+ transport and extraction by this calixarene to a differential solvation effect, rather than to special "cation -π" interactions. Finally, we report a MD simulation of the Cs^+ $\mathbf{L}^-$complex at a water / chloroform interface, where it remains "adsorbed" like a surfactant.

INTRODUCTION

Solid state structures of host-guest ionophoric systems, although invaluable for studying the geometrical features of their lock-and-key complementarity, [1] may not be relevant in solution state where complexation, extraction or transport processes take place. [2]. In solution, however, experimentation alone cannot provide precise pictures concerning the nature of the complexes and their conformation. NMR spectroscopy provides information on time averaged conformational ensembles which may not correspond to low energy instant structures. Computer simulations in supramolecular chemistry [3] and on ionophores [4] become a powerful tool to investigate structures *in solution*. Monte-Carlo and molecular dynamics techniques, using explicit representations of the solvent molecules provide individual microscopic pictures, as well as statistical representations. One advantage of computer simulations over experiments is the lack of physical constraints which allows to consider, for instance, hydrophobic species in water, and ionic species in organic solvents without achieving the total electroneutrality of the system, intimate like-ion pairs in different solvents, etc.... In a recent paper, we reported several computational studies of uncomplexed hosts (18-crown-6, the 222 cryptand, calixarenes) and their complexes with ions and ion pairs, whose nature and conformation changed from the solid to the simulated solution states, and from water to non-aqueous solutions. [3]

Here, we describe a simulation study of the M^+ salts ($M^+ = Li^+$ to Cs^+) of the *p-tert*-butylcalix[4]arene anion (referred to as $\mathbf{L}^-$, see Chart I) in the gas phase, and in aqueous and non-aqueous solvents, and compare the results to those of the solid state structures. The structural and energy analysis illustrate the role of dynamics, and of solvent and environment effects on the *endo* / *exo* / dissociated nature of the "complexes". In addition, we extend these structural features to the question of selectivity in complexation and extraction processes.

Finally, in relation to the ionophoric behaviour of calixarenes, we report a simulation at a water / organic interface explicitly represented.

Calixarenes, cyclic oligomers based on phenolic units [5, 6] represent an interesting platform to design ionophores. With suitable combination of their topological features and functionalization at their lower rim, it has become possible to synthetize neutral hosts which selectively bind and transport cations between two aqueous phase through a non-aqueous solution.[7-9] The unsubstituted *p-tert*-butylcalix[n]arene ligands (n = 4; 6; 8) display the following features desirable as membrane carriers: low water solubility, formation of neutral complexes with cations through the loss of a proton, and the potential for coupling cation transport through the reverse flux of protons. [10] Indeed, these ligands transport metal hydroxides through an organic phase (25% v/v CH_2Cl_2 - CCl_4), with a marked selectivity for Cs^+, increasing in the order [8] < [6] < [4]. [10] The pH dependent flux of the source phase indicates that transport takes place via the deprotonated host, which presumably complexes the cation; the precise nature of this complex is unknown. On the basis of transport experiments of Cs^+ complexed by calixarenes bridged at upper rim, Böhmer *et . al.* [56] suggested that the cation is inside the cone.

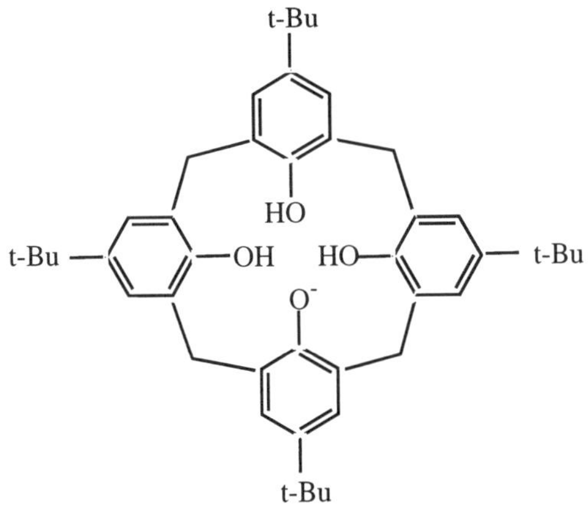

*Chart I. t-Butyl-Calix[4]arene monooxyanion (**L⁻**)*

The solid state structure of the Cs^+ salt (1:1 stoichiometry) of the *p-tert*-butylcalix[4]arene monooxyanion **L⁻** displays an inclusive structure, where Cs^+ sits inside the cone, closer to the aromatic carbon atoms than to the phenolic oxygens, and coordinated to one MeCN molecule (Figure 1). [11]

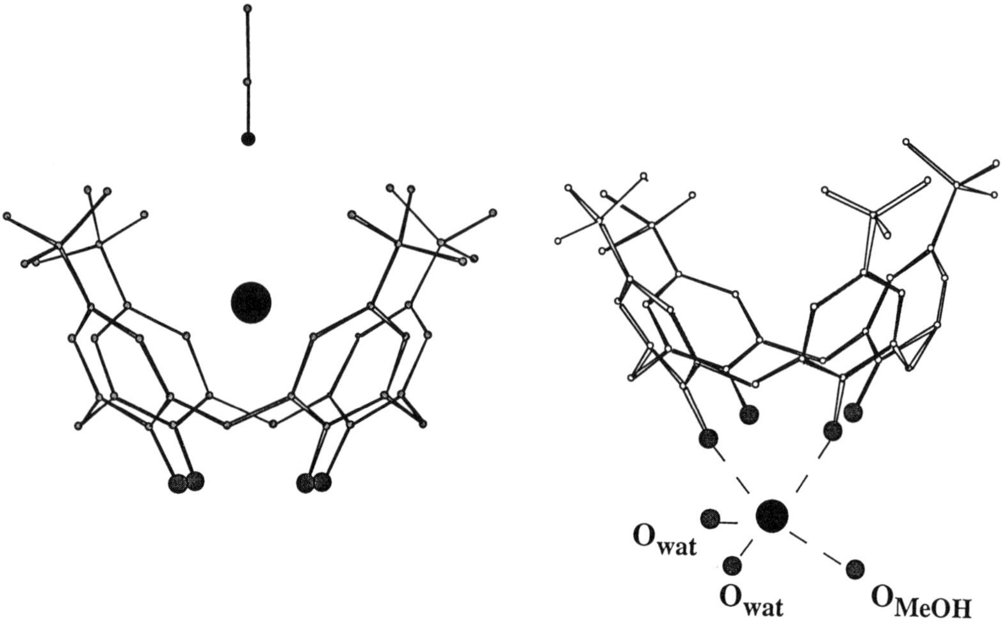

Figure 1. X-ray structures of the Cs^{+}**L**$^{-}$ *endo* complex (left) and of the Na^{+}**L**$^{-}$ *exo* complex (right). From references [11] and [12].

It is not clear to what extent this solid state structure is relevant for Cs^{+} complexation and transport in solution. In an apolar solvent such as chloroform, NMR data indicate that this structure is retained, *i.e.* the host adopts a cone conformation and the salt is undissociated, with Cs^{+} inside the cone.[11] In a polar non-aqueous solvent such as acetone, only external coordination of Cs^{+} to phenolic oxygens has been observed. [11] Other alkali salts of oxyanions of *p-tert*-butylcalix[4]arene have been prepared and crystallized, with Li^{+}, Na^{+} and K^{+} as metals.[12] In contrast to the Cs^{+} complex structure, Na^{+} and Li^{+} are *externally coordinated* to the ligand, as well as to MeOH and H_2O solvent molecules. The stoichiometry is 1:1 for Na^{+} and K^{+} but 2:1 for Li^{+}. The structure of the K^{+} salt could not be solved. [11] It seems, therefore, that the inclusive / non inclusive nature of metal salts of **L⁻** depends on the size and nature of the cation, as well as the nature of environment.

Early computational studies on calixarenes involved molecular mechanics calculations on their conformational properties *in vacuo* . [13-16] The dynamical exchange between the cone / partial cone / alternate forms of calix[4]arenes was recently studied by MD in the gas phase. [17] In solution, there have been several molecular dynamics studies and free energy perturbation simulations. Grootenhuis *et al.* investigated the acid-base and structural properties of calixarenes. [18] Simulations on cation complexes include the studies of Miyamoto and Kollman on calixspherands in water, [19] of Wipff *et al.* on calix[4]tetraamides in water [20] and in acetonitrile, [21] and of Guilbaud and Wipff on the UO_2^{2+} complexes of calix[5]arene^{5-} and calix[6]arene^{6-} anions. [22] Alkali cation complexes of calix[4]crowns [23] and calix[4]-bis-crowns [24, 25] were compared in water, methanol, acetonitrile and chloroform, to address the question of possible counterion effects. In all of these complexes, there was limited choice on the possible location of M^+, since M^+ was either sitting more or less inside a cavity or pseudo-cavity, or decomplexed. In the case of the *p-t-butyl*-calix[4]arene anion **L⁻**, it is not clear whether the ligand acts as a host (with *endo* binding of M^+), or simply as a counterion (with M^+ *exo*, forming an ion pair with the phenolate anion). We therefore consider *endo* (inclusive) and *exo* (exclusive) alkali cation (Li^+ , Na^+ , K^+ , Rb^+ , Cs^+) complexes of **L⁻** *in vacuo* and in water, acetonitrile and chloroform solutions. A 1:1 stoichiometry is assumed throughout the study. Although these salts have low solubility in water, it is important to investigate structures and stabilities in the source phase, as well as in a typical receiving organic phase used in transport or extraction experiments. We first focused on *structural features* concerning the precise location of the cation: (*i*) do the *endo* cations occupy a similar position in solution as *in vacuo* ? (*ii*) how do the solvent and phenolic groups compete in attracting M^+ ? (*iii*) for the *exo* complex, is the ion pair intimate or dissociated ? (*iv*) is the average structure of fourfold symmetry, as observed in the solid state, or unsymmetrical ? (*v*) how does the hydrogen bonding network at the lower rim of the cone evolve as a function of time, and vary from one complex to the other ? and (*vi*) what happen at the water / chloroform interface when M^+ is extracted by **L⁻**. Another important issue concerns the question of *binding selectivity.* It is important to understand why Cs^+ is extracted and transported more efficiently than Na^+, K^+ and Rb^+ cations.

METHODS.

We used the AMBER software [26] for molecular mechanics and molecular dynamics simulations, with the following representation of the potential energy:

$$E_T = \Sigma_{bonds} K_r (r-r_{eq})^2 + \Sigma_{angles} K_\theta (\theta-\theta_{eq})^2 + \Sigma_{dihedrals} V_n(1+\cos n\phi)$$

$$+ \Sigma_{i<j} (q_i q_j/R_{ij} - 2\varepsilon_{ij} (R_{ij}^*/R_{ij})^6 + \varepsilon_{ij}(R_{ij}^*/R_{ij})^{12})$$

The bonds and bond angles were treated as harmonic springs, and a torsional term was associated to the dihedral angles. The interactions between atoms separated by at least three bonds were described within a pairwise additive 1-6-12 potential. The host was built from one phenolate and three phenol residues, connected by explicit CH_2 groups. Parameters were taken from the AMBER force field. [27] The atomic charges of the phenolate and of the three phenol residues were derived by Grootenhuis *et al.* from electrostatic potentials, [27] and used without special scaling factor for 1...4 interactions. The cation parameters were fitted by Aqvist to reproduce relative and absolute free energies of hydration. [28] For the solvent, we used the TIP3P model for water [29] and the OPLS models for chloroform [30] and for acetonitrile, [31] where CH or CH_3 are represented in the united atom approximation. All complexes were first energy minimized *in vacuo* prior to any solution studies.

The solvent box was defined as a "cube" of 28-30 Å length, containing 829 to 972 H_2O TIP3P water molecules, or 253 to 299 chloroform molecules or 374-399 acetonitrile molecules (see Table 2), with periodic boundary conditions. The simulation at the water / chloroform interface were performed with 891 H_2O and 310 $CHCl_3$ molecules. The solute was placed at the center of the box and all solvent molecules within 2 Å and beyond 12 Å from the solute were deleted. In solution, the O-H and C-H bonds were constrained to constant values with SHAKE, in conjunction with a time step of 2 fs. In the gas phase, the time step was 1 fs, without SHAKE. After 1000 steps of conjugate gradient energy minimization, the MD simulations were run for 50-200 ps at 300K (N, P, T) ensemble in water, and (N, V, T) ensemble in chloroform, acetonitrile and a two-phases system) using the Verlet algorithm, starting with random velocities. A residue based cut-off of 10 Å was used for non-bonded interactions, taking the ion pair as a single residue. The temperature was controlled by velocity scaling in the gas phase, and by coupling to a thermal bath in solution. The "FEP" (Free Energy Perturbation) calculations were performed with the windowing technique, changing the

ε, R* parameters of M^+ linearly with λ, as suggested in reference: [32] $\varepsilon_\lambda = \lambda \cdot \varepsilon_{M1^+} + (1 - \lambda) \cdot \varepsilon_{M2^+}$; $R^*_\lambda = \lambda \cdot R^*_{M1^+} + (1 - \lambda) \cdot R^*_{M2^+}$. We successively mutated Na^+ -> K^+ -> Rb^+ -> Cs^+. The mutation of a given cation M_1^+ (free or complexed) to the next M_2^+ was achieved in 11 windows. At each window, 1 ps of equilibration was followed by 4 ps of data collection, and the change of free energy ΔG was averaged from the forward and backward cumulated values. The change in ΔG from Na^+ to Cs^+ was obtained from the cumulated free energies involved in the intermediate states. In order to evaluate the polarization energy contribution to the *exo* /*endo* preference in the gas phase, additional molecular mechanics calculations were performed on the M^+L^- complexes (M^+ = Li^+, Na^+, K^+, Rb^+ and Cs^+) using the MM2IS software. [33, 34] The applied force field combines the MM2 method [35] for the intramolecular energies of the host and the modified Claverie method [34, 36] to compute the cation-ligand interaction energies. The latter include electrostatic, polarisation and van der Waals terms. In these calculations we used the same set of charges as in simulations with AMBER4.

Analysis of results. The interaction energies cation-ligand ($E_{M^+...L^-}$), cation-solvent ($E_{M^+...s}$), ligand-solvent ($E_{L^-...s}$), solvent-solvent (E_{ss}), and intrinsic energies of the ligand (E_{L^-}) have been recalculated from the MD trajectories using MD_DRAW program. [37] The energy fluctuations are typically 3-5 for $E_{M^+...L^-}$, $E_{M^+...s}$, $E_{L^-...s}$, E_{L^-} and 50-80 kcal/mol for E_{ss} and for the total energy of the system. The radial distribution functions (*rdf*) of the **L⁻** phenolic oxygen atoms and of the solvent's donor atoms around M^+ were calculated in the gas phase, in chloroform and in acetonitrile. The coordination number (CN) of M^+ was obtained by integration of the first peak of the *rdf*.

RESULTS.

1. The M^+L^- salts in the gas phase. Typical structures of these complexes are represented in Figure 2. During simulation, the host retains its cone form, and the Na^+, K^+, Rb^+ and Cs^+ complexes remain either *endo* or *exo*, as in the starting form. The case of Li^+ is different, since both *exo* and *endo* starting forms converge to the same structure, where Li^+ sits near the plane of the phenolic oxygens, slightly more *endo* than *exo*. This Li^+ complex is therefore described as *endo* (Table 1).

The average total energies of the Li^+ - Cs^+ complexes are reported in Table 1. For a given cation, comparison of the *exo* / *endo* complexes indicates a small preference for *endo* structures, which is largest for Cs^+ (about 5 kcal/mole) and smallest for Na^+ (about 3 kcal/mole). These numbers are comparable to, or smaller than the statistical fluctuations (about 5 kcal/mole). The energy component analysis shows that this *endo* preference results from the balance between the cation - ligand interaction energy $E_{M^+...L^-}$ (which is more attractive in the *endo* complexes), and of the energy E_{L^-} of the ligand (which favours the *exo* complexes).

Table 1. M^+L^- complexes (*exo/endo*) in the gas phase. Average energies[a] (kcal/mol) and distances[b] (Å).

	free	Li^+	Na^+	K^+	Rb^+	Cs^+
$-E_{M^+L^-}$		275 / 276	243 / 246	228 / 231	222 / 227	215 / 220
$-E_{L^-}$	121	104 / 101	118 / 109	119 / 113	119 / 115	119 / 118
$-E_{M^+...L^-}$		171 / 175	125 / 137	109 / 119	103 / 112	96 / 102
$-E_{M^+...AR1}$		7 / 9	-4 / 6	-3 / 6	-3 / 5	-3 / 4
$-E_{M^+...AR2}$		131 / 131	111 / 113	99 / 96	95 / 93	89 / 88
$-E_{M^+...AR3}$		20 / 19	15 / 9	10 / 11	9 / 9	7 / 6
$-E_{M^+...AR4}$		14 / 16	3 / 9	3 / 6	3 / 5	2 / 4
$d_{M^+...<O>}$		0.8 / 0.8	2.5 / 1.5	2.8 / 2.8	3.0 / 3.1	3.2 / 3.4
K_{asym}[c]	1.04	1.09 / 1.09	1.13 / 1.05	1.16 / 1.43	1.13 / 1.33	1.17 / 1.16

[a] Average intrinsic energy of the complex ($E_{M^+L^-} = E_{L^-} + E_{M^+...L^-}$); intrinsic energy of the ligand (E_{L^-}); interaction energies between M^+ and ligand ($E_{M^+...L^-}$); M^+ and the residues 1 - 4 ($E_{M^+...AR1}$ - $E_{M^+...AR4}$). [b] Average distances between M^+ and the center of mass of four phenolic oxygens ($d_{M^+...<O>}$). [c] Asymmetry index of the average cone's shape (K_{asym}), calculated as the ratio between the largest average distance between opposite t-Bu carbons and the smallest one.

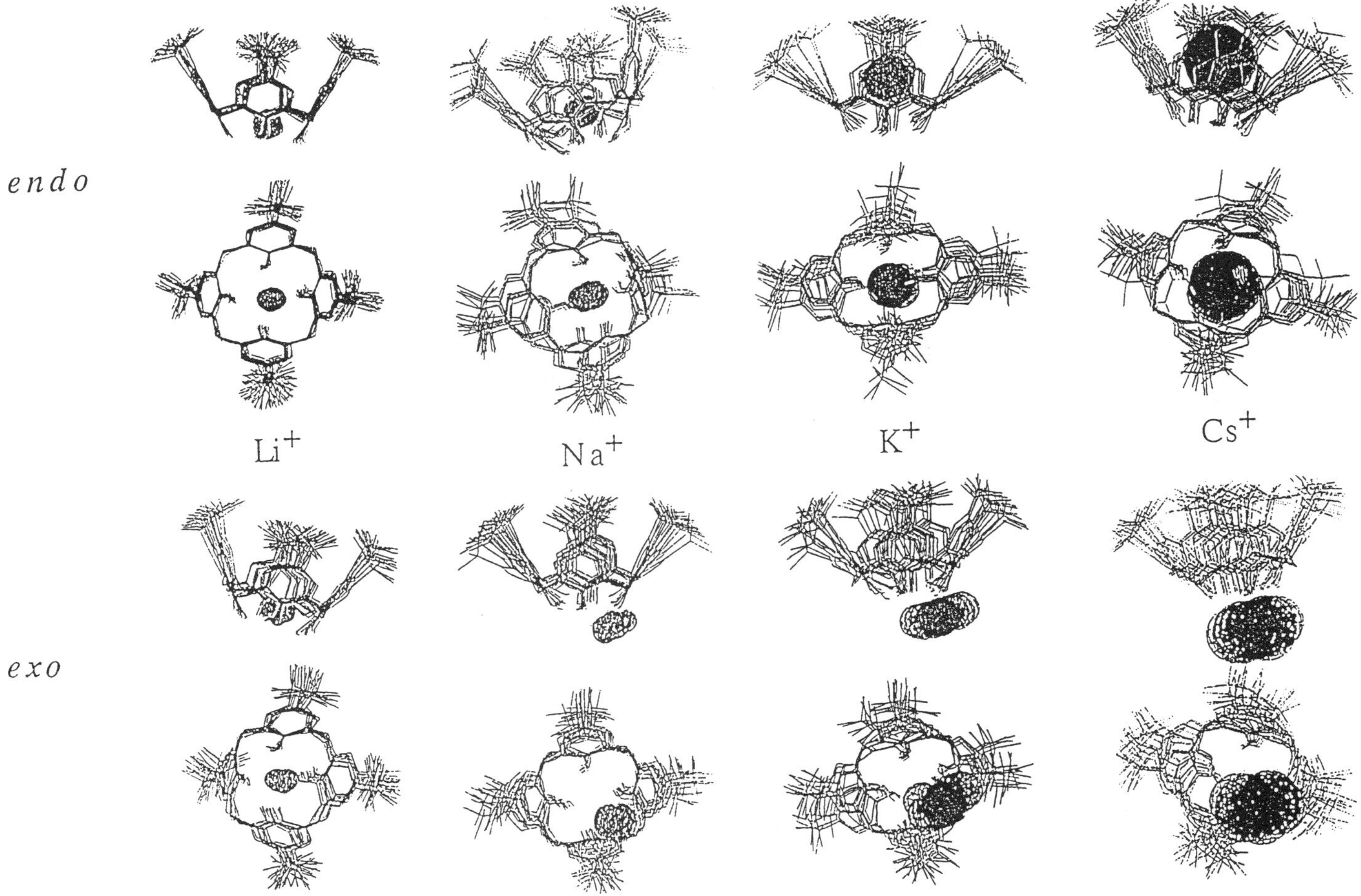

Figure 2. The M^+L^- complexes. From left to right: $M^+ = Li^+$, Na^+ , K^+ and Cs^+ *endo* and *exo* simulated *in vacuo*. Cumulated views (orthogonal representations).

A comparison of one cation complex to the others reveals much larger differences in total energies, than those for the *endo* /*exo* complexes, with stabilities increasing from Cs^+ to Li^+. The Li^+ / Cs^+ energy difference amounts to 60 and 55 kcal/mole for *exo* and *endo* complexes, respectively. This is primarily due to the $E_{M^+...L^-}$ contribution (about 75 kcal/mole), which is most attractive for Li^+. The energy of the host within the complex is also cation dependent. From Cs^+ to Li^+ complexes, the host is destabilized by about 16 kcal/mole for *exo* and *endo* complexes.

A further dissection of the $E_{M^+...L^-}$ energy into that contributed by the three phenols and one phenolate residues shows two features: (*i*) the phenolate provides the largest contribution (about 76% for Li^+ *endo* or *exo* , to 93 % for Cs^+ *exo*), and (*ii*) the average contribution of the three phenols differs from one to the other, indicating that *the* M^+ *cation lies asymmetrically* with respect to these fragments.

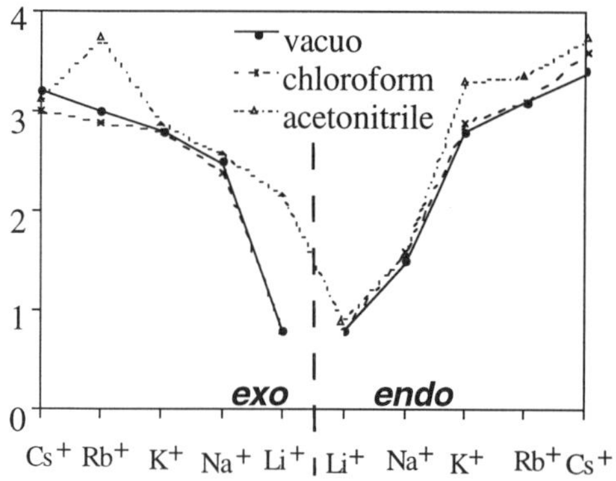

Figure 3. The M^+L^- complexes in the gas phase, in chloroform and in acetonitrile: average "altitude" of M^+ over the four phenolic oxygens (see text).

The altitude of the M^+ cations fluctuates more or less as a function of time, for the *endo* and *exo* complexes, as shown in Figure 2. The corresponding time average distances, $d_{M^+...<O>}$ (Table 1 and Figure 3) increase from Li^+ to Cs^+, in relation with the size of the ion.

The $M^+...O_{ph}$ *rdfs* (Figure 4) characterize the coordination of M^+ by the oxygen atoms of $\mathbf{L}^-$. For the *exo* Na^+ - Cs^+ complexes, these *rdf*s display two peaks, indicating that the M^+ is closer to the phenolate and one phenolic oxygen than to the others. This location is related to the asymmetry of the hydrogen bond network at the lower rim where the O^- oxygen acts as a single proton acceptor in all complexes, as well as in the free $\mathbf{L}^-$. Simulations starting

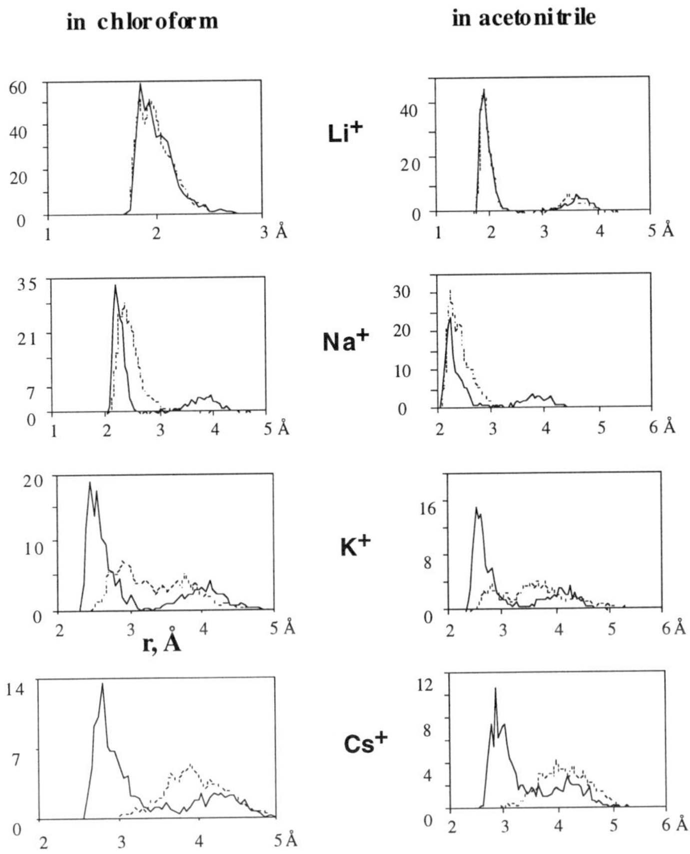

Figure 4. The M^+L^- complexes *exo* (solid line) and *endo* (dashed line): $M^+...O_{ph}$ *rdf*'s in chloroform (left) and in acetonitrile (right). The *rdf*'s in vacuo (not shown) are nearly identical as in chloroform.

with O^- acting as a double proton acceptor converged immediately to the same structure for the free and complexed host.

The Li^+ and Na^+ *endo* cations are approximately equidistant from the four phenolic oxygens (one peak on the $M^+...O_{ph}$ *rdfs*). This is not the case for the K^+, Rb^+ and Cs^+ *endo* complexes, where broad peaks on *rdfs* show that the $M^+...O$ distances to one pair of opposing phenolic oxygens are smaller than to the other pair of oxygens (Figure 5). Because of the mechanical top-bottom coupling, [20] such coordination introduces some asymmetry of the cone; the asymmetry index K_{asym} is larger for K^+ - Cs^+ *endo* complexes than for the others (Table 1).

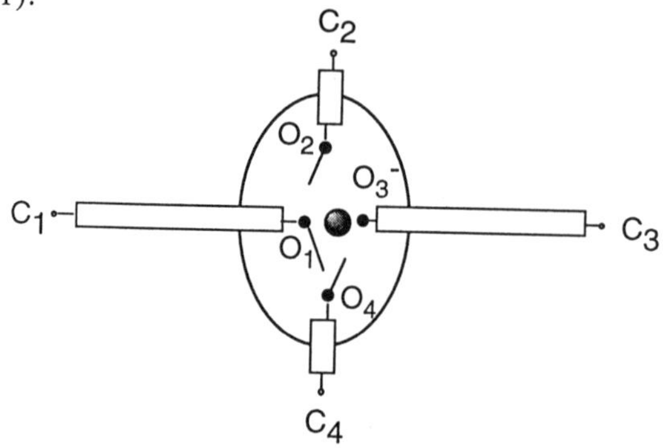

Figure 5. Schematical representation of the coordination pattern at the lower rim of the *endo* K^+, Rb^+ and Cs^+complexes .

2. The Li^+ to Cs^+ complexes in chloroform. In chloroform solution, the complexes remain similar to those observed in the gas phase (Figure 2). The *endo* structures remain of inclusive type, while the *exo* ones form intimate ion pairs. Only the Li^+ complexes converge, as seen *in vacuo*, to a unique structure, where Li^+ is slightly *endo*, but in a more symmetric position with respect to the "C_4" axis.

Other structural features in chloroform display the same trends as *in vacuo*, in particular as far as the altitude and position of the cation, and the overall asymmetrical shape of the calixarene host are concerned (see the asymmetry index K_{asym} and $d_{M^+ \ldots <O>}$ distances; Table 2 and Figure 3). In solution, we cannot simply compare the stabilities of *endo* / *exo* complexes based on total energies. Because the simulation boxes contain different numbers of particles the solvent-solvent interaction energies are not consistent. The other energy components (Table 2) show interesting features. As *in vacuo*, for one given M^+ cation, the average energy of the M^+L^- solute slightly favours the *endo* complexes (by about 3 - 7 kcal/mole for Na^+ to Cs^+). On the other hand, the solute - solvent interaction energies are attractive and are larger for *exo* complexes (by 12, 13, 11, 12 kcal/mole for Na^+, K^+, Rb^+ and Cs^+, respectively). This stems from two additive contributions. First, the cation - solvent interactions. The weakly polar $CHCl_3$ solvent molecules interact repulsivly with the M^+ *endo* (3 to 2 kcal/mole for Na^+ to Cs^+), but attractively with the M^+ *exo* (-3 to -2 kcal/mole for Na^+ to Cs^+). Second, the ligand interacts to a greater extent with chloroform in the *exo* complexes (-64 to -67 kcal/mole for

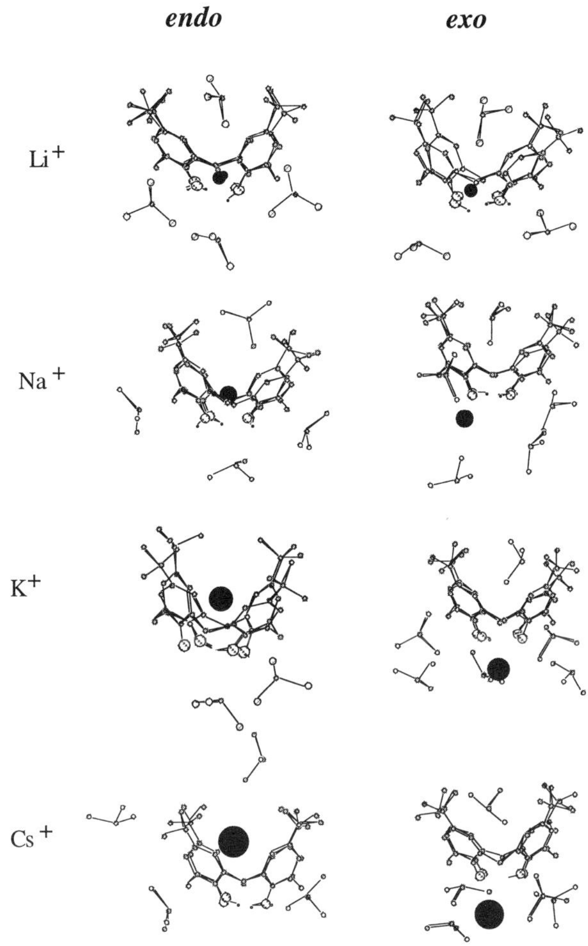

Figure 6. The M^+L^- complexes in chloroform. $M^+ = Li^+$, Na^+ , K^+ and Cs^+ *endo* and *exo* . Snapshots after 50 ps of MD.

Na^+ to Cs^+ complexes) than in the *endo* complexes (-65 to -58 kcal/mole). This difference is related to the filling of the cone cavity of *exo* complexes by one $CHCl_3$ guest molecule (Figure 6), which adds about 3-5 kcal/mole extra stabilization. In the *endo* complexes, such solvent uptake is prevented by the encapsulated cation. Finally, comparison of the solute - chloroform interaction energies with the free host **L**$^-$ and with its **L**$^-$ M^+ complexes shows that the former is favoured by 10-15 kcal/mol.

3. The L$^-$ host free and its Li^+ to Cs^+ complexes in acetonitrile. As in chloroform solution, the free host **L**$^-$ is better solvated than the complexed one (Table 2).

One important contribution comes from two MeCN molecules which fill the cone, oriented with their methyl groups pointing toward the phenolic oxygens. This is different than for the *t*-butyl-calix[4]arenetetraamide in acetonitrile where only one MeCN molecule was found in the cone, [21] presumably because the dipole - dipole interactions between MeCN and the cone are larger in **L**$^-$ than in the neutral host.

As in chloroform and in the gas phase, for a given cation, the M^+**L**$^-$ solute is intrinsically more stable *endo* than *exo* (Table 2). The structures of the M^+**L**$^-$ complexes are also similar to those in the gas phase or in chloroform except that of Li^+ which forms distinct *endo* and *exo* complexes. In the latter, the cation occupies an asymmetrical position, coordinated to only two phenolic oxygens (Figures 3 and 7). For the other *endo* or *exo* complexes, the altitude of a given cation is larger, and the asymmetry of the cone is smaller in acetonitrile than in the gas phase or in chloroform. This is because the cation is somewhat pulled out by the solvent attractions (Figure 7 and Table 2).

Again, *exo* complexes are solvated much better than the *endo* ones, but the difference in solute - solvent interaction energies is larger (-20 to -40 kcal/mole) than in chloroform. Cations Na^+, K^+, Rb^+ and Cs^+ are coordinated respectively to 3.0, 3.5, 4.5 and 5.0 (*exo*), and 1.0, 1.8, 1.8 and 1.8 (*endo*) solvent molecules (Figures 4 and 7). In the *endo* complexes, the MeCN molecules, bound to M^+ by their N terminus, sit more or less in the cone, with their dipole parallel to the dipole of the cone. This cone /MeCN arrangement is therefore somewhat repulsive, compared to the antiparallel arrangement found in the *exo* complexes. The larger solvent stabilization of the *exo* complexes results from direct coordination to M^+, and from the filling of the cone by MeCN as in the free host.

Table 2. M^+L^- *Exo* (first line) and *Endo* (second line) complexes in chloroform and in acetonitrile. Average energies [a] (kcal/mol) and distances [b] (Å).

	in chloroform						*in acetonitrile*					
	free	Li^+	Na^+	K^+	Rb^+	Cs^+	free	Li^+	Na^+	K^+	Rb^+	Cs^+
N_s [c]	299	266 / 264	271 / 263	276 / 256	278 / 256	286 / 253	374	399 / 374	399 / 374	399 / 374	399 / 374	399 / 374
$-E_{M^+L^-}$		271/ 272	238/ 241	223/ 227	218/ 223	209/ 216		263/ 271	234/ 240	220/ 221	213/ 214	209/ 210
$-E_{L^-}$	115	96 / 97	112 / 104	114 / 109	114 / 113	113 / 116	114	114 / 98	113 / 104	114 / 111	116 / 108	116 / 111
$-E_{M^+..L^-}$		175/ 175	126 / 137	109 / 118	104 / 110	96 / 100		149 / 173	121 / 136	106 / 110	97 / 106	93 / 99
$-E_{M^+L^-...S}$		67 / 64	70 / 62	69 / 56	68 / 57	68 / 56		147 / 126	149 / 114	147 / 122	148 / 121	145 / 116
$-E_{M^+...S}$		-1 / 1	3 / -3	3 / -4	4 / -2	2 / -2		64 / 30	60 / 16	54 / 27	56 / 28	48 / 22
$-E_{L^-...S}$	83	68 / 63	67 / 65	66 / 60	64 / 59	66 / 58	144	83 / 96	89 / 99	93 / 95	92 / 92	97 / 95
$-E_{S...S}$ [d]	1513	1364 / 1356	1356 / 1349	1347 / 1351	1319 / 1332	1292 / 1312	2500	2525 / 2508	2519 / 2518	2515 / 2516	2521 / 2511	2518 / 2513
$d_{M^+...<O>}$		0.8 / 0.8	2.4 / 1.6	2.8 / 2.9	2.9 / 3.1	3.0 / 3.6		2.2 / 2.1	2.6 / 1.6	2.9 / 3.3	3.7 / 3.4	3.1 / 3.7
K_{asym}	1.05	1.05 / 1.05	1.11 / 1.04	1.05 / 1.45	1.07 / 1.17	1.07 / 1.00	1.06	1.08 / 1.04	1.05 / 1.18	1.08 / 1.28	1.07 / 1.33	1.06 / 1.26

[a] Average intrinsic energy of the complex ($E_{M^+L^-} = E_{L^-} + E_{M^+...L^-}$); intrinsic energy of the ligand (E_{L^-}); interaction energies between M^+ and ligand ($E_{M^+...L^-}$); M^+L^- and solvent ($E_{M^+L^-...s}$); L^- and solvent ($E_{L^-...s}$); M^+ and solvent ($E_{M^+...s}$); solvent-solvent interaction energies ($E_{s...s}$). [b] See Table 1 for definitions. [c] Number of solvent molecules in the simulation box. [d] Scaled to 299 (in chloroform) or to 399 (in acetonitrile) solvent molecules

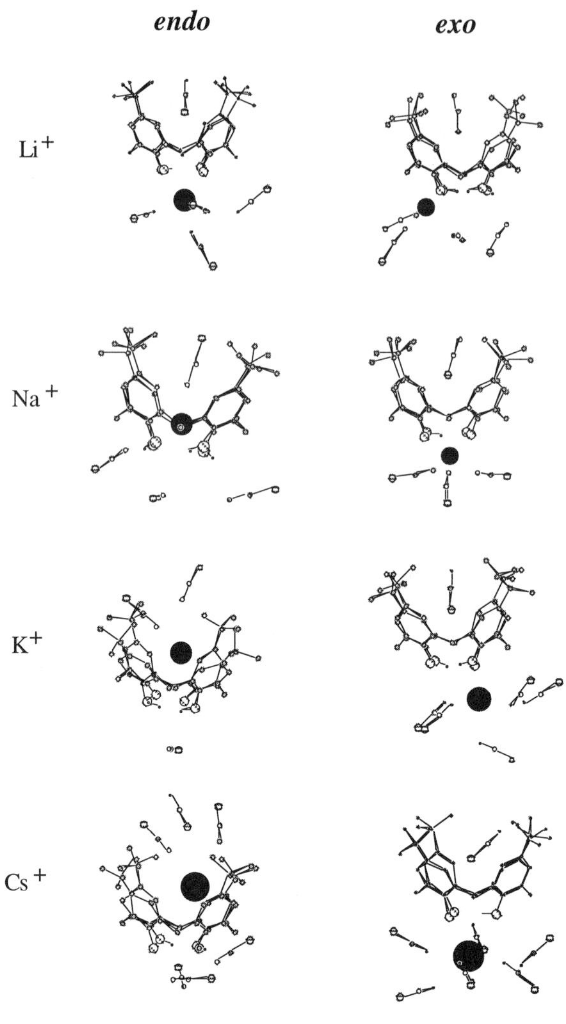

Figure 7. The M^+L^- complexes in acetonitrile. $M^+ = Li^+$, Na^+, K^+ and Cs^+ *endo* and *exo*. Snapshots after 50 ps of MD.

4. The Li^+ to Cs^+ "complexes" in water. In aqueous solutions, all M^+L^- complexes, excepted Li^+L^- display different behaviour, compared to that in chloroform or acetonitrile because of the larger cation - water attraction.

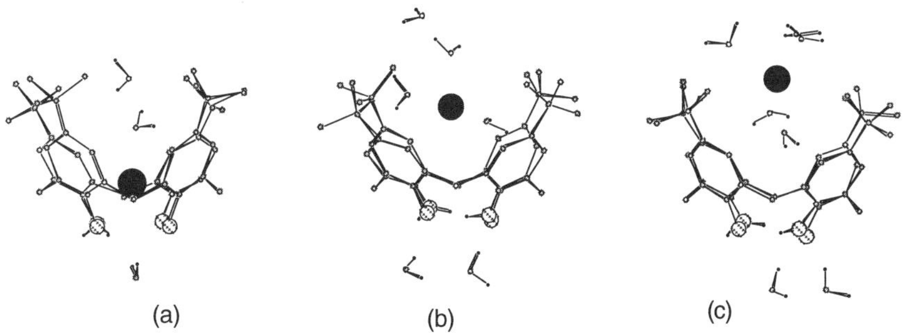

Figure 8. The Na^+L^- *endo* complex in water after 5 ps (a), 50 ps (b) and 70 ps (c) of MD.

The *exo* or *endo* Li^+ complexes remained bound after 250 ps of simulations. Li^+ *endo* is located close to the plane of the phenolic oxygens, as seen in chloroform or in the gas phase. The calixarene coordinates one water molecule sitting deeply in the cone. In the *exo* complex the Li^+ coordination resembles that seen in acetonitrile, *i.e.* loose binding to the three phenol oxygens and direct coordination to the O^- phenolate oxygen. Integration of the $Li^+...O_W$ *rdf* gives a coordination number of 2.8 water molecules.

For the *exo* and *endo* Na^+, K^+, Rb^+ and Cs^+ "complexes", complete dissociation takes place to form separated ion pairs. Repeating the MD simulations on the Cs^+ complex in the (N,V,T), instead of (N,P,T) ensemble leads also to dissociation. In the *exo* complexes, the lifetime of the initial intimate ion pair increases regularly as the cation becomes smaller (2, 10, 40, 130 ps respectively for Cs^+, Rb^+, K^+ and Na^+). All *endo* Na^+ - Cs^+ complexes dissociate in 50 - 100 ps. This dissociation takes place stepwise as described in the following for the Na^+ complex. In the first 50 ps, the cation, attracted by water, moves about 0.5 Å higher in the cone than in chloroform. At this stage, it is coordinated to two water molecules, which sit between the *t-butyl* groups (Figure 8a). However, these groups do not shield adequately the Na^+ from water to prevent its decomplexation. Indeed, 30 ps later, decomplexation of Na^+ is observed, followed by migration of one H_2O molecule inside the cone (Figure 8b,c). This water migration is assisted by the partial H_2O coordination to M^+ , and by its dipole - dipole attractions with **L^-**.

Although the M^+**L^-** complexes (except Li^+) displayed a relatively short lifetime in water, we analyzed their solvation shells before dissociation in order to gain insight into the number of water molecules bound to the complexed M^+. Such water molecules form a "supermolecule" with the complex and thus could migrate indeed to the organic phase upon extraction from an aqueous phase. As a consequence, for *endo* complexes, integration of the water *rdf's* gives coordination numbers of 1.1, 1.4, 1.8, 2.1 respectively for Na^+, K^+, Rb^+ and Cs^+. In contrast with what was observed in chloroform, the M^+ / solvent interactions are attractive in water (-17, -28 and -27 kcal/mole, respectively for K^+, Rb^+ and Cs^+). For Na^+, this average energy is only -6 kcal/mole for the first 50 ps, because of its initial deep position. However, in the next 30 ps, it increases to -137 kcal/mol, due to the Na^+ migration to water. In this predissociated *exo* complex, Na^+ coordinates to 3.2 water molecules.

In order to elucidate the role of t-butyl groups on the Cs^+ binding, we first performed an energy component analysis on the Cs^+ complex. The four *t-butyl* substituents slightly stabilize the complex: the Cs^+ / t-butyl interaction energy is weak (about -7 kcal/mole per group). They also play an essential role in preventing the solvent assistance for decomplexation, as shown by the following computer experiment: we replaced the butyl groups by H atoms to generate the unsubstituted calix[4]arene$^-$ anion, and simulated the *endo* Cs^+ complex in water as described above. Dissociation was rapid (less than 3 ps), compared to the Cs^+**L^-** complex (about 100 ps). The "gating" hydrophobic groups are therefore essential to enhance any kinetic stability of the complex. Conversely, they may prevent inclusion of the cation into the cone.

Concerning the *solvent content of the cone*, we find that the cone of the free host, or of the *exo* "complex", contains from one to two water molecules whose dipole is more or less antiparallel to that of the cone (Figure 8c). The deepest water molecule is connected to the bulk medium via a relay of hydrogen-bonded water molecules, through the hydrophobic gate delineated by the four *t-butyl* groups.

5. pH modulation of the binding properties of the calixarene. It follows from experiments [38] that calixarenes unsubstituted at the lower rim act as *anionic* carriers of alkali cations through liquid membranes. When the pH is not adequately high, no transport takes place. In order to investigate the binding ability of the *neutral* ligand (L-H) for cations, we performed 50 ps of MD on the complex L-H·Cs^+ (*endo* and *exo*) in the gas phase, in chloroform and in water. We find that *in vacuo* and in chloroform only *endo* complexes are kinetically stable, whereas in water both complexes dissociate. In chloroform, the interaction energy of Cs^+ with **L**-H (-20 kcal/mol) is about 5 times smaller than the Cs^+ / $\mathbf{L}^-$ interaction energy. The binding affinity of the neutral host is therefore very weak, compared to the anionic host, in agreement with the pH-effect on the transport properties of this family of calix[n]arenes. [38]

6. The Cs^+L^- *endo* complex at the water / chloroform interface. In order to gain microscopic views of the extraction of cations at a water / organic interface, and of transport processes, we initiated MD studies of typical ionophores (18-crown-6, 222 cryptand, calix[4]arenetetraamide, calix-crowns and calix-bis-crowns). Structures uncomplexed, and complexed with several cations, at a water / chloroform interface, including counterions were considered. [39]

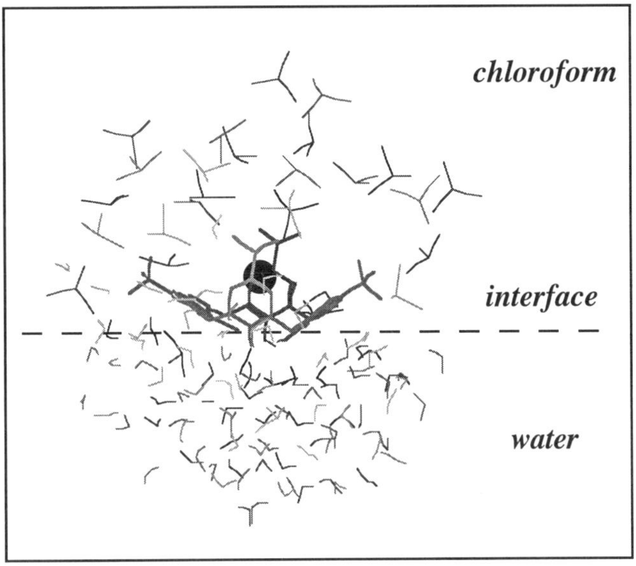

Figure 9. The Cs^+L^- *endo* complex at the chloroform/water interface. Snapshot after 350 ps of MD.

The MD simulation starts with a rectangular box composed of one cubic box of water adjacent to one box of chloroform equilibrated separately. The Cs^+ *endo* salt of $\mathbf{L}^-$ is placed at the interface, symmetrically exposed to the two solvents, and the time evolution is followed for 350 ps. We observe in the first 50 ps, that the *endo* complex reorients, in such a way that the hydrophobic alkyl groups are immersed in chloroform, while the polar lower rim remains in the aqueous phase. Then the complex, almost completely immersed in chloroform, stays in contact with the interface, without achieving further diffusion into the bulk of the organic phase. It undergoes several rotations, accompanied by "capture" of water molecules at the upper rim by the polar Cs^+, and the lower rim by the oxygen atoms (Figure 9). Such a situation can be viewed as the early stage of migration to the organic phase. Whether this migration will take place or whether the present situation is close to a thermodynamic equilibrium require additional time-consuming experiments. Further investigations involving the questions of starting configuration and protocol are presently in progress in our laboratory. [39]

7. Calculated relative binding selectivities upon complexation and extraction: Free energy simulations in the gas and in solution.

The M_1^+ / M_2^+ binding selectivity of the ionophore $\mathbf{L}^-$ in solution measured as $\Delta G_c = \Delta G_1 - \Delta G_2$ was obtained computationally [40-43] as $\Delta G_c = \Delta G_3 - \Delta G_4$, using the thermodynamic cycle:

$$\begin{array}{ccccc} & & \Delta G_1 & & \\ & \mathbf{L} + M_1^+ & \text{------------------>} & \mathbf{L}M_1^+ & \\ \Delta G_3 \downarrow & & & & \downarrow \Delta G_4 \\ & \mathbf{L} + M_2^+ & \text{------------------>} & \mathbf{L}M_2^+ & \\ & & \Delta G_2 & & \end{array}$$

Table 3. Relative Free Energies of Mutation (kcal/mol) of Li^+, Na^+, K^+, Rb^+ and Cs^+ Uncomplexed and Complexed by L^-.

	Na^+->K^+	K^+->Rb^+	Rb^+-> Cs^+
	ΔG_4 (complexed M^+)		
gas phase, endo	16.3	4.6	6.7
gas phase, exo	15.6	4.8	6.6
chloroform, endo	14.9	4.7	6.7
chloroform, exo	16.4	5.2	7.2
water, endo	15.7	4.9	7.5
acetonitrile, endo	19.7	5.4	6.8
acetonitrile, exo	18.5	6.5	9.6
	ΔG_3 (free M^+)		
chloroform, calc	2.3	2.1	0.9
water, exp	17.6	5.1	7.7
acetonitrile, exp	16.2	4.9	7.2
acetonitrile, calc	13.3	5.6	5.3
	$\Delta\Delta G_C$ (complexation of M^+)		
chloroform, endo	-14.1	-3.1	-6.3
chloroform, exo	-12.5	-2.6	-5.8
water, endo	2.0	0.2	0.3
acetonitrile, endo [a]	-6.4 / -3.5	0.2 / -0.5	-1.5 / 0.4
acetonitrile, exo [a]	-5.2 / -2.3	-0.9 / -1.6	-4.3 / -2.4
	$\Delta\Delta G_{ext}$ (extraction of M^+ by L^- from water to chloroform)		
endo	2.8	0.4	1.0
exo	1.2	-0.1	0.5

[a] the first value uses the calculated ΔG_3, the second value uses experimental ΔG_3.

In water, our calculated difference in solvation energies ΔG_3 is identical to the experimental value [44] and to the value calculated by Aqvist [28] using a somewhat different protocol and water representation. No experimental results are available in chloroform, due to the insolubility of the salts. In acetonitrile, there exist substantial differences between the "experimental" (extracted from free energies of transfer [21]) and calculated energies ΔG_3's. The free energy results of the Na^+ to Cs^+ complexes are reported in Table 3 for both *endo* and *exo* complexes, when stable. In water, no value is reported for the *exo* complexes because they rapidly dissociate. We report only results for the *endo* complexes, obtained for the first 50 ps of simulation, *i.e.* before these complexes dissociate. Indeed, as noted above, all cation complexes dissociate after 50-100 ps.

<u>Complexation selectivity in pure solvents.</u>

As for most alkali cation complexes, [3] the ΔG_4 free energy component favours the smallest ion Na^+ over Cs^+. In the *endo* or *exo* series, values of ΔG_4 follow the same order: $Na^+ < K^+ < Rb^+ < Cs^+$. The Na^+ - Cs^+ differences in ΔG_4 are similar for both the *endo* / *exo* series *in vacuo* (27.6 / 27.0 kcal/mol), in chloroform (26.3 / 28.8 kcal/mol), and in acetonitrile (31.9 / 34.6 kcal/mol). Taking into account the differences in free energies of solvation of free cations, ΔG_3, leads however to complexation selectivities $\Delta G_c = \Delta G_3 - \Delta G_4$, which are markedly solvent dependent.

In chloroform, the free cations have weak interactions with the solvent, and the ΔG_3 differences in solvation free energies are small (5.3 kcal/mol from Na^+ to Cs^+). As a result, the order of binding selectivities remains the same for *endo* or *exo* complexes as *in vacuo, i.e.* Na^+ is preferred ($Na^+ > K^+ > Rb^+ > Cs^+$).

The changes in ΔG_3 are larger in water than in chloroform, and close to the changes in ΔG_4 (30.4 kcal/mol from Na^+ to Cs^+). As a result the binding sequence differs from water to chloroform. In water, the order of calculated stabilities would be $Cs^+ > K^+ = Rb^+ > Na^+$. Thus, Cs^+ is clearly better complexed to $\mathbf{L}^-$ than Na^+.

The binding sequence, calculated with the experimental values of ΔG_3, is the same for both types of complexes in acetonitrile as in chloroform ($Na^+ > K^+ > Rb^+ > Cs^+$). Using the calculated instead of the experimental ΔG_3 free energies results in a slight difference for *endo* complexes ($Na^+ > Cs^+ \geq Rb^+ \approx K^+$ and Na^+) while for *exo* complexes, the sequence is unchanged. Since in acetonitrile *exo* complexes are more stable than *endo* complexes (because of the their larger interactions with the solvent), the order of binding selectivity is not critically affected by the differences between experimental and calculated ΔG_3 free energies. In particular, the Na^+ complex is more stable than the Cs^+ complex, in contrast with to what is observed in water.

Extraction selectivity from water to chloroform.

In order to model the extraction phenomenon, we consider the competition between cations M_1^+ and M_2^+, extracted from an aqueous phase to an organic phase by ligand **L**.

$$
\begin{array}{ccc}
 & \Delta G_1' & \\
L_{org} + M_1{}^+{}_w & \xrightarrow{\hspace{3cm}} & LM_1{}^+{}_{org} \\
\Delta G_{3w} \downarrow & & \downarrow \Delta G_{4\,org} \\
L_{org} + M_2{}^+{}_w & \xrightarrow{\hspace{3cm}} & LM_2{}^+{}_{org} \\
 & \Delta G_2' &
\end{array}
$$

Competitive extraction of M_1^+ / M_2^+ by **L** can be represented by the above cycle, which assumes that in water the ions are in the free state only, while in the organic phase they are completely complexed by **L**. The $\Delta G_1'$ and $\Delta G_2'$ changes in free energies may involve complicated mechanistic events at the water / organic interface and relative partition coefficients of the free and complexed ions. If one assumes, however, that the concentration of the complexed cation in water, and of the free cations in the organic phase, are negligible, the extraction selectivity is given by the difference

$$\Delta G_{ext} = \Delta G_1' - \Delta G_2' = \Delta G_{3w} - \Delta G_{4org}.$$

The corresponding values are reported in Table 3. In fact, since the ΔG_{4org} energies are nearly identical to the ΔG_{4w} values, the sequence of relative free energies of extraction ΔG_{ext} is the same as the order of binding selectivities calculated in pure water, *i.e.* $Cs^+ > Rb^+ > K^+ > Na^+$. Thus, we calculate that Cs^+ is selectively extracted, as has been observed experimentally.[10, 45]

DISCUSSION

1.- Comparison of *endo* / *exo* M^+L^- complexes. Intrinsic stabilities and solvation effects.

It follows from ^{133}Cs NMR spectra in chloroform that the cation is situated inside the cone [11] as seen in the solid state structure of the Cs^+**L**$^-$ salt. [11] This is in contrast to the situation in polar solvents such as acetone, where NMR data suggest that Cs^+ is *exo*. In our calculations we considered all *endo* and *exo* complexes of **L**$^-$ in the gas phase, pure chloroform, acetonitrile and water solutions. In the gas phase and in chloroform, there is a clear preference, using the AMBER 1-6-12 potentials for non-bonded interactions, for *endo* complexes M^+**L**$^-$. In the case of Cs^+, the difference is about 5 kcal/mole. On the other hand, our calculations show a significant difference in solute - solvent interaction energies, in favour of the *exo* form. Another computational proof of this solvation effect is the fact that in water, none of the Na^+ to Cs^+ complexes remain bound, however, they remain associated in the other solvents.

Before discussing these solvation effects in more detail, it is important to assess the importance of the energy representation on the intrinsic gas phase *endo* / *exo* stabilities. This is why we addressed the question of polarization effects on host - guest interaction energies. Indeed, the polarization energy E_{pol}, not explicitly included in our standard AMBER 1-6-12 non-bonded potentials might be a possible source of differentiation between *exo* and *endo* forms of the of M^+**L**$^-$ complexes. We therefore ran additional Molecular Mechanics calculations *in vacuo*, using the MM2IS software [33], which includes polarisation energy. Now, for the Na^+ to Cs^+ complexes, we found a greater stability of the *endo* form for a given ion, mostly because of differences in E_{pol} (by about 10 kcal/mol). Since all *endo* complexes behave similarly in chloroform and acetonitrile as *in vacuo*, they can be estimated to be intrinsically more stable than the *exo* ones by 13-15 kcal/mole, when the polarization contribution is included.

In solution, the solute-solvent interaction energies are in opposition to the intrinsic *endo* / *exo* stabilities, since they are found to always favour the *exo* complexes. In fact, other contributions such as solvent-solvent interactions E_{SS} also differ from one form to another (Table 2). However, given the very large statistical fluctuations (about 50 - 80 kcal/mole), changes in E_{SS} may not be significative. Our discussion on the *exo* /*endo* preference in solution is, therefore based on the comparison of the intrinsic energies of the solute and of solute - solvent interaction, assuming "everything else being comparable". This analysis reveals a difference between behaviour in chloroform and acetonitrile solutions. In chloroform, the *endo* / *exo* difference in solvation energies is about 2 - 5 kcal/mole weaker than the intrinsic stabilities, when E_{pol} is included. Therefore, we would predict that, in chloroform, *endo* complexes remain preferred because solvation effects are not large enough to overcome the intrinsic *endo* preference. This is in agreement with experimental results for the Cs^+ complex in this solvent. [11] For the other cations, using the same arguments, we would predict the *endo* forms to be also most populated in chloroform.

The case of acetonitrile solutions is different: due to the polarity of this solvent, the *exo* / *endo* difference in solute - solvent interaction energies is quite large (-20 to -40 kcal/mole), *i.e.* larger then the intrinsic preference of *endo* forms (10 - 15 kcal/mole, including the polarization effects). Thus, in polar non-aqueous solvents such as acetonitrile, solvation effects overcome the intrinsic stabilities, and favour the *exo* complexes. We would therefore predict all *exo* complexes to be preferred over the *endo* ones in acetonitrile as well as in polar solvents. This is in agreement with the experimental observation of *exo* Cs^+ in acetone. [11] The results of simulations in water are also consistent with this trend: the *exo* complexes are preferred, but they dissociate, in relation with the larger dielectric constant of this solvent.

Thus, our calculations demonstrate that the *endo* or *exo* nature of the complexes results from an interplay between intrinsic stabilities of the solute, and solvation effects.

As far as the "cation-π" interactions [46] are concerned, it is clear that all cations display attractive interactions with the phenol groups (Tables 1 and 2). These interactions are largest for Li^+ and decrease regularly with "standard" 1-6-12 potentials from Li^+ to Cs^+. The polarization term is also larger for the smallest cation. For instance, MM2IS gives for Li^+ / Cs^+ complexes a $\Delta E_{pol} \approx -19.5$ kcal/mol for *endo* complexes and $\Delta E_{pol} \approx -29.1$ kcal/mol for *exo* complexes. This trend follows that for the gas phase alkali cations - benzene interaction energies. [46] As discussed in more detail in reference [24], the preference for Cs^+ complexation in aqueous and acetonitrile solutions results mainly from differential solvation effects, rather than from special "cation-π" interactions which would favour the largest cation.

2.- Structure of the Na^+ and Cs^+ complexes in the solid state *vs* gas phase and solution.

In the solid state only two structures of M^+L^- complexes are available: the *endo* complex of Cs^+ [11] and the *exo* complex of Na^+, [12] and it is interesting to compare some of their structural features with our calculations. First, let us emphasize that the *endo* / *exo* nature of these complexes in the solid state cannot in principle be compared to our simulations, which correspond to another state. It is known indeed from experiments that the complex which crystallizes may not correspond to the most stable one in solution. For example, the t-butyl-calix[4]arenetetraamide binds selectively Na^+ over K^+ in methanol or acetonitrile solutions, but only the K^+ complex is found to date in the solid state [47]. In the complexation of chiral substrates by Cram's crown ether macrocycle, the enantiomeric form present in the crystal [48] is the *least stable* one in solution. Packing and environment effect clearly differ from the solution to the solid states. Here, however, in the case of Na^+L^- complex observed in the solid state, it is noticeable that Na^+ is *exo* and coordinated to one MeOH and two H_2O molecules. This *exo* structure is consistent with the "solvent -induced" *exo* stabilization discussed above. Conversely, the *endo* Cs^+ structure in the solid state is consistent with the calculated intrinsic stability, not overcome by solvation effects. Only one MeCN molecule is coordinated to Cs^+ *endo*. Stabilization of the *exo* complex would require the presence of more solvent molecules.

Structure and symmetry of the Cs^+L^- *endo* complex. In the crystal, this complex appears to be of C_{4v} symmetry, where Cs^+ lies on the C_4-symmetry axis, equidistant from all phenolic oxygens, and coordinated to one MeCN molecule. However, as quoted in reference, [11] the phenolic protons are disordered, and could not be located. Our simulated structure takes these protons explicitly into account and differentiate the three phenol from the phenolate residue. This leads to a calculated time averaged structure of approximate C_2-symmetry in the gas phase and in solution. In fact, deviation from C_{4v} is small, as reflected from the asymmetry index K_{asym} which is environment dependent (1.0 in chloroform, 1.16 in the gas phase and 1.26 in acetonitrile). The small differences between the X-ray and calculated symmetries may be therefore related primarily to environmental effects. Since the phenolic protons were disordered in the X-ray structure, it is likely that other effects such as thermal motion, possible transitions of the phenolic protons from one oxygen to another one, and different orientations of the complex within the unit cell also contribute to the "observed" structure.

In order to simulate a statistical delocalisation of the phenolate charge over all fragments of the **L⁻** anion, we calculated average distances between carbon atoms of aromatic rings and the Cs^+ cation, where the average is determined over the four aromatic residues of **L⁻**. Indeed, the six calculated average Cs^+...C distances are in excellent agreement with the X-ray published distances (Figure 10). This, therefore, strongly indicates that *the X-ray C_{4v} structure observed in the solid state does not correspond to an energy minimum, but to a time and statistical average of more stable and less symmetrical forms.*

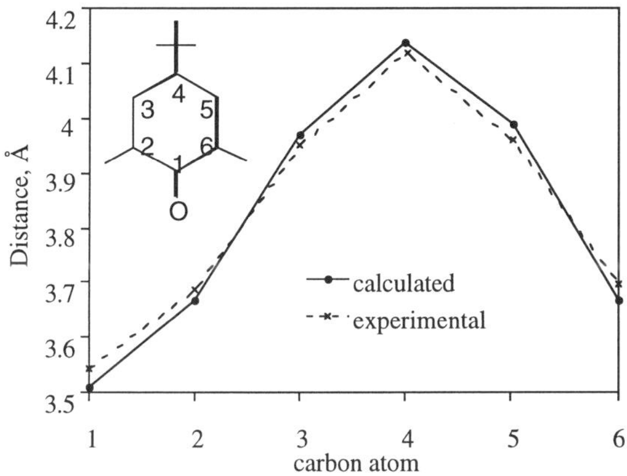

Figure 10. The Cs^+**L⁻** endo complex: average distances between Cs^+ and the aromatic carbons calculated in the gas phase and obtained experimentally in the solid state [11].

<u>The Na^+**L⁻** *exo* complex.</u> For the Na^+ *exo* complex, the solid state structure, and the structures simulated nicely complement each other, with regard to the cone shape, the binding of Na^+ to **L⁻**, and its additional coordination to solvent molecules. Firstly, all Na^+ structures are less symmetrical than the Cs^+ ones, since Na^+ is positioned away from the "C_4" symmetry axis, and coordinated to two phenolic oxygens of the host (Figures 2, 6 and 7). Additional coordination to solvent molecules is observed in the solid state as well as in the simulations. In the crystal, Na^+ is , indeed, coordinated to two H_2O (at about 2.4 Å) and one MeOH molecule. This correlates well with the hydration pattern simulated in water: the Na^+...O_W *rdf* calculated before the Na^+ dissociates peaks at 2.4 Å, and integrates to 3.2 water molecules. Such coordination of solvent molecules to Na^+ *exo* compensates for the partial loss of Na^+ coordination to two of the phenolic oxygens, which are, therefore, no longer prevented from

being hydrogen bonded to water. As far as possible proton transfer is concerned, it can be speculated that the *exo* cation, closest to the phenolate oxygen, prevents any proton from jumping to this oxygen. Such inhibition of proton transfer may contribute to the asymmetry of the *exo* complexes, as observed in the solid state of Na^+**L**$^-$. The above considerations show that structures of the complexes in the solid state with included solvent molecules may shed light on the M^+**L**$^-$ geometry in solution.

3.- On the preferred Cs^+ *extraction* and *transport* by L$^-$.

Based on free energy simulations, we first predict that in the gas phase, pure chloroform or acetonitrile solutions, the smallest cation Na^+ is complexed better than Cs^+. In water, the predicted trend is reversed, *i.e.* Cs^+ preferred over Na^+, due mainly to the dehydration cost of the cations in their free states. There are, to our knowledge, no experimental data directly related to these results. The calculations however clearly point out the importance of solvent environment on the binding selectivity, as documented in macrocyclic chemistry.[49] They also provide detailed microscopic pictures of the complexes in solution.

Two experimental studies on Cs^+ selectivity have been reported. The first one concerns the *transport* of cations from an aqueous phase through a thick CH_2Cl_2/CCl_4 liquid membrane. The relative fluxes of ion transported are 260, 5.6, 1.5 and < 0.9, respectively for Cs^+ , Rb^+, Na^+ and K^+. [10] The second study concerns cation *extraction* from water to dichloroethane.[45] Individual ion extraction gives the same sequence as obtained in transport experiments ($Cs^+ > Rb^+ > Na^+ > K^+$, with relative extraction constants K_{ex} = 340, 2.5, 0.9 and < 0.01, respectively), whereas *competitive* extraction gives a somewhat different order ($Cs^+ > Na^+ > Rb^+ > K^+$). The whole sequence is therefore somewhat dependent on the way the experiment is performed, but the important result, which prompted our study, was the high Cs^+ selectivity observed in all cases. At the microscopic level, such experiments are not clearly rendered. As a theoretical approach to these experiments, we considered a simple model, which deals with the free ions in pure water, and the complexes in pure chloroform. Based on this scheme, we predict that Cs^+ is best extracted ($Cs^+ > Rb^+ > K^+ > Na^+$), in agreement with experimental results. The conclusion holds to whatever *exo* or *endo* complexes are considered. Thus the calculations demonstrate that the peak observed for Cs^+ does not correspond to the largest host-guest interactions, but results, instead, mostly from solvation effects. At a more quantitative level, some discrepancies should be noted. The first one concerns the Na^+ / K^+ inversion. The second concerns the Cs^+ / Rb^+ selectivity, which is underestimated in the calculations (1.1 instead of 2.9 kcal/mole). From the computational side, force field effects cannot be ruled out, with regard to the host - guest interactions, as well as the solvent - solute interactions. The physical mimic used to model the extraction may also be questioned. Important features such as ionic strength and ion concentration in the source phase, pH effects, and water dragging to the organic phase, not represented computationally, may also play an important role in precisely determining the sequence of extraction or transport. It should be also noted that the extraction selectivities may not correspond to the selectivities in transport experiments, as was shown for instance by Kirch and Lehn [50] for the alkali cation cryptates. In particular, a too high binding affinity for M^+ in the organic phase may prevent its release from the organic phase to water. Taking into account such complexity at the molecular level is beyond our present computational possibilities. As far as the Cs^+ extraction, transport or complexation by the calixarene **L**$^-$ are concerned, our analysis strongly indicates that no "special $M^+...\pi$" interactions in favour of Cs^+ over the smaller alkali cations are required to account for the experimental results. Instead, solvation effects play a major role in these processes.

CONCLUSION.

Molecular Dynamics and Free Energy perturbation simulations were performed on the alkali salts ($M^+ = Li^+$, Na^+, K^+, Rb^+ and Cs^+) of *t*-butyl-calix[4]arene monooxyanion **L**$^-$ in the gas phase, chloroform, acetonitrile and water, assuming a 1:1 stoichiometry for the complex and a cone conformation for the host. We compared the relative stabilities and the structural characteristics of *exo* / *endo* complexes. In the gas phase, chloroform and acetonitrile the *endo* and *exo* cations remain as intimate ion pairs with **L**$^-$. This is in contrast to the results in water, where all complexes (except Li^+) dissociate. A detailed structural analysis suggests that the C_{4v} structure observed in the solid state for the Cs^+ *endo* complex does not correspond to an energy minimum, but results from the average of more stable asymmetrical forms. We also provide a detailed picture of the complexes in solution, which could not be obtained directly from experiment.

As far as the *endo* / *exo* forms are concerned, we find that intrinsically, in the gas phase and in solution, the *endo* forms are more stable than *exo* for a given cation. This result, found with 1-6-12 potentials, is strengthened by calculations including a polarization term in the host

- guest interactions. In chloroform and acetonitrile solutions, solute -solvent interactions favour *exo* coordination. Taking into account the solvent, we predict that in chloroform *endo* complexation is preferred because solvation effects are too weak to overcome the intrinsic *endo* preference. This contrasts with acetonitrile and polar non-aqueous solvents where *endo / exo* solvation effects are dominant, and lead to an *exo* preference, as observed experimentally in acetone. [11]

Based on free energy perturbation calculations, we predict that the order of binding selectivity is $Na^+ > K^+ > Rb^+ > Cs^+$ in pure chloroform or in pure acetonitrile, as well as in the gas phase. Thus, intrinsically, **L**$^-$ displays the weakest attractions with Cs^+. However, the calculated order of *extraction* selectivity from water to chloroform is reversed: $Cs^+ > Rb^+ > K^+ > Na^+$. These results agree qualitatively with the preferred Cs^+ extraction or transport from an aqueous phase. No specific "cation-π" interactions are required to account for preferred Cs^+ extraction from water to an organic phase. In fact, Cs^+ has the lowest interactions with the host: its binding or extraction selectivities are ascribed mostly to differential solvation effects.

Finally, we report a 350 ps MD simulation of the Cs^+**L**$^-$ complex at the water / chloroform interface. Although it is almost completely immersed in the organic phase, the complex does not migrate further, but remains "anchored" to the interface like a surfactant, accompanied by a few water molecules. Further studies involving the question of ion capture and transport at the interface are in progress in our laboratory. It is hoped that these microscopic pictures and related energy features will provide insights into the complex process of ion extraction and transport in solution, [51, 52] as well as with membrane mimics. [53-55]

Acknowledgements. The authors are grateful to the CNRS for allocation of computer time on IDRIS, to E. Engler for assistance in software modifications and to Dr N. Branda for linguistic assistance. AV is grateful to NATO and INTAS (94-3249) for supporting grants.

REFERENCES

(1) Dobler, M. *Ionophores and their Structures*; **1981**, Wiley Interscience: New York.

(2) Lehn, J.-M. *Angew. Chem., Int. Ed. Engl.* **1988**, *27,* 89-112.

(3) Wipff, G.; Troxler, L. in *Computational Approaches in Supramolecular Chemistry*; **1994**, Wipff, G.; Kluwer Academic Publisher, Dordrecht, Boston, London pp 319-348.

(4) Wipff, G. *J. Coord. Chem.* **1992**, *27,* 7-37.

(5) Zinke, A.; Zingler, E. *Chem. Ber.* **1944**, *77,* 264.

(6) Gutsche, D. *Acc. Chem. Res.* **1983**, *16,* 161 *and references cited therein.*

(7) Gutsche, C. D. *Calixarenes*; **1989**, J. Fraser Stoddart, U. K.; The Royal Society of Chemistry: Cambridge.

(8) *Calixarenes: a versatile class of macrocyclic compounds*; **1991**, Vicens, J. Böhmer, V.; Kluwer Academic Publisher: Dordrecht, Boston, London.

(9) Böhmer, V. *Angew. Chem., Int. Ed.* **1995**, *34,* 713.

(10) Izatt, R. M.; Lamb, J. D.; Hawkins, R. T.; Brown, P. R.; Izatt, S. R.; Christensen, J. J. *J. Am. Chem. Soc.* **1983**, *105,* 1782-1785.

(11) Harrowfield, J. M.; Olden, M. I.; Richmond, W. R.; White, A. H. *J. Chem. Soc., Chem. Comm.* **1991**, 1159-1161.

(12) Hamada, F.; Robinson, K. D.; Orr, G. W.; Atwood, J. L. *Supramol. Chem.* **1993**, *2,* 19-24.

(13) Royer, J.; Bayard, F.; Decoret, C. *J. Chim. Phys.* **1990**, *87,* 1695-1700.

(14) de Mendoza, X.; Prados, P.; Campillo, N.; Nieto, P. M.; Sachez, C.; Fayet, J.-P.; Vertut, M. C.; Jaime, C.; Elguero, J. *Recl. Trav. Chim. Pays-Bas* **1993**, *112,* 167.

(15) Lipkowitz, K.; Pearl, G. *J. Org. Chem.* **1993**, *58,* 6729.

(16) Groenen, L. C.; van Loon, J. D.; Verboom, W.; Harkema, S.; Casnati, A.; Ungaro, R.; Pochini, A.; Ugozzoli, F.; Reinhoudt, D. N. *J. Am. Chem. Soc.* **1991**, *113,* 2385.

(17) Fischer, S.; Grootenhuis, P. D. J.; Groenen, L. C.; van Horn, W. P.; van Veggel, F. C. J. M.; Reinhoudt, D. N.; Karplus, M. *J. Am. Chem. Soc.* **1995**, *117,* 1611-1620.

(18) Grootenhuis, P. D. J.; Kollman, P. A.; Groenen, L. C.; Reinhoudt, D. N.; van Hummel, G. J.; Ugozoli, F.; Andreetti, G. *J. Am. Chem. Soc.* **1990**, *112,* 4165-4176.

(19) Miyamoto, S.; Kollman, P. A. *J. Am. Chem. Soc.* **1992**, *114,* 3668-3674.

(20) Guilbaud, P.; Varnek, A.; Wipff, G. *J. Am. Chem. Soc.* **1993**, *115,* 8298-8312.

(21) Varnek, A. A.; Wipff, G. *J. Phys. Chem.* **1993**, *97,* 10840-10848 *and references cited therein* .

(22) Guilbaud, P.; Wipff, G. *J. Mol. Recogn. Inclus. Phenom.* **1993**, *16,* 169-188.

(23) Wipff, G.; Lauterbach, M. *Supramol. Chem.* **1995**, 000.

(24) Varnek, A.; Wipff, G. *J. Mol. Struct. (THEOCHEM)* **1995**, 0000.

(25) Varnek, A.; Wipff, G. *J. Comp. Chem.* **1996**, 0000.

(26) Pearlman, D. A.; Case, D. A.; Cadwell, J. C.; Seibel, G. L.; Singh, U. C.; Weiner, P.; Kollman, P. A. *AMBER4*; **1991**, University of California: San Francisco.

(27) Weiner, S. J.; Kollman, P. A.; Nguyen, D. T.; Case, D. A. *J. Comput. Chem.* **1986**, *7,* 230-252.

(28) Aqvist, J. *J. Phys. Chem.* **1990**, *94,* 8021-8024.

(29) Jorgensen, W. L.; Chandrasekhar, J.; Madura, J. D. *J. Chem. Phys.* **1983**, *79,* 926-936.

(30) Jorgensen, W. L.; Briggs, J. M.; Contreras, M. L. *J. Phys. Chem.* **1990**, *94,* 1683-1686.

(31) Jorgensen, W. L.; Briggs, J. M. *Mol. Phys.* **1988**, *63,* 547-558.

(32) Singh, U. C.; Brown, F. K.; Bash, P. A.; Kollman, P. A. *J. Am. Chem. Soc.* **1987**, *109,* 1607-1614.

(33) Varnek, A. A.; Glebov, A. S. *MM2IS Program Package*; **1991**, Mendeleev Institute of Chemical Technology: Moscow.

(34) Tsivadze, A. Y.; Varnek, A. A.; Khutorsky, V. E. *Coordination Compounds of Metals with Crown-Ligands (Russ)*; **1991**, Nauka Publ. House: Moscow.

(35) Allinger, N. L. *J. Am. Chem. Soc.* **1977**, *99,* 8127-8134.

(36) Gresh, N.; Claverie, P.; Pullman, B. *Int. J. Quant. Chem.* **1979**, *13,* 243.

(37) Engler, E.; Wipff, G. *MD_DRAW. A Program of graphical representation of a molecular trajectories*; **1992**, Université Louis Pasteur: Strasbourg.

(38) Izatt, S. R.; Hawkins, R. T.; Christensen, J. J.; Izatt, R. M. *J. Am. Chem. Soc.* **1985**, *107,* 63-66.

(39) Wipff, G.; Engler, E.; Guilbaud, P.; Lauterbach, M.; Troxler, L.; Varnek, A. *New. J. Chem.* **1996**, 0000.

(40) Beveridge, D. L.; DiCapua, F. M. *Ann. Rev. Biophys. Biophys. Chem.* **1989**, *18,* 431-492.

(41) Bash, P. A.; Singh, U. C.; Langridge, R.; Kollman, P. A. *Science* **1987**, *236,* 564-568.

(42) Jorgensen, W. L. *Acc. Chem. Res.* **1989**, *22,* 184-189.

(43) Straatsma, T. P.; McCammon, J. A. *Ann. Rev. Phys. Chem.* **1992**, *43,* 407.

(44) Marcus, Y. *Ion Solvation*; **1985**, Wiley, J.; Chichester.

(45) Natatou, L.; Burgard, M., Azfari, Z.; Vicens, J. *J. Inclusion Phenom.* **1995**, 0000.

(46) Kumpf, R. A.; Dougherty, D. A. *Science* **1993**, *261,* 1708-1710.

(47) Arduini, A.; Ghidini, E.; Ponchini, A.; Ungaro, R.; Andreetti, G. D.; Calestani, G.; Ugozzoli, F. *Journal of Inclusion Phenomena* **1988**, *6,* 119-134.

(48) Goldberg, I. *J. Am. Chem. Soc.* **1977**, *99,* 6049-6057.

(49) Izatt, R. M.; Bradshaw, J. S.; Nielsen, S. A.; Lamb, D. *Chem. Rev.* **1985**, *85,* 271-339.

(50) Kirch, M.; Lehn, J.-M. *Angew. Chem., Int. Ed. Engl.* **1975**, *14,* 555-556.

(51) Benjamin, I. *J. Chem. Phys.* **1992**, *96,* 577-585.

(52) Benjamin, I. *Science* **1993**, *261,* 1558-1560.

(53) Roux, B.; Karplus, M. *J. Am. Chem. Soc.* **1993**, *115,* 3250.

(54) Guba, W.; Kessler, H. *J. Phys. Chem.* **1994**, *98,* 23-27.

(55) Guba, W.; Haessner, R.; Breipohl, G.; Henke, S.; Knolle, J.; Santagana, V.; Kessler, H. *J. Am. Chem. Soc.* **1994**, *116,* 7532-7540.

(56) Goldman, H.; Vogt, W.; Paulus, E., Böhmer V. *J. Am. Chem. Soc.* **1988**, *110,* 6811-6817.

Solid State NMR in Inclusion Compounds

Synoptic Approaches to Structure Determination

John A. Ripmeester and Christopher I. Ratcliffe
Steacie Institute for Molecular Sciences
National Research Council of Canada
Ottawa, Ontario, Canada K1A 0R9

1. Introduction and Scope

Over the years NMR spectroscopy has evolved to become a major tool in just about all areas of chemistry, and without a doubt it is the principal tool for structure elucidation in solution. With chemical shifts and J-coupling as primary information, there is now a wealth of techniques to trace out through-bond connectivities, non-bonded distances and conformational information. Most chemists will have some familiarity with at least some of these techniques, which are well-reviewed in a number of articles and books. It is beyond the scope of this article to give an overview of this rapidly advancing field.

For studies of structure in the solid state, the most powerful technique is diffraction, with single crystal work the technique of choice, although there have been many recent advances in powder methods. So what is the place of solid-state NMR? In this contribution we hope to show how solid-state NMR techniques can be used both independently and as an adjunct to diffraction to derive structural and dynamic information which often cannot be obtained easily by other techniques. It must be kept in mind that diffraction tends to be a "snap-shot" technique where static and dynamic disorder usually aren't distinguishable, and all types of disorder are projected back into a single unit cell. Thus diffraction methods give structural information which is space and time averaged, whereas most NMR parameters are sensitive especially to local structure. Also, in X-ray diffraction studies the thermal ellipsoids which locate the atomic positions usually are drawn to account for a large fraction of the total electron density. It can be of some importance to find some of the missing electron density, however, for locating secondary guest sites which could play an important role in topologically controlled reactions. In the case of severely disordered systems sometimes attempts are made to locate guest species by empirical modelling techniques. Solid state NMR offers a means to test potential functions and parameters, as reorientational barriers may be calculated and compared with experimentally derived values.

G. Tsoucaris et al. (eds.), Crystallography of Supramolecular Compounds, 101–125.

From a more fundamental point of view, it is important to show that solution NMR techniques do not easily carry over to the solid state, and to identify some of the important NMR parameters unique to solid state studies. This account will serve as a non-rigorous introduction to solid state NMR and will also present some recent results where NMR techniques have brought new insights pertinent to various aspects of inclusion chemistry. The serious student should consult some of the appropriate specialized literature cited. For the NMR fundamentals (section 2) a number of reference texts and articles are given [1-7], whereas for the applications some specific references are presented.

2. NMR Spectroscopy of the Solid State

2.1. THE FUNDAMENTAL PHENOMENON

As a fundamental physical phenomenon, the observation of a nuclear magnetic resonance signal has three basic requirements, a polarizing magnetic field H_o to remove the degeneracy of the Zeeman levels, a resonant radio-frequency field H_1 which drives the transitions, and a relaxation mechanism so that the transitions aren't saturated. We will be concerned primarily with the nature of the resonance absorption itself rather than relaxation phenomena [1].

In order to visualize the relationship of the different fields and the generation of a signal, it is useful to introduce the well-known rotating frame of reference. In a fixed frame of reference the spins with a magnetic moment m and gyromagnetic ratio g precess about the magnetic field H_o along the z -axis at the Larmor frequency $\omega_o = \gamma H_o$, and the radio-frequency field H_1 is an oscillating field in the x-y plane at a similar frequency, $\underline{H}_1 = H_1$ (cos ωt + sin ωt) (fig. 1a). If now a frame of reference is chosen which rotates at the Larmor frequency ω_o, the net spin magnetization is now aligned along the z-axis. H_o must be replaced by $H_o - \omega/\gamma$, so that at exact resonance H_o vanishes, and H_1 becomes a stationary field in the x-y plane (fig. 1c,d). If H_1 is turned on for a time t_w the spin magnetization vector is rotated by an angle $\theta = \gamma H_1 t_w$ (fig. 1d). For an angle of 90° (90° pulse) ,the magnetization vector ends up in the x-y plane, and generates a signal in the sample coil. Under the influence of various spin interactions the spins dephase in the x-y plane and the signal decays with a characteristic time T_2, the spin-spin relaxation time. Normally this time-domain signal, after phase-sensitive detection and digitization is Fourier transformed to give a frequency spectrum. The spins return to their equilibrium position along the z-direction with a characteristic time constant T_1, the spin-lattice relaxation time. Because of the use of phase sensitive detection, the signal digitized and stored in computer memory is actually a "rotating frame" video signal.

The power of the NMR technique as traditionally used in solution work is that there is a one-to-one correspondence between resonance lines in a spectrum and chemically distinct nuclei in a molecule. This is, of course, true only if the spectrum can be obtained under quantitative conditions, and spin couplings can be identified, or suppressed by decoupling.

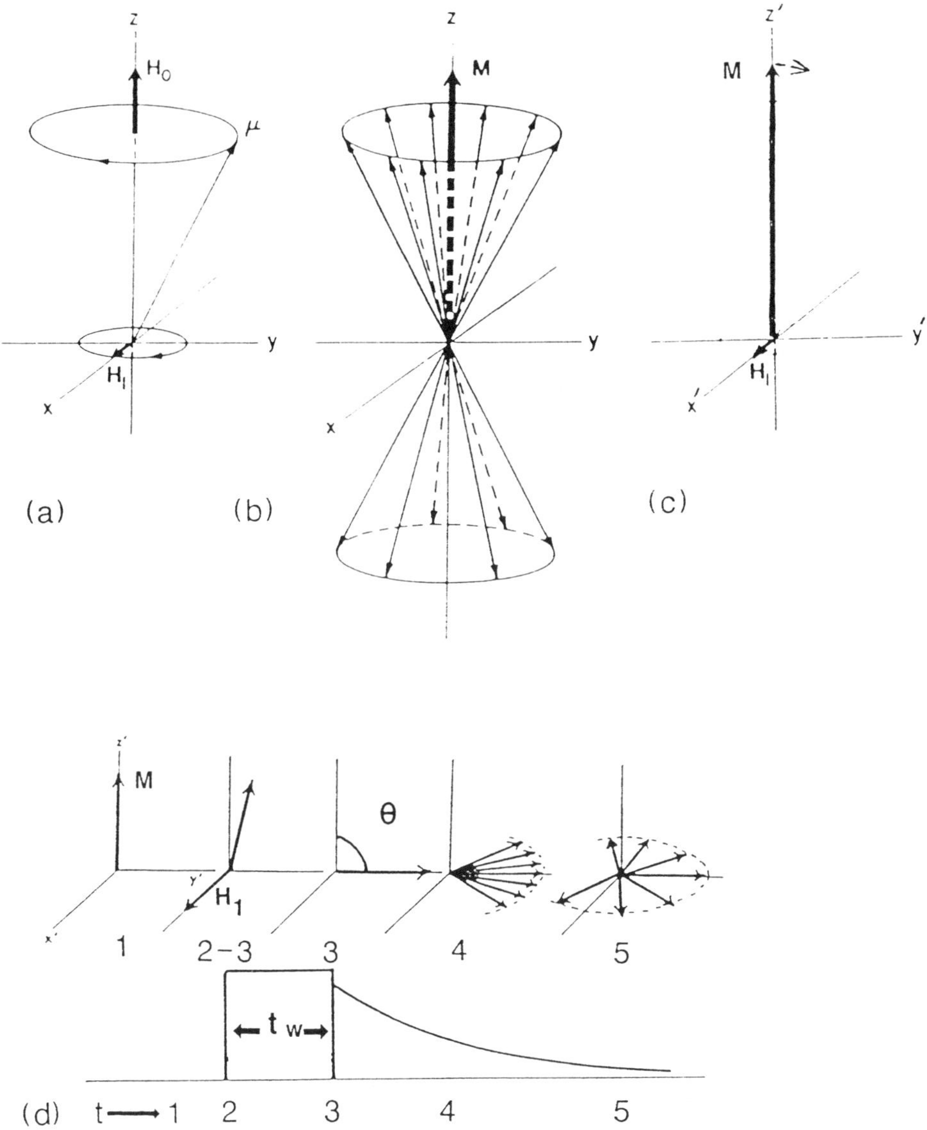

Figure 1 (a) Laboratory reference frame showing spin magnetic moments μ precessing about the magnetic field H_o, and the rotating rf field H_1; (b) laboratory reference frame showing excess spins aligned along the H_o field, giving rise to a resultant spin magnetization vector M; (c) rotating reference frame with magnetization vector and the now static field H_1; (d) Timing sequence and spin vector diagram for a simple NMR pulse experiment (see text).

Is it possible to obtain a high resolution spectrum in the solid state to obtain similar information? The answer is yes, but it will become clear that the experimental methods required are quite different from solution techniques, and that in the solid state the nuclear spin energy levels are affected by a number of interactions which do not manifest themselves directly in solution NMR lineshapes. Although sometimes one must go to considerable effort to remove the effects of these interactions to observe a simple chemical shift spectrum, they often give very useful information in their own right.

2.2. THE SOLID STATE NMR SPECTRUM

The Hamiltonian governing the energy levels in the general case is [1,2,4,6]

$$\mathcal{H} = \mathcal{H}_Z + \mathcal{H}_Q + \mathcal{H}_D + \mathcal{H}_\sigma + \mathcal{H}_J \qquad (1)$$

where $\mathcal{H}_Z$ is the Zeeman interaction, $\mathcal{H}_Q$ is the quadrupolar interaction, $\mathcal{H}_D$ is the nuclear dipolar interaction, $\mathcal{H}_\sigma$ is the chemical shielding interaction and $\mathcal{H}_J$ is the scalar or J coupling. It must be noted that the Hamiltonians all should be considered as tensor interactions, rather than scalar interactions as for solutions. This is due to the fact that the rapid molecular motions which average the interactions to their isotropic values usually are absent in the solid state, and that all of these interactions therefore depend on the orientation of the interaction tensor with respect to the magnetic field direction. The different interactions will now be examined in more detail. All take the general form: $\mathcal{H} = \underline{A}.\underline{\underline{X}}.\underline{B}$ where $\underline{\underline{X}}$ represents the interaction tensor and can be represented by a 3x3 matrix. $\underline{A}$ and $\underline{B}$ are row and column vectors which usually describe spin vector operators.

In all of the cases of interest here, the Zeeman interaction which lifts the degeneracy of the 2I+1 nuclear spin energy levels, will be taken to be much larger than the other interactions.

2.2.1. *The Nuclear Quadrupolar Interaction*

The lineshape We choose to start with this interaction as it often is the only one which needs to be considered to explain lineshapes for nuclei such as ^2H (spin I = 1)[6]. The quadrupolar Hamiltonian represents the interaction between the nuclear quadrupole moment Q and the electric field gradient at the nucleus $\underline{\underline{q}}$, and takes the form

$$\mathcal{H}_Q = \underline{I}\, e^2Q\, \underline{\underline{q}}.\, \underline{I}\, /(2I(2I-1)) \qquad (2)$$

The electric field gradient tensor with its 3 principal components q_{xx}, q_{yy} and q_{zz}, is defined in a coordinate system attached to the molecule (fig. 2a). e^2qQ/h is known as the quadrupole coupling constant (qcc) and the asymmetry parameter is defined as $\eta = (q_{yy}-q_{xx})/q_{zz}$. $\nu_Q = 3/2\, e^2Qq/h$ is known as the quadrupole frequency. For nuclei such

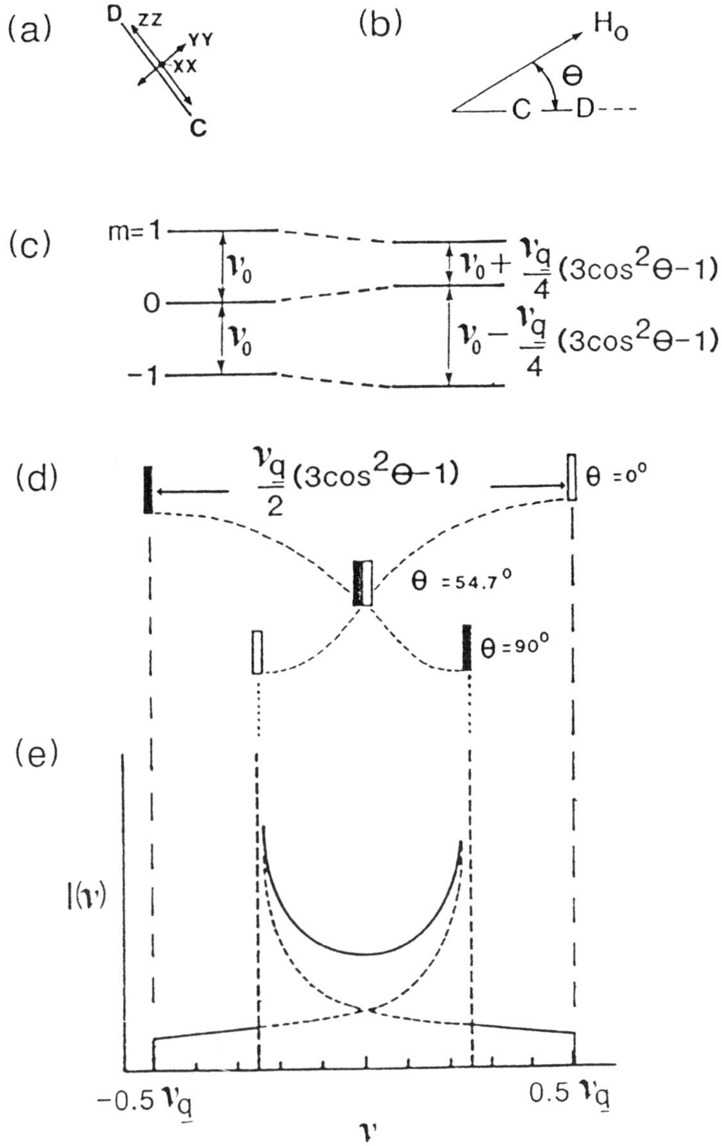

Figure 2 (a) C-D bond fragment showing the orientation of the three electric field gradient components; (b) orientation of the C-D bond with respect to the magnetic field; (c) quadrupole perturbation of spin I=1 Zeeman levels; (d) orientation dependent line splittings for C-D bond in single crystal; (e) axially symmetric powder pattern.

as ^2H, the quadrupolar interaction is small enough to be no more than a perturbation on the Zeeman levels. This also requires the definition of the interactions in terms of a coordinate system defined by the magnetic field (fig. 2b). The resulting energy level scheme is shown in fig. 2c. There are two allowed transitions which give rise to a doublet of lines which is symmetric about the Zeeman frequency. The energy levels (fig. 2c), and hence the spectrum (fig. 2d) now have an orientation dependence. For single crystal samples both the orientation and the magnitude of the qcc tensor principal components can be determined by observing the doublet splitting as a function of rotation angle, usually about three orthogonal axes, and diagonalization of the general qcc tensor. For powder samples all angles with respect to the field are present, and the much broader powder pattern I(n)= sinq dq/dn results (fig. 2e) because the frequency at which certain angles occur in the powder sample must be taken into account . In figures 2d and 2e the relationship between the single crystal and powder spectra are clearly evident. It also should be noted that the powder lineshape consists of the sum of many individual single crystal lines. This kind of lineshape is known as an inhomogeneously broadened lineshape.

The sharp transitions as shown in fig. 2 are also broadened by interactions with nuclear dipoles, and this can be accounted for by convolving the basic powder lineshape with a Gaussian or Lorentzian function. (fig. 3a,b). The three splittings in the general powder pattern (fig.3b) still relate directly to the quadrupole coupling constant and asymmetry parameter as follows:

$$\Delta\nu_3 = \nu_Q, \;\; \Delta\nu_2 = (1+\eta)\,\nu_Q/2, \;\; \Delta\nu_1 = (1-\eta)\,\nu_Q/2 \qquad (3)$$

DynamicEffects Although nuclear quadrupole coupling constants can be calculated to yield information on the electronic structure of molecules, the main use of ^{2}H NMR has been to study dynamics in the solid state[6]. Since the qcc's depend on the electric field gradients at the nucleus, molecular motions with a characteristic rate and amplitude average the gradients and hence also the observable qcc's. This also necessitates the introduction of a timescale, just the same as for fine structure averaging in solution due to chemical exchange. Motions start to affect the lineshape when the rate becomes similar to $1/\pi T_2$, the linewidth of the individual powder lineshape components, with a maximum effect when the rate is similar to the quadrupole splitting. The lineshape, although narrowed becomes independent of the motional rate in the fast motion limit. The fast motion limit lineshape is again relatively easy to calculate. In general, the qcc's must be calculated in a frame of reference where the averaged tensor is diagonal, with the qcc for each site connected by the motion calculated by the use of rotation matrices. For relatively simple motions, it is possible to illustrate motional effects explicitly. Lineshapes for a para disubstituted benzene molecule rotating about the molecular symmetry axis either by two-fold flips or an n-fold reorientation ($n \geq 3$) are shown in figs. 3a,b,c.

For intermediate motional rates, the calculation is considerably more involved, and, unlike the case for the fast motion limit, the lineshape also is able to give information on the details of the motion such as the order of the rotation axis and the motional rate. The complex nature of the problem can be envisioned if we first consider a single crystal with a single type of C-D bond. At a certain orientation of the crystal, the motional averaging depends on the number of sites for the C-D vector connected by the motion and their angular disposition. The lines will broaden, and reappear as a doublet with a different splitting in a well defined manner as the motional rate changes with temperature. In a powder, all doublet splittings are present,

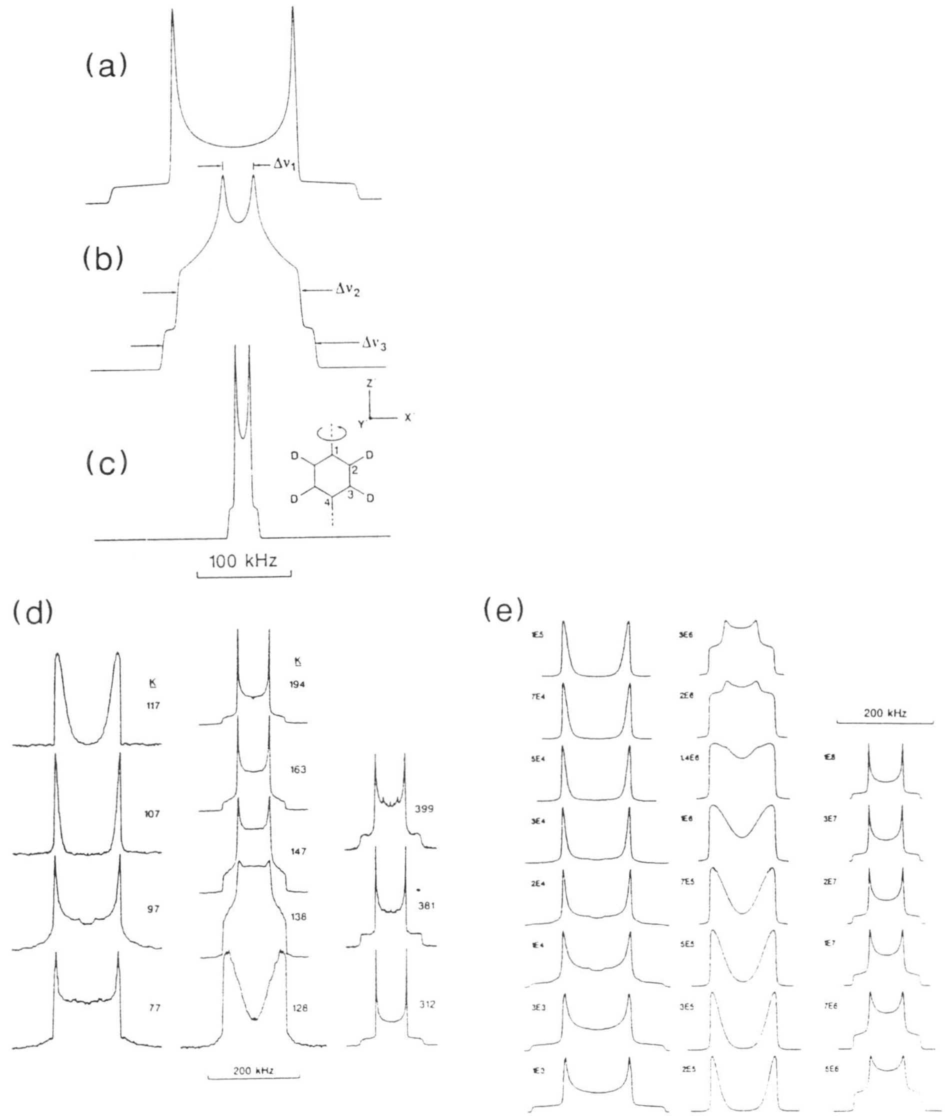

Figure 3 (a, b ,c) calculated ^{2}H NMR powder patterns for para disubstituted benzene ring; (a) rigid lattice; (b) fast motion limit for two-fold flip; (c) fast motion limit for axial rotation; (d) experimental ^{2}H NMR lineshapes for enclathrated benzene ring in $Cd(NH_3)_2Ni(CN)_4.2C_6D_6$ as a function of temperature; (e) calculated ^{2}H NMR lineshapes for in-plane 6-fold benzene ring rotation as a function of jump rate [8].

and the degree of maximum broadening for each splitting will occur for different rates, yielding a lineshape which varies with temperature in a very complex manner. Experimental and calculated lineshapes for in-plane rotation of an enclathrated benzene molecule in $Cd(NH_3)_2Ni(CN)_4.2C_6D_6$, are shown in figs. 3d,e [8].

2.2.2 *The Nuclear Dipolar Interaction*

The nuclear dipolar interaction $\mathcal{H}_D$ represents the through space interaction between nuclear dipoles according to $\underline{\mathcal{H}}_D = \underline{I}_1.\underline{\underline{D}}.\underline{I}_2$, and in a high magnetic field again is treated as a perturbation on the Zeeman energy levels. The secular part of the Hamiltonian, the part responsible for lineshapes can be expressed as [1,2,4]:

$$\mathcal{H}_D = \frac{\gamma^2 \hbar^2}{r^3} (1\text{-}3\cos^2\theta) [I_{1z}I_{2z} - 1/4 (I_{1+}I_{2-} + I_{1-}I_{2+})] \qquad (4)$$

where q is the angle between the internuclear vector and the magnetic field direction. This expression holds for like spins. For unlike spins, the expression is usually written with I for I_1 and S for I_2, and the last terms in I_+ and I_- must be omitted. The inverse cube dependence on internuclear distance makes this interaction most important for directly bonded nuclei, although the interactions are summed over all nuclear pairs in the lattice. For small isolated clusters of nuclei (up to 6 or so) the interactions give rise to specific lineshapes which can be calculated, yielding the internuclear distances [9]. It must also be kept in mind that a sharp line spectrum is obtainable only for a single crystal, and that for a polycrystalline sample again a powder pattern results.

More often, the nuclear dipolar interactions from nuclei both near and further away act to obscure fine structure information by line broadening. Strategies to remove the effect of dipolar coupling from the spectrum will be discussed later.

Some other features of the dipolar interaction must be mentioned at this time. As pointed out above, the dipolar interaction is different for homo and heteronuclear interactions. In the case of homonuclear interactions the extra term in the Hamiltonian, often referred to as the flip-flop term, causes efficient transfer of magnetization throughout the sample by a process known as spin diffusion, and it is the reason why all ^1H nuclei in a uniform single phase solid sample usually have the same spin-lattice relaxation time. The other property of most dipolar broadened lines is that they are homogeneous, that is irradiation of the line at a specific frequency causes uniform saturation of the entire line rather than the " hole burning" which would be observed for an inhomogeneously broadened line.

2.2.3 *The Chemical Shift*

The chemical shift interaction in the solid state also is an anisotropic interaction and can be written as: $\mathcal{H}_S = \underline{I}.\underline{\sigma}.\underline{H}$ [2,4,5,6]. The anisotropic chemical shift therefore has an orientational dependence, which becomes evident when studying the spectrum of a single crystal where the principal axis of the chemical shift tensor may be oriented at different angles with respect to the magnetic field. The line position can be expressed in terms of the the direction cosines cos θ_{ii} of the tensor components with respect to the field:

$$\nu = (\gamma H_o/2\pi) [1 - \sigma_{xx} \cos^2\theta_{xx} + \sigma_{yy}\cos^2\theta_{yy} + \sigma_{zz}\cos^2\theta_{zz}] \quad (5)$$

In a polycrystalline sample, a powder pattern results (fig.4a,c) as previously discussed for the quadrupolar interaction, with the tensor principal components σ_{xx}, σ_{yy}, and σ_{zz} obtainable from the features of the lineshape. For an axially symmetric lineshape there are just two parameters $\sigma_{\|}$ and $\sigma_{\perp}$, the shielding tensor components along and at right angles to the bond direction. Note the relationship between powder patterns for the anisotropic chemical shift with a single transition, and the quadrupolar coupling with two transitions.

Next, it is necessary to distinguish "rare" spins such as ^{13}C which represent only about 1% of all C, from "abundant" spins such as ^1H and ^{19}F which are close to 100% naturally abundant. If rare spins are to be observed, it is relatively easy to get a useful spectrum, as the nuclear dipolar broadening due to like spins is very small, and broadening due to unlike spins can be removed by decoupling. This kind of decoupling, known as dipolar decoupling (DD), needs to be at least of the order of the dipolar coupling in strength to be effective, and is described theoretically as a "coherent averaging " process [5,6,10]. For directly bonded ^1H-^{13}C pairs the decoupling field strength needs to be of the order of 50 kHz (≈ 100 W). The resulting spectrum will contain the anisotropic chemical shifts, as well as dipolar and J coupling information from undecoupled nuclei.

The final step in obtaining an isotropic chemical shift spectrum is to remove the effect of anisotropy by magic angle spinning. If the expression for the orientationally dependent chemical shift is expanded to allow explicit time dependence due to sample rotation at angular frequency w_r, the direction cosines become (fig.4a) :

$$\cos\theta_{ii} = \cos\beta \cos\chi_{ii} + \sin\beta \sin\chi_{ii} \cos (\omega_r t + \phi_{ii}) \quad (6)$$

where the ϕ_{ii} are the azimuth angles at t=o,and

$$\nu = (\gamma H_o/2\pi) [1\text{-}3/2\, \sigma_i \sin^2\beta + 1/2 (3\cos^2\beta - 1) \Sigma_{ii} \sigma_{ii} \cos^2\chi_{ii}] \quad (7)$$

The anisotropic terms vanish when β = the magic angle (tetrahedral angle divided by two), leaving an isotropic chemical shift spectrum. The chemical shift traced out by one of the single crystal lines in the powder pattern on magic angle rotation is shown in fig. 4b. Again, magic angle spinning is a coherent averaging process, yielding rotational echoes which result in the presence of spinning side bands in the Fourier transformed spectrum. Fig. 4c shows the effect of spinning speed on the isotropic spectrum.

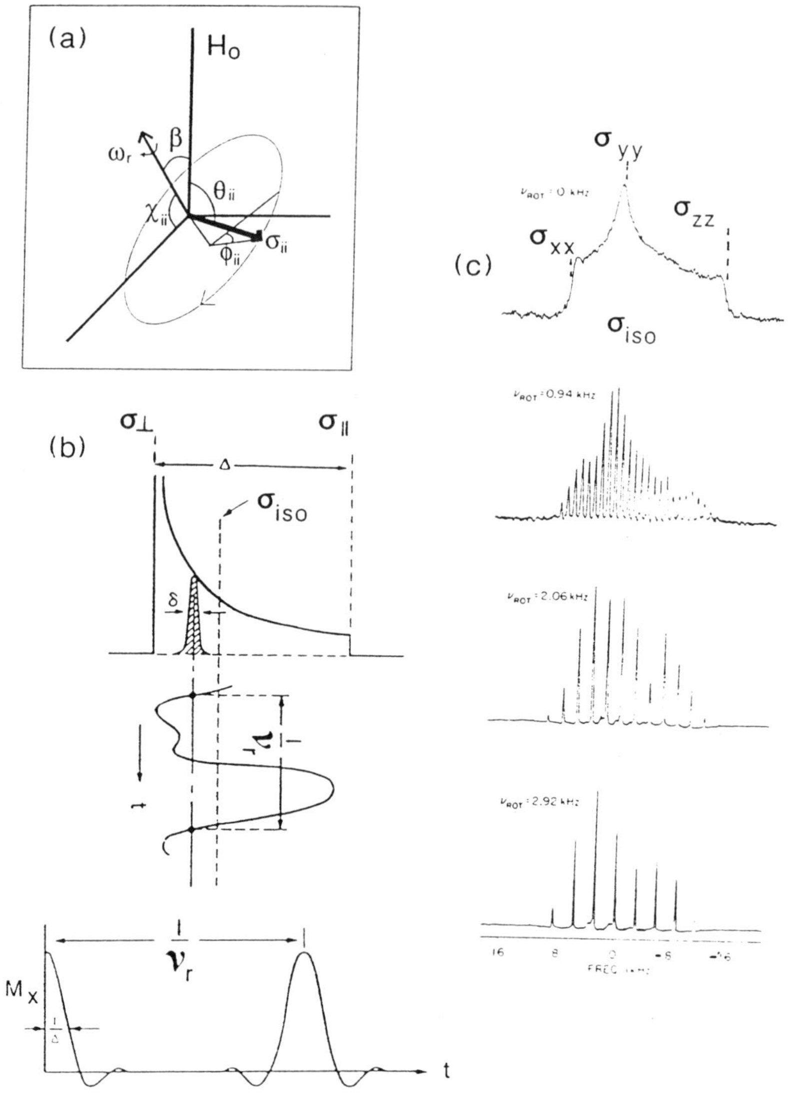

Figure 4. (a) Relationship of chemical shift tensor components with respect to magic angle rotation axis and magnetic field direction; (b) powder lineshape for axially symmetric chemical shift tensor; also shown is the chemical shift path traced out by one of the single crystal component lines and the generation of a rotational echo (after ref.[11]); (c) powder pattern for a general shielding tensor and the spectra obtained under magic angle spinning at various rates (adapted from ref.[12]).

The effect of molecular motion on the chemical shift anisotropy pattern is described in the same way as for the quadrupole powder lineshapes, and again a " chemical exchange" description can be employed [2,5,6]. It is relatively simple to calculate " fast motion limit" lineshapes, but considerably more difficult to calculate lineshapes for motions with intermediate rates. There are also two kinds of motional effects unique to solid state spectroscopy where coherent averaging techniques such as dipolar decoupling and magic angle spinning are employed. Since molecular motion causes random averaging, it interferes with the coherent averaging, and produces anomalous spectral broadening for specific motional rates. For magic angle spinning the maximum broadening occurs for spinning rates such that the motional correlation time $t_c \sim 1/\omega_r$, whereas for dipolar decoupling the maximum broadening occurs when the motional correlation time $\tau_c \sim 1/\gamma H_1$.

For nuclei which are abundant, eg ^{1}H, the chemical shift information must be extracted from a spectrum in the presence of strong homonuclear dipolar couplings. In this case the spin part of the Hamiltonian is averaged by application of complex multiple pulse cycles, yielding anisotropic chemical shift information, or a combination of this with magic angle spinning (CRAMPS) to yield an isotropic chemical shift spectrum [10].

2.2.4 *J Coupling*

This interaction also is a tensor interaction in the solid state and can be written as: $\mathcal{H}_J = \underline{I}_1 \cdot \underline{\underline{J}} \cdot \underline{I}_2$. In solid state NMR spectroscopy, J coupling has not been used extensively. In order to extract J coupling information, one must use multiple-pulse decoupling techniques rather than the standard dipolar decoupling which removes the effects of J coupling from the spectrum completely. In obtaining distance information from nuclear dipolar coupling data, often anisotropic J coupling plays a complicating role as these two interactions cannot be separated.

2.3 SOME BASIC PULSE TECHNIQUES

2.3.1 *Echo Spectroscopy*

The fundamental, Fourier transform pulse experiment was described in section 2.1 and this carries over to the solid state without any modifications if the resonance lines aren't too broad. The broadest lines recorded routinely are ^{2}H spectra which are 200-300 kHz wide. Then there are pulse power and digitizer speed requirements so that the pulse bandwidth and the spectral sweep width are sufficiently broad to cover the spectrum. In addition, it is necessary to resort to echo techniques in order to obtain reliable lineshapes. This is due to the fact that after a high voltage rf pulse there is a short but finite time necessary for the probe circuit to recover before it can receive the nuclear induction signal. This means that critical information is lost during this "dead" time, which causes serious distortions in the spectrum, and which can be recovered by means of echo techniques. The quadrupole echo experiment as carried out for ^{2}H cannot be described in terms of spin vector diagrams [2], however examination of the Hahn echo [1],is also informative (fig.5). The experiment is again described in the rotating frame of reference and starts with a 90° pulse (step 1-2) which rotates the spins along the y' axis. During the echo delay time the spins, due to anisotropic spin interactions, lose step with each other, the faster and slower spins going in different directions in the x'y' plane (step 3). The 180° pulse applied (step 4-5) a time t after the

90° pulse flips the spins about the x' axis, so that after another time t they get back into step (step 6), thus producing an "echo" signal (step 7). The timing diagram (fig. 5) indicates how this allows resolution of the signal at 0 time, not possible after the first pulse because of probe deadtime. Almost all ^{2}H NMR lineshapes are recorded by the related quadrupolar echo technique, which gives the important 0 time information necessary to minimize lineshape distortions.

2.3.2 *Cross polarization*

Another one of the most important solid state NMR experiments is the cross-polarization experiment [2,3,6]. This is commonly used to enhance the weak signal of rare spins such as ^{13}C and ^{29}Si, and also avoids the often long relaxation times of these nuclei which govern the pulse repetition times in time averaging experiments. The experiment starts with the generation of spin magnetization for abundant spins such as ^{1}H with a 90° pulse (fig. 6 step 1-2). This is followed by a phase-shifted pulse (x' direction in the rotating frame) on the protons which locks the magnetization along the field direction (step 3). At the same time a pulse is applied to the rare spins, and its amplitude is adjusted so that the following condition holds (known as the Hartman-Hahn matching condition)

$$\gamma_a H_{1a} = \gamma_r H_{1r} \tag{8}$$

The precession frequencies of both rare and abundant spins about their respective H_1 fields are then the same, and the spins can exchange energy via the flip-flop term of the dipolar interaction. After optimum transfer, the H_1 field on the rare spins is turned off, resulting in a free precession signal which may then be digitized (step 4), meanwhile keeping the abundant spin H_1 field on for decoupling purposes. The delay between successive sequences is now governed by the usually much shorter relaxation time of the abundant spins. There is then an overall gain in signal/noise ratio achieved over the standard Fourier transform experiment both from the cross polarization (maximum gain is γ_a/γ_r), and the shorter delay times allowed by the more efficient relaxation of the abundant spins.

2.4 HIGH RESOLUTION SOLID STATE NMR SPECTROSCOPY

The above label usually refers to the fact that use has been made of either MAS or CP/MAS techniques, and this kind of spectroscopy has found many applications in modern research. A brief summary of its features will be presented.

One important feature of solid state spectroscopy is that nuclei which are normally considered to be chemically equivalent may not be equivalent in the solid state, and hence unexpected multiplicities may be observed. In fact, all crystallographically inequivalent nuclei in the crystal asymmetric unit should give rise to separate resonances, although there often are accidental degeneracies. Examination of a solid state spectrum therefore can give a quick "snapshot" of the assymmetric unit contents, and should be considered as a routine step before carrying out diffraction work.

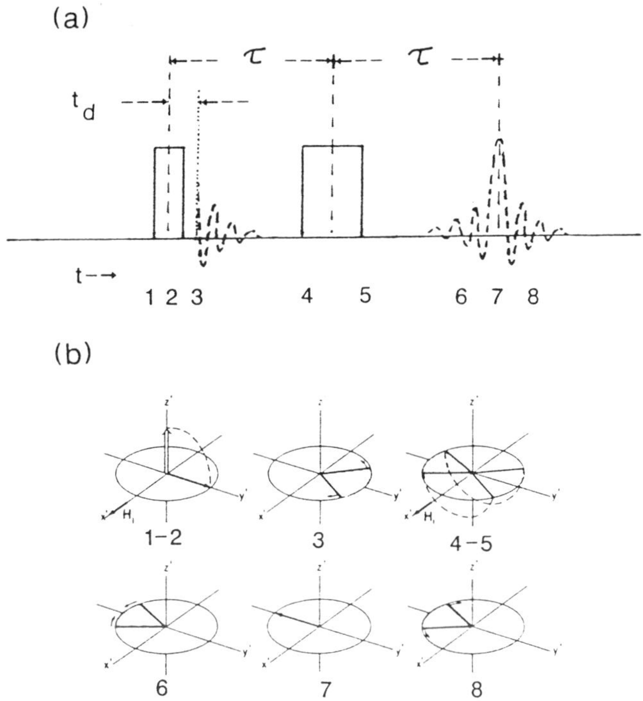

Figure 5. Timing sequence (a) and spin vector diagram (b) for a Hahn echo pulse sequence.

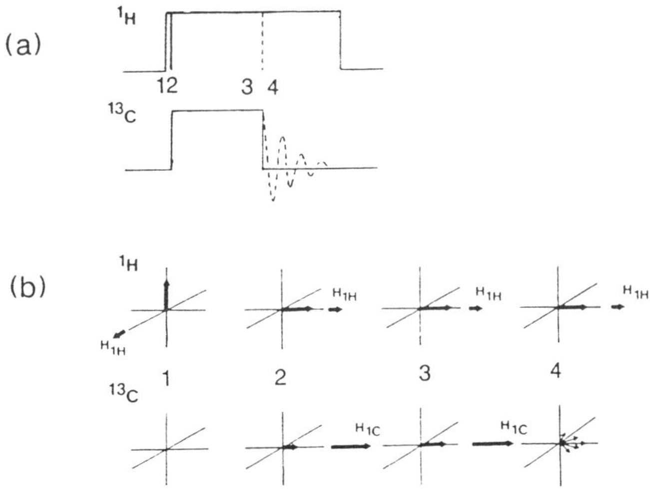

Figure 6. Timing sequence and spin vector diagram for cross-polarization sequence.

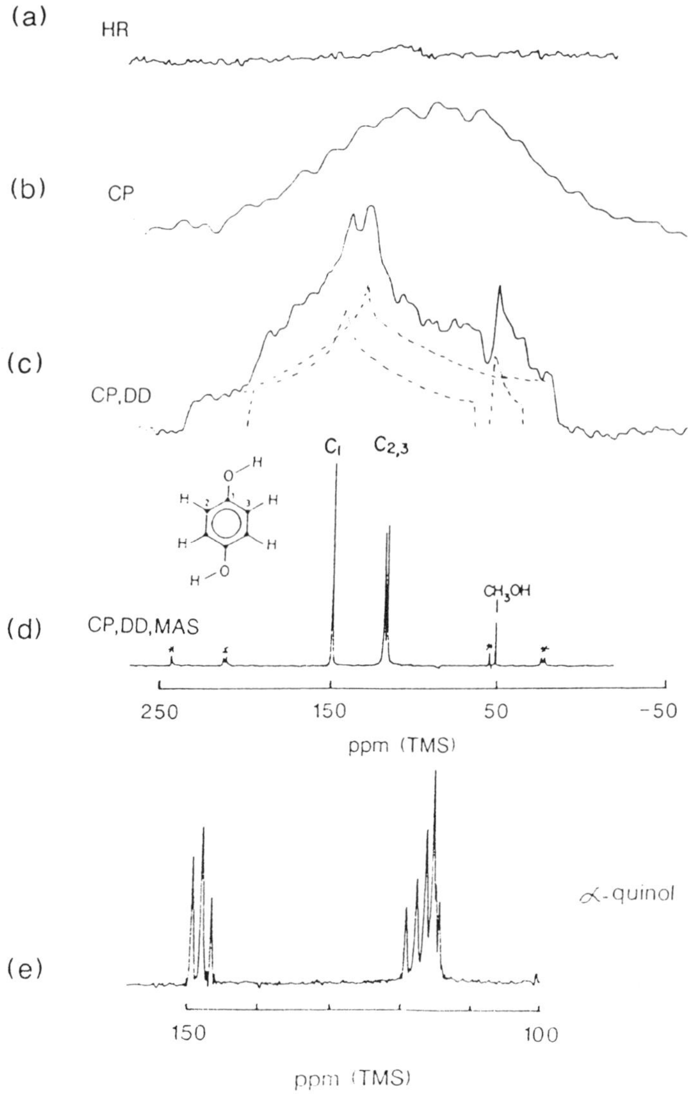

Figure 7 (a-d) Development of ^{13}C CP/MAS spectrum of solid methanol β-quinol clathrate; (a) spectrum obtained by using high resolution solution techniques; (b) spectrum obtained with cross-polarization, no decoupling; (c) spectrum obtained with cross-polarization and ^{1}H dipolar decoupling; (d) spectrum obtained with cross-polarization, dipolar decoupling and magic angle spinning (CP/MAS); (e) ^{13}C CP/MAS spectrum of α-quinol

The development of a CP/MAS (cross-polarization/magic angle spinning spectrum) is shown in fig. 7 as the various components of the technique are implemented. The top spectrum, obtained by using solution spectroscopy techniques on the solid sample shows little intensity. The application of cross-polarization shows considerable intensity, but no resolution. Dipolar decoupling gives the recognizable features of a number of overlapping chemical shift powder patterns which have been sketched in, and finally magic angle spinning gives the high resolution spectrum. The inequivalence of the quinol ring carbons 2 and 3 in the crystal is indicated clearly. For comparison, the spectrum of the empty quinol also is given. By carefully noting intensities, one can conclude that in this case there are 6 inequivalent phenolic carbons and 12 other carbons in the asymmetric unit. In fig.7c, the small chemical shift anisotropy for the methanol carbon also is indicative of considerable motional averaging. The utility of the solid state techniques in studying weak guest-host interactions is clearly indicated, as both the empty host and the clathrate would have identical solution spectra.

Since polarization transfer spectroscopy depends on the nuclear dipolar coupling between the rare and abundant nuclei, spectra recorded without special experimental considerations should not be taken as quantitatively accurate. For carbon compounds, quaternary carbons and carbons which have considerable motional averaging cross-polarize much more slowly than rigid methylene and methine carbons. This, in fact, forms the basis for a number of spectral sorting techniques. Other useful signatures are given by nuclei coupled to quadrupolar nuclei [2].

Another feature unique to the solid state is the motional rate dependent broadening associated with random averaging processes mentioned in section 2.2.3 which may cause specific lines in the spectrum to disappear completely in a particular temperature range. In favourable cases the linewidth may be used to derive motional rate information [2].

How about the transferability of the numerous 2-dimensional solution NMR techniques to the solid state? In general, this is not feasible for solids which have strong dipolar couplings, as many of the 2-D techniques depend on the much weaker J coupling. However, COSY and INADEQUATE experiments have been performed on systems with very weak dipolar coupling such as plastic crystals, and silicates [13]. Instead, a new array of experiments has been developed to utilize some of the unique solid state couplings.

3. Some Applications

3.1 INORGANIC LATTICES

3.1.1 *Zeolites and Lattice Silicates*

One area of great interest is the design, synthesis and study of new framework host lattices. Zeolites represent one such group in which there is not only a scientific, but also a considerable economic interest [14]. One traditional problem in studying zeolites, and aluminosilicates in general, is the presence of Si/Al disorder, and also, the difficulty experienced in distinguishing these two atoms by X-ray diffraction. Hence the discovery that the ^{29}Si MAS NMR spectrum chemical shifts are sensitive to the nearest neighbour Si/Al distribution proved to be extremely valuable both for solving a number of standing problems and for the design of new materials by altering the ordering schemes [14].

For materials with many atoms which are chemically equivalent, but crystallographically inequivalent such as lattice silicates where there may be many types of SiO_4 units, NMR methods can give rapid and detailed information on the asymmetric unit in the crystal. An example is the clathrasil Dodecasil-3C, which is formally isostructural with the cubic clathrate hydrate known as Structure II [15]. However, this particular material undergoes a number of symmetry-lowering transitions [16], and correspondingly, the ^{29}Si CP/MAS NMR spectrum changes to reflect the increasingly complicated asymmetric unit in the crystal. The NMR, and synchrotron powder diffraction data, together with group - maximal subgroup relationships, were used to derive the following scheme for the D3C phases: cubic (Fd3m) $\rightarrow$ tetragonal (I4/amd) $\rightarrow$ orthorhombic (Imma) $\rightarrow$ monoclinic (C2/m).

Often it is impossible to assign any one of a number of closely-spaced resonance lines to a specific site in the asymmetric unit of known structures. Two-dimensional NMR techniques such as COSY and INADEQUATE can then be helpful at arriving at an assignment [13]. An example is that given by the zeosil ZSM - 12, which has 7 different T sites. The Si atoms are relatively far apart and the natural abundance of ^{29}Si is relatively low so that dipolar couplings are weak. The J coupling between ^{29}Si atoms in $^{29}Si-O-^{29}Si$ pairs is small (about 10Hz), but it is sufficient to tranfer magnetization between coupled spin pairs for a COSY spectrum, or to generate the double quantum coherence required for an INADEQUATE spectrum [13].

3.1.2 *Lattice Cyanides*

Recently, a new group of versatile framework hosts based on tetrahedral metal centers joined by cyanide links has become of interest [17-19]. Crystal structures have been obtained for a number of these, and one outstanding problem is that X-ray diffraction normally cannot be used to distinguish C from N in these materials [18,19]. Crystal structure refinements often were carried out in a number of different space groups, with the final choice based on results of the refinement [18,19]. By using this approach, ordering schemes were picked for $Cd(CN)_2$, a $Cd(CN)_2$-tBuOH clathrate and a hydrated $Cd(CN)_2$ clathrate of cyclohexanol [20 -22].

The Cd based lattices lend themselves very well to study, as Cd is a good NMR nucleus [20,21]. For guest-free $Cd(CN)_2$, the structure consists of two interpenetrating but unconnected cristobalite lattices, with tetrahedral metal centers connected by cyanide links (fig. 8a). Three different possibilities exist for these lattices a) static ordered CdC_4 and CdN_4 centers, b) dynamic disordered cyanide groups which make all Cd centers equivalent, and c) static disorder of the cyanide groups which should give rise to all five $CdC_nN_{(4-n)}$, n=0-4 centers. The ^{113}Cd MAS spectrum for

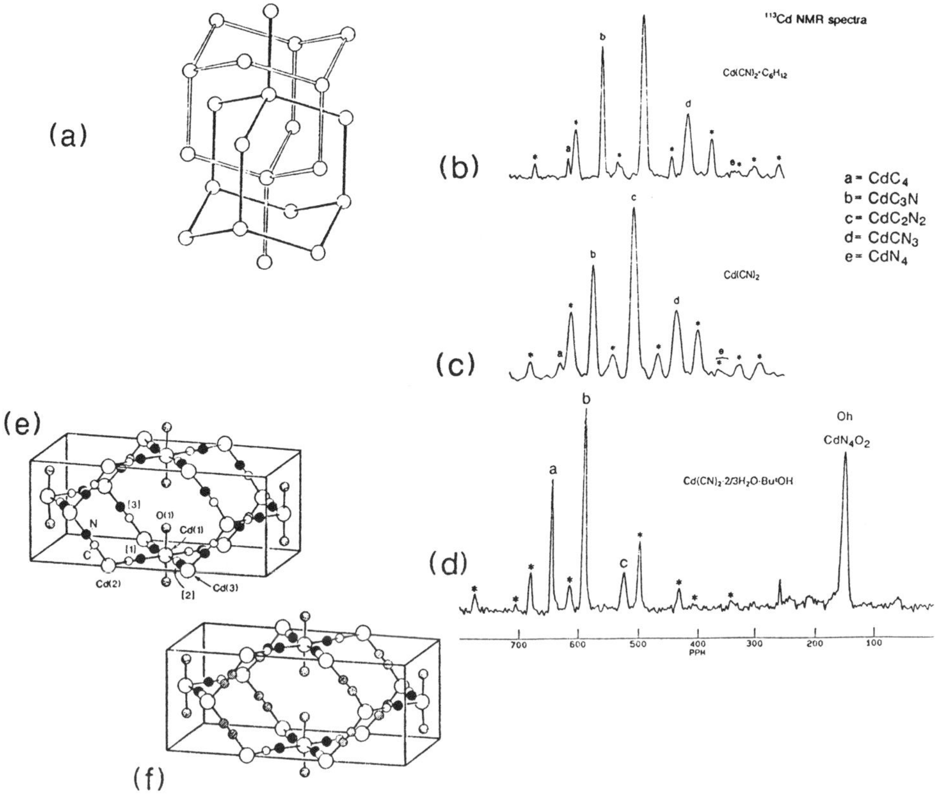

Figure 8 (a) Structure of cadmium cyanide; ^{113}Cd MAS spectra of (b) the cyclohexane clathrate of cadmium cyanide (c) cadmium cyanide, and (d) the hydrated t-butanol clathrate of cadmium cyanide; (e) structure of hydrated t-butanol clathrate of cadmium cyanide based on single crystal diffraction; (f) structure as in (e) modified from ^{113}Cd NMR data.

$Cd(CN)_2$ is shown in fig. 8c. After identifying all of the spinning side band manifolds, 5 Cd resonance lines can be identified, showing that situation c) is the correct one. The integrated intensities show that the intensity distribution departs slightly from a statistical one, with a slight tendency to favour the CdC_2N_2 center. Other points of note include the fact that, of course, only the CdC_4 and CdN_4 are very nearly true tetrahedral centers, whereas the diffraction data would indicate that all metal centers are tetrahedral on *space average*. Cadmium cyanide clathrates based on cristobalite lattices also exist, as in fig. 8a, but with one of the two interpenetrating networks missing. The resulting adamantane-like cavities are then available for occupancy by guests such as cyclohexane. The ^{113}Cd spectrum for this material shows that it also has a near-statistical disorder of cyanide groups (fig 8b) [20].

A second example is that of the t-BuOH clathrate. The lattice now has a honeycomb structure, with an octahedral Cd center coordinated by 4 CN's and two water oxygens, and two tetrahedral Cd's (19). The ordering scheme of the cyanides obtained from the refinement is given in fig. 8e, and the refinement itself was again the basis of choice for the spacegroup, of which five were considered (19). Examination of the ^{113}Cd CP/MAS NMR spectrum (fig. 8d), shows that in fact the cyanides have quite a different ordering scheme (21). The octahedral Cd is ordered as CdN_4O_2, and this limits the disorder on the rest of the centers, with lines for a statistical distribution of $CdC_{(2+n)}N_{(2-n)}$, n=0-2 (fig. 8f). This then also precludes the choice of space group based on the refinement of diffraction data, and three of the others originally considered must be correct. It is also worth noting that the resonance for the only true tetrahedral center, CdC_4, has no spinning side bands, whereas, the other lines do, in keeping with large chemical shift anisotropies for the centers with low local symmetry.

By using a synoptic approach, that is by combining single crystal X-ray diffraction and ^{113}Cd NMR data, it was possible to derive the crystal structure, including the ordering scheme of the cyanides around the metal centres for a hydrated cadmium cyanide framework lattice with cyclohexanol as guest ($Cd_5(CN)_{10}.(H_2O)_4.4C_6H_{11}OH$) [22].

3.1.3 *Hydrates*

In host-guest systems such as the clathrate hydrates, guest species often are distributed over more than one site [9,23]. Both structure I and II hydrates have two kinds of cages, one large and one small (I - 5^{12}, $5^{12}6^2$, II - 5^{12}, $5^{12}6^4$) and structure H hydrate has been proposed to have 2 kinds of small cage and one large (5^{12}, $4^35^66^3$, $5^{12}6^8$) (fig. 8). For a complete description, one needs to specify not only the structure, but also the cage occupancy, as these materials are known to be non-stoichiometric. Guest chemical shifts now are known to vary with cage size, and the best examples of this are hydrates which contain xenon as one of the guests. The smaller cages induce chemical shifts further away from the dilute gas reference at 0 ppm than the larger cages. In addition, the presence of chemical shift anisotropy indicates the presence of a non-spherical guest site (I - $5^{12}6^2$, II - 5^{12}, H - 5^{12} and $4^35^66^3$, xenon apparently does not occupy the $5^{12}6^8$ cage in structure H (fig. 8). It is clear that the xenon spectrum not only gives an indication of the distribution of xenon atome over the guest sites, but it also is an excellent indicator of structure type. Many new guest species have been identified with the use of ^{129}Xe NMR spectroscopy [23, 36].

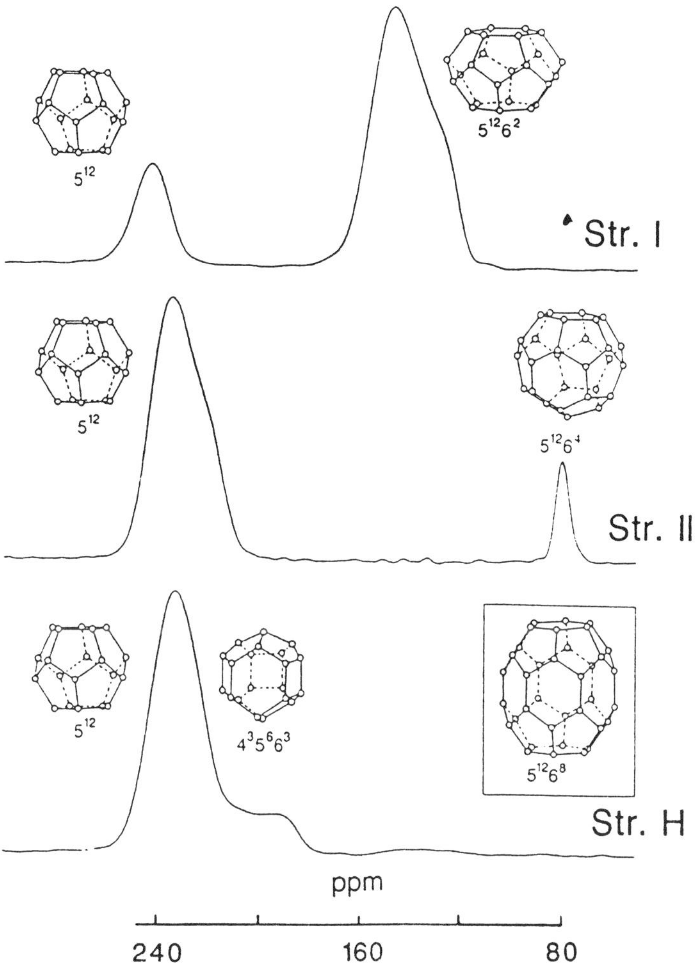

Figure 9 ^{129}Xe NMR spectra (obtained with cross-polarization) of clathrate hydrates : top - structure I xenon hydrate; middle- structure II double hydrate of xenon and benzene; bottom - structure H double hydrate of xenon and cis-cyclooctene. The cage types present in each structure are indicated.

3.2 ORGANIC LATTICES

3.2.1 *Crown Ethers*

Early work [24] on dibenzo and dicyclohexano-18-crown-6 structures indicated that ^{13}C CP/MAS NMR could be useful in giving information on the content of the asymmetric unit and also on the geometry of torsional networks in these solids. Subsequent work on 18-crown-6 itself and some of its complexes (fig. 10a) suggested the presence of molecular motion [25,26]. For 18-crown-6 itself there was averaging

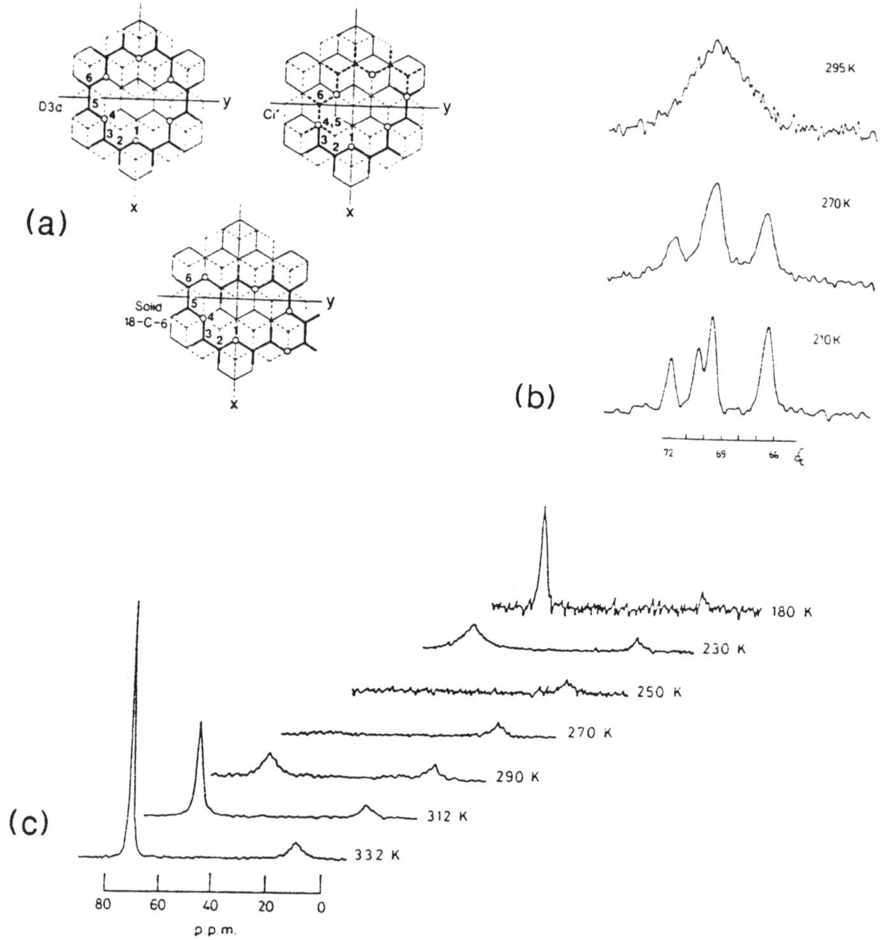

Figure 10. (a) Crown ether conformations shown on a "diamond lattice " (b) ^{13}C CP/MAS NMR spectra of solid 18-crown-6 as a function of temperature showing chemical shift averaging for the various inequivalent CH_2 carbons; (c) ^{13}C CP/MAS spectra for the malononitrile - 18-crown-6 complex as a function of temperature, illustrating the "dipolar fadeout". Spectra were normalized on the malononitrile peak.

of ^{13}C chemical shifts associated with distinct methylene groups at low temperature, followed by considerable line broadening at room temperature (fig. 10b) [25]. For the malononitrile complex, there was a complete broadening, disappearance and reemergence of the ^{13}C spectrum (fig. 10c)[26]. This was assigned to a dipolar fadeout phenomenon, as discussed in section 2.2.3. The presence of a large amplitude motion presents a considerable problem, as the diffraction structures show very well ordered structures. The problem was resolved by recording ^{2}H NMR lineshapes for a ^{2}H-labelled crown ether complex [27]. A large amplitude motion involving reorientation of the entire macrocycle was indeed found to be present. Reorientation of the crown takes place so that each OCCO fragment rotates to an adjacent site, and that during this rotational jump the local crown configurations readjust so that the overall crown conformation is exactly the same as before the jump. Comparison of calculated with experimental ^{2}H NMR lineshapes at different temperatures has yielded motional jump rates, which in turn were used to derive a motional activation energy [27]. The NMR results reveal that 18-crown-6 both as a pure phase and in its complexes is indeed very flexible, and that the motion which takes place relatively easily is completely diffraction invisible in the complexes studied.

3.2.2 *Calixarenes*

Examination of a number of t-Butyl Calix-[4]-arenes with simple aromatic guests has shown highly symmetrical structures [28]. For instance, the toluene clathrate shows a 2-fold disorder of the guest, consistent with 4-fold overall symmetry of the clathrate structure [29]. One question that then comes to mind is, is the high symmetry due to dynamic disorder, or is there a static disorder which gives only a high symmetry on space average? Examination of the ^{13}C CP/MAS spectrum (fig. 11a) [30] shows that the local symmetry is indeed 4-fold, suggesting the presence of dynamic disorder. From the multiplicities of the ^{13}C lines, there is one-quarter calix molecule in the asymmetric unit at room temperature. ^{2}H NMR on the deuterated toluene guest confirms that the toluene behaves like an axial rotor in the calixarene cone.

On lowering the temperature, a fairly sharp change in ^{13}C spectral features (fig. 11b,c) shows that there is a transition to a lower symmetry phase. It was surprising, therefore, that a redetermination of the structure showed exactly the same space group as at room temperature, and the residuals were large (0.094) as was the case for the structure determined previously at room temperature [29]. In order to try and derive a better model for refinement, additional work was carried out on the low temperature phase with ^{2}H NMR on the partially deuterated toluene compound inorder to derive information on the local symmetry of the calix units [31]. The spectrum indicated that the toluene performed two-fold flips, suggesting that the local symmetry was only two-fold rather than four-fold as derived from diffraction data. Once this was incorporated into the model, and it was realized that the disorder of the calix involved not only the t-butyls, but also a portion of the attached aromatic ring, a model with much better residuals resulted (< 0.05). The t-butyl (and attached aromatic ring) disorder now comes out as 1:1, rather than 77:23 as found previously [29]. Additional structural features that can be derived from this synoptic approach include the observation of about 5% of the toluene guests which are inserted in the calix upside down, and the presence of an off-axis tilt of the guest [31].

3.2.3 *Tri-ortho-thymotides*

Tri-o-thymotide forms clathrates which show the very interesting property of self-resolution when recrystallised from some racemic solvents (ref. [32] gives an overview of TOT work). Each crystallite contains either right or left handed propellor-shaped molecules, and shows preference for one of the enantiomeric guests. Structural studies by diffraction have been hampered by considerable guest disorder, both static as dictated by cage symmetry, and dynamic [32]. A number of NMR studies have now

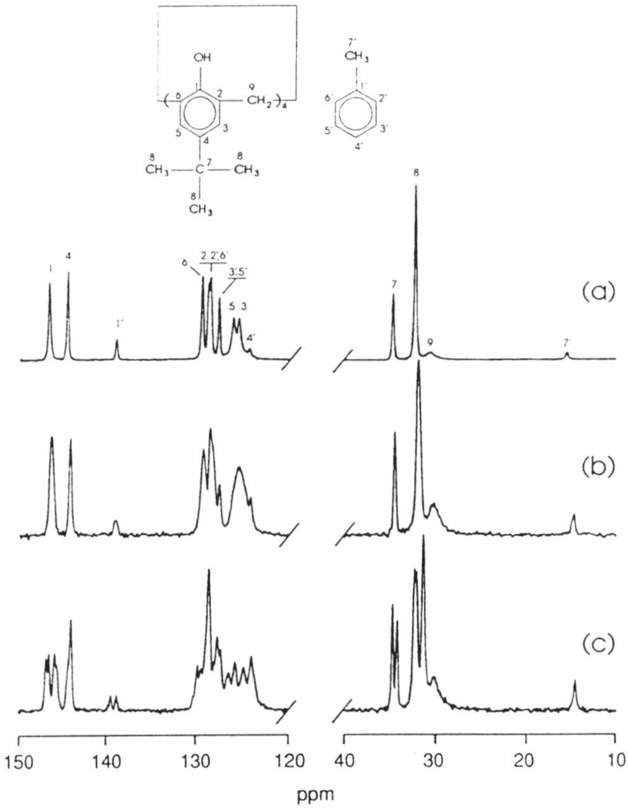

Figure 11. ^{13}C CP/MAS NMR spectra for the t-butylcalix[4]arene toluene compound at several temperatures (a - 295 ,b - 250 ,c - 238K); note the change in multiplicity of both guest and host signals on going from room temperature to 238K.

been carried out to help define the structure and dynamics of these materials. First of all, it was found that for a number of enantiomeric guests some of the guest peaks were split into doublets (fig. 12a-d), and this was interpreted in terms of separate signals for the diastereomeric guest-host combinations P-R, M-S on the one hand, P-S and M-R on the other (fig. 12, inset) [33]. This then gives a unique new way of measuring the different guest-host populations, previously obtainable only from the enantiomeric excess determined by wet chemistry techniques on single crystals. The difficulty in defining the less favoured enantiomer by diffraction prompted us to look for significant differences in dynamic behaviour of the two guest populations. Examination of the ^{2}H

NMR lineshape of racemic 2-bromobutane-1,1,1d_3 (fig. 12f) indeed showed that the

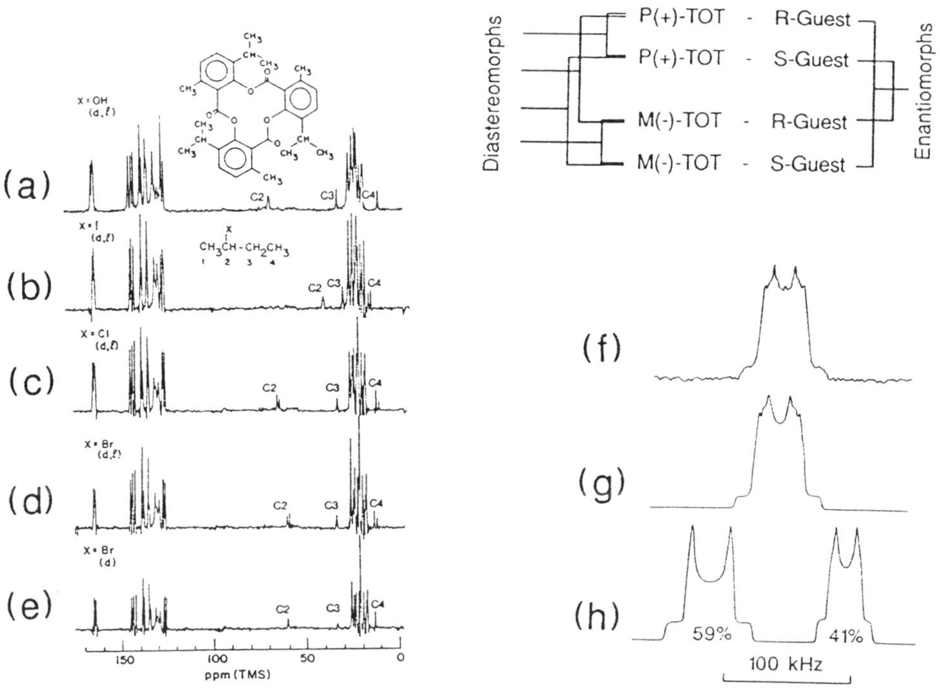

Figure 12. (a-d) ^{13}C CP/MAS NMR spectra for TOT clathrate of racemic 2-substituted butanes; note doubling of all C2 and some C4 guest resonances; (e) ^{13}C CP/MAS NMR spectrum of TOT clathrate of d-2-bromobutane; note single guest resonances; (f) ^{2}H NMR lineshape of the TOT clathrate of racemic 2-bromobutane-1,1,1d_3; (g) simulated lineshape consisting of the two contributions shown in (h) for the major and minor enantiomers;

two guest populations had quite different dynamic properties, and that this also gives another quantitative way of studying chiral recognition (fig.12g,h) [34]. Modelling calculations have shown, in fact, that there should be considerable differences in guest conformations for the major and minor enantiomers. [35]. There is a distinct possibility that it may be possible to link chiral recognition to guest mobility which should be related to guest-host binding.

Other work on NMR applications to clathrates, inclusion compounds and host-guest chemistry has been reviewed in refs. [9, 36,37]

Acknowledgments: The authors would like to thank S. Nishikiori, G. A. Facey, G. D. Enright and E.Brouwer for their contributions to the work reported here.

4. References

1. Farrar, T. C. and Becker, E. D. (1971) *Pulse and Fourier Transform NMR*, Academic Press, New York

2. Fyfe, C. A. (1983) *Solid State NMR for Chemists,* CFC Press, Guelph

3.Jelinski, L. W., and Melchior, M. T., (1987) in *NMR Spectroscopy Techniques*, Ed. Dybowski, C., and Lichter, R. L., Marcel Dekker, New York

4. Harris, R. K., (1983) *Nuclear Magnetic Resonance Spectroscopy: a Physicochemical View*, Pitman, London

5. Mehring, M., (1983) in *NMR Basic Principles and Progress*, Springer-Verlag, Berlin 2nd., Ed'n

6. Grimmer, A.-R, and Blumich, B., (1993), in *NMR. Solid State NMR I, Methods, Basic Principles and Progress*, Vol. **30**., Springer-Verlag, Berlin

7. Clayden, N. J. (1992) , *Ann. Reports NMR Spectr.* **24**, 2

8. Nishikiori, S, Ratcliffe, C. I., and Ripmeester, J. A., (1991) *J. Phys. Chem.* **95**, 1589

9. Davidson, D. W., and Ripmeester, J. A., (1984) in *Inclusion Compounds* Vol. **III**, Ed. Atwood, J. L., Davies, J. E. D., and MacNicol, D. D., Academic Press, N. Y.

10. Gerstein, B. C. and Dybowski, C (1985) *Transient Techniques in NMR of Solids* , Academic Press, London

11. Waugh, J. S, Maricq, M. M., and Cantor, R.. (1978) *J. Magn. Reson.* **29**, 103

12. Herzfeld, J., and Berger, A. E. (1980) *J. Chem. Phys.* **73**, 6021

13 Fyfe, C. A., Feng, Y., Grondey, H., Kokotailo, G. T., Gies, H. (1991) *Chem. Rev.* **91**, 1525

14 Engelhardt, G. and Michel, D. (1987) *High Resolution Solid State NMR of Silicates and Zeolites,* John Wiley, Chichester

15. Gies, H. (1991) in *Inclusion Compounds*, Vol. **V**, Ed. Atwood, J. L., Davies, J. E. D., MacNicol, D. D., Oxford University Press, Oxford

16. Tse, J. S., Desando, M., Ripmeester, J. A. and Handa, Y. P. (1993) *J. Am. Chem. Soc.* **115**, 281

17 Kitazawa, T., Nishikiori, S., Kuroda, R. and Iwamoto, T. (1988) *Chem. Lett.* 459 ; 1729

18 Hoskins, B. F. and Robson, R. (1990) J. *Am. Chem. Soc.* **112**, 1546

19 Abrahams, B. F., Hoskins, B. F. , Liu, J. and Robson, R. (1991) *J. Am. Chem. Soc.* **113**, 3045

20. Nishikiori, S., Ratcliffe, C. I. and Ripmeester, J. A. (1990) *Can. J. Chem.* **68**, 2770

21. Nishikiori, S., Ratcliffe, C. I. and Ripmeester, J. A. (1991) *J. C. S. Chem. Comm.* 735

22. Nishikiori, S., Ratcliffe, C. I. and Ripmeester, J. A. (1992) *J. Am. Chem. Soc.* **114**, 8590

23. Ripmeester, J. A., Ratcliffe, C. I., Klug, D. D., Tse, J. S. (1994) *Ann. N. Y. Acad. Sci.* **715**, 161

24. Buchanan, G. W., Ripmeester, J. A., Bovenkamp, J. W. and Rodrigue, A. (1986) *Tetrahedron Lett.* **27**, 2339

25 Buchanan, G. W., Kirby, R. A., Ripmeester, J. A. and Ratcliffe, C. I. (1987) *Tetrahedron Lett.* **28**, 4783

26 Buchanan, G. W., Morat, C., Ratcliffe, C. I. and Ripmeester, J. A. (1989) *J. C. S. Chem. Soc.* 1306

27. Ratcliffe, C. I., Ripmeester, J. A., Buchanan, G. W. and Denike, K. A., (1992) *J. Am. Chem. Soc.* **114**, 3294

28. Perrin, M. and Oehler, D. (1990) in *Topics in Inclusion Science*, Ed. Vicens, J. and Boehmer, V., Kluwer, Dordrecht

29. Andreetti, G. D., Ungaro, R. and Pochini, A. (1979) *J. C. S. Chem. Comm.* 1005

30. Facey, G. A., Dubois, R. H., Zakrzewski, M., Ratcliffe, C. I., Atwood, J. L.and Ripmeester, J. A. (1993) *Supramol. Chem.* **1**, 199

31. Brouwer, E. B., Enright, G. D., Ratcliffe, C. I. and Ripmeester, J. A. , unpublished results

32. Gerdil, R. (1987) in *Topics in Current Chemistry* V. **140**, Springer-Verlag, Berlin

33. Ripmeester, J. A. and Burlinson, N. W. (1985) *J. Am. Chem. Soc.* **107**, 3713

34. Facey, G. A. and Ripmeester, J. A. (1990) *J. C. S. Chem. Comm.* 1585

35. Facey, G. A. (1990) PhD Thesis, Carleton University, Ottawa

36 Ripmeester, J. A. and Ratcliffe, C. I. (1991) in *Inclusion Compounds* Vol. **V**, Ed. J. L. Atwood, J. E. D. Davies and D. D. MacNicol, Oxford University Press, Oxford

37.Ripmeester, J. A. and Ratcliffe, C. I. (1992) in *Spectroscopic and Computational Studies of Inclusion Compounds*, Ed. J. E. D. Davies, Kluwer, Dordrecht

THE SINGLE CRYSTAL AS A SUPER MOLECULE

CLAUDINE PASCARD
Centre National de la Recherche Scientifique
Institut de Chimie des Substances Naturelles
91198-Gif-sur-Yvette Cedex France

1.What constitutes a crystal ?

-molecules
-solvent
-ions

We can define two levels of organization: the first one will assemble the initial nucleus called the asymmetric unit. It is constituted by one atom (ion), one molecular fragment, one whole molecule or several molecules. The second level of organization reproduces this nucleus in three dimensions by symmetry operations thus forming a three-dimensional regular pattern.

The asymmetric unit: Several fragments, non-covalently bonded, may exist in the asymmetric unit: they may be different (anionic and cationic groups for instance), or chemically identical: n exemplaries of the same molecular fragment, which may present conformational differences [1], very seldom different configurations [2] (Figure 1).

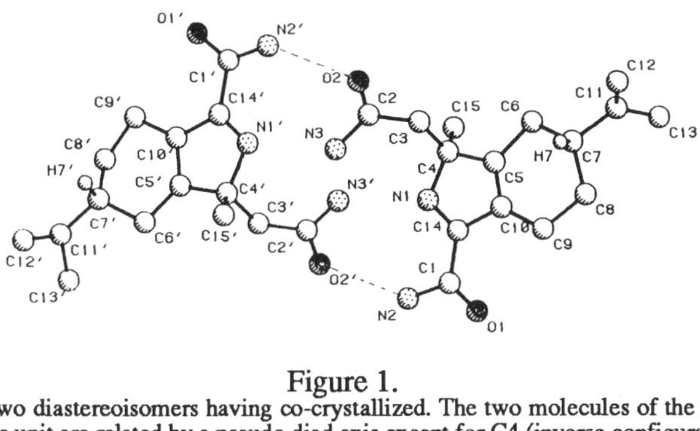

Figure 1.
Rare case of two diastereoisomers having co-crystallized. The two molecules of the alkaloid Manicolin B of the asymmetric unit are related by a pseudo diad axis except for C4 (inverse configuration).

They can display an association presenting a certain chemical significance [3] (Figure 2).

The symmetry elements observed in the crystal are of two types:

-real: they belong to the crystal space-group and repeat the asymmetric unit exactly.

G. Tsoucaris et al. (eds.), Crystallography of Supramolecular Compounds, 127–141.

- pseudo: these elements make correspond parts of the asymmetric unit with symmetry operations which can be exact or not. They do not belong to the crystal class.

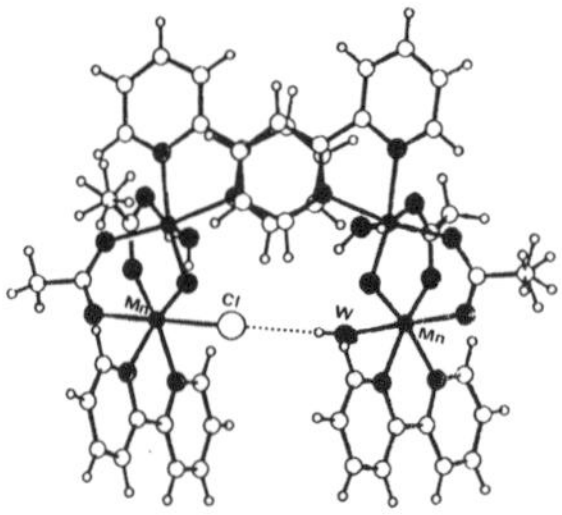

Figure 2.
A supermolecule: these two manganese complexes $Mn_2O(CH_3CO_2)_2bpy_2Cl/H_2O[H_2O]$ are identical except for one substituent: chlorine on one manganese cation, and water molecule on the other one. The interactions holding this asymmetric unit are a hydrogen bond between Cl and Water (3.2Å) and an aromatic stacking between two pyridines (3,8Å).

2. How the molecular objects are held in the crystal ?

The long range interactions are:

- electrostatic interactions, H-bonding,
- aromatic stacking, and dipole-aromatic
- hydrophobic interactions;

These interactions are the same binding all the fragments within the asymmetric unit, as those linking together the asymmetric units repeated by real crystal symmetry. The resulting pattern is an infinite 3D array of molecules all linked together by such interactions:

Mol -medium- Mol or Mol-Mol
(solvent-ions).

For macromolecules, we speak of secondary, tertiary and quaternary structure. Secondary structure is represented by helices and β–sheets of peptidic chains; tertiary structure by the internal folding of these units. Quaternary structure is said of the assembly of several domains or macromolecules associated to form a constant 3D shape.

We can figure out a single crystal as a quaternary construction, an assembly, extending in space, of molecular objects, linked together in a regular fashion. This organization is based on molecular recognition. A crystal needs molecular association for nucleation and growth. It expands by binding new molecular objects through a network of long range interactions which repeat indefinitly. A good example is offered by experiments on crystal growth by Leiserowitz [4]. In the crystallizing medium, the approching molecule will have to realize the same hydrogen bonds as the crystallized molecules in order to be accepted. If a molecule, having a similar shape than the crystallized ones, but non-identical binding sites, is introduced as an impurety in the solution and wants to bind on the crystal face, the balance in the hydrogen bonding network will not be fulfilled, the incoming molecule will bind with lesser ease and the surface of this crystal face will grow more slowly.

3. Importance of the medium:

The cohesion of the molecular array is often due to bonding through the solvent molecules. The shape of flexible amphiphilic molecules is influenced by their environment. If binding is possible with the solvent (polar), then the molecules will adopt a certain shape with their polar heads turned outward. If the environment is hydrophobic, the molecules will present their hydrophobic side to the medium.

Cyclosporin (Figure 3) is a good example.

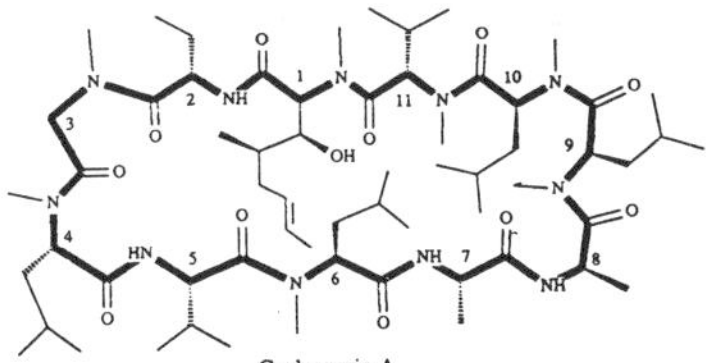

Figure 3.
Cyclosporine A..

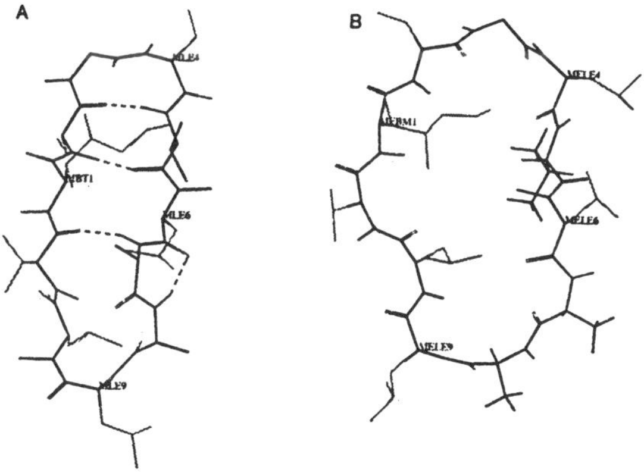

Figure 4.
Cyclosporine A. (a) conformation observed in apolar solvent (RMN) and in the crystal state. (b) Conformation observed in macromolecular complexes.

The conformation of this cyclopeptide has been established in the solid state in 1976 [5], then in 1985[5] as being conformation I (Figure 4a). Further analyses in solution by NMR in 1985 [6] confirmed this conformation: four intramolecular hydrogen bonds tighten the elongated cyclic molecule. A *cis* peptidic bond was observed. However, in 1991, a NMR study of a cyclosporin-cyclophilin complex [7], followed by the X-ray structure of a Cyclosporin A-Fab fragment complex [8], exhibited both the conformation II, blown out (Figure 4b): all hydrogen bonds bind externally to the receptor, and the formerly *cis* peptidic bond is observed *trans*. It appears that conformation I was observed in apolar environments (acetone, $CDCl_3$), whereas, in the presence of H-donors and acceptor sites, the functional groups are turned inside out, binding with the surroundings.

Medium organization: Molecules can associate differently according to the medium organization. Figure 5 shows a macrobicycle bearing carboxylate groups fixed on a hydrophobic skeleton, sufficently flexible to adapt its shape. The sodium salt derivative was crystallized in two hydrated forms (14 and 18% of water content) [9].

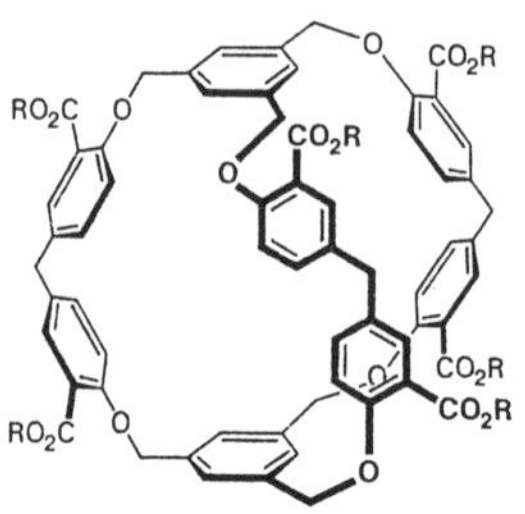

Figure 5.
Molecule **1**.

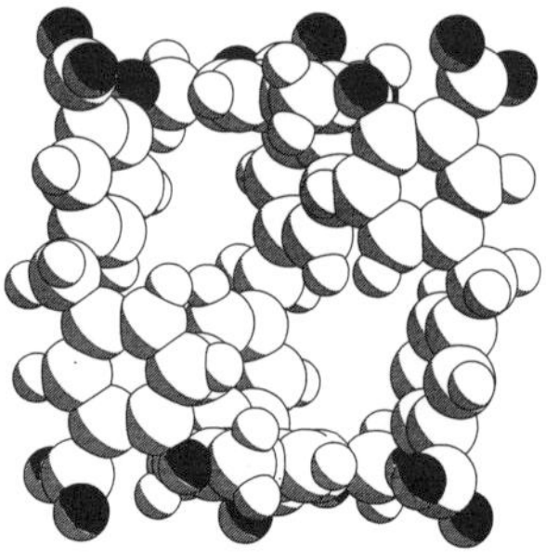

Figure 6.
Molecule **1** and molecular shape I.

We observe two types of molecular associations: in crystal I (space-group P63) (Figure 6), the macrobicycle has a cylindrical shape (barrel) with the acidic groups pointing outwards, and forming two negatively charged crowns around each barrel lid. These barrels are piled one above the other along the **c** direction, with a π-stacking interaction. The resulting cyclinders, immersed in the Na^+-containing solvent, form a hexagonal close-packing (Figure 7) [10,11].

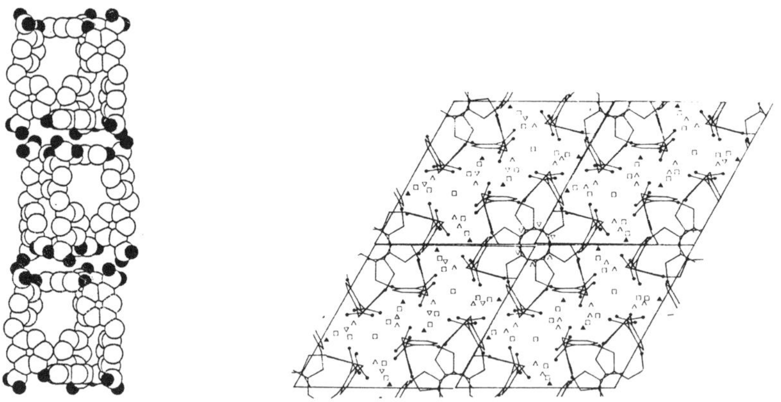

(a) (b)

Figure 7.
(a) "Polymer "of molecule **1**(form I) . (b) Crystal structure of Molecule I (form I.): projection perpendicular to the hexagonal axis. white squares = Na^+, white triangle s= water, black triangle s=MeOH..

In crystal II (space-group Cc), the molecule adopts a flattened shape (Figure 8) so as to keep its charged groups turned toward the medium. The latter is ordered in layers, and the cyclophane skeletons form a monomolecular hydrophobic layer, coated on each side by an anionic layer of oxygens bordering the solvent layer (Figure 9). This behavior is typical of amphiphilic molecules [12][13].

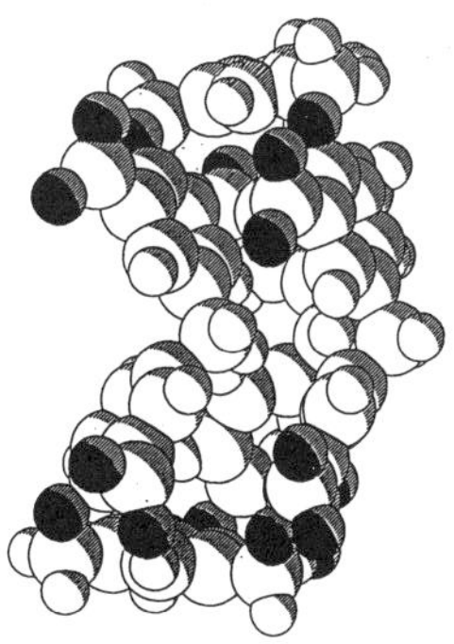

Figure 8.
Molecule 1 and molecular shape II.

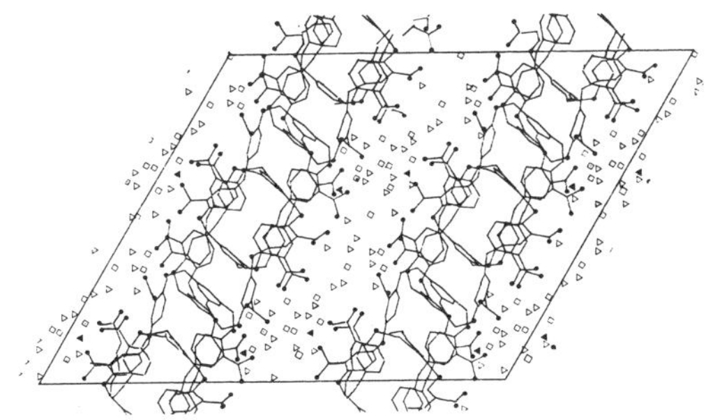

Figure 9.
Crystal structure of form II. Monoclinic projection.
Same labels as in figure 8.

Important role of the solvent
solvent is a stabilizing agent for a self-assembly
In function to its binding possibilities, it links to its neighbors, dragging them more closely, thus stabilizing a molecular assembly. The structure of a Copper(I) dinuclear complex (Figure 10) [14] shows two superimposed molecular rings sandwiching two copper cations.

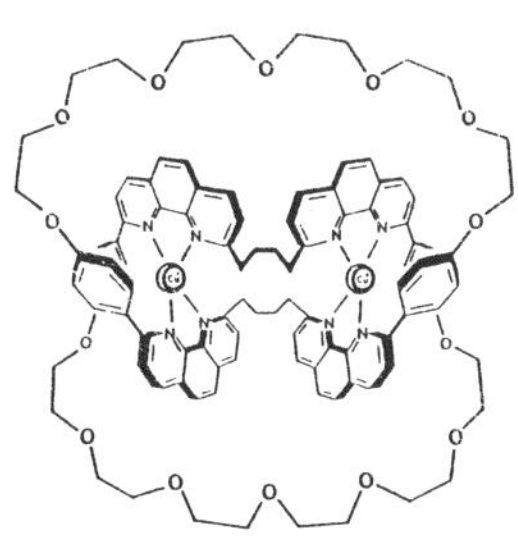

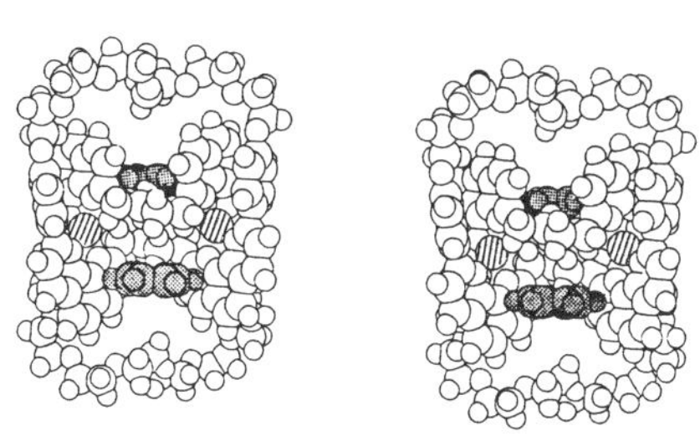

Figure 10.
(a) Molecule **2** = Copper dinuclear complex.
(b) Stereoview of the molecular structure of **2**.
Hatched=copper cations, dotted=benzene.

This association is not observed in solution, although it has well crystallized. The clue is the presence of two benzene molecules which offer very beautiful aromatic interactions:

a) the benzene rings are parallel to the central alkyl chains (short contacts: 3.6Å)

b) the benzene rings are pointing two opposite hydrogens towards the centers of perpendicular phenanthrolins at 3.6Å (carbone to phen. center).

By means of these interactions, the solvent molecules are the agent of cohesion of this supermolecule.

solvent as an hydrogen-donor or acceptor can act as a crystallizing agent

When the molecular shape is not suitable for close-packing or when all the binding sites can not be fulfilled, the solvent molecules step into the molecular association process to supply for the missing links.

water may interefer by its mere bulk

Water plays a very important role in biologial macromolecules. It has a structural role, included into the tertiary structure (see proteins, nucleic acids), and also surrounds each molecule, filling all the surface irregularities, and allowing packing by acting as a bridge [15].

4. Different modes of molecular associations.

The following exemples will illustrate the different types of non-covalent interactions linking all the pieces of the crystal architecture:

electrostatic interactions-hydrogen bonds

Short and direct H-bonds between, for example, protonated nitrogens and anionic oxygens, impose the geometric features of the molecular association. The formation of extracavity cryptates is a good example: the structure of the complex between squaric acid and a macrocyclic cyclophane (Figure 11a) [16] shows a squarate dimer, linked by an internal OH..O bond, imprisoned by three molecules of the ligand; the latters distort their shape to present their protonated nitrogens to the free anionic oxygens of the dimer (Figure 11b).

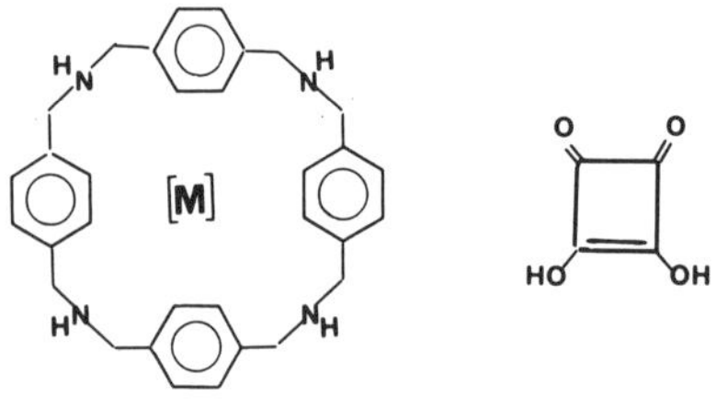

Scheme 1

(a)

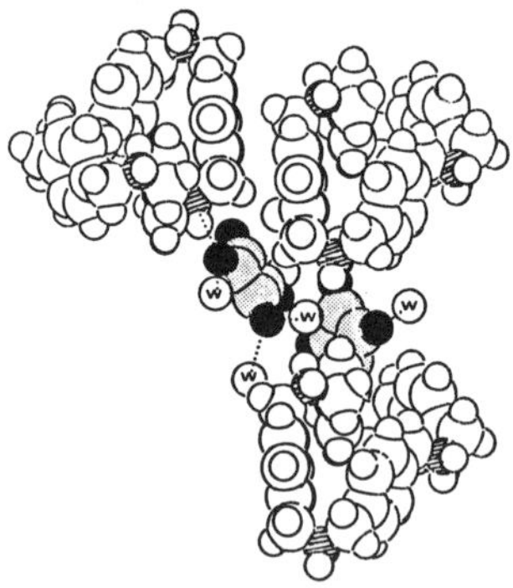

(b)

Figure 11.

(a) Cyclophane **3**. (b) Three ligands **3** surround two squarate anions linked together by a OH..O short bond thus sheltering the monoanion entity from the environnement..

Looking for the control of the building of a defined superstructure by molecular chirality features, Lehn and his group (Collège de France) crystallized a bis-lactam **4** (Figure 12) as a racemate and as the (-) enantiomer. [17]

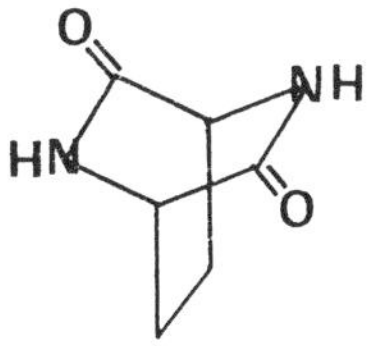

Figure 12.
bBs-lactam **4.**

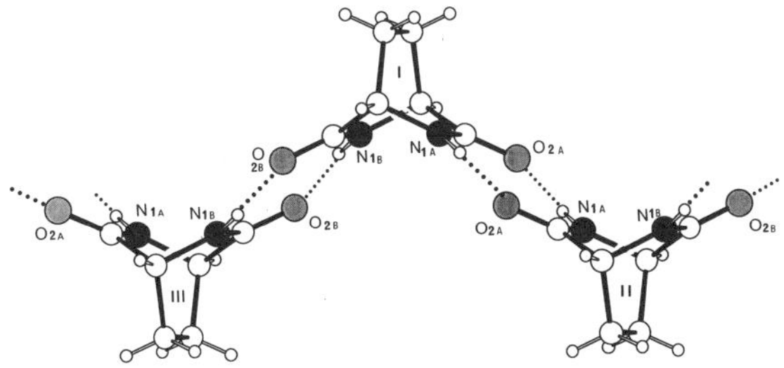

Figure 13.
Packing of the racemate: infinite chain of alternating. enantiomers.

In the racemate crystal (Figure 13), the molecules are assembled into infinite undulating chains of alternating enantiomers of opposite chirality related by different symmetry operations through a centrosymmetrical hydrogen bonding network.

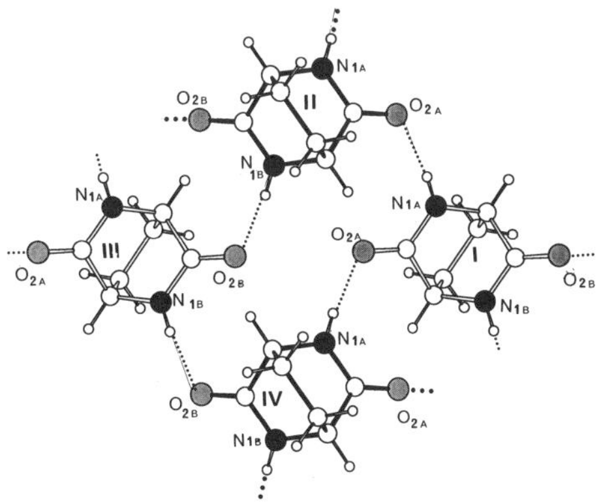

Figure 14.
Packing of the enantiomer: formation of a tetrameric array of chiral molecules.

In the enantiomer structure (Figure 14) with no center of symmetry, the chiral molecules, related by crystallographic symmetry operations, are assembled in cyclic tetrameric arrays, repeated indefinitely in space through hydrogen bonding network. The molecules are closely packed, showing an alternation of polar and hydrophobic layers. Here, the chirality of the molecule has determined the supramolecular architecture generated by the hydrogen bonding networks.

Aromatic interactions.

Although weaker than an H-bond, this interaction with an aromatic implies a large surface, and therefore, has a large "active" area. This favorizes a certain molecular shape, or a certain intermolecular association[18]

In the crystallized triple catenane **5** (Figure 15) [19], the two small rings rotate around the interlocking strand of the large ring, in order to bring their phenanthrolines into a parallel contact with a large overlapping of the rings. This association is sufficently strong to be observed in solution. Furthermore, the phenanthrolines of the large rings are superimposed with the homologous nuclei of centrosymmetrically-related catenands. Consequently, throughout the crystal medium, all molecules are interacting between the aromatic rings.

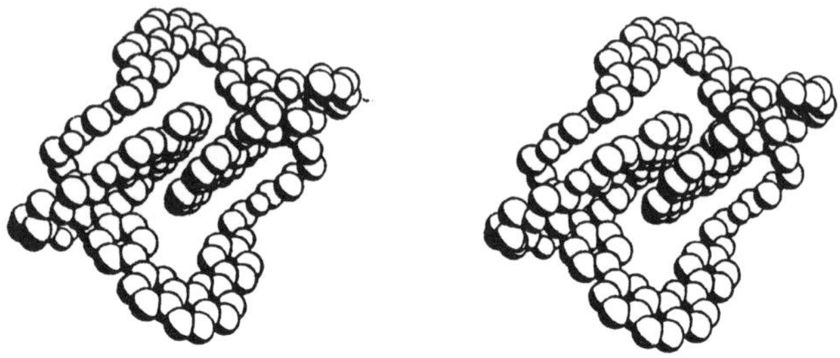

(b)

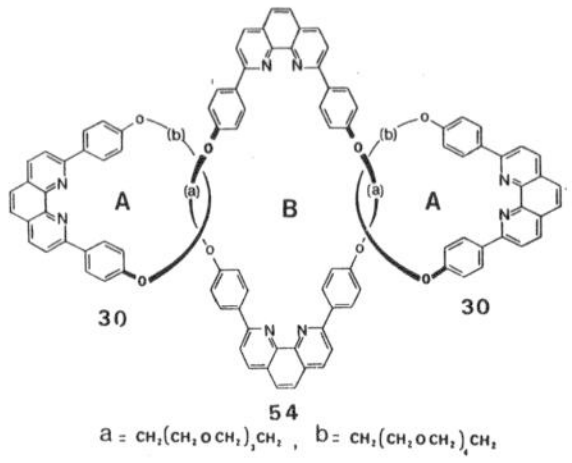

(a)

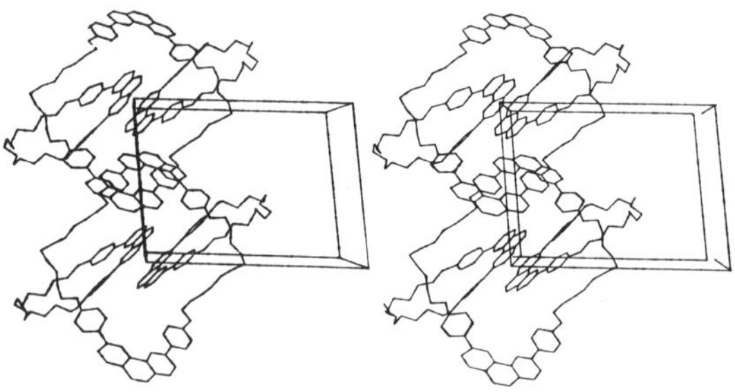

(c)

Figure 15.
(a) Triple catenand **5**. (b) Stereoview of the molecular structure. (c) Stereoview of the superimposition of two molecules.

The aromatic stacking is the tool with which the crystal structures of oligonucleotides are built. DNA oligomers crystallize in double helices piled in infinite columns by the same base-stacking effect as the intra-helix base pairs (Figure 16a) [20]. The hexagonal close-packing (Figure 16b) form a classical array for long helical rods.

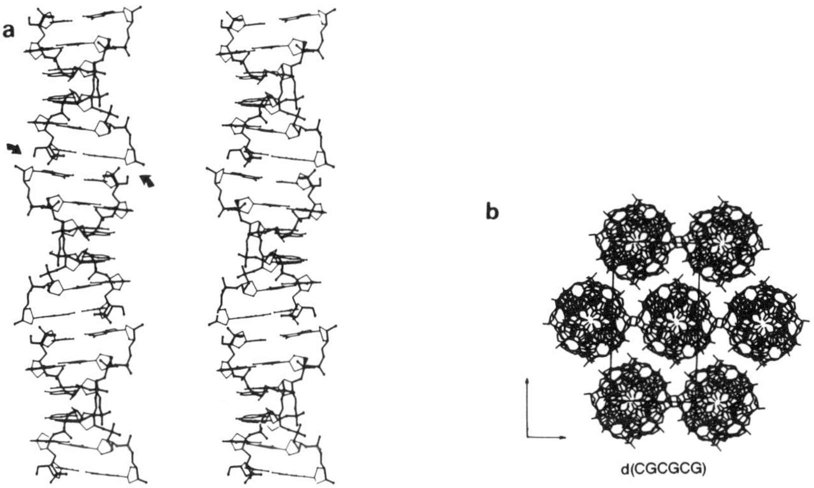

Figure 16.
(a) Head-to-tail stacking of hexanucleotide double helices. The arrows indicate the interruption in the phosphate chain. (b) Hexagonal close-packing of the helices.

Aromatic stacking has been a ingenious tool used by Stoddart in the interlocking of his catenanes [21] after he has observed how in the crystal structure of a catenane, the molecules superimpose their aromatic rings from one cell to another, forming a new association by π-stacking.

Hydrophobic contacts

They are numerous and essentially between neutral sp3 and sp2 carbon atoms (CH3, =CH2, =CH groups). Their large number may often impose a certain shape to the molecule, and induce a certain steric behaviour in the molecular assemblies. I.Karle showed that apolar oligopeptidic helices formed parallel or antiparallel assemblies with tightly imbricated hydrophobic residues [22].

5. Assemblies types

The asymmetric unit contains several fragments which can form in the lattice different assemblies with different chemical significance.

Symmetry: real and pseudo.

Centrosymmetry

The above cited complex of squaric acid and a cyclophane macrocycle crystallizes in P-1 space-group [16]. The asymmetric unit contains 2.5 squarate anions, one ligand, and nine water molecules (Figure 11). Two guest-specific spaces corresponding to two different anionic entities are observed in the crystal structure: aside two dimeric monoanions, inserted between three ligands (see above and Figure 11b), one squarate dianion, situated on the center of symmetry, is encapsulated within four macrocycles (Figure 17).

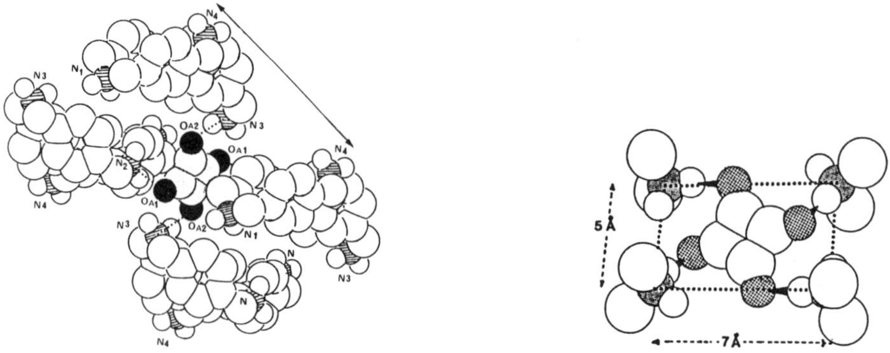

(a) (b)

Figure 17.

(a) Supermolecule formed by four ligands **3** encapsuoating one squarate dianion. (b) The links are $NH^+...O^-$ bonds. The monomer sits at a center of symmetry.

The latter association is realized by strong hydrogen bonds between NH(+) of each macrocycle and the four anionic oxygens. The squarate monomer has selected its suitable environment to build this supermolecule. Two different chemical species were thus isolated in the solid state.

binary axes

The cyclophane macrobicycle **1** in its convex form I (Figure 6) is built around a hexagonal axis 6_3. The molecule has a internal ternary symmetry, and repeats itself in a helix with a half-turn and a displacement of **c**/2. Then, the aromatic tops are parallel and close enough to π-interact, the staggered carboxylate side-chains forming a tight crown (Figure7a).

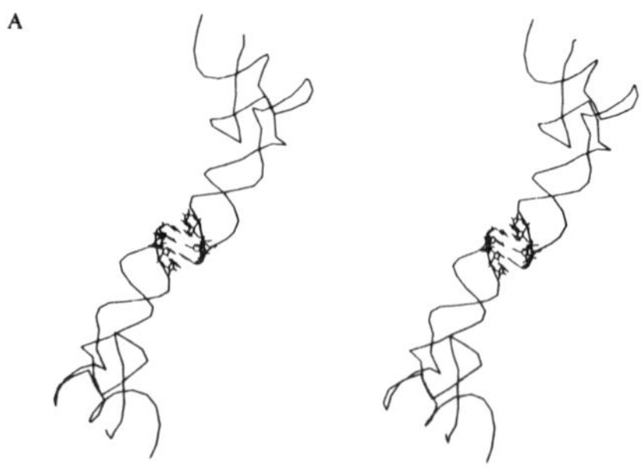

Figure 18.

$tRNA^{asp}$: two tRNA molecules related by a diad axis at their anticodon terminals, are base-paired.

The diad axis relating two $tRNA^{Asp}$ in the crystal state [23] allows intermolecular base-pair stacking and Watson-Crick linking, thus mimicking perfectly RNA-RNA recognition through their anticodons(Figure 18).
The bis-zinc porphyrin 1,10-Phenanthroline (Figure 19) crystallized in $P2_1/a$ space-group with two molecules in the asymmetric unit and 30% solvent (ethanol and water in disorder) [24]. Molecule I is repeated by the screw axis parallel to **b**, and Molecule II by the center of symmetry (Figure 20)

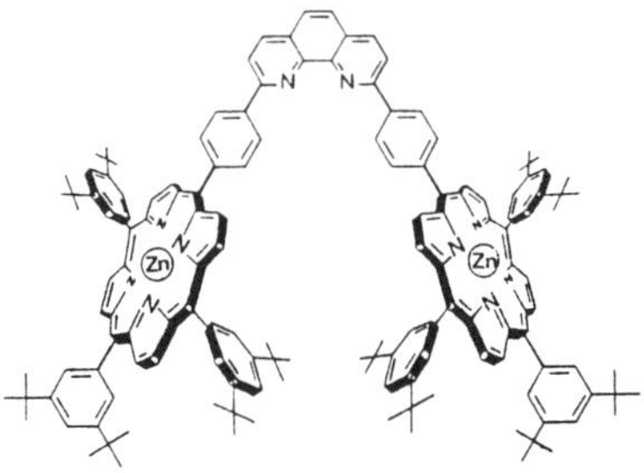

Figure 19.
Molecule 5 = bis-zinc porphyrine-1,10-phenanthroline.

The pseudo mirror conformation of the two independent molecules (comparable torsion angles of the porphyrins around their link to their respective spacer) creates new elements of symmetry : a pseudo screw axis passing by the real symmetry center, and a pseudo center situated on the real screw axis

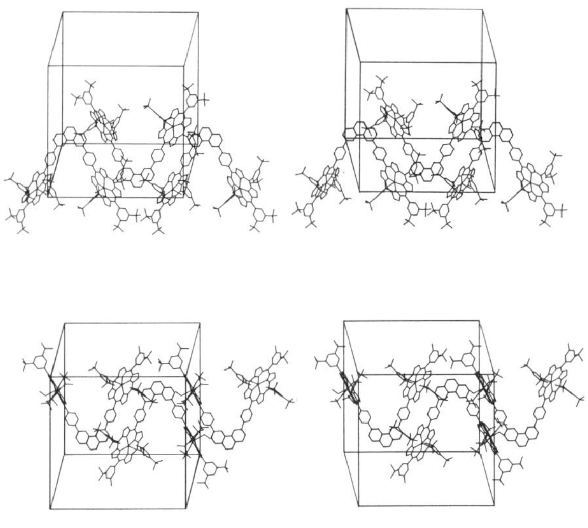

Figure 20.
Stereoview of the unit cell with (top) chain of molecules I and (bottom) chain of molecules II.

Therefore, two independent but very similar sets of helices parallel to the **b** axis exist throughout the crystal, one formed by a real screw axis and a pseudo-center of symmetry, (Figure 20, top) and the second one by a real symmetry center and a pseudo screw axis (Figure 20, bottom). In each helix, the molecules are held together through interactions between parallel aromatic rings and hydrophobic contacts (figure 20). This packing of parallel helices with a hydrophobic surface, leaves large channels which explains the flowing of solvent molecules.

This astonishing self-organization presents, as a consequence, an interesting disposition of the porphyrins. Inside a given helix, or between helices, the large flat rings are regularly spaced, with only two possible orientations. The comparison with the disposition of the hemes in the cytochrome and of the bacteriochlorophyls and bacteriopheophytins in the transmembrane part is striking (Figure 21). The organic crystal structure builts itself with porphyrin centers (Zinc atoms) no closer than 10Å.

When two-or more- molecules exist in the a.u., it may happen that they do not associate together, but form two - or more- intermolecular "homo" associations, completely independent, assembling each individual family. The resulting assemblies may be different, if molecules differ, or identical.

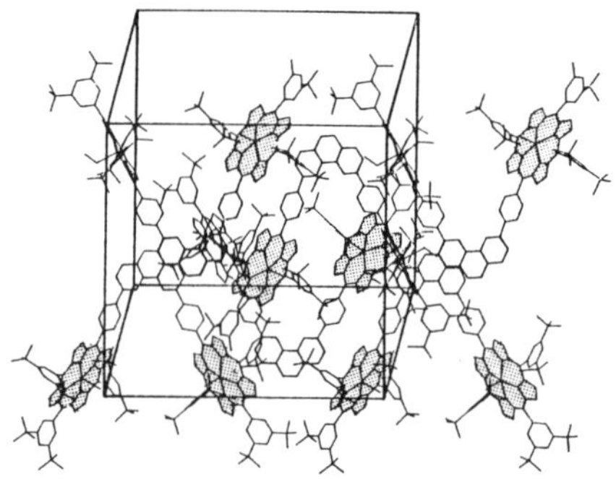

Figure 21.
Unit cell with the spatial distribution of the porphyrins.

Pseudo-symmetry: biological relevance ?

This property is of major importance. Frequently, molecular assemblies have internal symmetry elements which may - or not- coïncide with the crystal elements. And it appears, more and more, that this property has biological significance for macromolecules: viruses and proteins polymers present a high degree of symmetry which is of utmost importance for their biological functions.

We have seen, on the previous example, how pseudo-symmetry elements are used to obtain a certain planified organization. Striking examples are provided by the viruses. The icosahedral shape of certain viruses is the assembly of three identical (T1)(ref. for explanation of the symbols, see the paper by Caspar and Klug, [25]) or different (T3) proteins to form the face of a icosahedron. This face is then repeated (60 times)(figure 22).

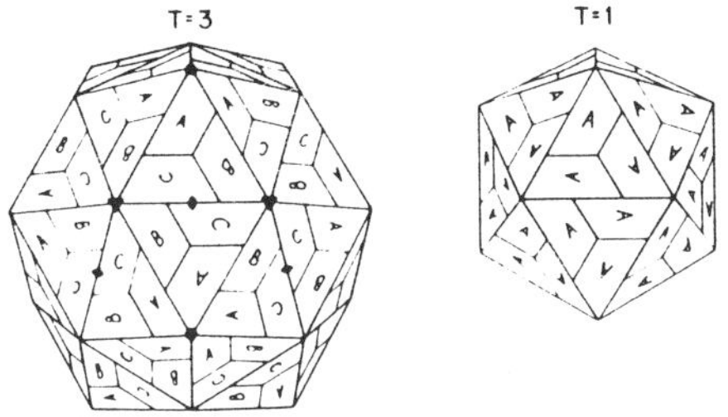

Figure 22.
Symmetries of viruses.

This symmetry which ordonates the coat proteins arrangement, does not govern the content of the capsid which may be a single nucleic acid strand.

Protein oligomers:

- cyclosporin A forms with its receptor, cyclophilin, an assembly which crystallized with five molecular complexes in the asymmetric unit, disposed around a pseudo-5 axis (26]. The pentamer, represented figure 23a is repeated by a crystallographic diad axis perpendicular to the pseudo 5-axis, giving a second pentamer slightly staggered with repect to the first one. The two protein discs sandwiched the cyclosporin molecules (figure 23b), thus forming a supramolecular complex.

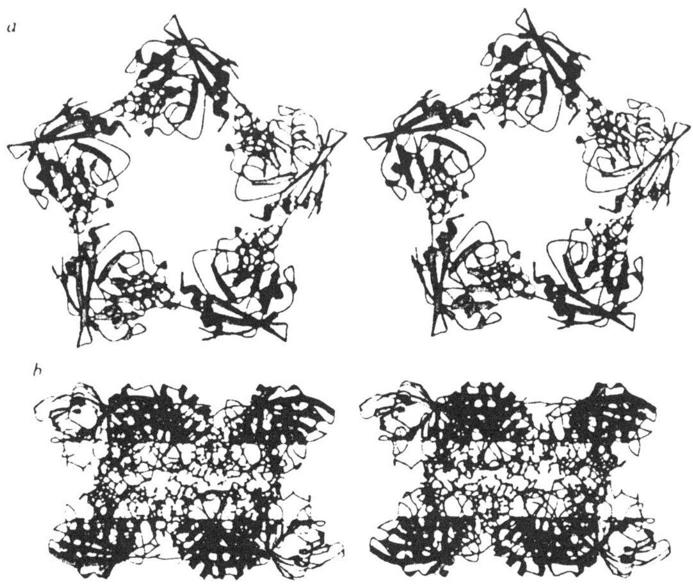

Figure 23.
(top)Pentamer of cyclophilin (asymmetric unit). (bottom) Two pentamers related by a crystallographic diad axis perpendicular to the pseudo 5-fold axis. (b) view perpendicular to (a) showing the cyclosporine molecules sandwiched between the two cyclophilin layers.

6. Flexibility-Adaptabilily-"thermal agitation"- disorder

The so-called "thermal agitation" is the factor which characterizes the local atomic motion. Some molecular fragments exhibit a larger motion amplitude than others, characterizing a portion of the molecule possessing a greater "flexibility", i.e. a better adaptability to the environment. Flexibility is a favourable property which allows local molecular recognition. Slight -or more pronounced- twisting occur to adapt molecular shapes in a complementary fashion. The above cited example of $tRNA^{Asp}$ crystal structure [23] demonstrates that the highest observed temperature factors (i.e. motion amplitude] are on the anticodons and at the helix portion where head-to-tail packing occurs.

When the entire molecule has no possibility to bind to its environment, then it can show disorder phenomenon. This disorder can keep a certain ordered packing as it is seen in fibers. In some oligonucleotides crystal structures, the helices, piled in infinite columns throughout the crystal, may show rotational disorder.

Complete disorder is frequently observed for ions such as ClO_4, BF_4, PF_6, which have a spherical behaviour. They are very much favored by the chemists as counter ions, and very much less by crystallographers.

7 Conclusion

In the first chapter, two levels of organization were distinguished in the crystal state. However, the above structural examples make obvious that the first molecular self-assembly, organized by pseudo and/or real symmetry and presenting a certain chemical significance, is maintained by the same intermolecular forces which will repeat this association throughout the crystal, thus forming a single supermolecule. The environment, the medium in which the crystal was obtained, is very important: it plays often the role of a cohesion agent, without it some molecular assembly would not exist. The medium own organization may even impose the spatial molecular association.

Special attention is drawn on aromatic interactions which play a major role in driving the molecules toward typical assemblies.

Self-assembly process exists throughout the whole crystal: the different fragments, molecules, ions, solvent, are packed by what can be considered a single network of comparable long-range interactions; their local motion, observed in the crystal state, is the witness of the molecular flexibility, a necessary quality for association.

To end, one should remember that "any ordered structure, whether it is a crystal or a virus, will have some type of well-defined symmetry" (Caspar and Klug, 25]

Acknowledgements.

I am grateful to Dr.D.Moras and Dr.J-C.Thierry for the $tRNA^{asp}$ and cyclosporine figures. I am indebted to Dr.Jean Guilhem and to Dr.Michèle Cesario who took a large part in all the structural examples given, and helped me in the preparation of the manuscript by many helpful discussions.

References

1 Pohl,E., Heine,A., Sheldrick,G.M., Dauter,Z., Wilson,K.S., Kallen,J., Huber,W., and Päffli,P.J., (1995) Structure of Octreotide, a Somatostatine xanalogue. Acta Cryst. **D51** 48-59

2 Polonsky,J., Prangé,T., Pascard,C., Jacquemin,H., Fournet,A. (1984) Structure (X-ray analysis) of manicoline B, a mixture of two diastereoisomers of a new alkaloid from Dulacia guianensis (Olacaceae). Tetrahedron Lett. **25** 2359-2362

3 Cesario, M., Guilhem,J., Tchertanov,L., Pascard,C., Frapart Y., Delfosse,M.,Philouze,C., and Girerd J-J., (1995) Modèles de clusters Mn-O du centre de dégagement d'oxygène des plantes Congrès National de Cristallographie Abstract A56.

4 Weissbuch,I., Popovitz-Biro,R., Lahav,M. and Leiserowitz L.(1995) Understanding and Control of Nucleation, Growth, habit,Dissolution and Structure of Two-and Three-dimensional Crystals using "Tailor-made" auxillaries. Acta Cryst. **B51** 115-148

5 Petcher,T.J., and Weber,H.-P., (1976) Crystal and molecular structure of a iodo-derivative of the cyclic undecapeptide Cyclosporin A. Helv.Chim.Acta **59** 1480-1489

6 Loosli,H.-R., Kessler,H., Oschinat,H., Weber,H.-P., Petcher,T.J., and Widmer,A., (1985) The conformation of Cyclosporin A in the crystal and in solution.Helv.Chim.Acta **68** 682-704

7 Weber C., Wider, G., von Freyberg,B., Traber, R., Braun,W., lidmer,H., Wüthrich,K., (1991) The NMR structure of Cyclosporin A bound to cyclophinin in aqueous solvent. Biochemistry **30** 6553-6574

8 Altschuh, D.,Vix,O., Rees, B., and Thierry, J.-C., (1992) A Conformation of Cyclosporin A in Aqueous Environment Revealed by the X-Ray Structure of a Cyclosporin A-Fab Complex. Science, **256** 92-94

9 Cesario,M., Guilhem,J., Lehn,J.-M., Meric,R., Pascard,C.and Vigneron,J.-P. (1993) Medium Effects in Action, Visualized by the Crystal Structures of Open and Closed forms of a Molecular Receptor J.Chem.Soc.Chem.Comm. **6** 540-543

10 Luzzati,V.(1968) X-Ray diffraction studies of Lipid-Water systems in Biological Membranes ed.D.Chapman, Academic Press, London 71-123

12 Ding,J., Steiner,T. and Saenger,W. (1991) Structure of g-cyclodextrin-1-propanol -17 H_2O inclusion complex. Acta Cryst.**B47** 731-738

13 Coleman,A.W., Bott,S.G., Morley,S.D., Means,C.M., Robinson,K.D.,Zhang,H., and Atwood,J.L.(1988) Novel layer structure of sodium Calix(4)arenesulfonate complexes - a class of organic clay mimics ? Angew.Chem.Int.Ed.Engl.**27** 1361

14 Dietrich-Bucheker,C., Sauvage,J.-P., Kintzinger,J.-P., Maltese,P., Pascard,C. and Guilhem,J. (1992) A Di-Copper Trefoil Knot and its Parent ring Compounds. New J.Chem. **16** 931-942

15 Salemme,F.R., Genieser, L., Finsel,B.C., Hilmer,T.M., and Wendoloski,J.J. (1988) Molecular factors stabilizing protein crystals J.of crystal Growth 90 273-282

16 Cesario,M., Guilhem,J., Pascard,C. (1993) Anion Complexes between a Macrocyclic Cyclophane and Squaric Acid: the Crystal Structure reveals the Ideal Receptor Topology ordered by the Substrate.Supramolecular Chemistry **2** n°4 331-336

17 Brienne,M.-J., Gabard,J., Leclercq,M., Lehn,J.-M., Cesario,M., Pascard,C., Chevé,M. and Dutruc-Rosset,G. (1994) Chirality directed self-assembly. Resolution of 2,5-diazabicycloooo[2,2,2]octane-3,6-dione and Crystal structures of its racemic and (-) enantiomeric forms. Tetrahedron Lett.**35** 8157-8160

18 Hunter,C.A., and Sanders J.K.M. (1990) The nature of π–π interactions.J.A.C.S. **112** 5525-5534

19 Guilhem,J., Pascard,C., Sauvage,J.-P., and Weiss,J. (1988) Solution study and Molecular structure of a [3]-catenand: intramolecular interaction between the two peripheral rings. J.A.C.S. **110** 8711-8713

20 Wang,A.H.-J., and Teng,M.-K., (1988) Crystallization and crystal packing analysis of DNA nucleotides J.of Crystal Growth **90** 295-310

21 D.B.Amabilino, P.Ashton, C.L.Brown, E.Cordova, L.A.Godinez,T.T.Goodnow, A.E.KAifer, S.P.Newton, M.Pietraszkiewics,. Philp,F.M.Raymo, A.S.Reder, M.T.Rutland,A .M.Z.Slawin, N.Spencer, J.F.Stoddart and D.J.Williams.(1995) Molecular Meccano: Self-Assembly of[n]Catenanes.J.Am.Chem.Soc. **117** 1271-1293

22 Karle,I., (1992) Folding, Aggregaiona dn Molecular Recognition in Peptides. Acta Cyst. .B48 341-356

23 Moras,D., Dock,A.-C., Dumas,P., Westhof,E., Romby,P., Ebel, J.-P. and Giégé,R. (1986) Anticodon-Anticodon interaction induces conformational changes in tRNA: yeast $tRNA^{Asp}$, a model for tRNA-mRNA recognition. P.N.A.S. 932-936

24 Pascard,C., Guilhem,J., Chardon-Noblat,S., Sauvage,J.-P. (1993) Molecular Structure of an Oblique Bis-Zinc Porphyrin 1,10-Phenanthroline: Model of a Fragment of the Photosynthetic reaction Centre.New J.Chem **17** 331-335

25 Caspar,D.L.D., and Klug,A., (1962) .Physical Principles in the Construction of Regular Viruses" Cold Spring Harbor Symp.Quant.Biol. **27** 1-24

26 Pflügl,G., Kallen,J., Schirmer,T., Jansonius,J.-N., Zurini,M.G.M. and Walkinshaw,M.D. (1993) "X-Ray Structure of a Decameric Cyclophilin-Cyclosporin CrystalComplex."Nature **361** 91-94

NEW LAYERED AND PILLARED-TYPE COMPOUNDS, THEIR INTERCALATION CHEMISTRY AND APPLICATIONS.

G. ALBERTI, C. DIONIGI, S. MURCIA-MASCARÓS
AND R. VIVANI
Chemical Department, University of Perugia, via Elce di Sotto 8,
06123 Perugia, Italy

1. Introduction

From a crystallographic point of view, lattice atom planes are present in all the crystalline solids that behave as a three-dimensional diffraction grating for X-ray radiation. However, we will consider that a solid exhibits a layered structure only when the bonds amongst atoms or molecules of a same plane are considerably stronger than the interactions amongst atoms or molecules of adjacent planes. In the greater part of typical layered solids there are covalent bonds between atoms of a same layer and van der Waals bonds between adjacent layers. In these compounds, a single layer can therefore be seen as a planar macromolecule, while the solid can be considered as a molecular crystal formed by the packing of these macromolecules. The planar macromolecule can be equipped with a large variety of functional groups that can interact more or less selectively with a large number of neutral or charged molecular species; thus, these interactions can be used as a driving force for the intercalation of many species (guest species) in the interlayer region of the solid (host).

G. Tsoucaris et al. (eds.), Crystallography of Supramolecular Compounds, 143–157.

A large number of different conformations and molecular aggregates of the guest molecules, depending on their reciprocal interactions in the interlayer region, on the nature of the active sites, on their density and distribution, as well as on the electrical conduction and flexibility of the layers, can be obtained. This explains why the intercalation chemistry of the layered hosts is considered an interesting aspect of the supramolecular chemistry. In this connection, it must be remembered that one volume of the series "Comprehensive Supramolecular Chemistry" has been dedicated to bidimensional and three dimensional compounds [1].

The structure, physical properties and intercalation chemistry of specific classes of layered solids such as cationic and anionic clays, graphite, metal chalcogenides, layered oxides and perovskites etc., are well enough established today [1-5], although, with advances in the field, many new problems are continuously arising so that the field is still growing in interest and very far to be exhausted.

In the present lecture there is no time to take into consideration, even superficially, the various classes of layered compounds. We will consider here only the class of layered metal phosphates and phosphonates. This choice was dictated not only because these layered solids are part of the research activity of the lecturer, but also because, as will be evident later, because of the presence of the organic groups covalently attached to the planar inorganic back-bone, many functional active groups can be easily introduced in the interlayer region. Furthermore, many different inorganic framework structures can be obtained with a proper choice of the metal atom and/or of the experimental conditions of the synthesis reaction [6]. Even mixtures of phosphate, monophosphonate and diphosphonate groups in different ratios can be introduced in these layered compounds. Thus many different arrangements of the organic groups in the interlayer regions can be obtained. Taking into account the different nature of the active sites attached to the organic groups, it is easy to imagine how many possibilities can be realised in the interlayer region and how many different interactions with the guest species can be obtained [6].

2. Layered Metal(IV) Phosphates

The chemistry of layered metal(IV) phosphates, having a general composition $M(IV)H_2X_2O_8{\cdot}nS$, where M(IV) = Zr, Ti, Sn, Pb, Ge, Ce, Th, etc; X = P, As; S = molecule intercalated in the interlayer region, can be considered to have begun in 1964, when the first member of this class, zirconium bis(monohydrogen phosphate) monohydrate, α-$Zr(HPO_4)_2{\cdot}H_2O$ was obtained and its layered structure (today known as α-layered structure) established [7]. Some years later another important member of composition $ZrH_2P_2O_8{\cdot}2H_2O$ (called γ-ZrP) was discovered [8] but its formula and structure remained unknown for many years. Only recently was it found that this compound is a zirconium phosphate dihydrogenphosphate dihydrate γ-$ZrPO_4H_2PO_4{\cdot}2H_2O$ whose structure was elucidated, too [9]. Although many other members with α- and γ- layer structures have been prepared up to date, the most widely investigated members always remain α- and γ-ZrP. These two compounds can be considered as the two archetypes of the α- and γ-layer structures. Our attention will therefore be essentially focused on these compounds and on their organic derivatives.

3. Layered Organic Derivatives of α- Zirconium Phosphate

A schematic view of the structure of $Zr(HPO_4)_2{\cdot}H_2O$ is shown in figure 1. Just after the synthesis of the first zirconium phosphonates and zirconium organo-phosphates of general formula $Zr(O_3P\text{-}R)_2{\cdot}nS$, and $Zr(O_3P\text{-}OR)_2{\cdot}nS$, where R is an organic radical (e.g. $-CH_3$, $-C_6H_5$, $-(CH_2)_n\text{-}CH_3$ etc.), S being a polar solvent intercalated in the interlayer region, it was realised, that

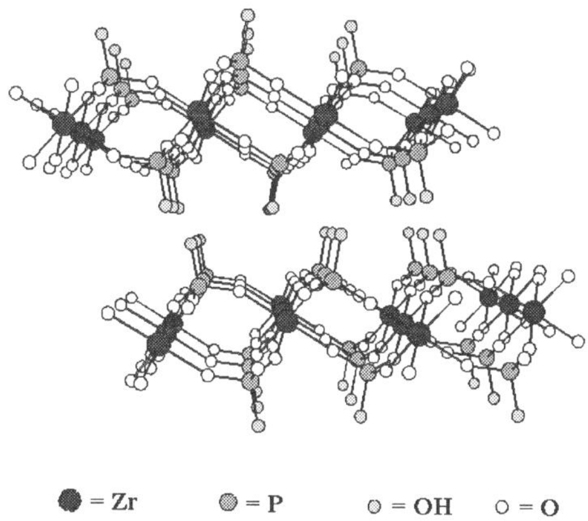

Figure 1. A schematic representation of the structure of α-$Zr(HPO_4)_2{\cdot}H_2O$. The water and hydrogens in the interlayer region are not shown.

these compounds can be considered as organic derivatives of α-Zr(O_3P-OH)$_2$·H_2O in which the inorganic -OH groups have been replaced by R groups, leaving the inorganic structure of the α-layer essentially unchanged [10] (fig. 2a). Recently, some zirconium phosphonate structures have been solved by X-ray powder diffraction data [11]. A projection down the *b*-axis of the structure of zirconium phenyl phosphonate is shown in figure 2b.

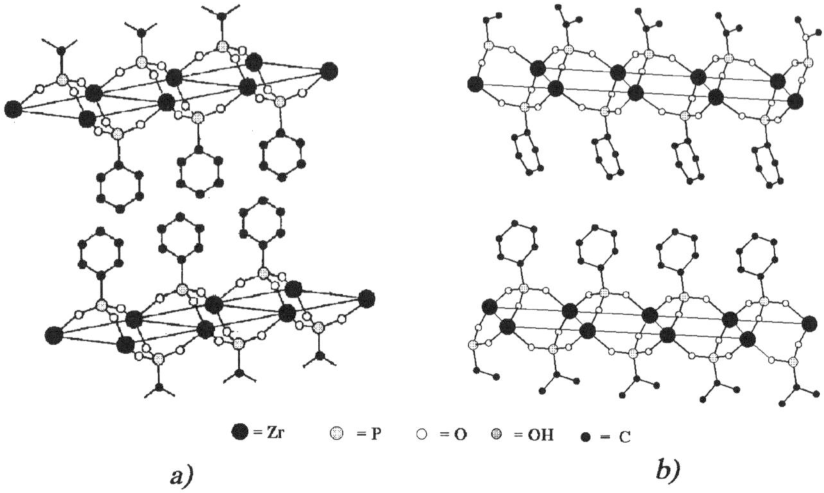

Figure 2. Structural model of a) α-Zr($O_3PC_6H_5$)$_2$ as proposed in 1978 by Alberti and c) the same compound drawn on the basis of its structural parameters (Clearfield, 1993)

The crystals belong to space group C 2/c with a = 9.0985(5), b = 5.4154(3), c = 30.235Å and, β = 101.333(5)°. The in-plane *a*- and *b*- dimensions are very close to the corresponding dimensions in α-ZrP. However, the phenyl groups are tilted relative to the layers by 30°. Apart from the orientation of the benzene rings, the recent study essentially confirms the model structure proposed 15 years before [10].

In recent years the chemistry of zirconium phosphonates with an α-type layered structure has grown considerably and a large number of compounds with a variety of organic groups have been prepared (see fig. 3). Note that in these compounds the inorganic polymeric matrix [M(IV)(O_3P-)$_2$]$_n$ of a single layer can be considered as a clothes-hook on which organic and inorganic R groups can be hung. The nature of the covalently attached groups depends only on the imagination and ability of the chemist to

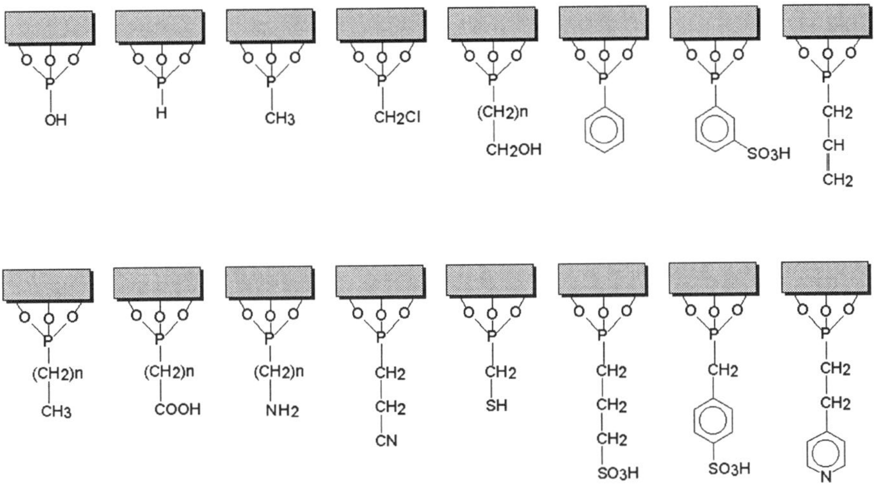

Figure 3. A view of some organic groups that have already been covalently attached to α-ZrP layers.

synthesise the appropriate phosphonate and organo-phosphate acids. The free area around each position of the α-layer is 24 $Å^2$; therefore two large R groups with cross sections greater than this value cannot occupy adjacent positions. Due to the short lateral distance among adjacent O_3P-R groups on each side of the α-layer, the interpenetration of R groups belonging to adjacent layers cannot occur for steric reasons; thus a double film of R groups, as shown in figure 2 for zirconium phenylphosphonate, must be expected for all the members of this class.

To illustrate how the interactions among adjacent layers of zirconium phosphonates can be influenced by the conformation of the R groups as well as by the nature of their terminal groups, compounds of the series $Zr(O_3P\text{-}(CH_2)_n\text{-}COOH)_2$ will be examined [12]. As shown in figure 4, the terminal -COOH groups take on a different orientation in compounds with an odd or an even value of n. The different interactions between -

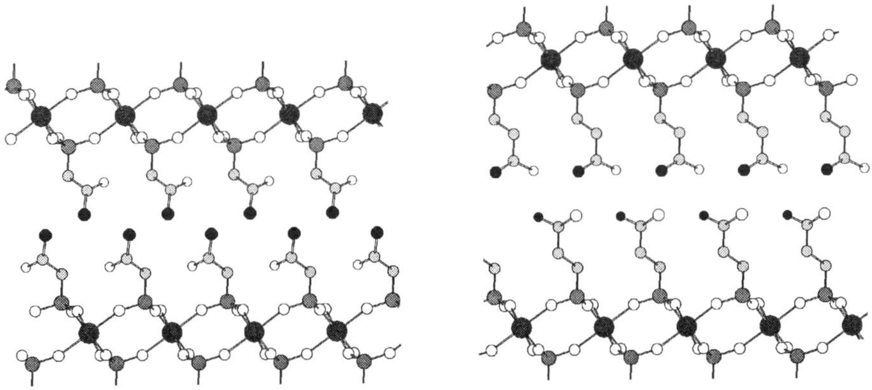

● Zr; ● P; ○ O; ◎ C; ● OH

Figure 4. Schematic view showing the different arrangement of carboxyl groups for even and odd

COOH groups of adjacent layers strongly influence the properties of the compounds (e.g., proton conductivity).

If a mixture of two different H_2O_3P-R acids is used in the synthesis, zirconium phosphonates with two different R groups are obtained. These compounds can be described by the general formula $Zr(O_3P\text{-}R)_x(O_3P\text{-}R')_{2-x} \cdot nS$.
In some cases asymmetrical layers may be formed in which one side of the layer is richer in R, the other in R' groups. The formation of compounds with asymmetrical layers has been investigated, in particular for the zirconium/phosphate/phosphite system [13]. In this case even compounds in which practically only phosphite groups are present in one side of the layer, while more than 65% of phosphate groups are present in the opposite side, have been found. The packing of the layers is that shown schematically in figure 5 in which there is an alternation of interlayer regions richer in phosphate and phosphite groups respectively. It is of interest to note that in this compound ion-exchange and intercalation processes can take place only in alternate interlayer regions where the acid O_3P-OH groups are present.

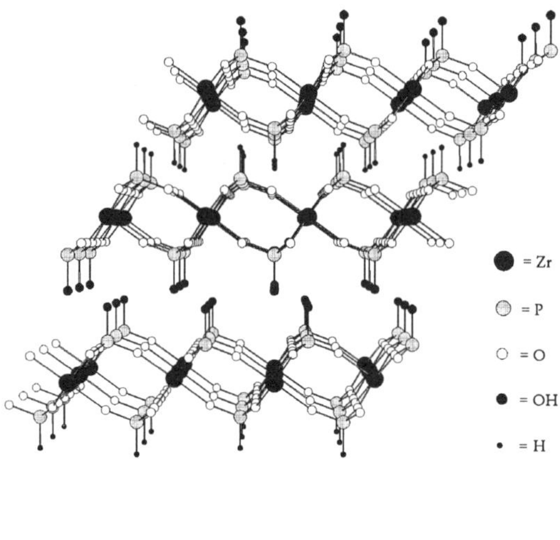

Figure 5. Schematic representation of the structure of α-zirconium phosphate phosphite, showing the packing of asymmetrical layers.

By choosing appropriate organic groups attached to the phosphorus atom, it is possible to vary the acid properties of the α-lamellae from a neutral (e.g. P-CH_3) or a weak acid (e.g. P-CH_2-COOH) to a strong acid (e.g. P-CH_2-SO_3H) or even a basic characteristic (e.g. P-CH_2NH_2). Now, an acid-base reaction is expected to take place if a colloidal dispersion of lamellae with acid groups is mixed with a colloidal dispersion of lamellae with basic groups. Such a reaction was of interest since interstratified compounds, in which there is an alternation of acid and basic layers, have been obtained [14].

Finally, it is of interest to point out that zirconium phosphonates with organic groups containing very strong acid sites (e.g. $-SO_3H$) have been investigated for their potential application as acid catalysts [15] and solid state protonic conductors [16]. A representation of the compound α-$Zr(O_3P\text{-}CH_2OH)_{1.27}(O_3PC_6H_4\ SO_3H)_{0.73}$ which exhibits a very high protonic conductivity (σ=1.6 $\cdot 10^{-2}$ S cm^{-1}) is shown schematically in figure 6.

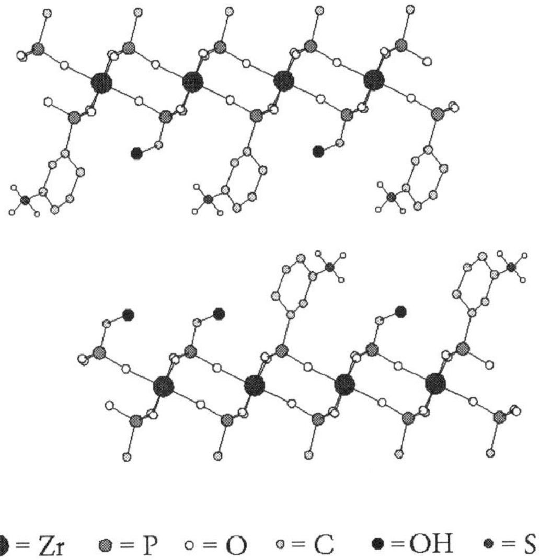

Figure 6. Schematic representation of the structure of α-$Zr(O_3P\text{-}CH_2OH)_{1.27}(O_3PC_6H_4SO_3H)_{0.73}$

4. Layered Organic Derivatives Of γ-Zirconium Phosphate

As previously discussed, only many years after its preparation was it realised that γ-ZrP, different from zirconium bis monohydrogen phosphate, must be formulated as zirconium phosphate dihydrogen phosphate, $ZrPO_4H_2PO_4 \cdot 2H_2O$, and that its interlayer dihydrogen phosphate groups

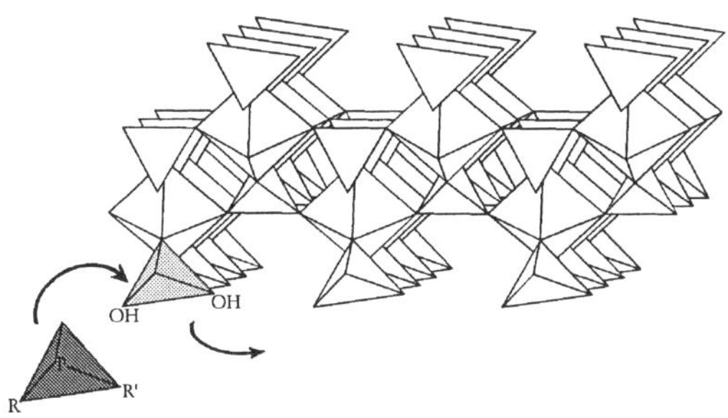

Figure 7. Schematic representation of a topotactic reaction of the exchange of $O_2P(OH)_2$ with O_2PRR' tetrahedra on the surface of a γ-ZrP lamella.

can easily be replaced, partially or completely, by monovalent phosphonate or phosphinate anions [17-19] by simple topotactic reactions (see fig. 7) of the type:

$$ZrPO_4\ O_2P(OH)_2 + x\ O_2P(OH)R^- \rightleftharpoons ZrPO_4\ (O_2P(OH)_2)_{1-x}\ (O_2P(OH)C_6H_5)_x + x\ O_2P(OH)_2^-$$

A schematic view of one of these, zirconium phosphate phenyl phosphonate of composition $ZrPO_4\ O_2P(OH)_2\ O_2P(OH)C_6H_5$ is shown in figure 8.

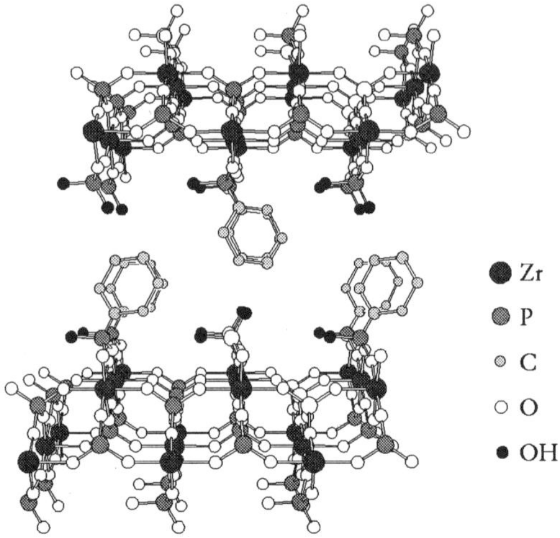

Figure 8. Computer generated structural model of γ-$ZrPO_4\ O_2P(OH)_2\ O_2P(OH)C_6H_5$.

Many organic derivatives with γ-layered structures have already been prepared [20-22]. In this laboratory, even some crown ethers have been attached as bi-layers in the interlayer region of γ-ZrP [23] (fig. 9).

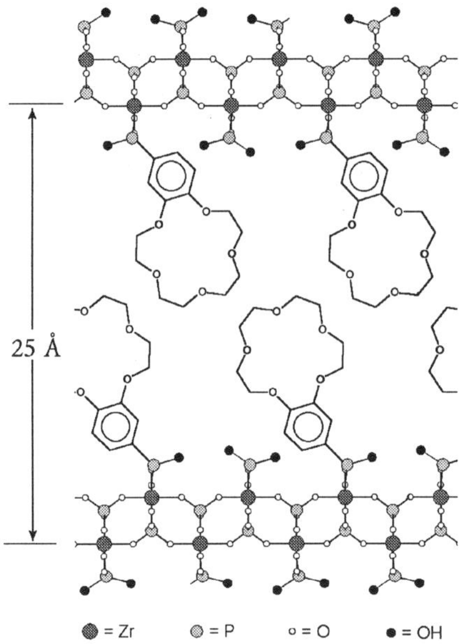

Figure 9. Schematic representation of a derivative of γ-ZrP containing a bilayer of crown ethers in its interlayer region.

These compounds could be of interest in the field of the molecular recognition, or, more in general, for the development of a supramolecular chemistry in the interlayer region, although it is too early for forecasting the real perspective of selective separations or other practical uses, such as the inclusion of lithium salts, for the preparation of lithium conductors.

5. Covalently Pillared Zirconium Diphosphonates

M(IV) diphosphonates, of general formula M(IV)(O_3P-R-PO_3), in which adjacent inorganic layers of the α-type are covalently joined to each other by divalent organic groups, may also be obtained. Due to the possibility that "tailor-made" materials with an interlayer porosity may be prepared by simply changing the length of the organic pillar, the acronym MELS (Molecularly Engineered Layered Structures) was given to M(IV) diphosphonates to emphasise their application potential [24]. A large number of organic derivatives of α-zirconium phosphate have been prepared and many attempts at preparing materials containing micro-cavities in the interlayer region, have been made. Unfortunately, the first MELS obtained in the 1980s were found to exhibit a low degree of crystallinity and a low number of micro-cavities in their interlayer region.

Very recently, it was found in our laboratory that MELS with α-layered structure, having a high degree of crystallinity and micro-cavities of uniform size, can indeed be obtained if the cross section of the pillars is not uniform, and in particular if their central cross-section is smaller than the terminal parts [25].

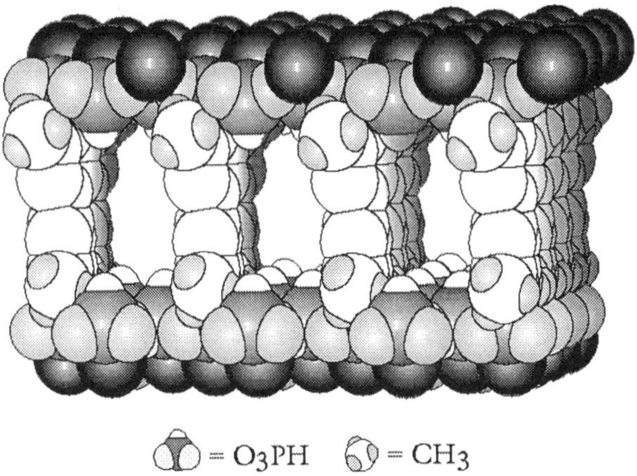

Figure 10. A computer generated view of a microporous α-zirconium phosphite diphosphonate in which pillars with bases are present.

Pillars with these characteristics are called pillars with bases. A computer generated structural model of a pillared compound based on one of these pillars (the 3,3', 5,5'-tetramethyl-biphenyldiphosphonate group) based on its unit cell parameters (a=15.286; b=9.365; c=28.48 Å; β=94.99°) is shown in figure 10. A good degree of crystallinity and an interlayer microporosity of 375 m^2/g (with a narrow pore distribution at about 5Å) have indeed been found [25]. It was furthermore found in our laboratory [26] that a strong excess of phosphorous acid in the synthesis of zirconium phosphites-diphosphonates may induce the formation of a high percentage of mesopores. E.g., compounds with high specific surface area (300-500 m^2/g of which 60-70% due to intercrystal mesoporosity and the remaining to interlayer microporosity), have been obtained. It was furthermore found that the dimension of the mesopores can be easily changed in the range 15-50 Å by simply changing the ratio and the concentrations of the reagents, while a fairly narrow distribution of the mesopores were obtained in the greater part of the syntheses investigated.

Diphosphonate groups can also be used for joining together adjacent layers of γ-zirconium phosphate, as shown in figure 11.

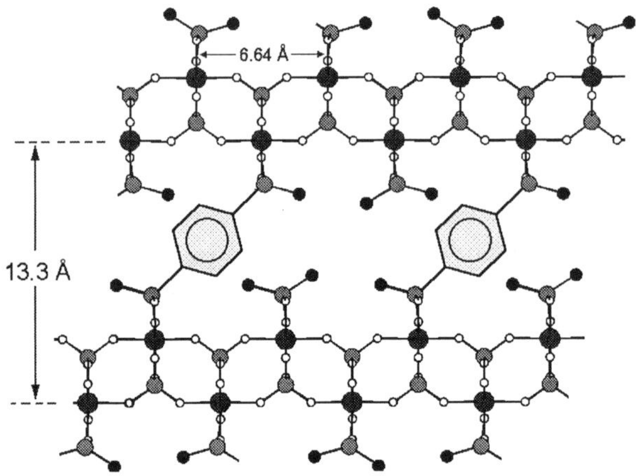

Figure 11. Schematic view of a γ-ZrP pillared with phenyldiphosphonic groups

In this case the pillaring can be performed simply, by a topotactic reaction [27]. Even pillared compounds with regular interlayer porosity have been obtained by a partial replacement of the dihydrogen phosphates with bivalent diphosphonate groups. A system studied in detail has been the pillaring with biphenyldiphosphonate groups [28]. Due to the rigidity of the γ-layer, a low percentage of pillaring can be obtained without a phase segregation. The pillared compounds so prepared possess a large specific surface area and a great micropore volume. A maximum of $\approx$320 m^2g^{-1} and $\approx$0.12 cm^3g^{-1}, respectively, and a narrow micropore distribution with an average diameter of 5.8 Å, were found for the sample at about 25% pillaring. A model structure of this interesting compound is shown in figure 12.

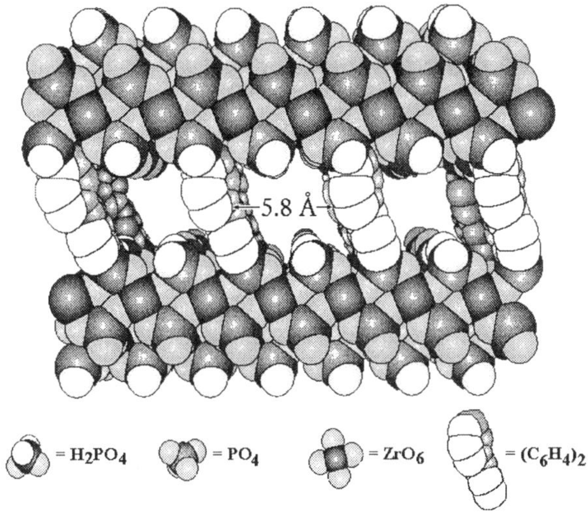

Figure 12. Computer generated view of a γ pillared microporous compound in which 25% of $H_2PO_4^-$ groups have been replaced by 4,4'-biphenyldiphosphonate groups.

The pillaring with covalently bonded diphosphonate groups, opens the way for the preparation of a large number of new inorganic or inorganic-organic solids with "zeolite-like" structures.

Although their thermal stability (230-250°C) is relatively low, modification of the length of the organic pillars and of their lateral distance should allow the engineering of compounds with predicted well-defined void spaces and, hence, tailored microporosities. Finally, taking into account that the original $H_2PO_4^-$ still present in γ-zirconium diphosphonates with a low degree of pillaring can be partially or completely replaced with more hydrophobic small groups such as $O_2P(OH)CH_3^-$, $O_2PHCH_3^-$, $O_2PH_2^-$, the interplay of hydrophilic and hydrophobic domains can be specifically designed to give desired properties to the cavities. Furthermore, the introduction of strong acid functions (i.e. $-SO_3H$) in the interlayer region opens new prospects to applications of the porous MELS in the field of acid catalysis.

Many other possibilities are opened. For example, crown ether diphosphonates have been used for the pillaring of γ-zirconium phosphate [22] (fig. 13) so widening the possibility of studying their behaviour in the interlayer regions. Non-rigid organic pillars (e.g., alkyl chains) can easily be introduced in the interlayer regions of γ-phases [29].

In this latter case, due to the possibility of different conformation of the pillars, the interlayer distance, and hence the pore size, will depend, within certain limits imposed

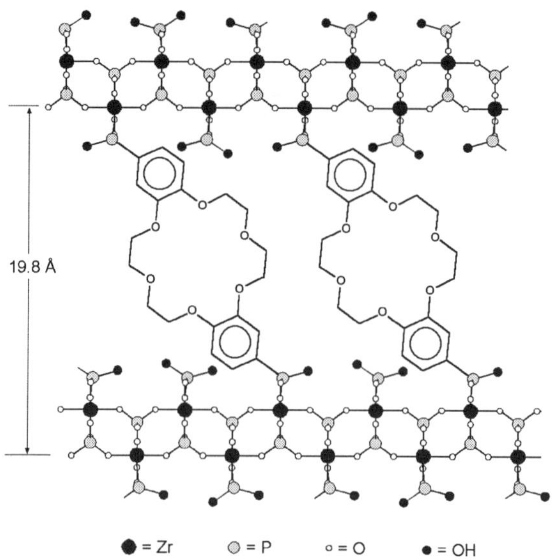

Figure 13. Schematic drawing of a γ-ZrP pillared with a diphosphonic group containing a crown ether in its organic radical.

by the nature of the particular pillar, on the amount and nature of the solvent included between the pillars. The change in the interlayer distance is reversible (see fig. 14).

The use of the interpillar regions as molecular vessels for specific reactions is under investigation in our laboratory. In this respect, the formation of some particular precipitates in the interpillar regions could also be of interest for quantum dot studies since the size and shape of the clusters can be modulated by the length and lateral distance of the pillars. Very recently, other particular pillars, such as polyethylene oxides of different length have been used for the pillaring of γ-ZrP [30].

These compounds too are expected to be of interest for supramolecular chemistry in the interlayer region (fig. 15).

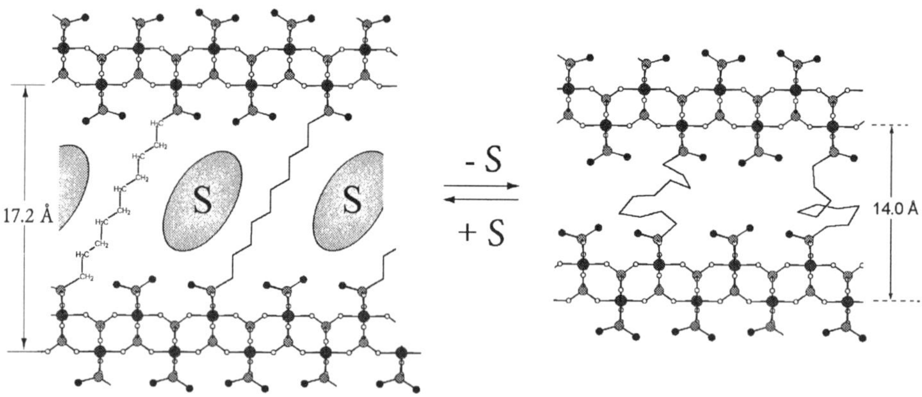

Figure 14. Schematic representation of reversible change of interlayer distance in a γ-pillared compound having non-rigid pillars at low (about 25%) pillaring percentage.

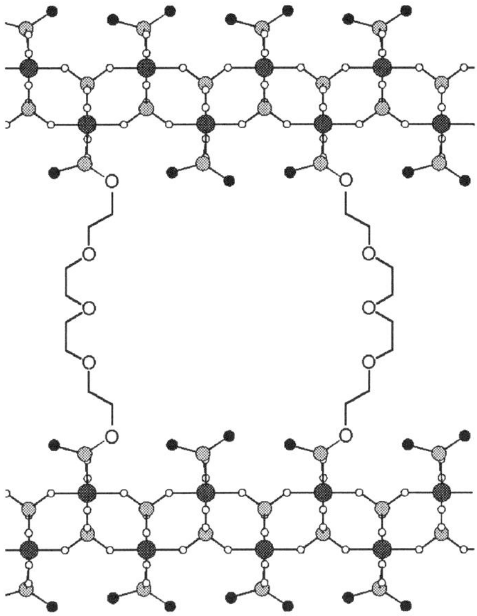

Figure 15: Schematic picture of a γ-ZrP pillared with polyethylene oxide columns.

Significant progress has been made in the preparation and characterization of molecular thin films on surfaces. Layered metal phosphonates are particularly suitable compounds for the layer-by-layer growth of very thin films which are of interest for their structural analogy with Y-type Langmuir-Blodgett films. The interested reader is referred to a review by Mallouk [31].

Finally, the recent experiments of Thompson and collaborators in photochromism via the photoreduction of zirconium phosphate-viologen compounds [32] seem to show exciting promise for their utilisation in the storage of solar energy.

6. Acknowledgements

We thank MURST (40%) for partial support.

7. References

1. Alberti, G. and Bein, T. (in press) Two and three-dimensional Inorganic Networks, in G. Alberti and T. Bein (eds.) *Comprehensive Supramolecular Chemistry*, Volume 7, Pergamon Press, Oxford, .
2. Alberti, G. and Costantino, U. (1991) Intercalation Chemistry of acid salts of tetravalent metals with layered structure and related materials, inM.S. Whittinghan and A.J. Jacobson (eds.),*Intercalation Chemistry,* Academic Press, New York, p. 147.
3. Alberti, G. and Costantino, U. (1991) Intercalation of zirconium phosphates and phosphonates, in J.L. Atwood, J.E.D. Davies and D.D. MacNicol (eds.), *Inclusion Compounds*, Oxford University Press, p. 136.

4. *Physics and Chemistry of Materials with Low-Dimensional Structures*, in W. Müller-Warmurth and R. Schöllhorn (eds.) (1994) *Progress in Intercalation Research*, Kluwer Academic Publishers, London, volume 17.
5. *Inorganic Ion Exchange Materials*, A. Clearfield, (ed.) CRS, Boca Raton, FL (1981).
6. Alberti, G., Murcia-Mascarós, S. and Vivani, R. (1994) Layered and Pillared Zirconium Phosphates with α- and γ-structures, in *Soft Chemistry Routes to New Materials* Proceedings of Int. Symposium *Mater. Sci. Forum* **87,** 115.
7. Clearfield, A. and Stynes, J.A. (1964) The preparation of crystalline zirconium phosphate and some observations on its ion exchange behaviour, *J. Inorg. Nucl. Chem.* **37,** 117
8. Clearfield, A. and Smith, G.D. (1969) The crystallography and structure of zirconium bis (monohydrogen orthophosphate) monohydrate, *Inorg. Chem.* **8,** 431.
9. Christensen, A., Andersen, E.K., Andersen, I.G.K., Alberti, G., Nielsen, N. and Lehmann, M.S. (1990) X-ray powder diffraction study of layer compounds. The crystal structure oα–$Ti(HPO_4)_2$ and a proposed structure for γ–$TiPO_4(H_2PO_4)\cdot 2H_2O$, *Acta Chem. Scand.* **44,** 865.
10. Alberti, G., Costantino, U., Allulli, S. and Tomassini, N. (1978) Crystalline $Zr(R\text{-}PO_3)_2$ and $Zr(R\text{-}OPO_3)_2$ compounds (R = organic radical). A new class of materials having a layered structure of the zirconium phosphate type, *J. Inorg. Nucl. Chem.* **40,** 1113.
11. Poojary, D.M., Hu, H.L., Campbell, F.L. and Clearfield, A. (1993) The determination of Crystal Structures from Limited Powder Data Sets: The Crystal Structure of Zirconium Phenylphosphonate,*Acta Crystallogr.* **B49,** 996.
12. Alberti, G., Costantino, U., Vivani, R., Casciola M. and Peraio, A. (1991) Proton conductivity of zirconium carboxy n-alkyl phosphonates with an α-layered structure, *Solid State Ionics* **46,** 61.
13. Alberti, G., Costantino, U. and Perego, G. (1986) Structural investigation on layered α-zirconium phosphite and zirconium phosphate-phosphite from X-ray powder data, *J. Solid State Chem.* **63,** 455.
14. Stein, E.W., Bhardwaj, C., Ortiz-Avila, C.Y., Clearfield, A. and Subramanian, M.A. (1994) Synthesis of novel metal phosphonate complex structures through soft chemistry, in *Soft Chemistry Routes to New Materials* Proceedings of Int. Symposium *Mater. Sci. Forum* **87,** 115.
15. Clerici, M.G., Alberti, G., Malentacchi, M., Bellussi, G., Prevedello, A. and Corno, C. (1989) Sulfonic group-containing zirconium phosphate-based catalysts for acidic catalysis and ion exchanger. Eur. Pat. Appl. EP 386,845.
16. Clearfield, A. (1992)Proton containing NASICON phases, in *Proton conductors. Solids, membranes and gels-materials and devices*, P. Colomban (ed.), Cambridge University Press, Cambridge, p. 224.
17. Alberti, G., Costantino, U., Vivani, R. and Biswas, R.K. (1992) Preparation and some preliminary investigation of ion exchange and intercalation properties oγ-zirconium phosphate phosphite,*Reactive Polymers* **17,** 245.
18. Alberti, G., Casciola, M. and Biswas, R.K. (1992) Preparation and some properties of zirconium phosphate hypophosphite and phosphate dimethylphosphinate with γ-layer structure, *Inorg. Chim. Acta* **201,** 207.
19. Alberti, G., Casciola, M., Vivani, R. and Biswas, R.K. (1993) Preparation and characterization of zirconium phosphate-phosphonates, $ZrPO_4(H_2PO_4)_{1-x}(RPO_2OH)_x\cdot nH_2O$, with γ-layer stucture (R= CH_3, C_3H_7, C_6H_{11}), *Inorg. Chem.* **32,** 4600.
20. Alberti, G., Vivani, R., Biswas, R.K. and Murcia Mascarós, S. (1993) Preparation and some properties of γ-zirconium phosphate benzenephosphonate, *Reactive Polymers* **19,** 1.
21. Yamanaka, S. and Koizumi, M. (1975) Structural consideration of zirconium phosphate and its organic complexes, *Clays Clay Miner.* **23,** 477.

22. Alberti, G., Boccali, L., Dionigi, C. and Kalchenko, V.I. To be published.

23. Yamanaka, S., Yamasaka K. and Hattori, M. (1984) Preparation and properties of the layered organic derivatives of γ–zirconium phosphate, Zr $(CH_3\text{-}(OCH_2CH_2)_n\text{-}OPO_3)(HPO_4)$ (n=1-3), *J. Inclusion Phenomena* **2,** 297.

24. Alper, J. (1986) Catalytica's successful sandwiches, *Chem. and Ind.*, 335.

25. Alberti, G., Costantino, U., Marmottini, F., Vivani R. and Zappelli, P. (1993) Zirconium phosphite biphenyl(3,3',5,5' tetramethyl) diphosphonate, a microporous, layered inorgano-organic polymer, *Angew. Chem. Int. Ed. Engl.* **32,** 1357.

26. Alberti, G., Costantino, U., Vivani R. and Zappelli, P. (1992) Composizione solida cristallina mesoporosa di difosfonato-fosfito di un metallo tetravalente con una stretta distribuzione di mesopori *Italian Pat.* 21213 .

27. Alberti, G., Murcia-Mascarós, S. and Vivani, R. (1993) Preparation and characterization of zirconium phosphate-diphosphonates with γ-structure: a new class of covalently pillared compounds, *Mater. Chem. and Phys.* **35**, 187.

28. Alberti, G., Marmottini, F., Murcia-Mascarós, S. and Vivani, R. (1994) Preparation and Preliminary Characterization of a Covalently Pillared γ-Zirconium phosphate-diphosphonate with Interlayer Microporosity, *Angew. Chem. Int. Ed. Engl.* **33,** 1594.

29. Alberti, G., Murcia-Mascarós, S. and Vivani, R. To be published.

30. Alberti, G., Dionigi, C. To be published.

31. Mallouk, E., Cao, G. and Hong, H. (1992) Layered Metal Phosphates and Phosphonates: From Crystals to Monolayers. *Acc. Chem. Res.* **25**, 420.

32. Thompson, M.E. and Vermeulen, L.A. (1992) Stable photoinduced charge separation in layered viologen compounds, *Nature* **358,** 656.

MOLECULAR TECTONICS.

Xavier Delaigue, Ernest. Graf, Frédérique Hajek,
Mir Wais Hosseini*and Jean Marc Planeix
Université Louis Pasteur
Laboratoire de Chimie de Coordination Organique
Institut Le Bel,
F-67000 Strasbourg, France

ABSTRACT. There is a true need for a rational approach to the synthesis of one-, two- and three-dimensional solids. Molecular tectonics based on iterative assembly of individual modules may be a viable way to the design and manufacture of large size molecules with strictly controlled structures. We describe in this contribution our approach and results dealing with the individual modules or tectons.

1. Introduction

The construction of large size molecules (10^{-6}-10^{-4} m scale) or hyper molecules with predicted and programmed structure may hardly be envisaged through stepwise classical synthesis using covalent linkages. However, the preparation of such higher-order materials may be attained through iterative process based on self-assembly [1] of individual tectons (from Greek TEKTON, builder) [2]. This particular area of research, called molecular tectonics [3], is currently under active investigation [4,5].

One of the most important aspect of molecular tectonics appears to be the fact that the assembling process is based on molecular recognition processes operating at the level of the complementary tectons. In other terms, the operational concepts in molecular tectonics are those defined for molecular recognition [6], plus topological features allowing iteration.

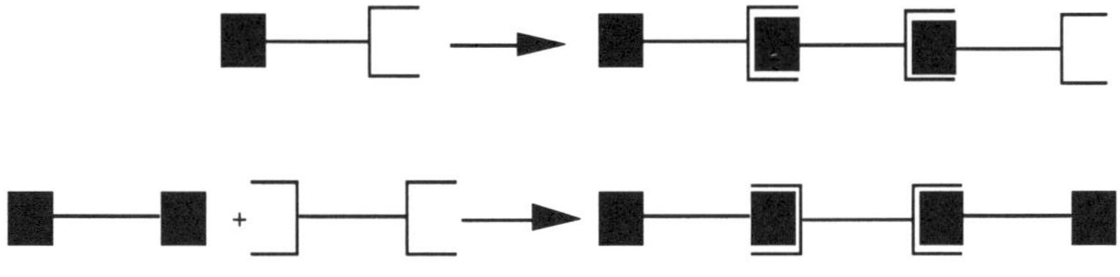

Fig. 1 Schematic representation of mono-dimensional hyper molecules based on iterative assembling of a self-complementary tecton (top) or of two complementary tectons (bottom)

G. Tsoucaris et al. (eds.), Crystallography of Supramolecular Compounds, 159–180.

The manufacture of hyper molecules may be achieved by an iterative assembling process engaging a self-complementary tecton (one component system composed of a unique motif), or several complementary modules (multi component system). In the later, case the most viable approach would engage two complementary tectons (Figure 1).

In order to achieve an iterative assembling process, the building blocks or tectons must fulfil both structural and energy criteria. In particular, the complementary tectons must be able to recognise each other (molecular recognition) and furthermore to allow the repetition of the recognition pattern (iteration). These two requirements lead to molecular modules possessing connecting points or interaction sites located in a divergent fashion. Although one may use the terms exo-receptors and exo-substrate to design the complementary modules (Figure 2), however, it is worth noting that, since the role played by both components is of the same importance, these terms are exchangeable. Never the less, an artificial hierarchy may be introduced based, quiet often, on the effort invested in the preparation of the component. One may simplify the terminology by using the terms exo- and endo-tectons. Whereas an endo molecular receptor (endo-receptor), by conception, is designed to form only molecular complexes with a selected exo-substrate, an exo-receptor may form hyper or giant structures with appropriate exo-substrates.

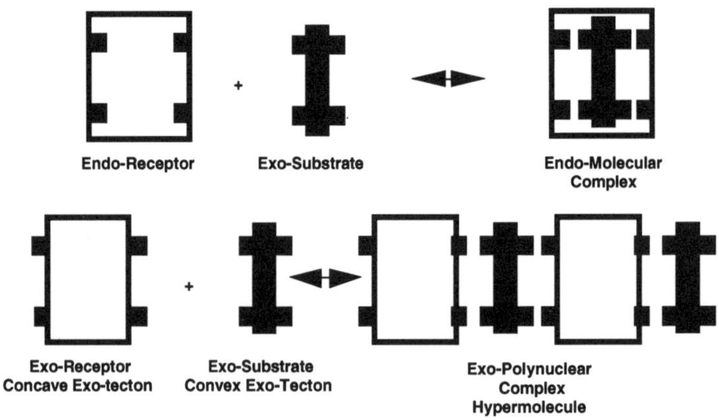

Fig. 2 Schematic representation of the formation of an endo-molecular complex (top) and of a one-dimensional hyper molecule resulting from an iterative assembly of one exo- and one endo-tectons

Dealing with the energetics involved in the assembling processes, weak interactions such as Van der Waals, electrostatic interactions, and hydrogen bonding may be used.

Design of linear assemblies in the solid state based on weak interactions:

The synthesis of solids based on iterative assembling of individual complementary tectons still remains a challenge for chemists. A strict

control of the self-assembly of molecular modules in the solid state should lead to structurally strictly controlled hyper molecules in the solid state.
Our approach, so far, focuses on the possibility of manufacturing one- and two-dimensional solids using complementary molecular tectons or modules.

In terms of energetics, our approach is based on both weak Van der Waals interactions and hydrogen bonding. In the latter case, using both electrostatic interactions and H-bonding one dimensional solids have been obtained (7). In this chapter, we only describe our results dealing with Van der Waals interactions. Although coordination bonds can not be classified as weak interactions, never the less, the same concepts of modular chemistry may be applied to coordination polymers. We also describe in this chapter our approach to one-dimensional coordination polymers *i.e.* polynuclear systems resulting from an iterative complexation of metals by exo-ditopic modules.

Approach to linear molecular arrays using Van der Waals interactions:

The chemistry of inclusion complexes, based on concave and convex molecules [8], *i.e.* the inclusion of a substrate within the cavity of a receptor molecule is an established area. One may extend the concept of inclusion in solution to the construction of new networks in the solid state. For example, a rigid and compact direceptor possessing two divergent cavities (linear *koiland* , from Greek koilos : hollow), and a linear connector, possessing two extremities capable each to be included within the cavities of the direceptor, may be assembled, by an iterative process, in the solid state through non-covalent Van der Waals interactions leading thus to a linear molecular array which may be called linear *koilate* [9,10] (Figure 3).

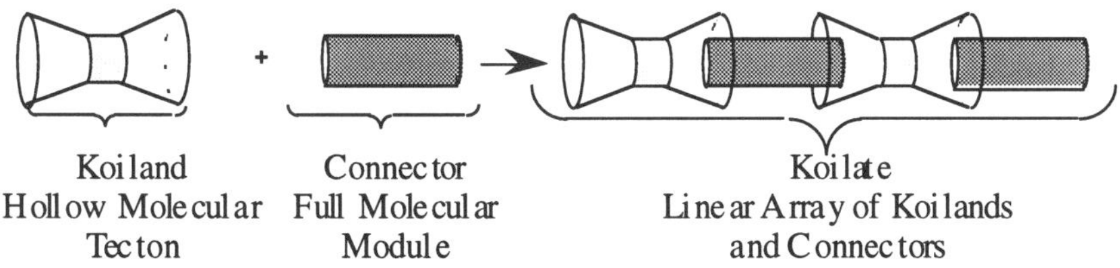

Fig. 3 Schematic representation of a linear array (koilate) resulting from interpenetration of a convex molecule (connector) into the cavity of a concave receptor (koiland)

For the construction of linear koilates, the design of a koiland possessing two divergent cavities with an angle of 180° between them is crucial. For this purpose, calix[4]arene seems to be an interesting unit since, it possess in its framework both hydrophobic and hydrophilic sites[11]. It has been shown that the simple *p-tert*-butylcalix[4]arene **1** (Figure 4), in its favoured cone conformation below 20°C in chloroform [12-14] and in the solid state [15,16] offers a genuine hydrophobic pocket surrounded by four *p-tert*-butyl aryl moieties. This basket type cavity was proved to accommodate in the solid state a variety of neutral substrates such as chloroform [14], toluene [15], anisole [16], xylene [17] and benzene [17]. Recently, complexation of Cs^+ through metal-π-aromatic ring interactions was reported for **1** [18] in its cone conformation.

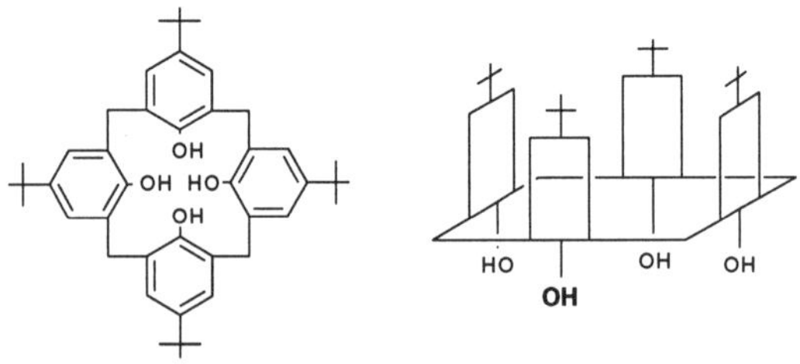

Fig. 4 *p-tert*-butylcalix[4]arene **1** (left) and a schematic representation of its cone conformation (right)

The fusion of two *p-tert*-butylcalix[4]arenes **1** in its cone conformation by two silicon atoms leads to a hollow brick (koiland) **2** [9,10] (Figure 5). Examples of fused *p-tert*-butylcalix[4]arene using titanium (19), niobium (20), aluminium (21) have been reported.

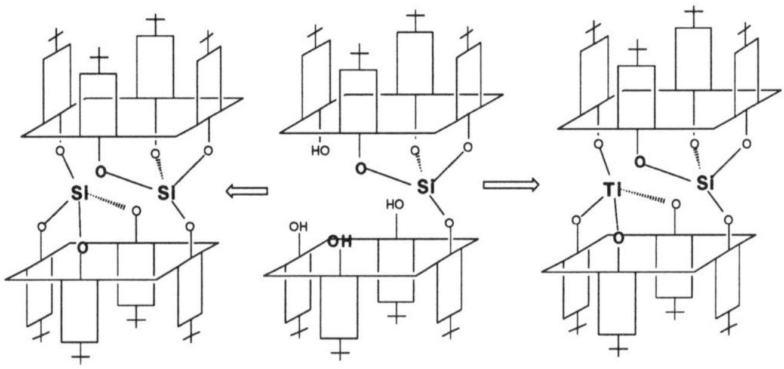

2 **3** **4**

Fig. 5 Schematic representation of koilands based on: the fusion of two *p-tert*-butylcalix[4]arene **1** by two Si atoms **2**, the fusion of two *tert*-butylcalix[4]arene by one Si atom **3**, and the fusion of two *tert*-butylcalix[4]arene by both Si and Ti atoms **4**.

In order to differentiate the two cavities, the synthesis of the heterobinuclear compound **4** containing both a Si and a Ti atoms was achieved [10]. The non-centro-symmetric nature of **4** is based on the difference between the two calix units induced by coordination to two different atoms. Indeed, one of the calix in **4** is triply coordinated to a Si atom whereas the other unit is triply coordinated to a Ti atom.

In order to prepare **4** by a directed strategy, we first prepared the monofused compound **3**. For the later, due to the possible free rotation around the Si-O bond, one may expect different rotamers. ROESY experiment showed that, due to possible intra plateau hydrogen bonding between hydroxy groups belonging to both calix units, the face to face (syn) rotamer was the preferred one (Figure 6).

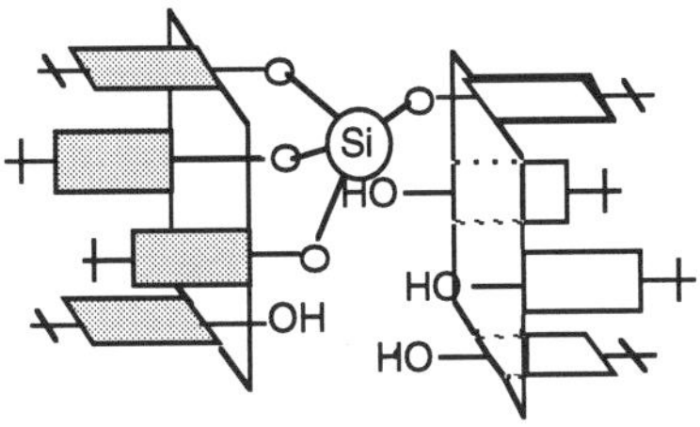

Fig. 6 Schematic representation of the syn rotamer of the monofused compound **3**

Treatment of the isolated **3** with $TiCl_4$ in CH_2Cl_2 produced, in high yield, the heterobinuclear compound **4**. The structure of **4** was established by mass spectrometry and by multidimensional NMR experiments.

In the solid state, the structure of **2** was proved by X-ray analysis. Compound **2** as shown in Figure 7 was indeed a centro-symmetric silicon aryloxy dimer consisting of 2 Si and 2 calixarenes. Both calixarene units were in cone conformation, thus, presenting two divergent cavities. The coordination geometry around both Si atoms was tetrahedral. Each of these two tetrahedrons was composed of 3 oxygen atoms belonging to one of the two calixarenes and the fourth one to the other calix unit.

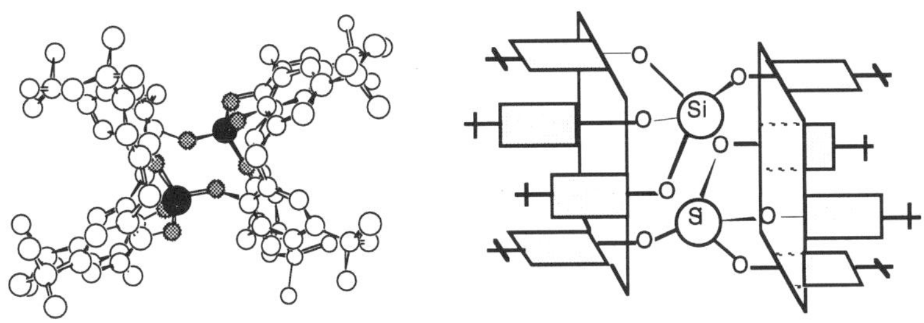

Fig. 7 Schematic representation of **2** composed of two calixarenes fused by two silicon atoms yielding a double basket type receptor (right), and its crystal structure (left).

Although 6 molecules of $CHCl_3$ were present in the solid sate, none of them was located within 3.3 Å distance of any atom belonging to **2**.

As shown in the figure 7, in the solid state, the hollow ditopic receptor **2** indeed possesses two divergent cavities. In order to ascertain the possibility of double inclusion, a binuclear complex in the solid state composed of one compound **2** and two anisol molecules was prepared and its structure investigated by X-ray analysis (Figure 8). The crystal study showed that both cavities in **2** accommodate, in a inclusive fashion, an anisol molecule. The latter penetrates the cavity of **2** through its CH_3 group. It is worth noting that anisol, due to its reduced dimension and its bent shape, is not a well suited connector for bridging the hollow modules **2**, and thus leading to the linear koilate.

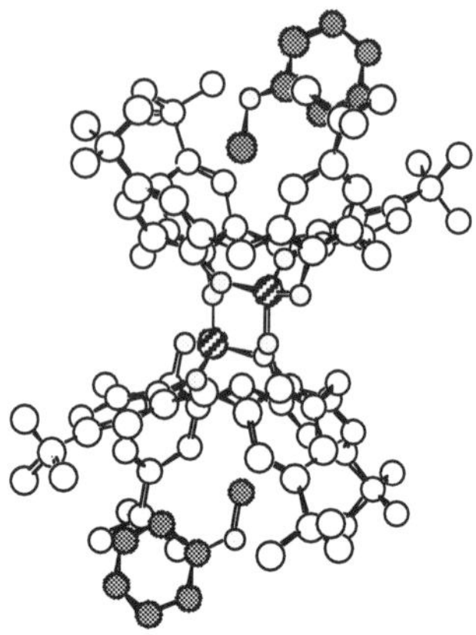

Fig. 8 X-ray structure of the binuclear inclusion complex formed between compound **2** and anisol

Our aim was the preparation of linear arrays in the solid state resulting from an iterative assembling process based on the inclusion of both extremities of connectors within the cavities of koilands (Figure 9).

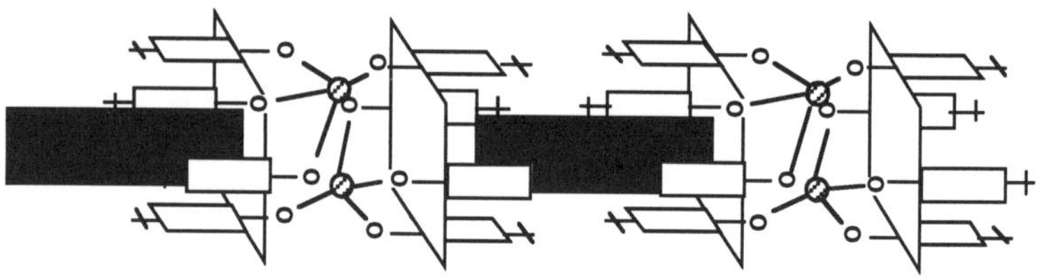

Fig. 9 Schematic representation of the formation of koilates resulting from an iterative assembling process engaging hollow molecular modules obtained by fusion of two *p-tert*-butylcalix[4]arenes with two tetrahedrally coordinated centres and connectors capable of penetrating the cavity of the ditopic receptors

With *para*-xylene as connector, again a discrete binuclear inclusion complex in the solid state was obtained (Figure 10). The complex was characterised by X-ray analysis which revealed that indeed both cavities of **2** form inclusive complexes with *para*-xylene. Each *para*-xylene, as in the case of anisol, penetrates into the cavities of **2** through its CH_3 group. Although, due to the discrete nature of the complex, *i.e.* the absence of interconnection between the bi-nuclear complexes, *para*-xylene as connector do not lead to linear koilate, nevertheless, the disposition, with respect to each other, of binuclear complexes appeared to be rather interesting and encouraging. Indeed, in the lattice, the different complexes are aligned indicating that with more suited connectors, one may succeed in preparing the linear koilates.

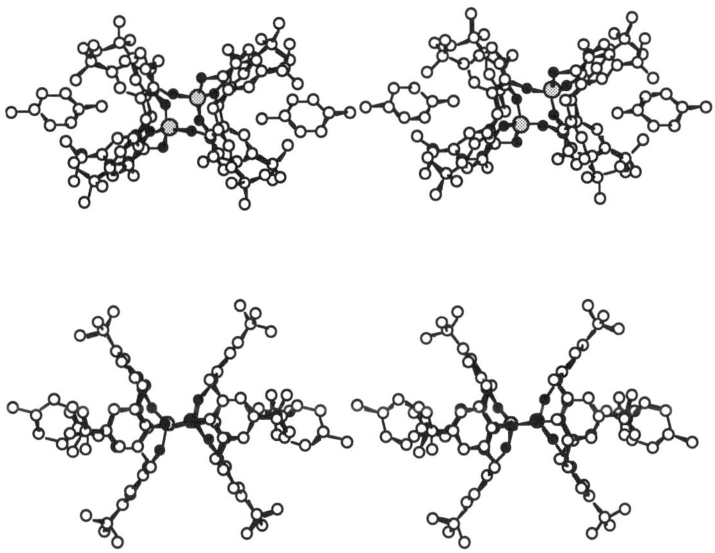

Fig. 10 Two views obtained from an X-ray study of the binuclear *para*-xylene complex with **2**. The analysis revealed that in the lattice the different modules are linearly arranged.

Another alternative to the formation of koilates may be based on the fact that one may use other calix[4]arene derivatives. Indeed, this class of molecules seems particularly suited to the design of the hollow modules since one may control both the entrance and the depth of their preorganised cavity by the nature of the substituant R at the *para* position, *i.e.* H, CH_3, $C(CH_3)_3$, Ph (Figure 11).

The synthesis of koilands **8-10** (Figure 12) based on the double fusion of two calix[4]arenes **5**, two *p*-methylcalix[4]arenes **6**, and two *p*-phenylcalix[4]arenes **7** with two silicon atoms was tried. Both, in the case of *p*-methylcalix[4]arenes **6**, and of *p*-phenylcalix[4]arenes **7** the corresponding koiland **9** and **10** could be prepared. However, we were unable so far to synthesise the koiland **8** (22).

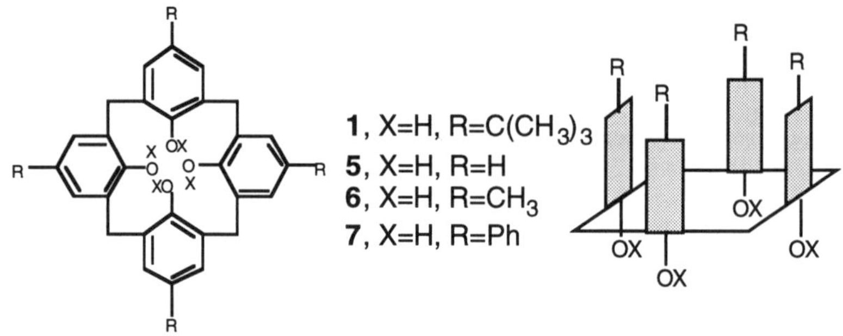

Fig. 11 Structure of calix[4]arene derivatives bearing different substituants at the *para* position (left) and a schematic representation of their cone conformation (right)

Attempts to prepare different koilates based on compound **9** and **10** using linear connectors are under current investigation.

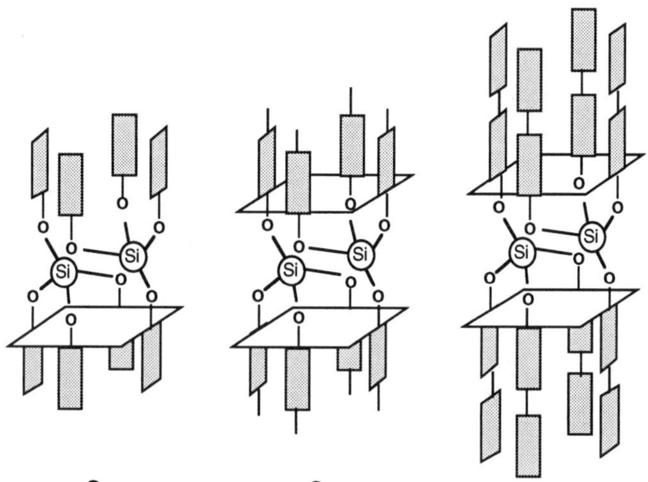

8 **9** **10**

Fig. 12 Schematic representation of koilands resulting from the fusion of two calix[4]arene **8**, *para*-methylcalix[4]arene **9** and *para*-phenylcalix[4]arene **10**

Design of linear assemblies based on coordination bonds :

Many binuclear complexes, based on ditopic receptors possessing convergent sets of binding sites (endoditopic receptors), have been reported. In these complexes intrinsic molecular proprieties such as magnetic coupling, redox activity and optical features may be tuned with remarkable precision. In principle, these molecular properties may be extended to polynuclear species. Although functionalised polymers (complexing polymers), bearing binding sites such as amino groups, carboxylate moieties and many other functional groups, capable of binding metal cations are well known, only in recent years much attention is focused on polymeric species (coordination polymers) in which the metal plays a structural role. This approach is interesting for two reasons. First of all, the metal, as a structural constituent, bridges other modules such as organic moieties and since, depending on its nature, the metal offers a large variety of coordination geometry (tetrahedral, octahedral, square planar etc.), one may reach polymeric species with special topology. Secondly, due to peculiar electronic, redox, photonic and magnetic properties of metals, the resulting coordination polymers may also exhibit pre-programmed or unexpected functional features. The preparation of such materials may be also envisaged through an iterative assembly of exoditopic ligands (exo-tectons) and metals. The design of the organic tecton and the choice of the interconnecting metal should allow the construction of materials with strictly controlled geometry and composition and thus properties.

One dimensional structures may be attained by the use of exo-ditopic tectons and tetra coordinated metals. As the organic module, one may envisage the tetracoordinating ligands in which the four coordination sites are either located at the summit of a pseudo-tetrahedron or occupying the four corner of a rectangle. Dealing with the metal, both the tetrahedron and square planar geometry of coordination may be used. A schematic representation of all four possibilities is given in the figure 13.

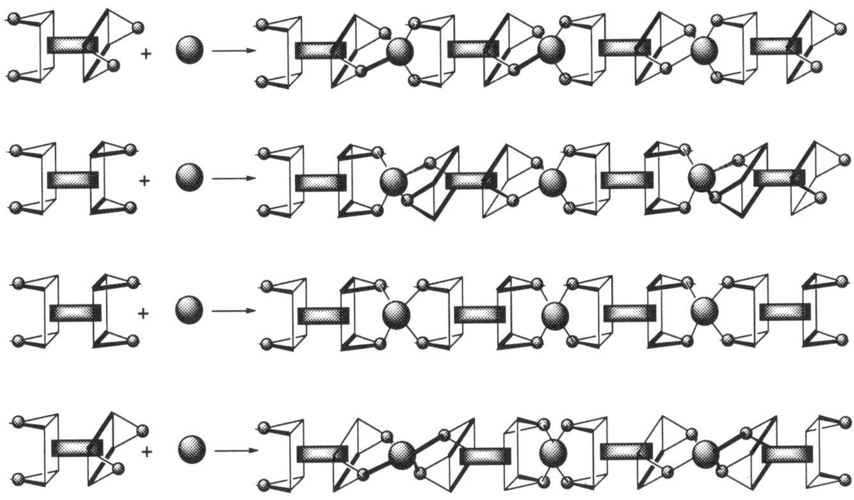

Fig. 13 Schematic representation of one dimensional coordination polymers resulting from an iterative assembling process engaging exo-ditopic tectons and tetra coordinated metals

The synthesis and binding ability of tetradentate exohomoditopic ligands based on bipyridine units have been reported (23, 24). As stated above, molecular units possessing four coordination sites pointing outwards and occupying the apices of a pseudo-tetrahedron may be of interest for construction of molecular wires. The design of such a ligand may be based on a preorganised backbone offering the possibility of anchoring four coordination sites in an alternating mode below and above the main plane of the backbone. In this regard, again *p-tert*-butylcalix[4]arene and paracyclophanes may be candidates of choice. In the case of calix[4]arene , unfortunately, due to strong intramolecular hydrogen bonding, the cone arrangement is the favoured conformation both in the solid state and in solution.

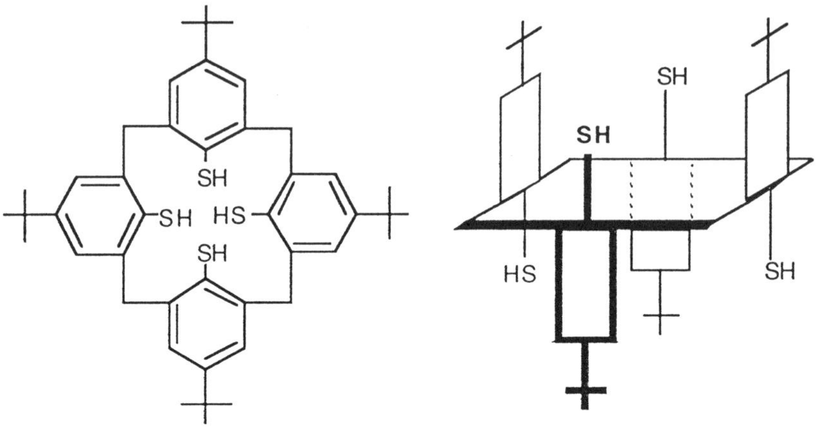

Fig. 14 Tetramercapto *p-tert*-butylcalix[4]arene **11** in a schematic representation of its 1,3-alternate conformation

One may predict that the disruption of the internal hydrogen bond pattern and the induction of steric hindrance should favour the 1,3-alternate conformation in calix[4]arene derivatives. The hydrogen bonding ability of thiol group in SH---H is roughly an order of magnitude less than the ability of alcohol to form OH---H hydrogen bond. Furthermore, the ionic radius of sulphur (1.85 Å) is substantially greater than that of oxygen (1.32 Å). Thus, replacement of all four OH groups in **1** by bulky SH moieties should lead to the tetramercapto *p-tert*-butylcalix[4]arene **11** in its 1,3-alternate conformation (Figure 14). Furthermore, due to its rich coordination chemistry, the thiolato group is an interesting coordination site for transition metals.

The synthesis of **11** in exceedingly poor yield but demonstrating that a non-cone conformation was involved was first reported by Gutsche et al (25). Our strategy (26) for the synthesis of **11** was based on the Newman-Kwart rearrangement (27) (Figure 15). This reaction is based on a thermal transformation of a *O*-phenyl-dimethythiocarbamate into a *S*-phenyl-dimethythiocarbamate. Subsequent to our success in finding an efficient synthesis of **11**, Gutsche et al reported their improved synthesis as a communication (28). The *p-tert*-butylcalix[4]arene **1** was first transformed into its tetra-*O*-phenyl-dimethylthiocarbamate **12** by reaction with dimethylthiocabamoyl chloride in the presence of NaH in diglyme. The thermal rearrangement of the tetra-*O*-phenyl-dimethylthiocarbamate **12** into the tetra-*S*-phenyl-dimethylthiocarbamate **13** was achieved in high yield under argon at 310-320 °C. Finally, the reduction of **13** by $LiAlH_4$ in dry THF afforded the desired compound **11**.

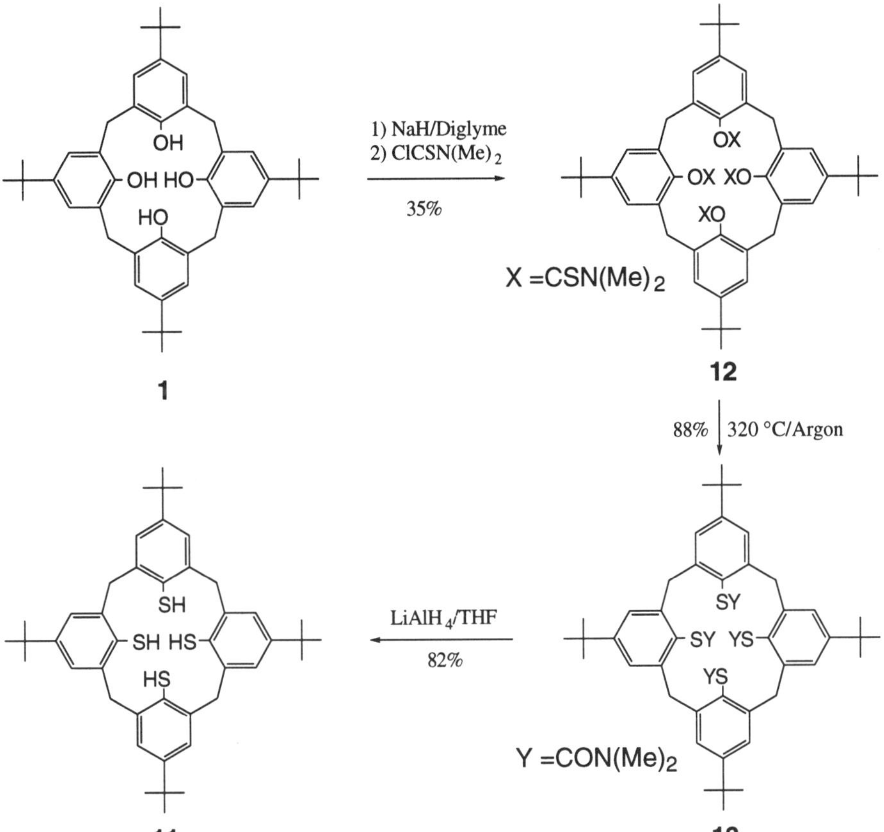

Fig. 15 Synthetic strategy employed for the synthesis of compound **11**

Although both ^{1}H and ^{13}C NMR data indicated **11** to have the 1,3-alternate conformation, its structure in the solid state was established by X-ray analysis (Figure 16). In the solid state, the four aromatic rings were located in a alternate fashion above and bellow the main plane defined by the four methylene groups. The two phenyl moieties disposed on the same side were almost parallel to each other, whereas the two sets of two aromatics were almost perpendicular. The average distance separating the two thiophenol derivatives appearing on the same side was *ca* 4.96Å.

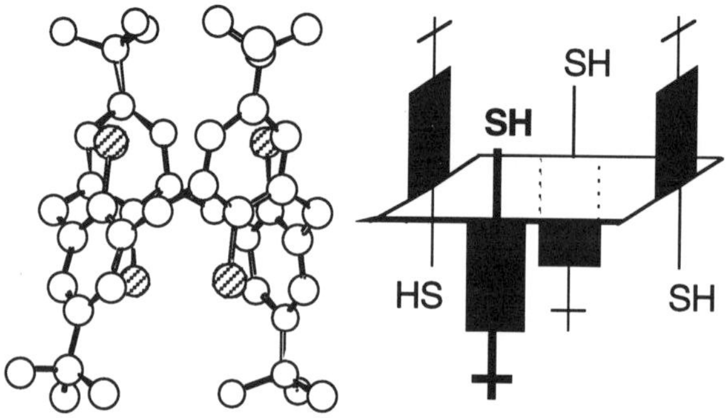

Fig. 16 X-ray structure (left) and schematic representation of compound **11** in its 1,3-alternate conformation. Sulphur atoms are presented by dashed circles

Compound **11**, as its parent *p-tert*-butylcalix[4]arene **1** may adopt the following four conformations: cone, partial cone, 1,2-alternate and 1,3-alternate (Figure 17).

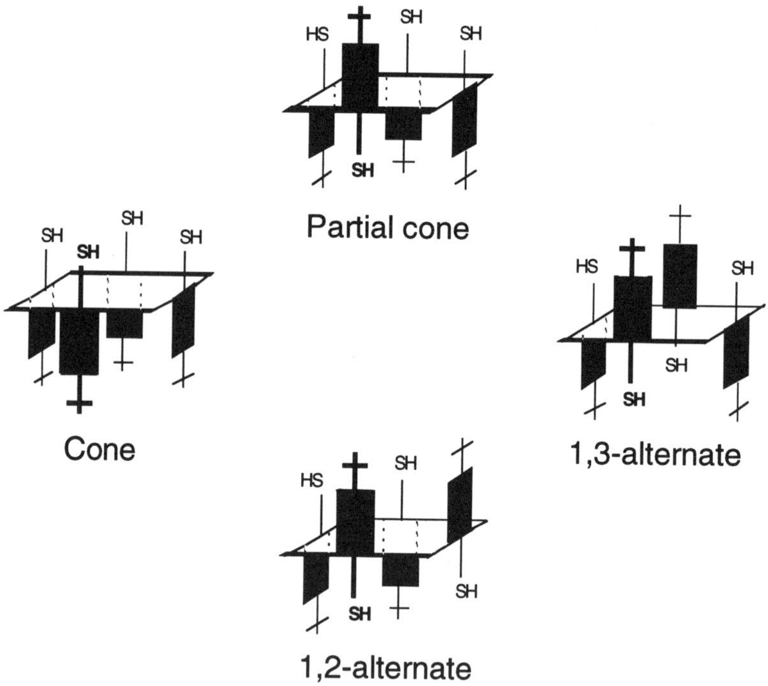

Fig. 17 Schematic representation of all four possible conformation of compound **11**

In order to compare the energy of all four conformers of **11**, they were modelled in the gas phase by molecular mechanics using both AMBER and Macromodel soft wares (29).

The results of the energy minimisation and its analysis in terms of components, obtained by AMBER, are given in figure 18.

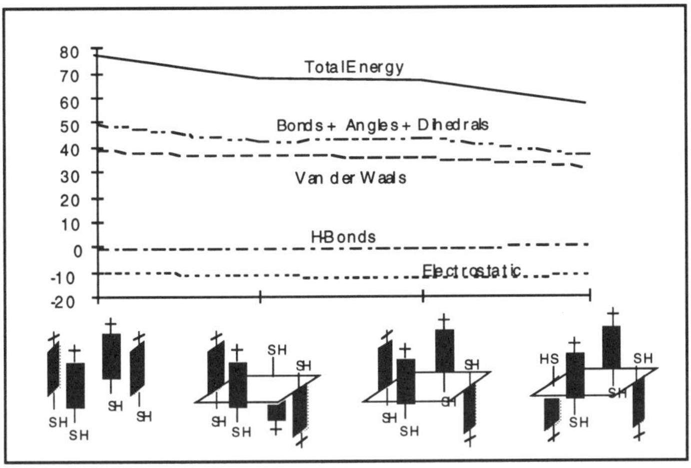

Fig. 18 Variation of total energy, Van der Waals energy component, and the sum of the bonds, angles and dihedrals energy components obtained by AMBER for compound **11**

In marked contrast with the results obtained for compound **1**, in the case of **11** the 1,3-alternate conformation was indeed found to be the most stable one.

In order to study the interconversion of the different conformations, a "simulated quenching", consisted of a single Molecular Dynamics (MD) simulation at high temperature (1000 K), from which structures were regularly extracted and their energy minimised (T = 0 K) was performed. The high kinetic energy at 1000 K, in principle, should allow to sample regions of the conformational space separated by high energy barriers. For both the cone and the 1,2-alternate conformations of **11**, "simulated quenching" was performed from a MD simulation at 1000 K. Every 5 ps a structure was extracted and minimised.

Starting from the cone conformation, a 500 ps MD simulation was performed, and 100 structures were extracted and minimised. In figure 19, the time evolution of the total energy for the minimised structures of **11** is shown. The analysis of the results showed that after 250 ps of MD simulation, the total energy decreased by *ca* 9 kcalmol^{-1} leading to the partial cone conformer. This interconversion resulted from the transannular passage of one of the SH groups. A second interconversion of the same type leading from the partial cone to the more stable 1,3-alternate conformation was observed after 20 ps.

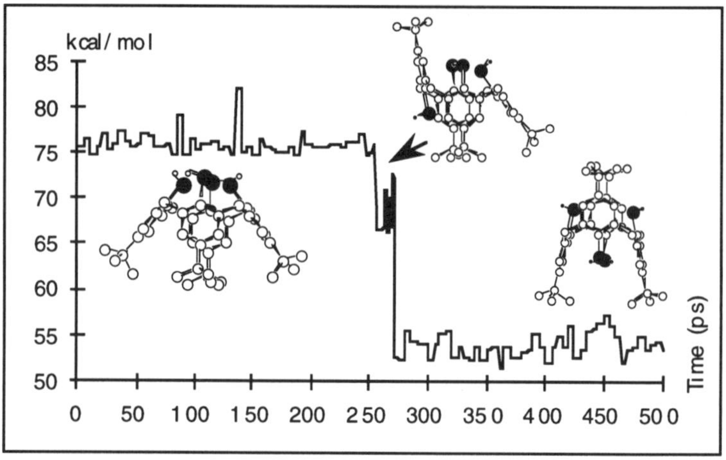

Fig. 19 Molecular Dynamics simulation of the interconversion of different conformers of compound **11** (see text)

Interestingly, starting from the 1,2 alternate conformation, by a 300 ps MD simulation 60 structures were extracted and minimised. As in the precedent case, two interconversions leading to the same lowest energy 1,3-alternate conformer were observed (Figure 20). After 130 ps, the first interconversion leading to the partial cone conformer was obtained. It is worth noting that the latter was also the intermediate conformer when starting from the cone conformation. As expected, the partial cone conformer was transformed into the stable 1,3-alternate conformer by a second interconversion after 40 ps and remained over a period of 130 ps.

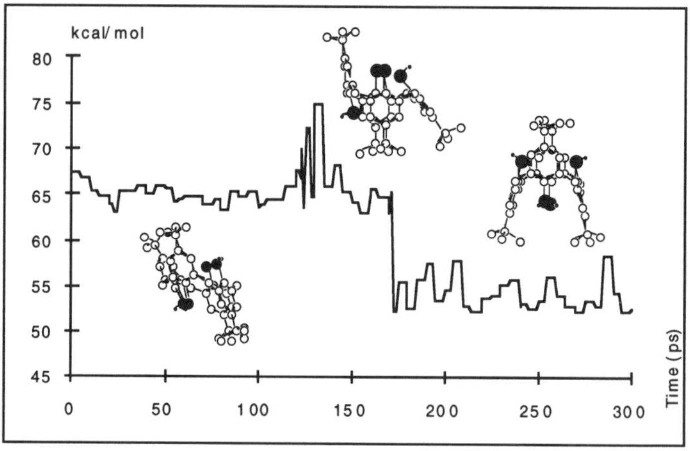

Fig. 20 Molecular Dynamics simulation of the interconversion of different conformers of compound **11** (see text)

Whereas in the case of tetramercaptocalixarene, the 1,3-alternate conformation was induced by replacing the OH groups by SH groups, one may also obtain the same type of molecule by modifying a precursor which adopts by construction the alternate disposition of binding sites. In this regard, metacyclophanes locked in 1,3-alternate conformation seem to be backbones of choice. The synthesis of **14** bearing 4 OH groups has been reported (30). Both in solution and in the solid state, the 1,3-alternated conformation of **14** (Figure 21) was established.

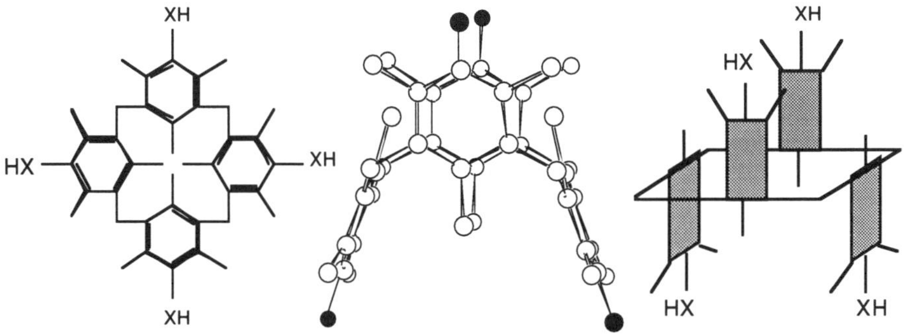

Fig. 21 The tetrahydroxy (X = O) **14**, and the tetramercapto (X = S) **15** paracyclophane (left), a schematic representation of their 1,3-alternate conformation (right) and the X-ray analysis for **14** (middle)

The synthesis of **15**, the tetrathio analogue of **14** was achieved (Figure 22). Starting from the preorganised metacyclophane **14**, a similar sequence as the one described above for the synthesis of the tetramercaptocalixarene **11** (Figure 15) affords the desired compound **15** in high yield (31).

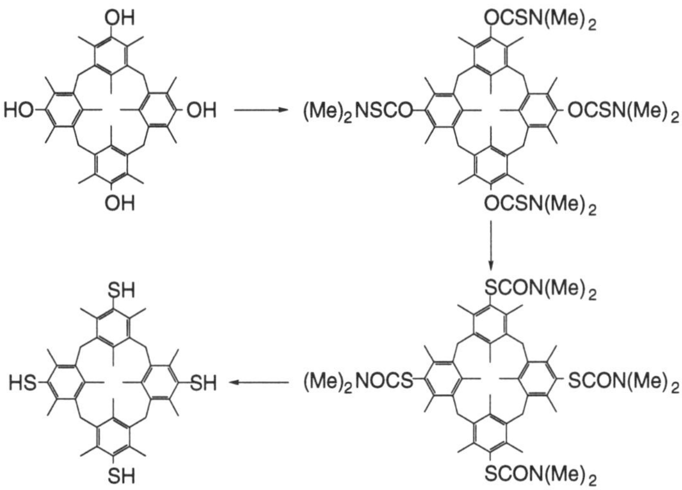

Fig. 22 Strategy for the synthesis of **15**

The tetrathiocalixarene **11** showed above presents several interesting binding features: due to its 1,3-alternate conformation, each side of the ligand is composed of two SH groups as "soft" and strong binding sites and two aromatic rings which may participate in the binding processes as "soft" and probably weak interaction sites. The thiophilicity of Hg(II) leads to the binding of two mercury ions by this ligand (23). In the binuclear complex, both Hg ions are indeed coordinated to two thiolate moieties with an average Hg-S distance of *ca* 2.34 Å and an average S-Hg-S angle of *ca* 178.5° (Figure 23). One essentially linear S-Hg-S unit projects upon the other at an angle of 89.3°, with the Hg-Hg distance being 3.66 Å. The average distance between two S atoms in the S-Hg-S units (4.59 Å) is shorter than that in the free ligand (4.96 Å), showing that the ligand must bow by 0.38Å in order to accommodate the metal ions. Linear two-coordination at Hg in $Hg(SR)_2$ compounds has been established, although, weaker further interactions are not excluded and the coordination chemistry of Hg(II) shows many aspects which may be rationalised in terms of a "2+4" model. In the binuclear complex the average distance between Hg ions and the centroids of the aromatic rings located on the same side is *ca* 3.1Å. Thus, due to the 1,3-alternate conformation of the ligand, it appears possible that a mercury(II) species may have its secondary coordination requirements satisfied by polyhapto (π) interactions with two of the phenyl rings.

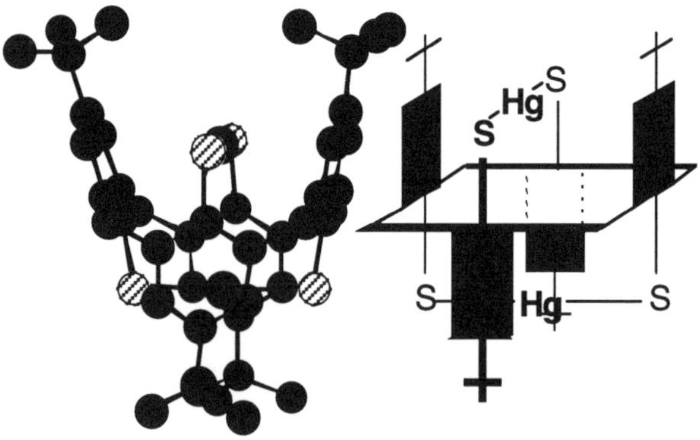

Fig. 23 X-ray structure of the mercury binuclear complex obtained with compound **11** (left) and its schematic representation (right)

Dealing with tetradentate exo-ditopic ligands based on calixarene derivatives in their 1,3-alternate conformation, differentiation of the two sides by introduction of two different sets of binding sites may be interesting for the formation of heterobinuclear complexes. We propose the substitution of two hydroxy groups in 1,3 positions by two mercapto groups on *p-tert*-butylcalix[4]arene **1**. The mixed ligand **16** possessing both "hard" OH and "soft" SH groups was synthesised (32).

The synthesis of **16** has been attempted previously (33). Our approach was inspired from the above mentioned publication (Figure 24). Compound **1** was first transformed into **17** by reaction with $(Me)_2NCSCl$ in the presence of NaH in DMF at 25 °C. Thermogravimetric analysis revealed that **17** decomposed above 260 °C, consequently, the thermal rearrangement of the latter into the di-*S*-phenyl-dimethylthiocarbamate derivative could not be achieved. In order to avoid the decomposition, the two hydroxy groups on **17** were protected by either methyl **18** or by benzyl groups . The thermal rearrangement of **18** to **19** was achieved by heating the solid at 360 °C under argon for 20-30 minutes. The reduction of **19** by $LiAlH_4$ in dry THF afforded compound **20**. Finally, the deprotection of **20** by BBr_3 in CH_2Cl_2 gave the desired compound **16**.

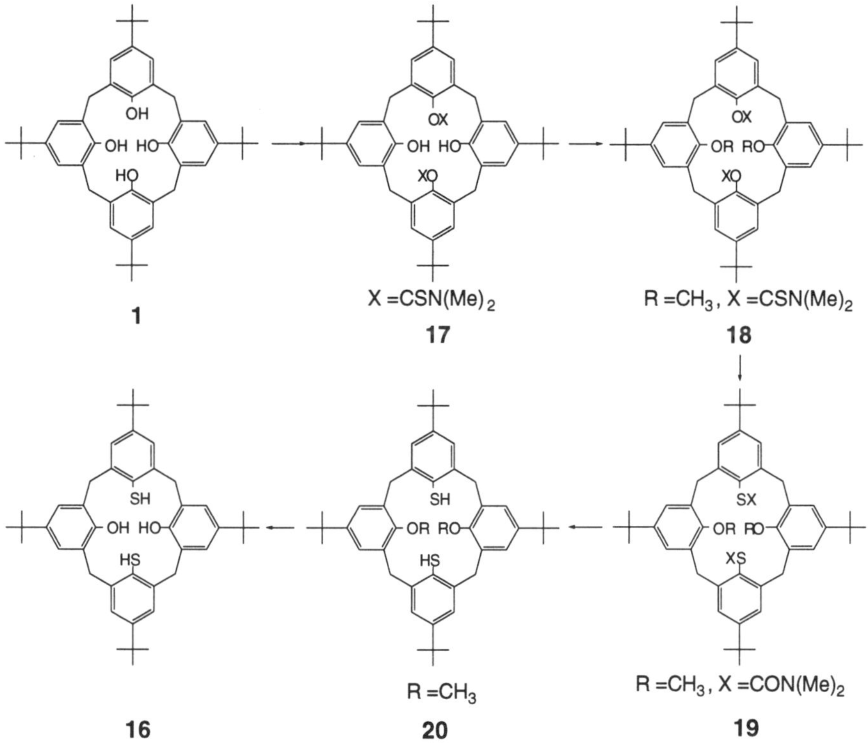

Fig; 24 Synthetic strategy for the preparation of the mixed compound **16**

For compound **16**, its ^1H-NMR spectrum in both $CDCl_3$ and in $C_2D_2Cl_4$ was composed of rather sharp signals at 25 °C demonstrating the presence of a unique conformer and a slow rate of interconversion. NOESY-NMR experiments revealed unambiguously the latter to be in a cone conformation. For the parent compound **1**, it has been proposed that the cone conformation was stabilised by intramolecular H-bonds. In the case of **16** the same explanation may also hold. In favour of this was the observation of a mixture of conformers at room temperature for **20** in which the two hydroxy groups were replaced by two methoxy groups, thus preventing the possible formation of OH···S hydrogen bonds.

Whereas for both compounds **1** and **11** only four different conformers are possible, for compound **16** five conformations may be envisaged (Figure 25).

The conformational mobility of **16** was studied over a range of 100 °C by variable ^{1}H-NMR in $C_2D_2Cl_4$. Heating to 130 °C caused coalescence of signals corresponding to the H_A and H_B protons of the CH_2 groups to a singlet. The coalescence occurred at *ca* 67 °C (at 200 MHz) with a rate constant of interconversion k_c of *ca* 350 s^{-1}, and a free energy of activation $\Delta G^{\#}$ of *ca* 16 Kcal/mole. For comparison purposes, the following characteristics were reported for **1** in $CDCl_3$: T_c=52 °C, $\Delta G^{\#}$=15.7 Kcal/mole, and k_c=150 s^{-1}. Although, for **1** a unique process was postulated for the cone inversion process leading to equivalence of the two protons, one can not assume an exchange process involving only two states for **16**. Indeed, above the coalescence temperature, the δ value of 4.10 ppm obtained for the singlet did not correspond to the average value of 3.99 ppm $((\delta H_A+\delta H_B)/2)$ for H_A and H_B protons.

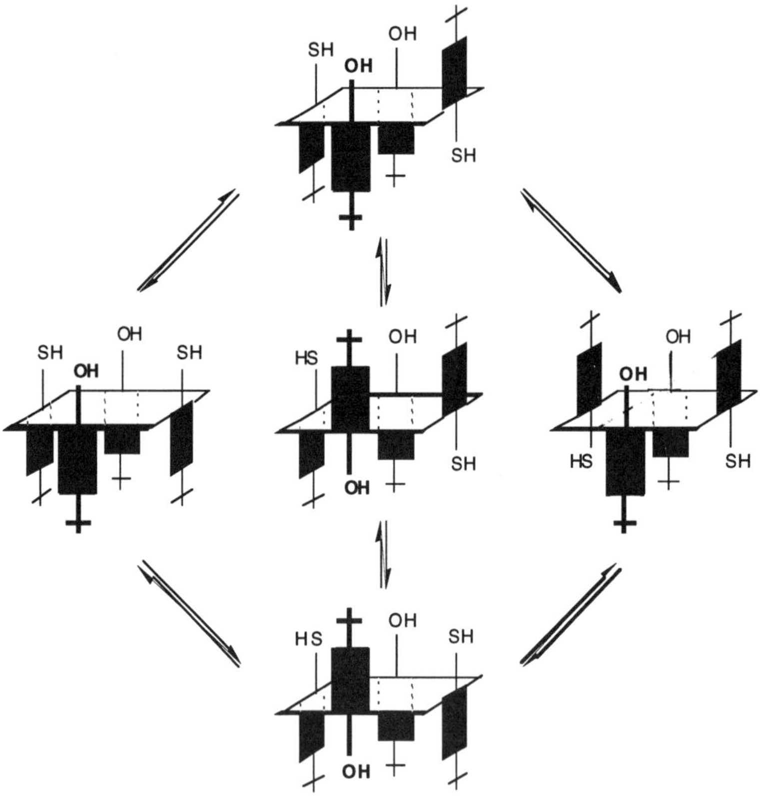

Fig. 25 Schematic representation of the five possible conformers for **16**

The structure of **16** was studied by X-ray analysis (Figure 26). Compound **16** also adopts a cone conformation in the solid state presenting the following features: i) the O1-O2 and S1-S2 distances are 3.469 Å and 5.317 Å respectively, ii) the average distance between O and S atoms is *ca* 3.29 Å, iii) with respect to a mean plane defined by all four CH_2 groups, the average angles between planes containing the aromatic moieties bearing SH or OH groups are *ca* 93° and 137° respectively showing that, whereas the thiophenol rings are almost perpendicular to the mean plane described above, the two phenol rings are considerably tilted towards the exterior of the calix, iv) the average O-H and S-H distances, using isotropically refined hydrogen atoms, are *ca* 1.03 Å and 1.29 Å respectively, v) the $d_{SH \cdots O}$ and $d_{OH \cdots S}$ distances and angles are O1H44···S1=2.374Å, 159.73°; S1H45···O2=2.203Å, 132.04°; O2H46···S2=2.249Å, 158.49°; S2H47···O1=2.200 Å, 145.87 °. For **16**, although the OH and SH hydrogen atoms are properly oriented, the average distance between O and S atoms of *ca* 3.29 Å and the observed angles indicate that if ever H-bonds exist, they should be rather weak.

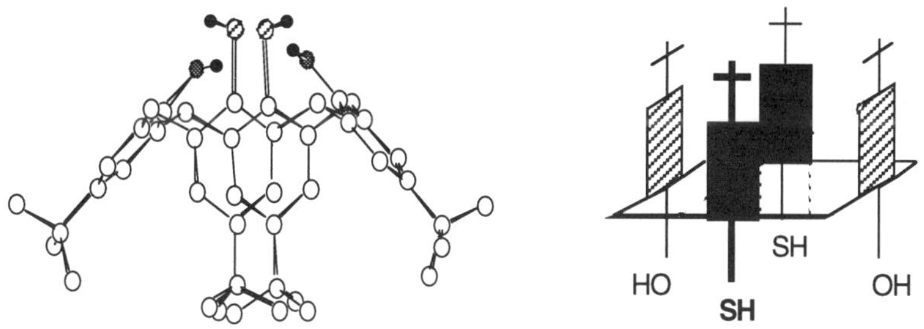

Fig. 26 X-ray structure of **16** (left) and its schematic representation (right)

Compound **16** possesses both "hard" OH groups and "soft" SH binding sites. Furthermore, as in the case of **11**, for its 1,3-alternate conformation, within the side bearing the two SH groups which may act as "soft" and strong binding sites, two aromatic phenolic rings are present as well and may also participate in the binding processes as "soft" and probably weak interaction sites. Compound **16** forms a mononuclear complex with mercury. The preferential binding of mercury at sulphur in **16** may be expected due to the thiophilicity of Hg(II). The structure of **16**-Hg complex was investigated by X-ray crystallography. This study revealed the following features (Figure 27): i) **16** adopts a 1,3-alternate conformation in the solid state in which the mercury ion is coordinated to two sulphur atoms, ii) in the mononuclear **16**-Hg complex, the average Hg-S distance of *ca* 2.35 Å and S-Hg-S angle of *ca* 176.0° are close to the values observed for the binuclear **11**-Hg_2 complex previously cited, iii) the distance between two S atoms in the S-Hg-S unit (4.710 Å) is shorter than that in the free ligand (5.317 Å), showing that the ligand must bow by *ca* 0.60 Å in order to accommodate the metal ion. Consequently, with respect to a mean plane defined by all four CH_2 groups, the average angles between planes containing the aromatic moieties bearing

SH or OH groups are *ca* 99.3° and 101.2° respectively. This shows that, both thiophenol rings and phenol rings are almost perpendicular to the this mean plane. Upon complexation of one mercury ion the ligand undergoes conformational changes and adopts a 1,3-alternate conformation in the solid state (Figure 27). In this complex, again, as in the case of the binuclear mercury complex with **11** mentioned above, the distance of *ca* 3.05 Å between Hg ions and the centroids of the aromatic rings located on the same side is observed. Although, in the case of **11**, one could argue that this was structurally imposed by the ligand, for the mixed ligand **16** mercury complex the fact that the ligand undergoes a conformational change in the complex is a clear cut argument in the favour of the "2+4" model.

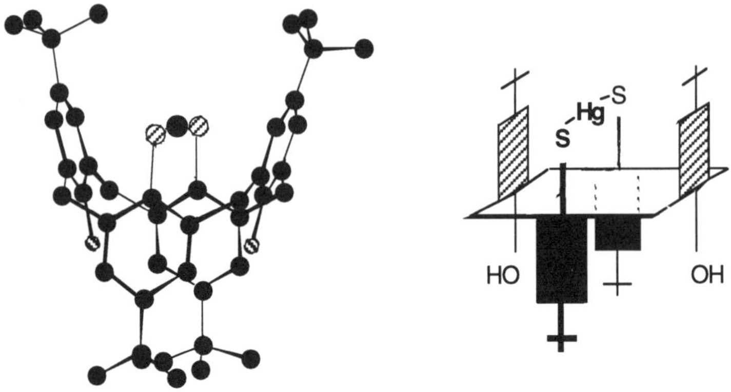

Fig. 27 X-ray structure of **16**-Hg complex (left) and its schematic representation (right)

Although we have so far only succeeded in preparing discrete bi- and mono-nuclear complexes with compounds **11** and **16** respectively, nevertheless these two exo-ligands were designed and prepared with the aim of forming metalloorganic one-dimensional solids (Figure 28). It is worth noting that in such linear arrays, the metal should be well encapsulated, and thus hidden, by four aromatic moieties.

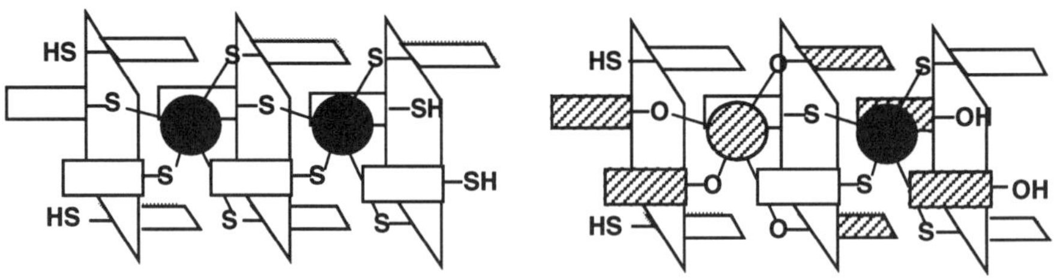

Fig. 28 Schematic representation of one-dimensional coordination polymers with **11** (left) and with **16** (right)

Furthermore, whereas in the case of **11** one may obtain homopolynuclear complexes engaging one type of tetrahedrally coordinated metal, for compound **16** heterobinuclear infinite complexes may be expected. Indeed, in the latter case, one may play with thiophile and oxophile metals (Figure 28). The preparation of the coordination polymers mentioned above is currently under progress.

References :

(1) Lindsey, J. S. *New. J. Chem.* **1991**, *15*, 153-180

(2) Simard, M. , Su, D. , Wuest, J.D. *J. Amer. Chem. Soc.*, **1991**, *113*, 4696.

(3) Mann, S. *Nature*, **1993**, *365*, 499.

(4) Lehn, J.-M. *Angew. Chem. Int. Ed. Engl.* , **1990**, *29*, 1304, Lehn, J.-M. in *Supramolecular Chemistry, Concepts and Perspectives,* VCH, Weinheim, 1995.

(5) Whitesides, G. M. , Mathias, J. P. , Seto, C. T. *Science*, **1991**, *254*, 1312

(6) Lehn, J.-M. , *Struc. Bonding.* , **1973**, *16*, 1; *ibid*, *Pure Appl. Chem.* **1978**, *50*, 871.

(7) M. W. Hosseini, R. Ruppert, P. Schaeffer, A. De Cian, N. Kyritsakas, J. Fischer, *J. Chem. Soc. Chem. Commun.* , **1994**, 2135; G. Brand, M. W. Hosseini, R. Ruppert, A. De Cian, J. Fischer, N. Kyritsakas, *New J. Chem.* , **1995**, *19*, 9.

(8) Lehn, J.-M. , *Leçon Inaugurale*, Collège de France, Paris, 1980.

(9) X. Delaigue, M. W. Hosseini, A. De Cian, J. Fischer, E. Leize, S. Kieffer, A. Van Dorsselaer, *Tetrahedron Lett.* , **1993**, *34*, 3285.

(10) X. Delaigue, M. W. Hosseini, E. Leize, S. Kieffer, A. Van Dorsselaer, *Tetrahedron Lett.* , **1993**, *34*, 7561; X. Delaigue, M. W. Hosseini, R. Graff, J.-P. Kintzinger, J. Raya, *Tetrahedron Lett.* **1994**, *35*, 1711.

(11) C.D. Gutsche,*"Calixarenes"*, Monographs in Supramolecular Chemistry, Ed. J.F. Stoddart, R.S.C., London, 1989; *Calixarenes. A Versatile Class of Macrocyclic Compounds,* Ed. J. Vicens and V. Böhmer, Kluwer Academic Publishers, Dordrecht, 1991; Böhmer, V. , *Angew. Chem. Int. Ed. Engl.*, **1995**, *34*, 713.

(12) Kämmerer, H.; Happel, G. ; Caesar, F. *Makromol. Chem.*, **1975**, *176*, 3317.

(13) Gutsche, C. D. ; Bauer, L. J. *J. Amer. Chem. Soc.*, **1985**, *107*, 6052.

(14) Gutsche, C. D. ; Dhawan, B. ; No, K. H. ; Muthukrishnan, R. *J. Amer. Chem. Soc.*, **1981**, *103* , 3782.

(15) Andreetti, G. D. ; Ungaro, R. ; Pochini, A. *J. Chem. Soc., Chem. Commun.,* **1979**, 1005.

(16) Ungaro, R. ; Pochini, A. ; Andreetti, G. D. ; Domiano, P. *J. Chem. Soc. , Perkin Trans. II*, **1985**, 197.

(17) Coruzzi, M. ; Andreetti, G. D. ; Bocchi, V. ; Pochini, A. ; Ungaro, R. *J. Chem. Soc. , Perkin Trans. II*, **1982**, 1133.

(18) Harrowfield, J. M. ; Ogden, M. I. ; Richmond, W. R. ; White, A. H. *J. Chem. Soc., Chem. Commun.,* **1991**, 1159.

(19) Olmstead, M. M. , Sigel, G. , Hope H. ; Xu, X. ; Power, P. P. *J. Amer. Chem. Soc.,* **1985**, *107,* 8087.

(20) Corazza, F. ; Floriani, C. ; Chiesti-Villa, A. ; Guastini, C. *J. Chem.*

Soc., Chem. Commun., **1990**, 1083.
(21) Atwood, J. L. ; Bott, S. G. ; Jones, C. ; Raston, C. L. *J. Chem. Soc., Chem.Commun.,* **1992**, 1349.
(22) F. Hajek, E . Graf, M. W. Hosseini, *unpublished.*
(23) C. Kaes, M. W. Hosseini, R. Ruppert, A. De Cian, J. Fischer, *Tetrahedron Lett.* , **1994**, *35*, 7233.
(24) C. Kaes, M. W. Hosseini, R. Ruppert, A. De Cian, J. Fischer, *J. C. S. Chem. Commun.* , **1995**, 1445.
(25) C. D. Gutsche, J.S. Rogers, D. Stewart, K. See, *Pure Appl. Chem.*, **1990**, *62*, 485.
(26) X. Delaigue, J. McB. Harrowfield, M. W. Hosseini, A. De Cian, J.Fischer, N. Kyritsakas, *J. C. S. Chem. Comm.* **1994**, 1579.
(27) M. S. Newman, H. A. Karnes, *J. Org. Chem.* , **1966**, *31*, 3980.
(28) C. G. Gibbs, C. D. Gutsche, *J. Amer. Chem. Soc.*, **1993**, *115*, 5338.
(29) J. M. Planeix, M. W. Hosseini, P. Guilbaud, G. Wipff, *unpublished.*
(30) Pappalardo, S. ; Ferguson, G. ; Gallagher, J. F. , *J. Org. Chem.* , **1992**, *57*, 7102.
(31) X. Delaigue, M. W. Hosseini, *Tetrahedron Lett.* **1993**, *34*, 8111.
(32) X. Delaigue, M. W. Hosseini, N. Kyritsakas, A. De Cian, J. Fischer, *J. Chem. Soc. Chem. Comm.* **1995**, 609.
(33) Y. Ting, W. Verboom, L. G. Groenen, J.-D. van Loon, D. N. Reinhoudt, *J. Chem. Soc. , Chem.Commun.,* **1990**, 1432.

SYMMETRY IN SPHEROALCANES, FULLERENES, TUBULES AND OTHER COLUMN-LIKE AGGREGATES

ANDRÉ RASSAT
Département de Chimie, Ecole Normale Supérieure (URA/CNRS 1679)
24, rue Lhomond, F-75231 Paris Cedex 05, France

Abstract : The point-group symmetry of spheroalcanes (ie $(CH)_N$ molecules, N even, of which the planar formula is a *simple, trivalent, planar graph*) (A.T. Balaban, in A.T. Balaban (Ed), *Chemical Application of Graph Theory*, Academic Press, Londres, 1976, p. 64) and other fullerene-like molecules is described (A. Rassat *et al., J. Chim. Phys.*, **91**, 1683-1710, 1994). Using the same graphical symbols, the well-known (L. Fejes Tϕth, *Regular Figures*, Pergamon Press, 1964) symmetry of friezes is presented. Because of the recent interest in column-like aggregates [tubules (S. Ijima, *Nature*, 1991, **354**, 56), peptide nanotube (M. Reza Ghadiri *et al., Angew. Chem. Int. Ed. English*, 1995, **34**, 93), "artificial nanotubes" (A. Ikeda, S. Shinkai, *J. Chem. Soc., Chem. Comm.* 1994, 2375) or polymeric one-dimensional aggregates (J. Perlstein, *J. Am. Chem. Soc.*, 1994, **116**, 455 ; R.P. Scaringe, S. Perez, *J. Phys. Chem.*, 1987, **91**, 2394)] the symmetry of infinite columns or rods (A.V. Shubnikov, V.A. Koptsik, *Symmetry in Science and Art*, Plenum Press, 1974) is presented and discussed in relationship with the symmetry of friezes and of some plane patterns : the 17 types of symmetry of infinite columns will be derived from the 7 axial point-groups or from the 9 two-dimensional patterns with rotation axes of order $n \leq 2$.

1. Introduction

There has been some recent interest in symmetric molecular and supramolecular aggregates such as fullerenes (1) and spheroidal molecules (2) or carbon nanotubes (3) and other tubular assemblies (4-16) (Figure 1, 2).

In this talk, I would like to review the symmetry of these chemical objects and of related geometric ones. In a first part, I shall present some symmetrical molecules and discuss their properties in relationships with those of rings and of friezes (17, 18a) or bands (18b). In a second part, I shall examine the symmetry properties of tubular aggregates, ie of infinite rods or ornamented cylinders of revolution.

G. Tsoucaris et al. (eds.), Crystallography of Supramolecular Compounds, 181–201.

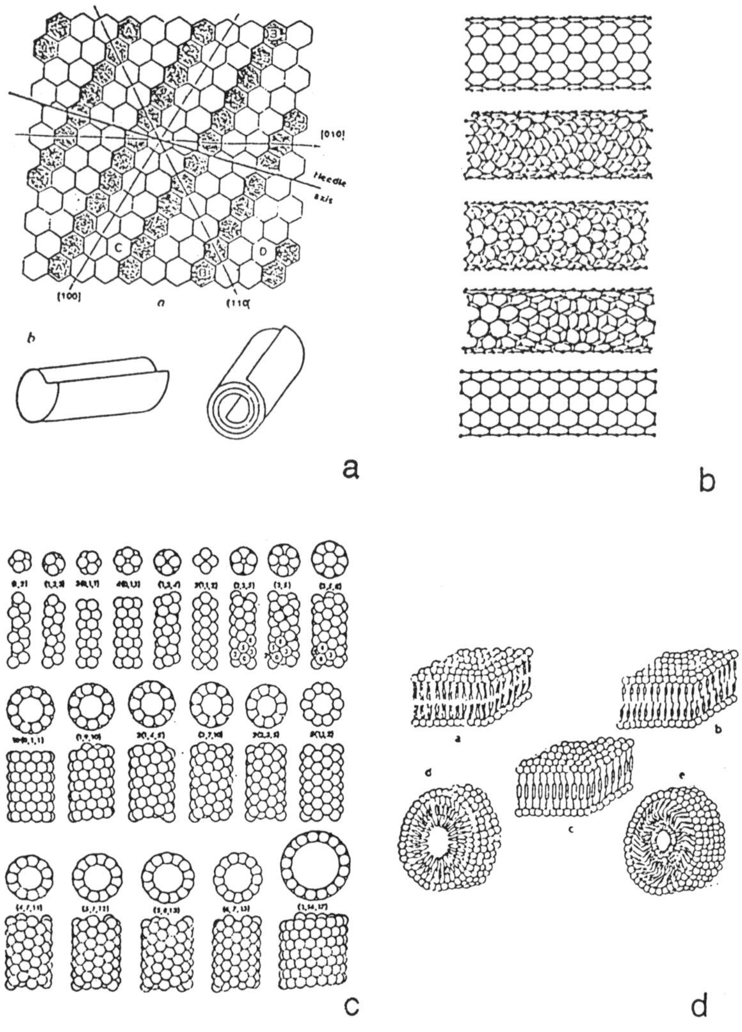

Figure 1 : Some examples of tubular assemblies : a) Ref 3 ; b) Ref 12 ; c) Ref 68 ; d) Ref 11. (Reproduced with permission).

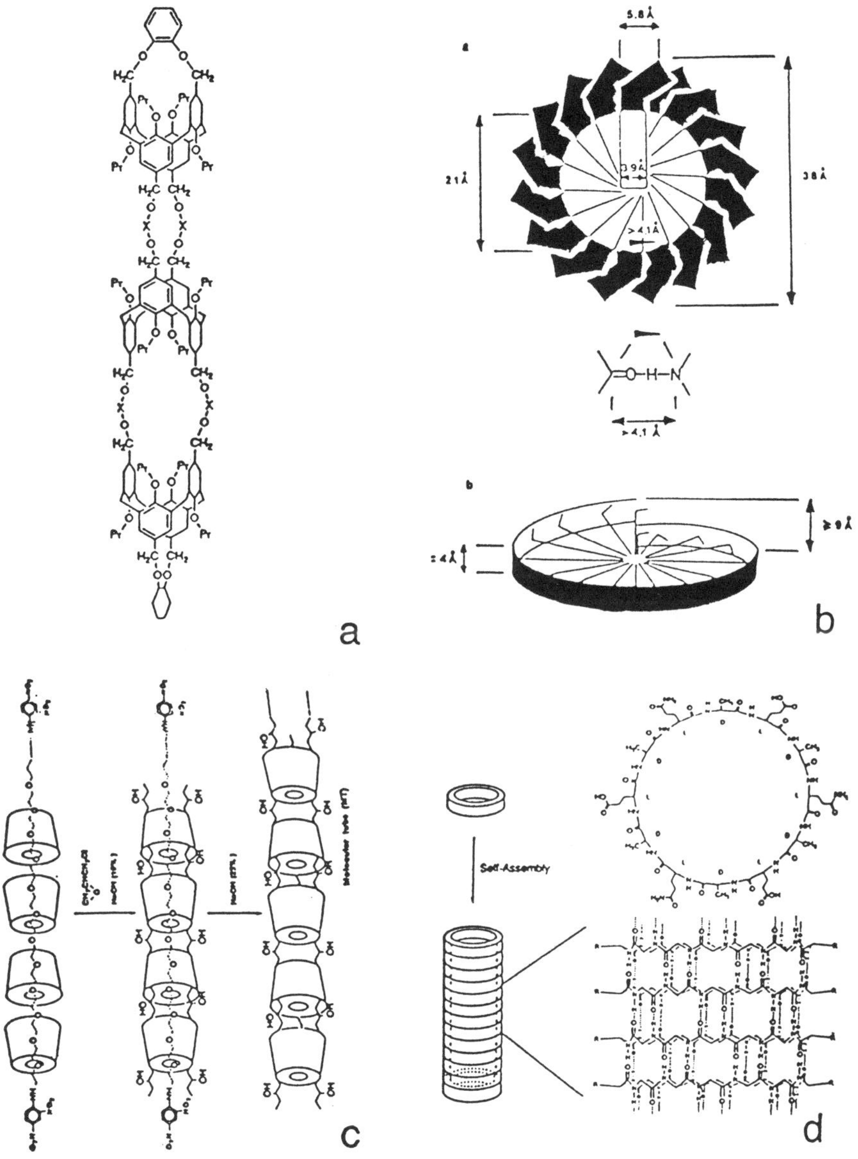

Figure 2 : Other examples of tubular assemblies : a) Ref 16 ; b) Ref 7a ; c) Ref 8 ; d) Ref 5c. (Reproduced with permission).

2 - Fullerenes and spheroalcanes

C_{60}-Fullerene (Figure 3a) is the archetype of symmetrical all-carbon cage molecules (1). According to Kroto (19) fullerenes are "*all closed carbon cages composed of twelve 5-membered and an unrestricted number of 6-membered rings*".

Various fullerenes have been isolated (20) (Figure 3b). Spheroidal all-carbon

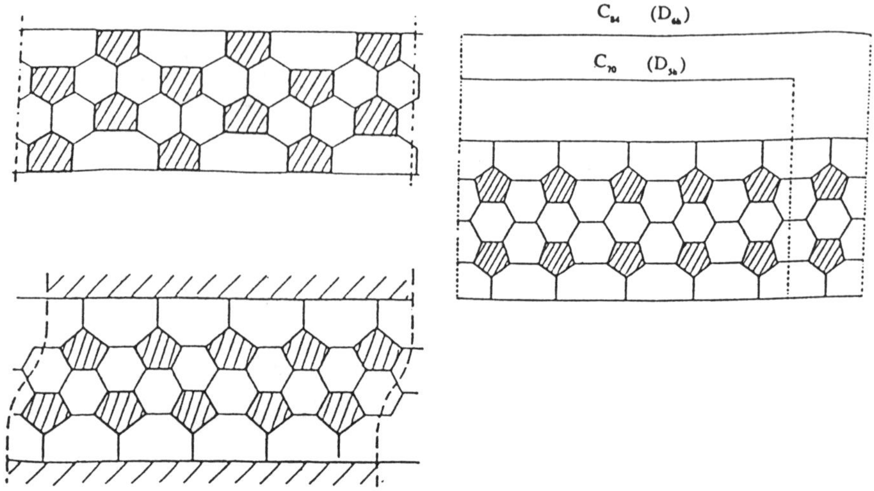

Figure 3 : Frieze-style representation of fullerenes : a) Mercator projections of C_{60} in $\mathbf{D_{3d}}$ and $\mathbf{D_{5d}}$ symmetries (subgroups of $\mathbf{I_h}$). b) Mercator projections of $\mathbf{D_{5h}}$ C_{70} and of the $\mathbf{D_{6h}}$ isomer of C_{84}.

molecules (including other fullerenes) have been studied theoretically (2). It is convenient to use the name *spheroarene* (21) for molecules made of carbon atoms only and such that their Schlegel diagram (22) is a *simple trivalent planar graph* (23a) : a "*trivalent*" or "*cubic*" graph is such that every point (C atom) is 3-coordinated ; it is "*simple*" if only this type of coordination occurs. ; a "*planar*" graph can be drawn in the plane so that no two lines cross each other. An enumeration of some simple trivalent planar graphs can be found in papers by Balaban (23). Among these graphs, type-3 graphs (24) (such that their graph cannot be separated into two distincts parts without "cutting" three bonds at least) may be considered as Schlegel diagrams of polyhedrons having only 3-coordinated vertices. For those polyhedrons, a modified Euler relationship exists (22a) between the number f_p of p-faces :

$$3f_3 + 2f_4 + f_5 = 12 + \sum_{p\geq 7}(p-6)f_p, \quad \forall f_p$$

(Fullerenes are those spheroarenes for which $f_5 = 12$ and $f_p = 0$ if $5 \neq p \neq 6$.) Similarly, the corresponding saturated hydrocarbons may be called *spheroalcanes* (21). "Normal" spheroalcanes are those where the hydrogens are "exo" (i.e the distance OC_i between the centre of gravity 0 of the spheroalcane and one carbon atom C_i is smaller than the distance OH_i between 0 and the hydrogen atom H_i to which C_i is bound).

We shall only deal with spheroarenes, but, except where mentionned specifically, remarks and conclusions are valid for normal spheroalcanes.

As any molecule, a given spheroarene may have different conformations, and these conformations may be of different symmetry. We shall always deal with their *topological symmetry* (2d), ie that of the most symmetrical conformation. (It is also the symmetry of their graph, ie the 3D-symmetry of their adjacency matrix (Hückel matrix (25) of the corresponding spheroarene). Of course, the most stable conformation of a molecule may not be the one of highest symmetry.

We can now examine some ways to design spheroarenes of all possible symmetries (24).

2.1. AXIAL POINT-GROUPS :

There is a one-to-one correspondance between the seven frieze-groups (17) and the seven axial point-groups :when a fragment (of n identical motifs) of an infinite frieze is wrapped arround a cylinder of revolution so that one C_n axis is conserved, a ring belonging to one of the seven axial point-group is obtained. (The C_n symmetry axis is taken as "vertical"). In view of a later use, we shall admit that these rings have a finite thickness *d* and that the horizontal mirror plane of the cylinder is the corresponding symmetry-element in the $\mathbf{C}_{nh}$, $\mathbf{S}_{2n}$ and $\mathbf{D}_{nh}$ groups. It may be of interest to notice that the *continuous* groups, $\mathbf{C}_{\infty}, \mathbf{C}_{\infty v}, \mathbf{C}_{\infty h}, \mathbf{D}_{\infty}, \mathbf{D}_{\infty h}$ may be considered (18c, 26a) as "limiting cases" of $\mathbf{C}_n, \mathbf{C}_{nv}, \mathbf{C}_{nh}, \mathbf{D}_n, \mathbf{D}_{nh}$ respectively, (the limiting case of $\mathbf{S}_{2n}$ and $\mathbf{D}_{nd}$ (18c) being $\mathbf{C}_{\infty h}$ and $\mathbf{D}_{\infty h}$) For Haln and Klapper (26b) "*the axes of order infinity... do not correspond to cyclic groups (as do the axes of finite order), because there is no smallest rotation from which all other rotations can be derived as higher powers, i.e. by successive application*" For Altman (26c), "*if the rotation angle can take any value, as for instance in the rotation around the axis of a cylinder, the group is a* continuous *one, i e., a group of infinite order the elements of which can be numbered with a continuous parameter*". In other words, in an axial group with a C_n axis, n is a integer and as it goes to infinity the cardinal is $\aleph_0$, while for continuous groups the cardinal is $\aleph_1$. If one considers a ring of radius R with n identical motifs separated by a $(2\pi R \approx na)$, there are two ways to obtain limiting cases :

1) R being constant, as $n \to \infty$, $a \to 0$ and the continuous axial point groups may be taken as limiting cases.

2) a being constant, as $n \to \infty, R \to \infty$ and the seven freize-groups are the limiting cases of the seven axial point-groups.

Figure 4 gives a representation of the seven frieze-symmetry types using Γ and Δ (Γ only and its transforms : $\hat{\sigma}_v\Gamma = \overline{\Gamma}, \hat{\sigma}_h\Gamma = L, \hat{C}_2'\Gamma = \overline{L}$, could be used but we found convenient to take $\Delta = \{\Gamma\overline{\Gamma}\}$ and $\Gamma_2 = \{\Gamma / \overline{L}\}$ as well as $\nabla = \hat{\sigma}_h\Delta$ and $\Delta_2 = \{\nabla / \Delta\}$). Other representations using commas (17), Greek keys (17b) or more artistic symbols (17, 18a) have also been used (Figure 5). The seven motifs can be named in the following way : $\mathbf{C}_n{:}\Gamma; \mathbf{C}_{nh}{:}\Gamma / L; \mathbf{C}_{nv}{:}\Delta; \mathbf{S}_{2n}{:}\Gamma, L; \mathbf{D}_n{:}\Gamma_2; \mathbf{D}_{nd}{:}\nabla, \Delta$ and $\mathbf{D}_{nh}{:}\Delta_2$. Notice that by setting only one elementary motif onto the cylinder we obtain rings of $\mathbf{C}_1$, $\mathbf{C}_{1h}$ and $\mathbf{S}_2$ symmetry but also of "$\mathbf{D}_1$", "$\mathbf{C}_{1v}$," "$\mathbf{D}_{1d}$" and "$\mathbf{D}_{1h}$", although these last rings

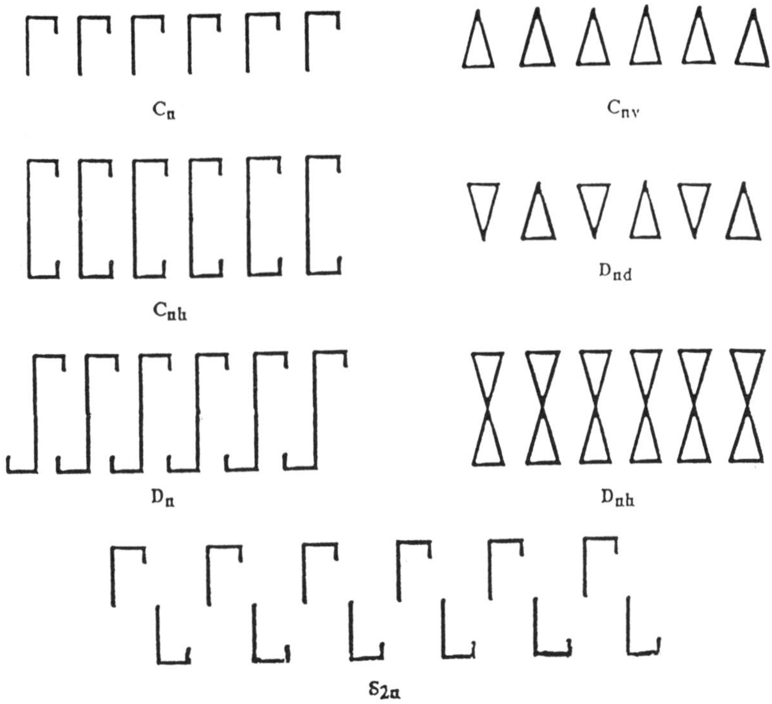

Figure 4 : The seven freeze-groups and their corresponding point-groups.

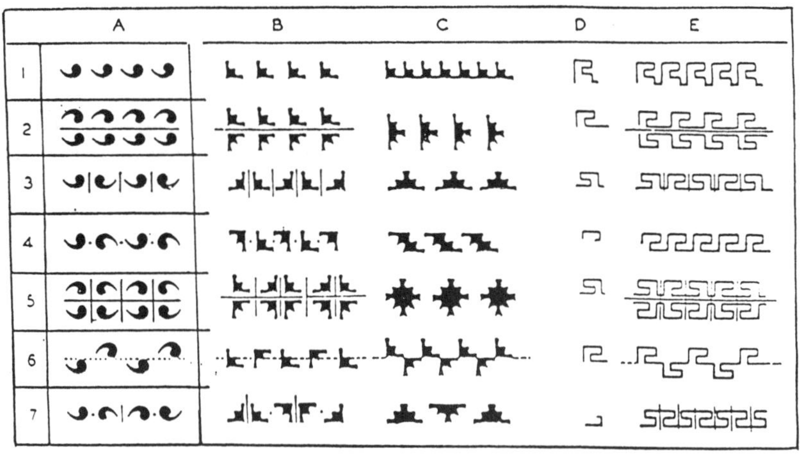

Figure 5 : Some representation of the seven freize-groups. (Reproduced from ref 17b, with permission).

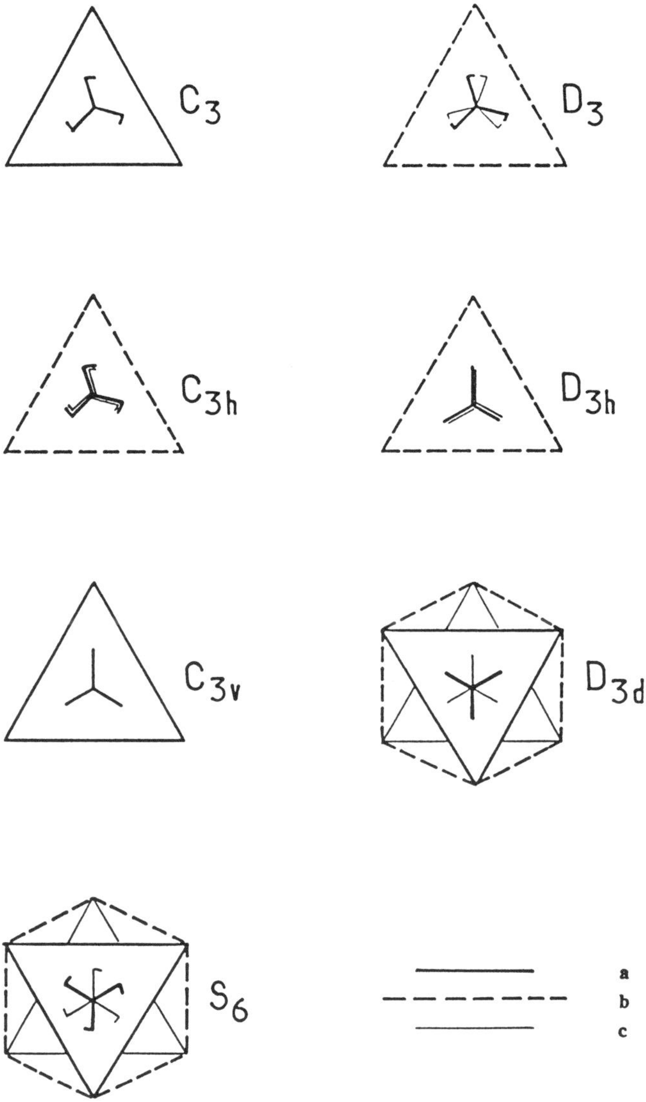

Figure 6: Seven symmetrical cardboard for n = 3. (b) is in the cardboard plane, (a) is drawn on the upper face and (c) on the lower one.

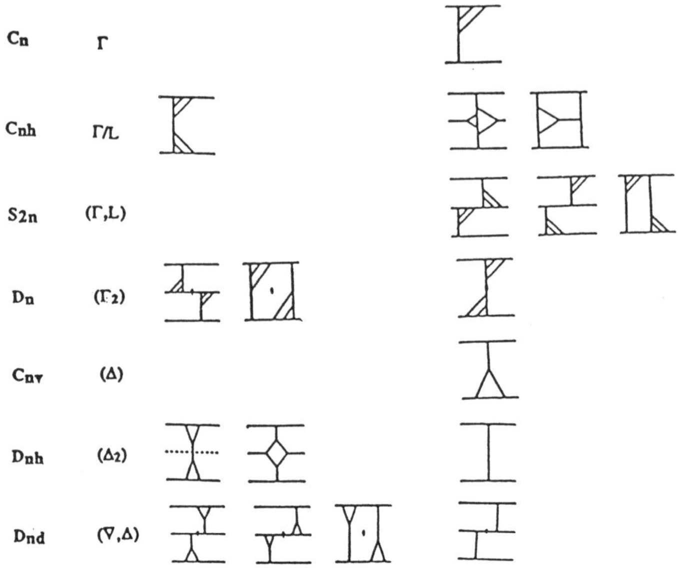

Figure 7 : Seven sets of "chemical motifs" corresponding to the motifs of Figure 2. Except for some cases (see text), a circular arrangment of n motifs generates a spheroarene of the appropriate topological symmetry, the motifs on the right generating the smallest.

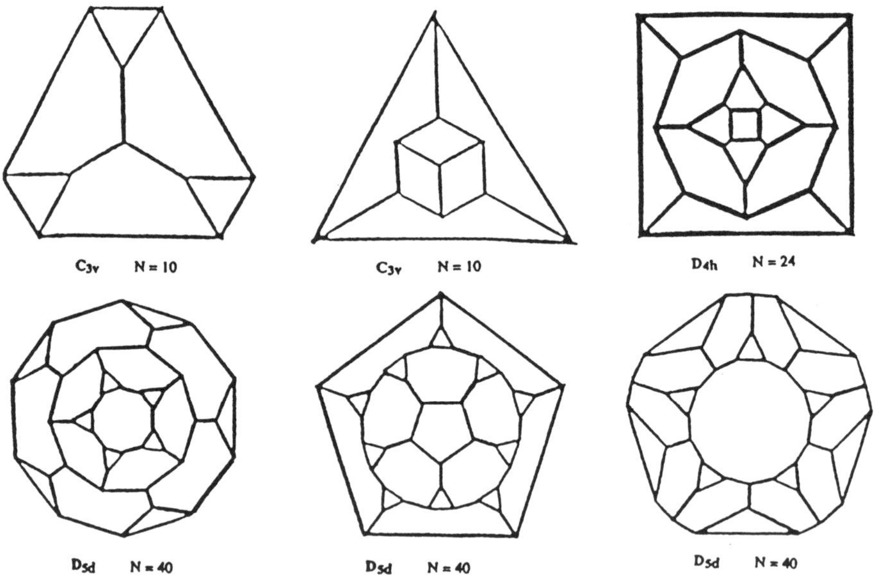

Figure 8 : Schlegel diagrams of the smallest spheroarenes of $\mathbf{C}_{3v}$, $\mathbf{D}_{4h}$, $\mathbf{D}_{5d}$ topological symmetry.

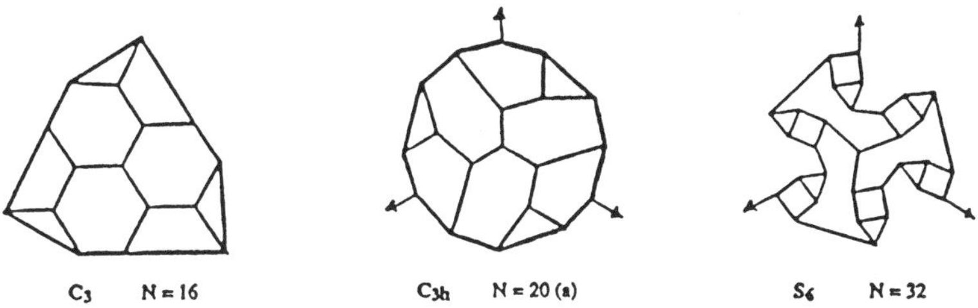

Figure 9 : Schlegel diagrams of the smallest spheroarenes C_N for n = 3.

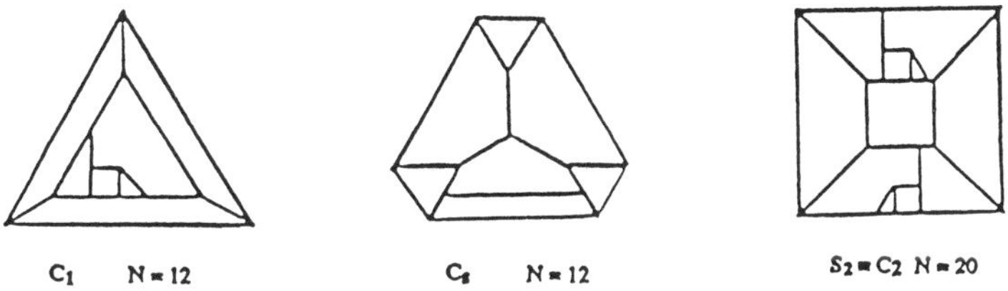

Figure 10 : Schlegel diagrams of the smallest spheroarenes C_N for n = 1.

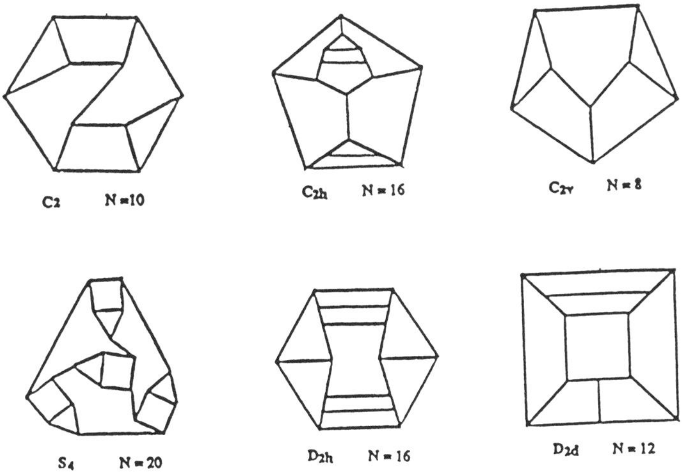

Figure 11 : Schlegel diagrams of the smallest spheroarenes C_N for n = 2.

belong to $\mathbf{D}_2$, $\mathbf{C}_{1h}$, $\mathbf{C}_{2h}$ and $\mathbf{C}_{2v}$ groups respectively. (This remark is also valid for symmetrical cardboards (18c) (Figure 6) but not for spheroarenes).

In order to obtain symmetrical spheroarenes belonging to any axial point-groups, it is sufficient to design seven "chemical motifs" having the symmetry properties of the seven basic motifs used before and to assemble them in such a way as to generate a $\mathbf{C}_n$ axis. Figure 7 gives sets of seven chemical motifs, chosen so that they be made of the smallest possible number of C atoms. They can also be named using the $\Gamma - \Delta$ nomenclature (For instance the two-carbon, ethylene-like motif is the Δ_2 motif). For $n \geq 6$ a molecule made by a "circular" arrangement of n chemical motifs has always the appropriate (topological) symmetry. For $n = 3,4$ or 5, three motifs $\Delta(n=3)$, Δ_2 $(n=4)$ and (∇,Δ) $(n=5)$ generate spheroarenes of a symmetry higher than the one expected : $\mathbf{T}_d$, $\mathbf{O}_h$ and $\mathbf{I}_h$ instead of $\mathbf{C}_{3v}$, $\mathbf{D}_{4h}$ and $\mathbf{D}_{5d}$ respectively. (The "expected" symmetry belongs to a subgroup of the actual symmetry). Figure 8 show the Schlegel diagrams of the smallest spheroarenes belonging to those three groups. For $n = 3$, except for $\mathbf{C}_{3v}$, molecules of the convenient symmetry are found as for $n \geq 6$, but some smaller spheroarenes may be obtained by replacing a 3-membered ring by a single trivalent carbon. (Figure 9). For $n = 1$ (Figure 10) and 2 (Figure 11) (except for $\mathbf{D}_2$), specific molecules have to be found, for instance by examination of the published lists of simple planar trivalent graphs (23).

Some spheroalcanes have been obtained by chemical synthesis (27-35) and others studied theoretically (36-45). See also (46-55).

2.2. SPHEROARENES OF CUBIC AND ICOSAHEDRAL SYMMETRY

$\mathbf{I}_h C_{60}$ is the smallest stable fullerene (20). Cubane (56-57), ($\mathbf{O}_h$), dodecahedrane (58-59) ($\mathbf{I}_h$) and substituted tetrahedranes (60-62) ($\mathbf{T}_d$) have been prepared or studied

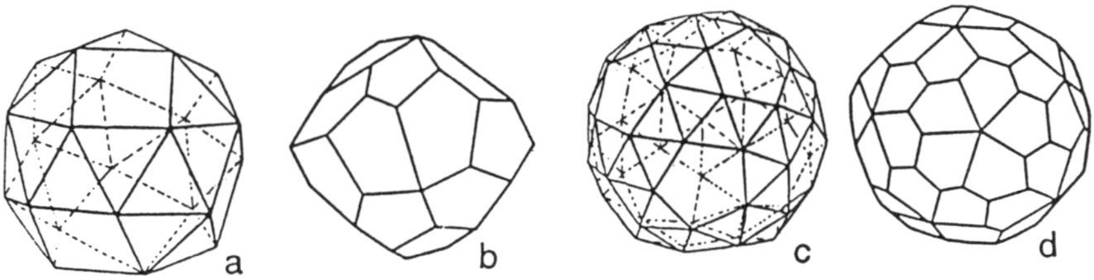

Figure 12 : Snub cube (a), snub icosahedron (c) and their duals (b,d).

theoretically (63-65). Valence graphs of spheroarenes belonging to the other high-symmetry groups may be obtained by truncation of selected vertices of the dodecahedron ($\mathbf{I}_h \rightarrow \mathbf{T}_h$ or $\mathbf{T}$) and of the duals of the snub cube ($\mathbf{O}$) and of the snub icosahedron ($\mathbf{I}$) (Figure 12). Figure 13 and 14 give the graphs of the smallest spheroarenes of cubic or

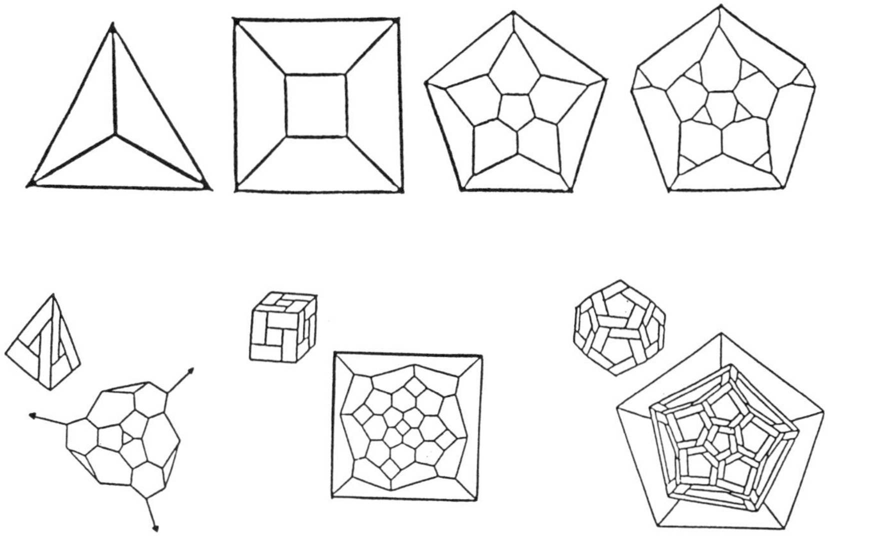

Figure 13 : Schlegel diagrams of the smallest spheroarenes of topological cubic or icosahedral symmetry.

icosahedral symmetry and steroviews of some spheroalcanes (66). It has already (65a) been noticed that a regular dodecahedron (I_h) labelled at "tetrahedral positions" (24) has **T** symmetry. Such labelling may be obtained with an endohedral tetrahedral carbon :. Figure 15 shows th corresponding C_{21} molecules of **T** symmetry (66).

3 - Symmetry of infinite columns :

By "infinite columns", we mean infinite cylinders of revolution on the external surface of which ornament of appropriate symmetry have been added. The revolution axis of the cylinder is considered to be vertical and becomes the C_n axis of the column. There are different ways to describe the generation of symmetrical columns :

3.1 RING APPROACH

In principle, symmetrical tubular molecular aggregates could be formed by piling identical spheroarenes belonging to one of the seven axial point-groups. In the same way, an infinite cylinder can be built up by piling (Figure 2d) a series of identical (or enantiomorphic) rings in a well defined manner. (This is essentially the approach used by Shubnikov and Koptsik (18d) to generate infinite rods made of cardboards belonging to one of the seven axial point-groups. Figure 6 gives a series of such cardboards). In particular, we can use the following processes, considered as one or more sucessive transformations of the original ring. (The first transformation is always a duplication of the original ring, the next transformation being applied only to the duplicated ring, and so on. After a certain number of these transformations, the ring obtained is exactly the one that would be obtained by applying a vertical translation of length *a* (we shall see that *a* may be infinite)).

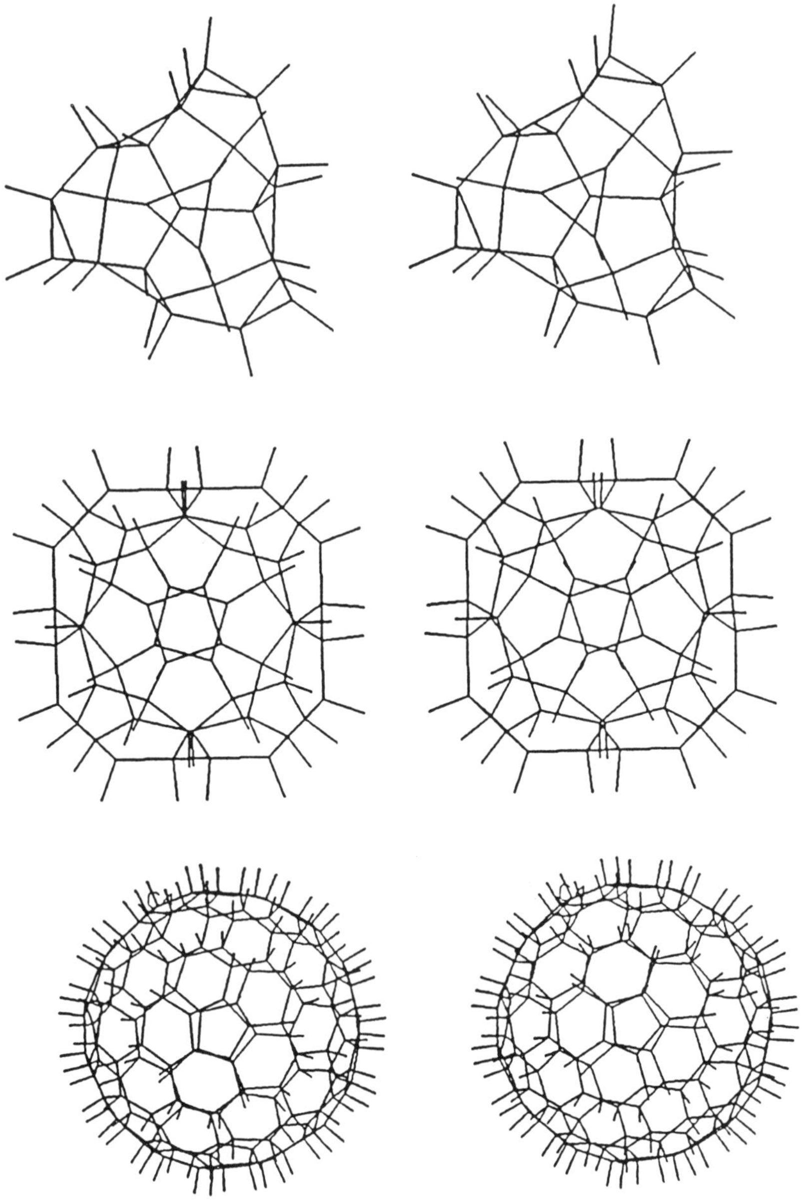

Figure 14 : Stereoview of $C_{28}H_{28}$, $C_{56}H_{56}$ and $C_{140}H_{140}$, smallest spheroalcanes of T, 0, and I symmetry. (66).

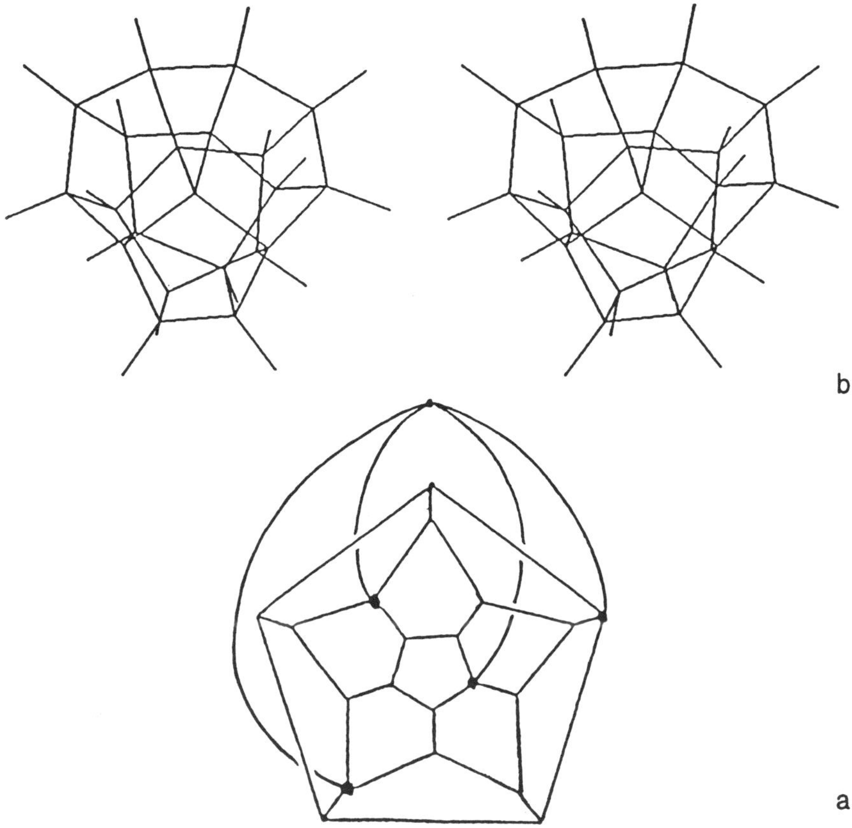

Figure 15 : a) Molecular formula (a non - planar graph) of C_{21} or $C_{21}H_{16}$ of topological T symmetry. b) Stereoview of $C_{21}H_{16}$ of T symmetry. (66).

Figure 16 : A fragment of a $D_{3h}T$ rod (p2mm pattern, C_3 axis).

$\hat{T}$: Vertical translation of length a, $a = d$, the thickness d of the ring (see also Figure 16).

$\hat{T}_2$: Rotation by $\alpha_2 = 2\pi/2n$ around the vertical axis, followed by a translation $\hat{T}$: the unit-cell is made of the two "staggered" rings, $a = 2\,d$.

$\hat{T}_q^p$: Rotation by $\alpha_q^p = 2p\pi / qn$ (p/q in its lowest terms and $p/q<1$) followed by a translation $\hat{T}$.($\hat{T}_2^1$ is $\hat{T}_2$,) The unit-cell is made of q rings, $\alpha = qd$. This process introduces a screw symmetry-axis $[nq_{np}]$. Except for q = 2, p = 1, this screw axis is chiral. By convention (18d), it is right-handed if $1 \le p < q/2$ and left-handed if $q/2 < p \le q-1$.

$\hat{t}$: Rotation by α ($|\alpha/2\pi|$ irrational and <1) : there is no finite translation a in the column. The corresponding screw axis is irrational. α is > 0 or < 0 for right- or left-handed screw axes, respectively. ($\hat{t}$ may be thought of as a limiting case of $\hat{T}_q^p$, when p and q become infinite with $O<p/q<1$).

$\hat{V}$: Reflection in a plane containing the vertical C_n axis followed by $\hat{T}$; $a = 2d$

$\hat{H}_2$:Reflection in the median horizontal plane of the ring followed by $\hat{T}_2$; $a = 2d$.

We may now examine the different symmetry types obtained : they are designated from the way they were generated : for instance C_nT means that the column is made from a C_n ring (Γ_n) to which the transformation $\hat{T}$ has been applied and the process repeated indefinitely.

a) $\hat{T}$ generates cylinders with rectangular type unit-cells. Seven cylinder-symmetry groups are obtained : $C_nT, D_nT, C_{nv}T, C_{nh}T, S_{2n}T, D_{nh}T$ and $D_{nd}T$. C_nT and D_nT are chiral groups, the other ones being achiral.

b) $\hat{T}_2$ generates groups with diamond-type unit-cells. Only five different group are now obtained : $C_nT_2, D_nT_2, C_{nv}T_2, C_{nh}T_2, D_{nh}$ the group from S_{2n} and D_{nh} being identical with $C_{nh}T_2$ and $D_{nh}T_2$ respectively.

c) $\hat{T}_q^p$ and $\hat{t}$ generate chiral groups with parallelogram-type unit-cells. Only two new different groups are obtained, $C_nT_q^p$ from C_{nv}, C_{nh} and S_{2n} and $D_nT_q^p$.from D_{nd} and D_{nh} .

d) $\hat{V}$ generates two new groups only, C_nV and $C_{nh}V$ with rectangle-type unit-cells and n vertical glide-planes.

e) $\hat{H}_2$ generates one new group only D_nH_2 with n vertical glide-planes and an infinity of reflection-rotation planes.

The Table gives these 17 non-redundant combinations of the seven ring-types (axial point-groups) and of the previous operations. Shubnikov and Koptsik's notations (18d) are also given.

(a)	(b)	(c)	(d)	(e)
p1	p1(R)	C_nT	(a).n	1
	p1(D)	C_nT_2	(a).2n$_n$	2
	p1(P)	$C_nT_q^p$	(a).n$_j$	3
p2	p2(R)	D_nT	(a).n:2	4
	p2(D)	D_nT_2	(a).2n$_n$:2	5
	p2(P)	$D_nT_q^p$	(a).n$_j$:2	6
pm	pm(h)	$C_{nh}T$	(a).n:m	7
	pm(v)	$C_{nv}T$	(a).n.m	8
pg	pg(h)	$S_{2n}T$	(a).$\widetilde{2n}$	9
	pg(v)	C_nV	(a).n.ã	10
cm	cm(h)	$S_{2n}T_2 = C_{nh}T_2$	(a).2n$_n$.$\widetilde{2n}$ = (a).2n$_n$:m	11
	cm(v)	$C_{nv}T_2$	(a).2n$_n$.ã = (a).2n$_n$.m	12
p2mm	p2mm	$D_{nh}T$	(a).m.n:m	13
p2gg	p2gg	D_nH_2	(a).$\widetilde{2n}$.ã	14
p2mg	p2mg(h)	$D_{nd}T$	(a).$\widetilde{2n}$.m	15
	p2mg(v)	$C_{nh}V$	(a).ã.n:m	16
c2mm	c2mm	$D_{nh}T_2$	(a).m.2n$_n$:m	17

Table : The seventeen symmetry types of infinite columns : *a*) the nine plane patterns with rotation axis $n \leq 2$; nomenclature of the column-symmetry types (*b*) originating from the plane patterns (see text § 3.2 and Figure 17), (*c*) from the "ring-approach" (see text § 3.1) ; *d*) : Shuknikov and Kopstsik's nomenclature (ref 18 d) ; e) : these numbers refer to the patterns of Figure 17.

3.2 PLANE-PATTERN APPROACH :

Ornamented infinite cylinders of revolution can also be obtained from some of the 17 plane-patterns by wrapping up an (infinite) fragment of such pattern on the cylinder. This fragment has to be finite in the "horizontal" direction and infinite in the "vertical" one.

The only horizontal symmetry-axis compatible with the column symmetry are C_1' and C_2'. This eliminates the plane-patterns having 3, 4 or 6 fold axes ; the only plane-patterns to be considered are *p1*, *p2*, *pm*, *pg*, *cm*, *p2mm*, *p2gg*, *p2mg* and *c2mm* (Table). *p1* patterns may have different plane-lattice types : parallelogram (*P*), rectangle (*R*) and diamond (*D*). Although these plane-lattice types do not change the *p1* or *p2* plane-symmetry type, they give rise to different types of column symmetry, because they generate $\hat{T}_q^p$ (or $\hat{t}$), $\hat{T}$ and $\hat{T}_2$ translations, respectively. In the table, they are designated by a combination of the plane pattern and of the plane-lattice type (eg *p1* (*P*)).

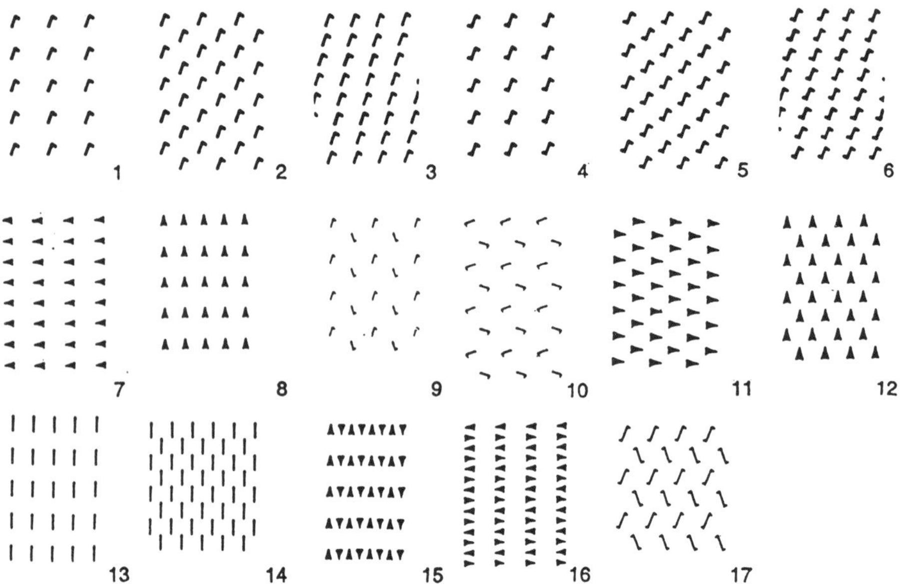

Figure 17 : Patterns generating the seventeen infinite-column (or rod) symmetry. The vertical axis is the C_n symmetry axis of the column.. The patterns are taken as infinite in the vertical direction and consisting of n motifs in the horizontal direction.

When there is a only one glide- or mirror-plane in the plane group, this plane may become vertical (*v*) or horizontal (*h*) (mirror planes remain mirror planes but only vertical glide-planes remain glide-planes, horizontal glide-planes becoming reflection-rotation planes).. Thus *pm, pg, cm* patterns may be wrapped on the cylinder in two different ways, generating *pm(h), pm(v), pg(h), pg(v), cm(h)*, and *cm(v)* symmetries. Similarly, *p2mg* generates two symmetry types : *p2m(v)g(h)* and *p2m(h)g(v)* or *p2mg(h)* and *p2mg(v)*. When there are two mirror- or two glide-planes (*p2mm, p2gg* and *c2mm*), there is only one way to wrap a band abstracted from the plane-pattern onto the infinite cylinder : there is then a one-to-one correspondance between the plane-symmetry and the column-symmetry.

Figure 17 give the plane patterns (67) corresponding to the seventeen column-symmetry types (the vertical direction is parallel to the cylinder axis).

Related problems, tubular arrangements of spheres (68) and some symmetrical conformations of linear polymers (69, 70) have been studied. Oligopeptides with topological $\mathbf{S}_{2n}$ symmetry have been prepared (71).

References :

1. Kroto, H.W., Heath, J.R., O'Brien, S.C., Curl, R.F., Smalley, R.E. (1985) C60 Buckminsterfullerene, *Nature* **318,** 162-163 ;
2. a). Schmalz, T.G., Seitz, W.A., Klein, D.J., Hite, G.E. (1988) Elemental carbon cages, *J. Am. Chem. Soc.* **110**, 1113-1127. b). Stankevich, I.V., Nikerov, M.V., and Bochvar, O.A. (1984) The structural chemistry of crystalline carbon ; geometry, stability, and electronic spectrum, *Russian Chemical Reviews*, **53**, 640-655. c). Fowler, P.W., Cremona, J.E., Steer, J.I. (1988) Systematics of bonding in non-icosahedral carbon clusters, *Theor. Chim. Acta* **73** , 1,26. d). Bakowies, D. and Thiel, W. (1991) MNDO Study of large carbon clusters, *J. Am. Chem. Soc.* **113**, 3704- 3714. e). Manolopoulos, D.E., Fowler, P.W. (1992) Molecular graphs, point groups, and fullerenes, *J. Chem. Phys.* **96**, 7603-7614.
3. Iijima, S. (1991) Helical microtubules of graphitic carbon, *Nature* **354**, 56-58.
4. Schnur, J.M. (1993) Lipid tubules : a paradigm for molecularly engineered structures, *Science* **262**, 1669-1676.
5. a). Ghadiri, M.R., Granja, J.R., Milligan, R.A., McRee, D. and Khazanovich, N. (1993) Self-assembling organic nanotubes based on a cyclic peptide architecture, *Nature* **366**, 324-327. b). Ghadiri, M.R., Granja, J.R. and Buehler, L. (1994) Artificial transmembrane ion channels from self-assembling peptide nanotubes, *Nature* **369**, 301-304. c). Khazanovich, N., Granja, J.R., McRee, D.E., Milligan, R.A. and Ghadiri, M.R. (1994) Nanoscale tubular ensembles with specified internal diameters. Design of a self-assembled nanotube with a 13-Å pore, *J. Am. Chem. Soc.* **116**, 6011-6012. d). Ghadiri, M.R., Kobayashi, K., Granja, J.R., Chadha, R.K. and McRee, D.E. (1995) The structural and thermodynamic basis for the formation of self-assembled peptide nanotubes, *Angew. Chem. Int. Ed. Engl.* **34**, 93-97.
6. Hamilton, E.J.M., Dolan, S.E., Mann, C.M., Colijn, H.O., McDonald, C.A., Shore, S.G. (1993) Preparation of amorphous boron nitride and its conversion to a turbostratic, tubular form, *Science* **260**, 659-661.
7. a). Fuhrop, J.H., Schnieder, P., Boekema, E. and Helfrich, W. (1988) Lipid bilayer fibers from diastereomeric and enantiomeric *N*-octylaldonamides, *J. Am. Chem. Soc.*

110, 2861-2867. b). Fuhrop, J.H., Svenson, S., Boettcher, C., Rössler, E., and Vieth, H.M. (1990) Long-lived micellar *N*-alkylaldonamide fiber gels. Solid-state NMR and electron microscopic studies, *J. Am. Chem. Soc.* **112**, 4307-4312. c). Fuhrop, J.H., Spiroski, D., and Boettcher, C. (1993) Molecular monolayer rods and tubules made of α-(L-Lysine), ω-(Amino) bolaamphiphiles, *J. Am. Chem. Soc.* **115**, 1600-1601.

8. Harada, A., Li, J. and Kamachi, M. (1993) Synthesis of a tubular polymer from threaded cyclodextrins, *Nature* **364**, 516-518.
9. Wang, Y. (1994) Encapsulation of palladium crystallites in carbon and the formation of wormlike nanostructures, *J. Am. Chem. Soc.* **116**, 397-398.
10. Georger, J.H., Singh, A., Price, R.R., Schnur, J.M., Yager, P., and Schoen, P.E. (1987) Helical and tubular microstructures formed by polymerizable phosphatidylcholines, *J. Am. Chem. Soc.* **109**, 6169-6175.
11. Okahata, Y., and Kunitake, T. (1979) Formation of stable monolayer membranes and related structures in dilute aqueous solution from two-headed ammonium amphiphiles, *J. Am. Chem. Soc.* **101**, 5231-5234.
12. Zhu, H.Y., Klein, D.J., Seitz, W.A., and March, N.H.(1995) BN alternants : boron nitride cages and polymers, *Inorg. Chem.* **34**, 1377-1383.
13. Kresge, C.T., Leonowicz, M.E., Roth, W.J., Vartuli, J.C., and Beck, J.S. (1992) Ordered mesoporous molecular sieves synthesized by a liquid-crystal template mechanism, *Nature* **359**, 710-712.
14. Brumlik, C.J. and Martin, C.R. (1991) Template synthesis of metal microtubules, *J. Am. Chem. Soc.* **113**, 3174-3175.
15. Roks, M.F.M. and Nolte, R.J.M. (1992) Biomimetic macromolecular chemistry : design and synthesis of an artificial ion channel based on a polymer containing cofacially stacked crown ether rings. Incorporation in dihexadecyl phosphate vesicles and study of cobalt ion transport, *Macromolecules* **25**, 5398-5407.
16. Ikeda, A. and Shinkai, S. (1994) Multiple connection of 1,3-alternate-calix[4]arenes. An approach to synthetic 'nano-tubes', *J. Chem. Soc. Chem. Commun.*, 2375-2376.
17 a). Feyes Tφth, L. (1964) *Regular Figures*, Int. Series of Monographs in Pure and Applied Math. **48**, Pergamon Press, Oxford. b). Shepard, A.O. *Ceramics for the archaeologist*, Publication 609, Carnegie Institution of Washington, Washington, D.C. c) Bhagavantam, S., Venkatara Yu Du, T. (1969) *Theory of Groups and its Applications to Physical Problems*. Academic Press, New York p. 13-15. d) Martin, G.E. (1980) *Transformation Geometry*. Springer Verlag, New York. p. 78-87.
18. Shubnikov, A.V. and Koptsik, V.A. (1974) *Symmetry in Science and Art,* Plenum Press, New York. a) p. 79-93 ; b) p. 95-101 ; c) p. 64 ; d) p. 103-127.
19. Kroto, H.W. (1987) The stability of the fullerenes C_n, with n = 24, 28, 32, 36, 50, 60 and 70, *Nature* **329**, 529-531.
20. Diederich, F. and Whetten, R. L. (1992) Beyond C_{60} : The higher fullerenes. *Acc. Chem. Res.* **25**, 119-126.
21. Fabre, C., Rassat, A. (1989) Geometrical partitions of C_{60} footballene into C_{15}, C_{30}, or C_{45} moieties. So called "coupe du roi" of the truncated icosahedron. Preparation of related C_{30} and C_{45} molecules. *C.R. Acad. Sci. Paris,* **308**, 1223-1228.
22. a). Wells, A.F. (1963) *The Third Dimension in Chemistry*, Clarendon, Oxford, b). Coxeter, H.M.S., (1963) *Introduction to Geometry*, Wiley, New York.

23. a)Balaban, A.T. in A.T. Balaban (ed) (1976) *Chemical Application of Graph Theory*, Academic Press, Londres, p. 64. b) Balaban, A.T. (1972) Valence isomers of [12]annulene. *Rev. Roumaine chim.* **17**, 865-881. c) Balaban, A.T. and Deleanu, C. (1987) Valence Isomers of [14]annulene. *Rev. Roumaine Chim.* **32**, *3*, 271-294.
24. Rassat, A., Seroussi, D., Coulombeau, C. (1994) Représentation des groupes de symétrie par des sphéroalcanes, *J. Chimie Physique* **91**, 1683-1710.
25. Günthard, Hs. H., Primas, H. (1956) Zusammenhang von Graphentheorie und MO-Theorie von Molekeln mit Systemen konjugierter Bindungen. *Helv. Chim. Acta* **39**, 1645-1653.
26. a) Landau, L., Lifchitz, E. (1967) *Mécanique Quantique*, Mir, Moscou p. 427 (§ 98). b) Hahn, Th., Klapper, H. (1983) *Point groups and crystal classes* in Th. Hahn (ed), *International tables for crystallography, Vol A, Space-groups Symmetry*, I.U.C., D. Reidel Publishing Company, Dordrecht p. 774-777. c) Altman, S.L., (1962) in O.R. Bates (ed), *Quantum Theory* Vol. II (*Aggregates of particules*), Academic Press, London p. 155.
27. Bosse, D., and De Meijere, A. (1977) A simple route to hexacyclo[4.4.O.O.^{2,4}.O^{3,10}.O^{5,8}.O^{7,9}] decane ("barettane") and pentacyclo[5.2.1.0^{2,6}.0^{3,5}.0^{4,8}]decane ("dihydrobarettane"), *Tetrahedron Letters* **13**, 1155-1158.
28. Cassar, L., Eaton, P.E., Halpern, J. (1970) Silver(I)- and palladium (II)-catalyzed isomerizations of cubane. Synthesis and characterization of cuneane, *J. Am. Chem. Soc.* **92**, 6366-6368.
29. a). De Meijere, A., Kaufman, D. Schallner, O. (1971) Hexacyclo[4.4.0.0.^{2,4}.0^{3,9}.0.^{5,7}.0^{8,10}]decan, ein überbrücktes *cis*-Trishomobenzol. *Angew. Chem.* **83**, 404-405. b). Kaufmann, D., Fick, Hs.H., Schallner, O., Spielmann, W., Meyer, L.U., Gölitz, P. et De Meijere, A. (1983) Darstellung und charakteristische Reaktionen einiger Tris-σ-homobenzol-Kohlenwasserstoffe. *Chem. Ber.* **116**, 587-609.
30. Prinzbach, H. and Stusche, D. (1971) Photochemische umwandlungen, 41, $3\sigma \rightarrow 3\pi$-isomerisierungen von monohetero-trishomobenzol-derivaten, *Helv. Chim. Acta* **54**, 755-759.
31. Allred, E.L., Beck, B.R. (1973) Concurrent thermolysis and photolysis of basketene. Formation and the interrelation of some new $(CH)_{10}$ isomers. *J. Am. Chem. Soc.* **95**, 2393-2394.
32. Katz, T.J., Acton, N. (1973) Synthesis of Prismane, *J. Am. Chem. Soc.* **95**, 2738-2739.
33. Lee, C.-H., Liang, S., Haumann, T., Boese, R. and de Meijere, A. (1993) *p*-[$3^2.5^6$]octahedrane, the $(CH)_{12}$ hydrocarbon with D_{3d} symmetry, *Angew. Chem. Int. Ed. Engl.* **32**, 559-561.
34. Eaton, P.E., Or, Y.S., Branca, S.J. (1981) Pentaprismane, *J. Am. Chem. Soc.* **103,** 2134-2136.
35. Shen, C.-C., Paquette, L.A. (1994) Development of a strategy for the synthesis of the spherical hydrocarbon *p*-[$4^2.5^8$]decahedrane, *Tetrahedron* **50**, 4949-4956.
36. Rassat, A., Scalmani, G., Seroussi, D., Berthier, G. (1995) Structure and stability of Spheroalcanes $(CH)_{10}$. *J. Mol. Structure* (THEOCHEM), in press.
37. Gordeeva, E.V., Katritzky, A.R., Shcherbukhin, V.V., Zefirov, N.S. (1993) Rapid conversion of molecular graphs to three-dimensional representation using the MOLGEO program, *J. Chem. Inf. Comput. Sci.* **33**, 102-111.

38. Iwamura, B.H., Morio, K., Kunii, T.L. (1971) Heats of formation and strain energies for $[CH]_8$ isomers calculated by all-valence-electron semi-empirical SCF-MO theories, *J. Chem. Soc. Chem.Com.* 1408-1409.
39. Schulman, J.M., Miller, M.A., Disch, R.L., (1988) Theoretical studies of triquinacene and diademane. *J. Mol. Structure* (THEOCHEM) **169,** 563-568.
40. Schulman, J.M., Disch, R.L. (1985) Ab initio heats of formation of medium-sized hydrocarbons. 3. The valence isomers of benzene. *J. Am. Chem. Soc.* **107**, 5059-5061.
41. Jonas, V., Frenking, G. (1992) Prismene : A theoretically predicted target for experimental studies. *J. Org. Chem.* **57**, 6085-6089.
42. Allinger, N.L., Eaton, P.E. (1983) The geometries of pentaprismane and hexaprismane insights from molecular mechanics, *Tetrahedron Letters* **24,** 3697-3700.
43. Reddy, V. P. Jemmis, E.D. (1986) A D_{4d} structure for [8]-prismane, *Tetrahedron Letters* **27**, 3771-3774.
44. Disch, R.L., Schulman, J.M. (1988) Ab initio heats of formation of medium-sized hydrocarbons,. 7. The [*n*]prismanes, *J. Am. Chem. Soc.* **110**, 2102-2105.
45. Engelke, R., Hay, P.J., Kleier, D.A., and Wadt, W.R. (1984) Theoretical study of dimeric forms of ground-state benzen molecules, *J. Am. Chem. Soc.* **106,** 5439-5446.
46. Schultz, H.P. (1965) Topological organic chemistry. Polyhedranes and prismanes. *J. Org. Chem.* **30**, 1361-1364.
47. Troischose A. (1970) Methin und seine sphäromerisierung. *Nach. Chem. Techn.* **18**, 127-128. DreidingA., *Personnal communication.*
48. Scott, L.T., Jones, M., Jr (1972) Rearrangements and interconversions of compounds of the formula $(CH)_n$. *Chem. Rev.* **72**, 181-202.
49. Garrat, P.J., White, J.F. (1977) Studies on $(CH)_{2n}$ hydrocarbons, alternative syntheses of [3]peristylane (triaxane), *J. Org. Chem.* **42**, 1733-1736.
50. a). Coulombeau, C., Rassat, A. (1987) Calculs de propriétés électroniques et des fréquences normales de vibration d'agrégats carbonés formant des polyèdres réguliers et semi-réguliers. *J. Chimie Physique* **84**, 875-882. b). Coulombeau, C., Rassat, A. (1988) Calcul par PSG des fréquences de vibrations d'alcanes ayant la structure de polyèdres réguliers et semi-réguliers. *J. Chimie Physique* **85**, 369-374.
51. Hassenrück, K., Martin, H.-D., Walsh, R. (1989) Consequences of strain in $(CH)_8$ hydrocarbons, *Chem. Rev.* **89**, 1125-1146.
52. Masumane, S., Darby, N. (1972) [10] Annulenes and other $(CH)_{10}$ hydrocarbons, *Acc. Chem. Research* **5**, 272-281.
53. Schriver, G.W., Gerson, D.J. (1990) Energies and isomerizations of $(CH)_{12}$ hydrocarbons, *J. Am. Chem. Soc.* **112**, 4723-4728.
54. Atovmyan, I.L., Shcherbukhin, V., Zefirov, N.S. (1993) Computer-assisted investigation of possible rearrangements of $(CH)_n$ (n = 4,6,8,10) annulenes. *Bull. Soc. Chim. Belg.* **102**, 247-257.
55. Jiang, Y., Shao, Y., Kirby, E.C. (1994) Topology and stability of trivalent polyhedral clusters.*Fullerene Science & Technology* **2**, 481-497.
56. Eaton, P.E., Cole, T.W. (1964) Cubane, *J. Am. Chem. Soc.* **86**, 3157-3158.
57. Griffin, G.W., Marchand, A.P. (1989) Synthesis and chemistry of cubanes, *Chem. Rev.* **89**, 997-1010.

58. a). Ternansky, R.J., Balogh, D.W., Paquette, L.A. (1982) Dodecahedrane, *J. Am. Chem. Soc.* **104**, 4503-4504. b). Paquette, L.A. (1989) Dodecahedranes and allied spherical molecules, *Chem. Rev.* **89**, 1051-1065.
59. Fessner, W.D., Prinzbach, H., (1990) The Pagodane Route to Dodecahedrane, in G.A. Olah, (ed.), *Cage hydrocarbons*, Wiley, New York, pp. 353-405.
60. Maier, G, Born, Pfriem, S., Schäfer, V., Matusch, R. (1978) Tetra-tert-butyltetrahedrane, *Angew. Chem. Int. Ed. Engl.* **17**, 520-521.
61. Maier, G., Born, D. (1989) Tri-tert-butyl(trimethylsilyl)tricyclo[1.1.0.0.^{2,4}]-butane - a second tetrahedrane derivative, *Angew. Chem. Int. Ed. Engl.* **28**, 1050-1052.
62. Maier, G. (1988) Tetrahedrane and cyclobutadiene, *Angew. Chem. Int. Ed. Engl.* **27**, 309-332.
63. Schulman, J.M., Venanzi, T.J. (1974) A theoretical study of the tetrahedrane molecule, *J. Am. Chem. Soc.* **96**, 4739-4746.
64 a) Veillard, A., Del Re, G., (1964) Cubane, *Theoret. Chim. Acta* . **2**, 55. b) Dewar, M.J.S., Worley S.D.,(1969) Cubane, *J. Chem. Physics* **50** 654. c) Bodor, N., Dewar, M.J.S., (1970) Cubane, *J. Am. Chem. Soc.* **92** 19. d) Schulman, J.M., Fischer, C.R., Solomon, P., Venanzi, T.J., (1978) Theoretical studies of the cubane molecule *J. Am. Chem. Soc.* **100** 2949-2953 and references therein.
65 a) Schulman, J.M., Disch, R.L. (1978) Theoretical studies of dodecahedrane. 2. Dodecahedrane, inclusion compounds, and fluorine derivatives. *J. Am. Chem. Soc.* **100**, 5677-5681.b) Disch, R.L., Schulman, J.M. (1981) Theoretical studies of dodecahedrane. 3. Ab initio studies of dodecahedrane and its inclusion compounds. *J. Am. Chem. Soc.*, **103**, 3297-3300. c) Schulman, J.M., Disch, R.L. (1984) Ab initio heats of formation of medium-sized hydrocarbons. The heat of formation of dodecahedrane, *J. Am. Chem. Soc.* **106**, 1202-1204.
66. Coulombeau, C,.Rassat, A., (1991) Unpublished work
67. Buerger, M.J. (1963) *Elementary crystallography,* Revised printing, John Wiley and Sons, Inc. New York.
68 Erickson, R.O. (1973) Tubular packing of spheres in biological fine structure, *Science,* **181**, 705-716.
69 Scaringe, R.P., Perez, S. (1987) A novel method for calculating the structure of small-molecule chains on polymeric templates, *J. Phys. Chim.* **91**, 2394-2403.
70. a). Perlstein, J. (1992) Molecular self-assemblies : Monte Carlo predictions for the structure of the one-dimensional translation aggregate, *J. Am. Chem. Soc.* **114**, 1955-1963. b). Perlstein, J. (1994) Molecular self-assemblies. 2. A computational method for the prediction of the structure of one-dimensional screw, glide, and inversion molecular aggregates and implications for the packing of molecules in monolayers and crystals, *J. Am. Chem. Soc.* **116**, 455-470.
71. Tomasic, L. Lorenzi, G.P. (1987) Some cyclic oligopeptides with S_{2n} symmetry, *Helv. Chim. Acta* **70**, 1012-1016.

STABILITY, SOLVENT PATTERNS AND MOLECULAR RECOGNITION IN CYCLODEXTRINS

G. TSOUCARIS
C.N.R.S.-UPR180
Univer. Paris-Sud
Centre Pharmaceutique
92290, Châtenay-Malabry, France

1. Introduction

Cyclodextrins (CD) have a long history as host molecules and supramolecular (SM) compounds. From a fundamental viewpoint, interesting questions arise from the existence of SM compounds both in solution and in the crystal state for the same host-guest system. For instance, does crystallisation proceed from a SM compound already preformed in solution, or, alternatively, a complete re-organisation of the host-guest-solvent system may occur upon crystallisation?

Such questions arose in our recent work on emission into the surrounding milieu of a volatile guest molecule encapsulated in a crystalline compound. The release process depends on thermodynamic stability and on kinetic factors; it involves both short range interactions (inclusion of a guest molecule in the host cavity) and long range effects (overall crystal cohesion). Thus we are led to considering molecular control versus crystalline control for the release process.

A problem of great practical importance is the controlled release of pheromones for large scale applications in agriculture, in order to reduce or even suppress the use of insecticides. Sex pheromones emitted from insects are active at very low concentration in the atmosphere, but basic problems related to their volatility have to be overcome first. The use of SM devices based on CD's has the advantage of compatibility with the increasing environmental demand, due to their vegetal origin and low toxicity.

The pheromone of *Bactrocera oleae* (also known as *Dacus oleae*) is a mixture of four components: 1,7-Dioxaspiro [5,5] undecane 1 (abbr. spiroacetal; 80% of the mixture; nonanal **2**, ethyl laurate **3**, α (-)-pinene **4**; related molecule: Z-9-dodecenol **5** (scheme 1). The pheromone of *Prays oleae* consists of Z-7-tetradecenal **6**.

G. Tsoucaris et al. (eds.), Crystallography of Supramolecular Compounds, 203–222.

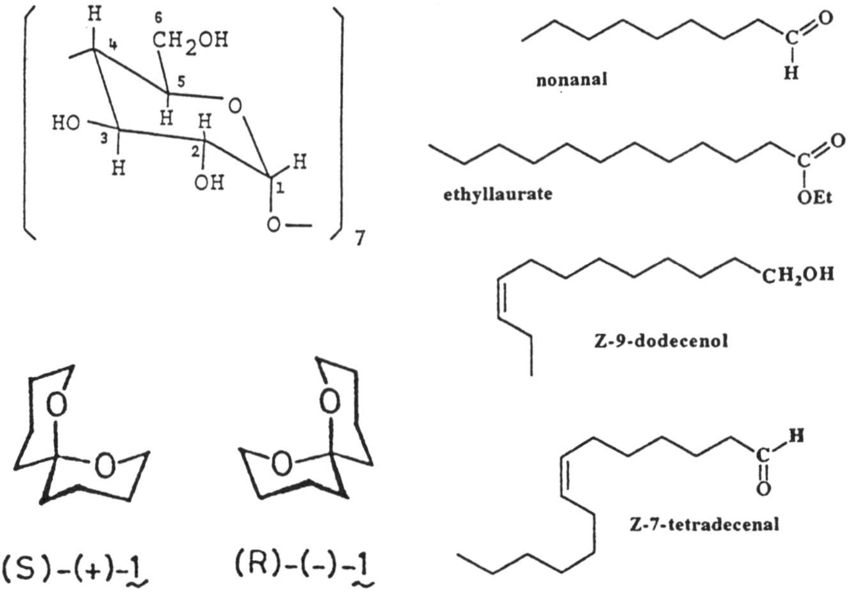

Scheme 1

The formation of the β-CD.spiroacetal crystalline complex is achieved by a quite straightforward process [1]. Laboratory and field experiments have shown at an early stage of our work that this complex is quasi-ineffective for spiroacetal emission [2]. From this point on, the research aimed at SM devices where the release rate approaches the required level, and it can be further adapted for various conditions by fine tuning the system. A particular difficulty of the task is the required versatility for the time period over which the emission of pheromone should be programmed; this time period can vary from several weeks to several months.
Clearly, the practical problem of guest release from a SM compound is linked to questions of fundamental interest, such as:

- What is the role of molecular inclusion versus the role of crystal packing in the overall cohesion and stability of a SM compound?
- How does the combination of several weak forces between hydrophobic entities, such as a pheromone molecule and the CD cavity walls, end up with a very stable, quasi-inalterable crystalline edifice?
- Are there parameters of molecular or crystalline nature which could be used in a fine tuning device in order to optimise this crucial property of SM Compounds i.e. the stability in the atmosphere over a long period of time?

Underlying these questions, we address the difficult problem of crystal structure prediction given the constituting molecular species. The approach to this problem involves the basic notions of molecular assembly and molecular recognition.

2. β-CD Assemblies and Hydration Patterns

α-, β-, γ-, ...CD are cyclic oligosacharides built up from 6, 7, 8, ... glucose units. A guest molecule located at the centre of the hydrophobic cavity 'sees' mainly the H(3), H(5) and the glucosidic O(4) atoms. The assembly of the CD crystalline complexes involves well-defined types. The first ones have been identified in the 70's by Saenger and named by evocative terms: Herringbone, Channel, Brick Type (p. 243, this book). In this lecture we will examine mainly β-CD complexes.

2.1. ASSEMBLY TYPES

β-CD Dimers and CD-CD Molecular Self-Recognition
An important characteristic of crystalline β-CD complexes is the formation of dimers where two CD molecules are linked head-to-head by seven hydrogen bonds between the secondary hydroxyls O(3)..H..O'(3) (Figure 1). This intermolecular hydrogen bond set is strengthened by 2x7 intramolecular hydrogen bonds O(2)..H..O(3). Clearly, the important feature of the dimer is the co-operative binding of this system, comprised of 21 hydrogen bonds. These bonds are embedded in an infinite hydration pattern thoroughly studied by G. Lebas.
We have here a case of strong molecular self-recognition achieved by the specific geometry in the mutual position and orientation of the two members of the dimer. The intermolecular vectors O(3)-O'(3) are oblique to the seven-fold axis in order to comply with the normal values of hydrogen bond lengths and angles.

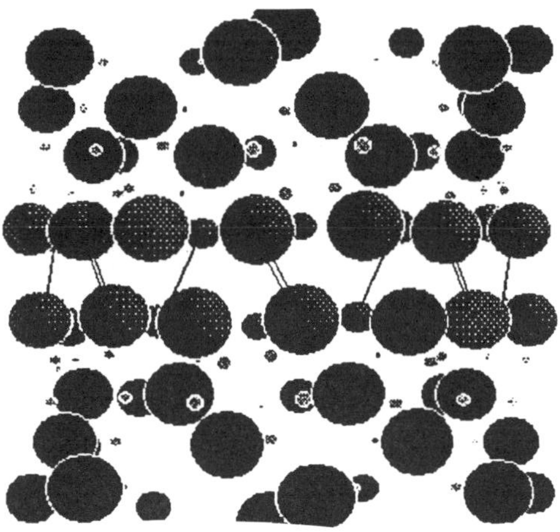

Figure 1. β-CD dimers. Van der Waals spheres centered at the cyclodextrin oxygen atoms illustrate the compactness of the dimer. The sphere centres are the experimental atomic positions, the radii are such as to provide a perspective view. Intradimer O(3)..H..O'(3) hydrogen bonds are depicted in lines oblique to the seven-fold axis. Double lines indicate hydrogen bonds at the front of the dimer, single lines, at the back.

Note that the two molecules are inter-related by a two-fold axis normal to the seven-fold axis. Comparatively, the so called duplex of the oligonucleotides $(TATA)_2$, involves eight hydrogen bonds; here also the two half-molecules are inter-related by a two-fold axis.
The dimer formation creates a larger cavity than the intramolecular CD cavity, presenting a strongly hydrophobic character despite the presence of 14 hydroxyls. The hydrophobic character results from the quasi-complete involvement of the hydrogen bond potentialities of the hydroxyls in CD-CD and CD-water hydrogen bonds. The chemical implication of this supramolecular architecture is similar to that of the hydrate clathrates where all possible hydrogen bonds among the water host molecules are actually established and only host-guest interactions of van der Waals type are allowed. The hydrophobic character of the dimer cavity will be further considered in section 2.3 and figure 4.

Dimers Stacked in Layers and Assembly Types
The dimers are stacked in infinite layers with their seven-fold axis approximately parallel. A remarkable fact is that the basic layer structure is invariant for all hydrated β-CD dimeric structures known so far (Figure 3). Systematic studies [3, 4] have led to a classification of the assembly types arising from different mutual position and orientation of the basic invariant layer: Channel CH, Chessboard CB, Intermediate IT, Screw Channel SC.

NOTE: α-CD complexes with neutral molecules usually crystallise as monomers, whereas the basic elements in β-CD crystals are dimers. Recently, however, a dimeric α-CD 2:1 inclusion compound with acetone has been described [5].

Host and Guest SM Columns in the Channel Type Assembly
A particular type of a clearly defined supermolecule in the crystal state is observed in the CH assembly. Infinite chains perpendicular to the layers are formed by piling up CD dimers via O(6)..H..O'(6) and O(6)..$H_2$0..O'(6) hydrogen bonds, creating thus a continuous channel.
The guest position and orientation relative to the CD cavity walls depend on the guest geometry and polarity. Small guest molecules are often located near the CD rims in two independent regions, respectively at the centre of the dimer and between dimers. They are also aligned along the channel. Figure 2 shows that the pheromone columns are completely secluded within the CD channel and they don't 'see' any other environment in the exterior of the channel.

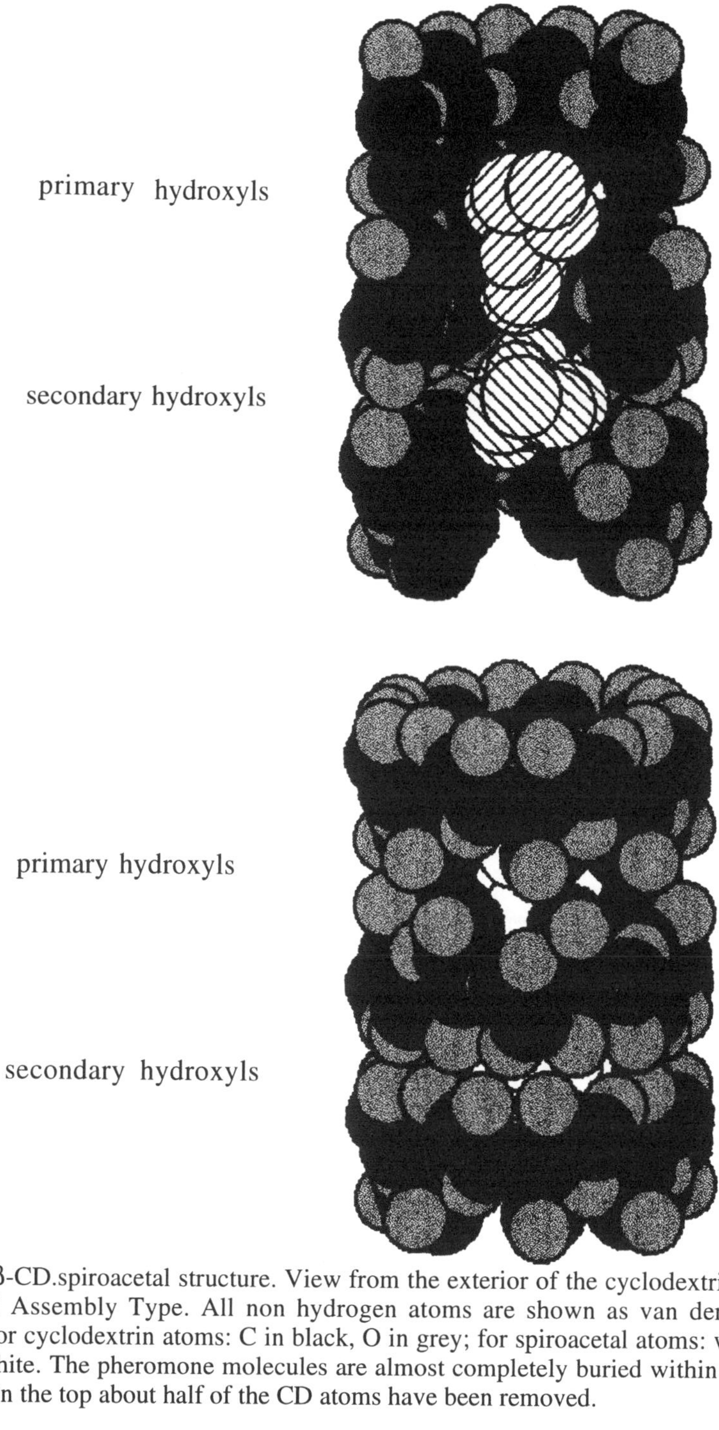

Figure 2. β-CD.spiroacetal structure. View from the exterior of the cyclodextrin chain in the CH Assembly Type. All non hydrogen atoms are shown as van der Waals spheres. For cyclodextrin atoms: C in black, O in grey; for spiroacetal atoms: white or hatched white. The pheromone molecules are almost completely buried within the CD channel. On the top about half of the CD atoms have been removed.

2.2. HYDRATION PATTERN IN DIMERIC COMPLEXES

Invariant 2D Hydrogen Bond Sub-Network

The hydrated dimeric complexes are associated with characteristic Hydration Patterns [3(a), 6]. Of particular interest is the existence of a strongly held quasi-invariant hydration sub-network in the sub-layer through the secondary hydroxyls, common to all Assembly Types. On the primary hydroxyl side, a part of the Hydration Pattern is also invariant, but the other part differs for each of the Assembly Types (Figure 3). Note that this studies have suggested crystallisation models for β-CD complexes [6].

Hydration patterns at a finer scale have been studied by neutron diffraction for several hydrated CD's by W. Saenger (p. 243, this book) and by G. Lebas [7]. The latter work has led to a plausible model for 'correlated host-guest disorder', extending thus the flip-flop model to host-guest interactions.

3D Hydration Network

The whole hydrogen bond network, comprising CD hydroxyls and water molecules, involves an oxygen atom framework reminiscent of the hydrate clathrates. The 'voids' of this framework are constituted precisely by the CD molecules and their guests.

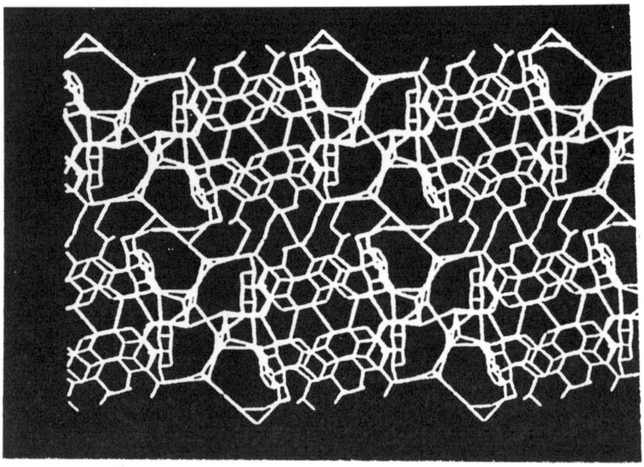

Figure 3. Hydrogen bond and hydration network in Channel Type dimeric structures (with Permission, the Journal of Supramolecular Chemistry, ref. [6]).

Proton Dynamic Aspects and Dielectric Behaviour

The description of hydration patterns based on X-ray diffraction has a static character. Other methods can provide access to dynamic aspects: NMR, neutron inelastic scattering, dielectric measurements, computer simulations. Dielectric methods reveal essentially movements of dipoles or charged structural elements, and their interpretation is often not straightforward. The existence of Hydration Patterns throughout the crystal has an incidence on its dielectric behaviour. The ε',ε' spectra, as well as the method of Thermally Stimulated Depolarisation Currents (TSDC) have been used to study hydrated CD's [8]. Among the results we mention:

- A distinction between the spectra of β-CD.11$H_2$0 and these of the dimeric complexes: the intracavity water in the former is less strongly bounded than the intermolecular water molecules in the latter.
- A series of peaks (200°-280°) in the dimeric crystals have been tentatively attributed to phase transitions resulting from the release of ionic conductivity via the hydrogen flip-flop mechanism .

2.3. DISORDER PROBLEMS

Difficult assessment of the exact guest position

We wish now to focus on the host-guest assembly for guest molecules bound to the host by the so called residual forces (van der Waals, weak dipoles, π–π interactions, very weak hydrogen bonds). The CD cavity having a quasi-cylindrical shape, we can expect the host-guest molecular recognition to occur only to a certain extent.

The crystals of β-CD with racemic spiroacetal present an accumulation of disorder sources. The spiroacetal enantiomers can fit into the β–CD channel at approximately the same positions, both in the intradimer and in the interdimer region (fig.2), and it has been experimentally observed that there is no chiral discrimination [17b]. For each of these two regions, a two-fold crystal axis generates from one molecular position another one, almost superimposed to the first. Consequently, in each region, there are four superimposed positions arising from the combination of enantiomeric and crystallographic disorder. From the four sets of guest coordinates (two enantiomers in each of the two regions) plausible structural models have been inferred [1b].

Disorder is an inherent phenomenon in many SM compounds and often impairs the determination of the exact position of the guest molecules. However, with the accumulation of data from several quasi-isomorphous complexes, more reliable results can be obtained for the guest preferential positions despite the lack of precision for individual crystals. Thus, we are led to seek additional information on the host-guest assembly by studying a series of related compounds.

Obtaining such information pertains to the general problem of crystal structure prediction from the molecular structure of its components. Here we consider the particular case of a given host associated with several guest molecules having in common a specific structural element.

Information'Despite Disorder'. Two Systematic Studies.

tert-butyl derivatives
t-Butyl being one of the most hydrophobic groups, one can expect to gather structural information on the host-guest geometry by grafting this group on various molecules having a more or less hydrophilic character [4,9] (compounds **1-5** in scheme 2). It is observed that the t-butyl group is most often inserted within or near the hydrophobic pocket at the centre of the dimer, whereas the less hydrophobic part is preferentially located in the interdimer region. But in the butylamine complex **2** (CH Type), on the opposite, the t-butyl is located in the primary region, and the amino group, at the dimer centre. Note, however, that in the CH-type the whole continuous channel has an hydrophobic character throughout the crystal.

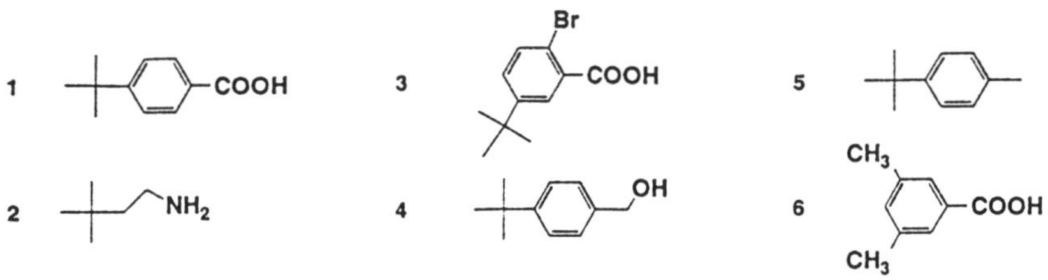

Scheme 2

The guest hydrophilic part plays an important role in the determination of the Assembly Type. As a typical example, we note that in **1** the guest-guest hydrogen bond geometry can account for the large lateral shift between successive layers in the IT Assembly Type adopted by this complex. The carboxyls of two guest molecules belonging to adjacent layers are interlinked by only one hydrogen bond. They are mutually shifted within their layer planes so as to fit the broken form of the IT channel.

long chains of pheromones and related molecules
The CH and, in special cases, the IT type are the most suitable for long linear molecules extended over more than one CD host (Figure 4): Z-9-dodecenol [10a], CH Type; Z-7-tetradecenal [11], CH Type; nonanoic acid [10b], IT Type. These crystals also show a propensity to accommodate the most hydrophobic part of the guest in the central dimeric pocket.
An interesting aspect to be further examined with new examples is the conformation adopted by the thin chain 'floating' within a larger channel.

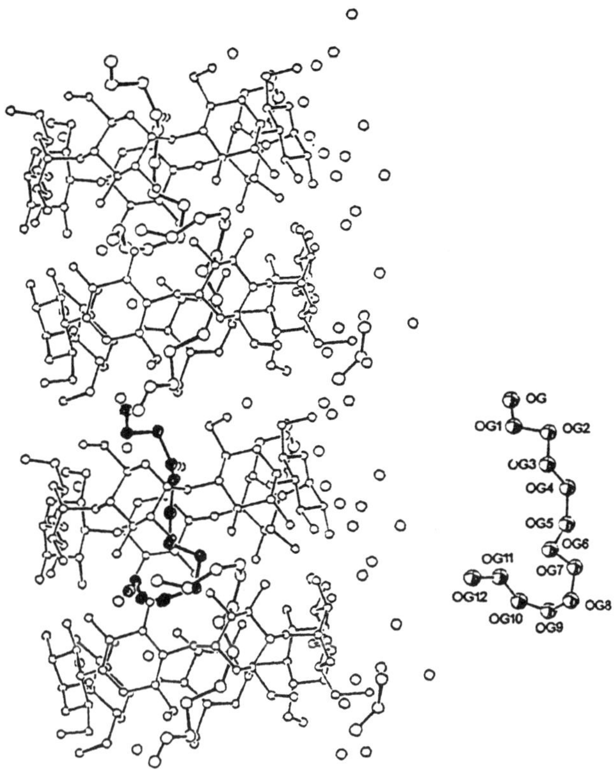

Figure 4. Z-9-dodecenol complex, CH-Type. The hydrophilic end lies in the primary inter-dimer region, the hydrophobic part makes a turn beyond the double bond and lies in the dimer centre perpendicular to the channel axis.

2.4. PHEROMONE RELEASE AND THE DOUBLE INCLUSION EFFECT

Many of the examined pheromone molecules crystallise in the dimeric CH type. Each CD molecule is linked via as many as 28 hydrogen bonds to the surrounding hydration network and to adjacent CD dimers. The hydration network, in which the channels containing pheromone are embedded, is likely to be a most important factor in the stability of the complex and the low release rate. We have indeed here a sort of double SM inclusion: pheromone columns included into CD channels, CD chains inserted into a strongly held hydration network. In order to enhance and control the release rate, we investigate in the following sections structures where the hydration pattern of natural CD's, and, subsequently, the host-guest supermolecules described above, are broken in various ways.

A further question is relevant: does disorder enhance guest mobility and guest release? Although no definite rules can be given at this stage, we mention that there are no indications of such correlation.

3. Methylated Cyclodextrins and Organic Solvents

The most usual derivatives are the seven-fold methylated 2-,6-, dimethyl-β-CD (DIMEB) and 2-,3-,6-, trimethyl-β-CD (TM-β-CD). The latter host forms channels and it has been used to entrap ethyl laurate [12] from aqueous solution. The interesting feature of this crystal is that the 15 atom long chain is extended within the channel over almost three host molecules.

The methylated host molecules are held together in the crystal by van der Waals contacts. A particularly interesting host molecule is the so-called 'overmethylated' DIMEB, differing from DIMEB in two points. It bears 15 methyls instead of 14 in DIMEB; the 'extra-methyl' is attached to an O(3) atom of a trimethylated ring. In addition, the ring adjacent to the latter is 3-,6-,dimethylated instead of the usual 2-, 6-, methylation. In the inclusion complex with spiroacetal the guest molecule is located at the centre of the 'overmethylated' DIMEB (Figure 5) [13].
On the other hand, the alkylated CD's have an increased solubility in organic solvents, which, in turn may enter in competition with the pheromone molecules or, alternatively, act as co-guests with them.

3.1. INDUCED FIT AND MOLECULAR ENCAPSULATION

Supermolecules can be further stabilised by the induced fit effect, a phenomenon well known in biological macromolecules. This effect stems from the flexibility occurring between various groups within the methylated CD molecule.
The flexibility between the glucosidic rings can be appreciated by the variation of the 'tilt' parameter, defined as the angle between the normals to the plane of the group: [C(2) C(3) C(5) O(5)] and the molecular axis (Figure 6). Note that an unambiguous definition of the latter is not obvious. Usually the molecular axis is identified as the normal to the O(4) mean plane. A better definition should take into account by a least squares procedure the glucosidic ring and the atoms directly attached to it. In the present calculations 10 atoms have been included in the calculation.

We observe that there is a progressive loss of both symmetry and rigidity in the order: β-CD dimers (21 hydrogen bonds); DIMEB.MeOH (7 hydrogen bonds); TM-α-CD.DMF and TM-β-CD.DMF (none).
The tilt angle variation, around a mean value of 80°-90°, is of a few degrees in the dimeric structures, but for TM the tilt angles cover a large range, from 55° to 110°. Note that in the usual trapezoidal CD drawings the primary edge is 'tilted in', while the secondary edge is 'tilted out' (Figure 7(b), left). A tilt angle of more than 90° means an 'inversion' of the usual tilt direction with the primary region outwards, and the secondary, inwards (Figure 7(b), right).

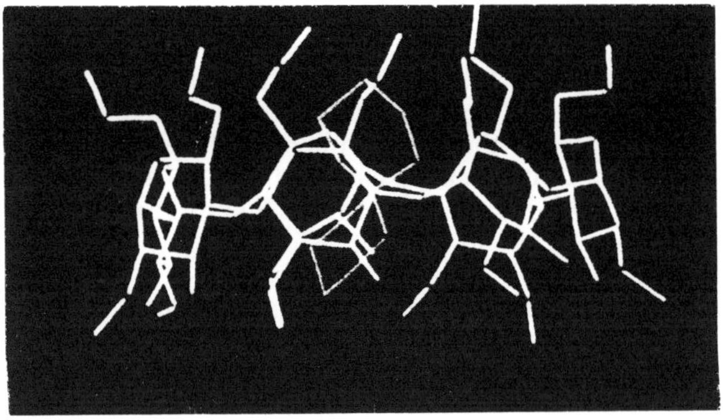

Figure 5. Molecular inclusion in the crystalline compound 'overmethylated' DIMEB.spiroacetal. The guest molecule (thin lines) is located at the centre of the CD cavity.

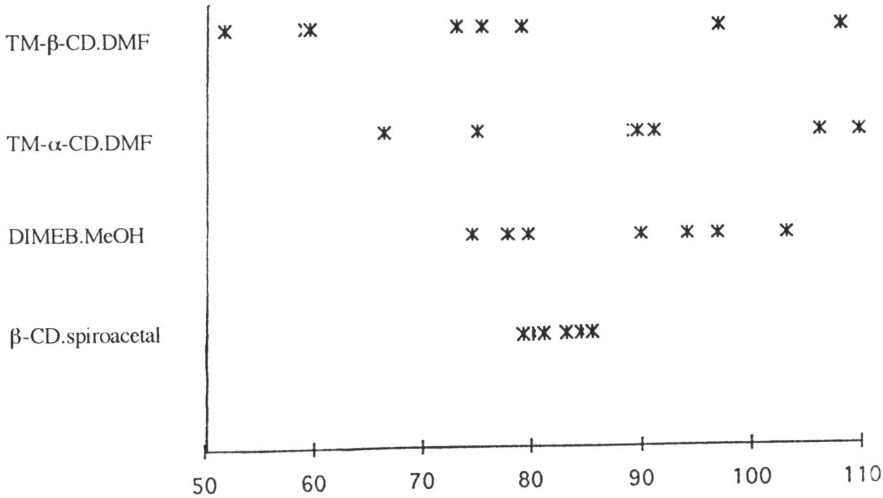

Figure 6. Tilt angles between the molecular axis and the normals to the mean planes through the groups [C2,C3,C5,O5] of the glucosidic rings for various cyclodextrin derivatives with decreasing number of intramolecular O(2)..H..O(3) hydrogen bonds.

We consider now the conformation of the methoxy group. The rotation barrier is so small that we would a priori expect the methoxy orientation to be submitted to unpredictable crystal proximity effects. In fact, this conformation parameter in certain structures is clearly correlated with the position and orientation of the guest molecule. Thus, in DIMEB.MeOH the MeOH position cannot be precisely assessed as a result of disorder. Yet, the orientation of two -OMe groups inwards (Figure 7a) can be related to the establishment of van der Waals contacts Me...guest.

Furthermore, in the 'overmethylated' DIMEB we have an interesting case where the induced fit stems from the combined effects of the macrocycle conformation (tilt angle) and the orientation of one methoxy group. These effects are here synergetic in establishing a large number of CD-spiroacetal contacts:

-The 'inversion' of the tilt angle for one of the seven residues enhances the adaptation of the cavity to the guest overall geometry. This 'inverted' residue is precisely that one which bears the 'extra-methyl' group on a O(3) secondary oxygen. Figure 7(b) shows two different views from directions perpendicular to the pseudo seven-fold CD axis. On the left one, the guest shape is similar to the usual trapezoidal CD cavity. On the right one, the 'inverted' residue confers to the CD cavity a particular shape, also adapted to that of the guest molecule.

-The 'extra methyl' group, already brought closer to the guest by the inversion of the tilt angle, is turned inwards and establishes contacts with the pheromone molecule, thus partially closing the cavity (Figure 7b, right) on the larger secondary side. The van der Waals sphere of the methyl appears as geometrically complementary to the local van der Waals contour of the guest molecule.

The overall effect is an excellent host-guest fit which can be appreciated in successive sections of van der Waals contours. Such a section through the centre of the O(4) heptagon is represented in Figure 8.

Another aspect of the induced fit is the significant ellipticity of 7% observed for the cavity form β-CD.spiroacetal, in relation with the guest shape.

3.2. GUEST-SOLVENT COMPETITION EFFECTS

The use of solvents such as DMSO and DMF is known to instantly achieve in solution the release of included molecules. The crystal structure of TM-α-CD.DMF and TM-β-CD.DMF reflects this host-solvent affinity [14,15]. The DMF molecules located within the CD channel form chains with a helical arrangement. The contacts between the DMF molecules along the chain further stabilise the structure.

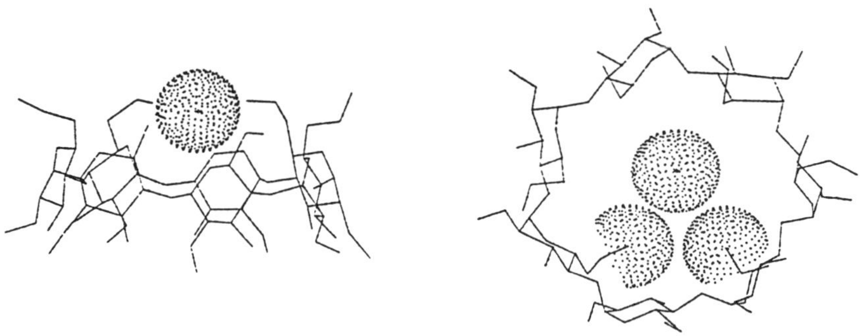

Figure 7(a). Induced fit in DIMEB.MeOH. Two methyl groups are directed towards a solvent molecule near the CD seven-fold axis; all others are oriented outwards.

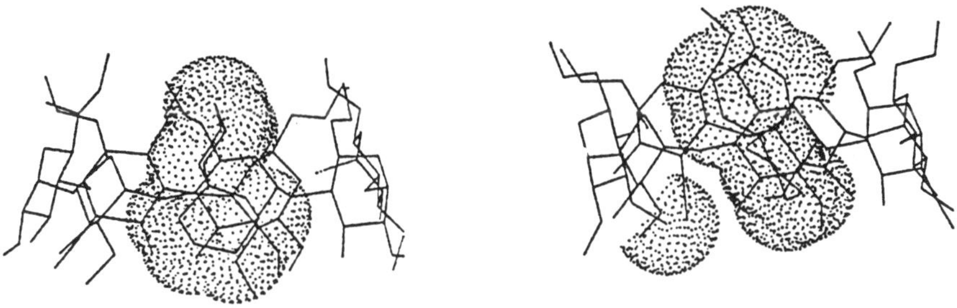

Figure 7(b). Induced fit from combined effects of the tilt angle and of the orientation of one methoxy group in the 'overmethylated' DIMEB.spiroacetal complex. On the right, the extra-methyl and the guest molecule are in contact, and their van der Waals contours have complementary shapes.

Figure 8. Host-guest fit illustrated by a section of the van der Waals contours, perpendicular to the cyclodextrin molecular axis, through the centre of the O(4) heptagon.

4. Inclusion in a Closed CD Cavity

The systems described so far are based on crystals where the inclusion/release mechanism greatly relies on the cohesion of the crystalline arrangement. The induced fit effect allows only a limited control by the host molecular structure on the geometry of intramolecular inclusion. We consider now a new device where the control arises mainly from a molecular mechanism.
The photoclosure ability of percinnamoylated cyclodextrins offers the possibility of such a control (Figure 9). This novel experiment relies on the complete closure of the β-CD cavity, achieving a very strong molecular encapsulation in three steps [16]:
- esterification of the CD hydroxyls by cinnamic acid
- inclusion of spiroacetal into the enlarged cavity of percinnamoylated CD
- photoclosure involving pairs of cinnamic moieties by irradiation at λ = 300 nm, entrapping the spiroacetal molecule inside.

Experimental evidence is in favour of efficient entrapping the pheromone molecule: only when the cavity is 'exploded' by high temperature the encapsulated pheromone is freed.

The inclusion/release control can be assured by incomplete closure during the third step. In turn, the control of the cavity closure can be achieved by incomplete esterification or by incomplete photoreaction between the existing cinnamates. A relevant question pertains to the number and mutual position of photoreacting cinnamates (primary or secondary, adjacent or distant) that are necessary to assure a permanent enclosure or a given release rate at room temperature. A further possibility relevant to the pheromone delivery is the reopening of the cavity by irradiation at λ= 254 nm.

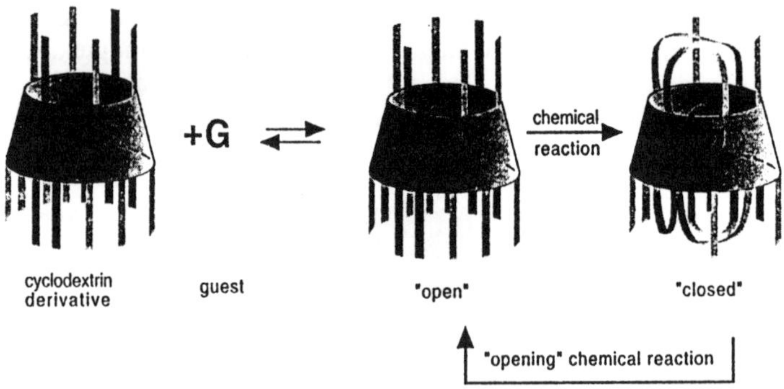

Figure 9. A schematic view of the reversible inclusion and imprisonment of guest molecule G in a CD derivative by closure of the cavity through intramolecular chemical bonds linking of the sealed CD substituents. A different chemical reaction opens the sealed CD and releases the guest.

Another route to the cavity closure is being investigated, based on the selective closure of the larger secondary hydroxyl face. One of the most difficult steps, the selective functionalisation of the least reactive hydroxyls O(3) is depicted in scheme 3, left part [17c]. An interesting aspect of this route is the use of exclusively natural reagents and the subsequent possibility of re-opening the cavity by a biochemical process.

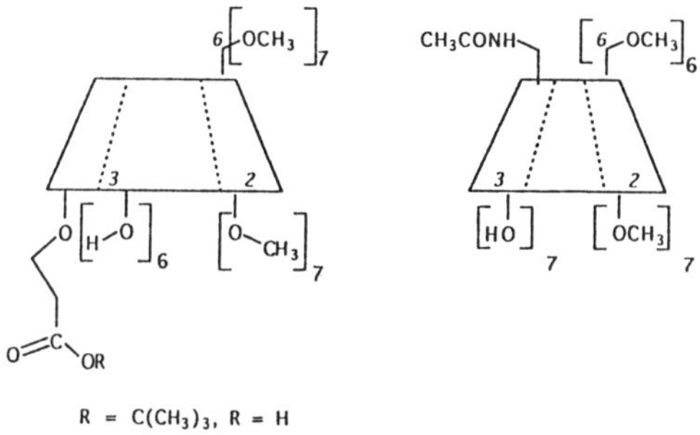

Scheme 3

5. Symmetry and Chirality

A high molecular symmetry of the host may have favourable or adverse effects on a desired property. Thus, the high axial symmetry of CD's contributes to a rather cylindrical shape of the cavity and to the subsequent poor molecular recognition, especially for guest molecules lacking hydrogen bond potentialities. However, we will examine below aspects where the symmetry effects have valuable consequences.

5.1. SYMMETRICAL MOLECULAR MODELS FROM EXPERIMENT

The CD structures present the complete set of crystallographic rotation axes, i.e. order 2,3,4,6. In certain crystal structures, a coincidence between the CD molecular axis and a crystallographic axis has been observed:

α-CD molecules may adopt a six-fold and a three-fold axis

γ-CD molecules may adopt a four-fold axis

β- and γ-CD structures present a two-fold axis relating two CD molecules within a dimer (symmetry controlled self-recognition, section 2.1).

An interesting consequence is the establishment from experiment of precise molecular models having the required symmetry. While a symmetrisation procedure by computer is necessary in the case of general crystallographic position, the occurrence of a special position reduces the number of parameters in γ-CD by a factor of 4 and even provides directly the whole symmetrical model for α-CD.

A further case of host-guest molecular recognition related to symmetry is provided by the crystal structure of the complex γ-CD.[12/4]crown ether, where the four-fold molecular axis of the guest coincides with the four-fold axis passing through the host [19].

5.2. CHIRALITY AND ENANTIOMERIC DISCRIMINATION

The general problem of molecular recognition through residual forces is a very difficult one for the today state-of-the-art. We focus now on a particular case, namely that of chiral recognition: we 'only' ask the system to discriminate between two enantiomeric configurations of the guest molecule.
Native CD molecules contain a large number of chiral centres, but enantiomeric discrimination (e.d.) upon complexation cannot be achieved in general. Native and modified CD's have widely been used to study e.d., mostly for polar guest molecules able to form hydrogen bonds. The chiral discrimination problem is considerably more difficult for molecules lacking functional groups such as spiroacetal, where the e.d. discrimination may only arise from residual forces. Yet, in favourable cases, the enantiomeric selectivity can be considerably enhanced by the induced fit phenomenon.

Induced fit effect and chiral recognition
It has been found that only S-spiroacetal is included in the overmethylated DIMB [13,17a]. This enantiomer perfectly fits into a chiral pocket formed at the CD centre and closed by the 'extra-methoxy' group.
This unsymmetrical 'overmethylated' DIMEB derivative is obtained in small quantities as impurity during the preparation of the (seven-fold symmetrical) DIMEB, but its synthesis with an acceptable yield is very difficult. However, the presence of the induced fit phenomenon in this type of compounds and the above efficient discrimination in the crystal state encouraged us to further investigate the use of modified CD's.

Experiments have been carried out with spiroacetal and ten modified CD's, symmetrically alkylated and/or monosubstituted on the primary hydroxyl. Only in two of them e.d. has been detected. In the mono-acetylamino derivative (scheme 3, right) a low but significant 12% enantiomeric excess was measured [17b].
In the TM-α-CD.spiroacetal complex [18] a total e.d. occurs: only the R-spiroacetal enantiomer is included, whereas only the S-enantiomer appears in the overmethylated DIMEB. Moreover, the use of TM-α-CD leads to a practical quantitative enantiomeric separation of spiroacetal.

Computer simulations
It would be instructive if we had a clear explanation on a structural basis of the above double e.d. of R- and S-enantiomers. In the present context of crystallographic results we mention the qualitative approach based on the host d guest van der Waals contours (rigid spheres centered at the atomic positions) for the 'overmethylated' DIMEB structure (section 3.1 and Figure 7b). The remarkable feature is the shape complementarity denoting an excellent host-guest fit for the experimentally observed S-enantiomer. The next step would be to examine the fit of the R-spiroacetal molecular model into the same host (p. 479, this book).

However, in order to convincingly account *ab initio* for the observed enantiomeric discrimination, we should simultaneously study the effect of all parameters describing the guest position, the CD conformation (induced fit effect), and furthermore the position of the solvent molecules as well as that of the neighbouring CD molecules. Clearly the simple examination of van der Waals contours, or crude energy calculations may not be sufficient for the present problem, and advanced methodologies such as these described by G. Wipff (p. 67, this book) are necessary.

In conclusion, e.d. does occur is certain CD systems. By chemically modifying the natural CD molecules, it is possible to create host-guest-solvent systems able to include either the R- or the S-guest molecule. The molecular recognition ability of the host may be enhanced by the induced fit phenomenon.

6. Concluding remarks

The study of various SM systems controlling the release rate of pheromones is linked to fundamental questions pertaining to the global scheme:

Mixing molecular species in solution ---> Supermolecules in solution --->
SM architecture in the crystal ---> Real crystal with defects --->
Global guest release rate

In this review we have encountered very different models, depending on the chemical nature of the host, guest and solvent. In certain crystals, a clearly intramolecular inclusion complex, most probably existing in solution with a dynamical in-out configuration, is actually observed in the crystal state, such as the overmethylated DIMEB complex. In other cases, a complete reorganisation occurs upon crystallisation and a novel SM entity appears in the crystal. For instance, in the β-CD.spiroacetal crystal the guest molecules are burried within the CD channel (figure 2). We can say that in the first case the release rate is mainly under molecular control, in the latter, under crystal control.

At this point it is important to comment on the relative importance of molecular versus crystalline factors of the stability of SM compounds. This problem is connected with the correlation between the structure of an SM entity in solution and in the solid state.
The nature of a crystal may merely yield a discontinuity of the stability (and of other properties as well) when shifting from one crystalline compound to another inside a given chemical family of host molecules: a small change in the molecular structure of the host may lead to a different crystalline structural type and a different symmetry, with a discontinuous variation of the properties in general.

However, it is also possible that small variations of the host chemical nature lead to a series of quasi-isomorphous crystals; in this case the variation of a property within this series may be very small.
Thus, crystallisation of SM compounds may provide a certain control of an inclusion/release process. The different crystal types achieve a very coarse screen. For each one of these structural types, isomorphous series may further allow a fine tuning.

On the other hand, a usual chemical modification of the host, such as substitution, may provide to some extent a molecular control both in solution and in the crystal state. In particular, a partial closure of the cavity can be achieved via the induced fit effect, even in the usual CD compounds where the cavity is open to the environment. The efficiency of this effect appears clearly with the enantiomeric discrimination and successful separation of the spiroacetal enantiomers. Here the combination of a large number of weak residual forces, together with the induced fit effect, may lead to an efficient chiral discrimination.

A further step towards molecular control can be provided by systems based upon mere imprisonment of the guest molecule in the host cavity, such as the photoclosed cinnamoylated cyclodextrins. Note, however, that their use as solid state devices, a form usually required for large scale applications, will again involve a new milieu and will require a further study.

Clearly, both aspects, molecular and crystalline, are necessary for an efficient control process by a solid state device, and a careful conjugation of their effects presides over the final output.

Acknowledgement. The study and practical evaluation of CD complexes with pheromones of olive tree pests has been undertaken in a collaborative project partially sponsored by the EC Value Program involving: the NCSR 'Demokritos', Greece, the CNRS-UPR180 Laboratory, France, and Agrigence-BCS Ltd, UK.

REFERENCES

1 (a) Tsoucaris, G., Rysanek, N., Le Bas, G., Villain, F., Hadjoudis, E., Moustakali-Mavridis, I. (1987) Chiral Interaction and Fine Structural Fit in Cyclodextrin Clathrates, *Acta Cryst. A,* **43**, Supplement, XIVth IUCr, Perth (Australie), N° 04.X-5, p.C-65.
(b) Rysanek, N., Le Bas, G., Villain, F., Tsoucaris, G., Inclusion of the main pheromone component of dacus oleae : 1,7 Dioxaspiro (5,5) Undecane in β Cyclodextrin, submitted to *Acta Cryst. C* .

2 (a) Mazomenos, B., Kondilis, P., Hadjoudis, E., Moustakali,I. and Tsoucaris, G.(1987) CD's as Dispensing System for Dacus Oleae Pheromone, *Proceedings of International Symposium on Controlled Release of Bioactive materials*, Chicago, Illinois.

(b) Kondilis, P., Mazomenos, B., Moustakali, I., Hadjoudis, E. and Tsoucaris, G. (1990) Inclusion complexes of cyclodextrins with the pheromone of the olive fruit fly *dacus oleæ*., *Minutes of the Fifth International Symposium on Cyclodextrins* ,Paris, Ed. Duchêne, D. p 578.

3 (a) Le Bas, G. (1985) Conformations chirales : leurs interactions avec des récepteurs chiraux. Etude des Cyclodextrines, Thèse de Doctorat d'Etat, Université Paris VI.

(b) Le Bas, G. and Rysanek, N. (1987) Structural Aspects of Cyclodextrins, in *Cyclodextrins and their Industrial Uses*, Chap.III, p.105 ; Ed. Duchêne, D., Editions de Santé, Paris.

4 Mentzafos, D., Mavridis, I.M., Le Bas, G. and Tsoucaris, G. (1991) Structure of the 4-tert-Butylbenzyl Alcohol-β-Cyclodextrin Complex. Common Features in the Geometry of β-Cyclodextrin Dimeric Complexes, *Acta Cryst. B* , **47**, 746-757.

5 Nicolis, I., Villain, F., Coleman, A.W., de Rango, C. (1994) The X-ray Crystallographic Structure of the two to one α-cyclodextrin - Acetone, 9 H_2O Compound, *Supramolecular Chemistry*, **3**, 251-259.

6 Le Bas, G. and Tsoucaris, G. (1994) Two dimensional hydrogen bond network in β-cyclodextrin complex, *Supramolecular Chemistry*, **4**, 13-16.

7 Le Bas, G., Mason, A., (1994) Neutron Diffration Structure of α-cyclodextrin cyclopentanone hydrate at 20K, *Acta Cryst.B*, **50**, 717-724.

8 Anagnostopoulou-Konsta, A., Apekis, L. and Tsoucaris G. (1992) Dielectric Investigation of β-cyclodextrin hydrates and inclusion complexes, *IEEE Transactions on Electrical Insulation*, **27**, 801-806

9 (a) Rontoyianni, A., Mavridis, I.M., Hadjoudis, E. and Duisenberg, A.J.M. (1994) The Crystal Structure of the Inclusion Complex of Cyclomaltoheptaose (β-cyclodextrin) with 4-*tert* -Butylbenzoic Acid, *Carbohydrate Research*, **252**, 19 - 32.

(b) Mavridis, I.M., Hadjoudis, E. and Tsoucaris, G. (1991) The crystal structure of the inclusion complex of cyclo-maltoheptaose (β-cyclodextrin) with 3,3-dimethylbutylamine, *Carbohydrate Res.*, **220**, 11-21.

(c) Hamilton, J.A., Sabesan, M.N., Steinrauf, L.K. (1981) Structure of cyclo-maltoheptaose inclusion complexes: crystal structure of substituted Benzoic acid and phenol derivatives, *Carbohydrate Res.*, **89**, 33-53.

(d) Mavridis, I.M. and Hadjoudis, E. (1992) The Crystal Structure of the Inclusion Complex of Cyclomaltoheptaose (β-cyclodextrin)with 4-*tert*-butyltoluene, *Carbohydrate Research*, **229**, 1-15.

(e) Rontoyianni, A. and Mavridis, I.M. (1994) The Crystal Structure of the Inclusion Complex of Cyclomaltoheptaose (β-cyclodextrin) with 3,5-Dimethylbenzoic Acid, *Journal of Inclusion Phenomena and Molecular Recognition in Chemistry*, **18**, 211 - 227.

10 (a) Mavridis, I.M., Mentzafos, D., Hursthouse, M.B., Structure of the β CD Z-9 dodecenol 2:1 complex, submitted to *Acta Cryst. C* .

(b) Rontoyianni, A. and Mavridis, I.M., Structure of the β CD nonanoïc acid 1:1 complex, submitted to *Acta Cryst. C* .

11 Mavridis, I.M., Yannakopoulou, K. and Papaioannou J. (1995) Structure and release properties of the complex between β-cyclodextrin and Z-7-tetradecenal, a pheromone of the *Prays* moth in solution and in solid state, *Abstracts of "Crystallography of Supramolecular Compounds"*, p.116.

12 Mentzafos, D., Mavridis, I.M. and Schenk, H. (1994) Crystal Structure of the 1:1 Complex of Heptakis (2,3,6-tri -*O* - methyl) cyclomaltoheptaose (permethylated β-cyclodextrin) with ethyl laurate, *Carbohydrate Research*, **253**, 39-50.

13 Rysanek, N., Le Bas, G., Villain, F. and Tsoucaris, G. (1992) Structure of the (1:1:1) Complex 2a,2b,2c,2d,2e,2f,3a,3g,6a,6b,6c,6d,6e,6f,6g-Pentadeca-*O* -methyl-β-cyclodextrin-1,7-Dioxaspiro [5.5] undecane - Methanol, *Acta Cryst. C*, **48**, 1466-1471.

14 Rysanek, N., and Tsoucaris, G. (1995) The crystal structure of heptakis (2,3,6-tri-O-methyl) α-cyclodextrin dimethylformamide complex, unpublished work.

15 Rysanek, N., Le Bas, G., Viswamitra, M.A., Villain, F., Tsoucaris, G. (1990) The crystal structure of heptakis (2,3,6-tri-O-methyl) β-cyclodextrin dimethylformamide complex), *Minutes of the Fifth International Symposium on Cyclodextrins*, Paris, Ed. Duchêne, D., p.150.

16 (a) Arad-Yellin, R., Green, B.S. (1994) Photochemical closing and opening of the guest-binding cavity of cyclodextrins, *Nature*, **371**, 320-322.
(b) Arad-Yellin, R., Tsoucaris, G. and Green, B.S. (1995) Trapping molecules in host-inclusion cavities, *Abstracts of "Crystallography of Supramolecular Compounds"*, p.86.

17 (a) Galons, H., Gnaim, J., Rysanek, N., Le Bas, G., Villain, F. and Tsoucaris, G. (1993) Inclusion, Complexation and Efficient Enantiomeric Discrimination by Regiospecific Overmethylation of a Dimethyl β-Cyclodextrin, *Tetrahedron Asymmetry*, **4**, 181-184.
(b) Bahaddi, Y., Galons, H., Rysanek, N. (1995) Enantiomeric discrimination in solid state cyclodextrin complexes, *Bull. Soc. Chem. Fr.*, **132**, 330-332.
(c) Bahaddi, Y., Galons, H. and Tsoucaris, G. (1995) New derivatives of cyclodextrins, *Abstracts of "Crystallography of Supramolecular Compounds"*, p.90.

18 Mavridis, I.M., Mentzafos, D., Yannakopoulou, K., Dandika, K. and Hadjoudis, E. (1995) Structure of the 2,3,6,-(Tri-O-methyl)-β-cyclodextrin/ 1,7-Diasaspiro(5,5)undecane inclusion complex. Resolution of the racemic mixture, *Abstracts of "Crystallography of Supramolecular Compounds"*, p.117.

19 Shigehiro, K., Hirotsu, K. and Higuchi, T. (1986) Crystal and molecular structure of the γ-cyclodextrin/12-crown-4 1:1 inclusion complex, *J. Chem. Soc., Chem. Com.*, **C48**.

CONSTITUTION AND STABILITY OF CLATHRATE HYDRATES

YU.A.DYADIN
Institute of Inorganic Chemistry, Russian Academy of Sciences, Siberian Branch, 630090, Novosibirsk, Russia

Abstract. Several aspects of clathrate formation in water systems, such as the stability of hydrates depending on the complementarity of the host and the guest subsystem , their stability at high pressures depending on the framework structure, the ability of water as a host component to adapt itself to the guest subsystem, including both very small guest molecules (like He) in the cavities of the frameworks of ices *Ih, Ic,* and *II,* and average-size molecules (**CH_4, THF**) in the cavities of clathrate frameworks, are considered from the structural-thermodynamic point of view. In the latter case there are two types of cavities: stable (which can be vacant) and unstable (which can not be vacant) ones. Inclusion of quite big molecules (like methylcyclohexane) is also possible, but this would require additional stabilization at the expense of the filling of the stable cavities by a help-component.

1. Introduction

At present there is an increasing interest in clathrate (gas) hydrates as a potential inexhaustible source of fuel. The amount of hydrocarbon raw material in the form of gas hydrates is evaluated to be approximately $2 \cdot 10^{16} m^3$ of methane [1]. (For comparison: Earth atmosphere contains $8 \cdot 10^{17} m^3$ of oxygen). Industrial exploiting of the deposits of gas hydrates will not only make it necessary to find a number of complicated technological solutions but undoubtedly cause enormous and grave ecological problems. The former as well as the latter requires fundamental knowledge and primarily structural and thermodynamic knowledge about water and its ability to construct one or other clathrate framework depending on the guest molecule dimension and shape. The guest-host interactions here are usually of van der Waals type only, and in this sense gas hydrates are typical clathrates and ideal representatives of supramolecular chemistry. However, it is well known that water as a substance has a number of wonderful properties. The properties which are most important from the point of view of clathrate chemistry are the following: several frameworks close in energy content with molecular size cavities can be build from its molecules; common ice can dissolve only very small molecules; the ice melting point decreases considerably under pressure due to high porosity of the structure which is not common for crystalline bodies. This results in the enormous variability

G. Tsoucaris et al. (eds.), Crystallography of Supramolecular Compounds, 223–241.

in clathrate formation, makes the chemistry of hydrates look nice and attractive, and shows vividly unique potentials of supramolecular chemistry.

So, in this paper we discuss the stability of clathrate hydrates which depends on their structure and the complementarity of the host and guest subsystems under widely varying conditions.

2. Water as a Host Component

Water as a host component possesses a number of unique peculiarities, which result from tetrahedral coordination of the oxygen atom in a water molecule and relatively variable *H*-bond lengths and angles.

(1) Solid water (ice *Ih*) is a quite porous structure (Fig. 1) [2]. The packing coefficient (k = 0.43) is too low for a crystalline body [3]. That is why pressure destabilizes this ice modification (dt/dp = -7.7 *K/kbar* for melting). When diminishing under pressure the ice crystallization field leaves the *P,T,X*-space, making the process of clathrate formation more distinct than it is under ordinary pressure. This is the reason why many hydrates melting incongruently begin to melt congruently, and in a number of systems where no clathrate formation occurs under ordinary conditions clathrate hydrates have been observed at higher pressures.

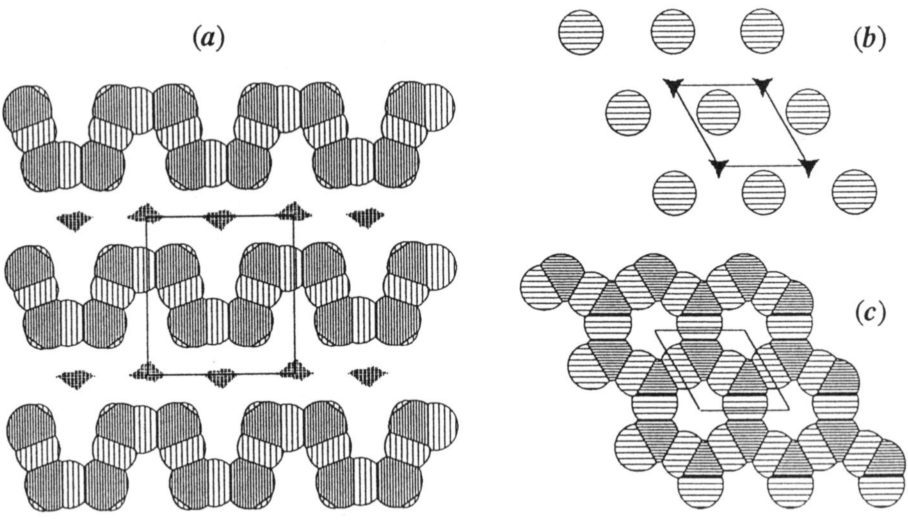

Figure 1 The dimensions and the shape of the channels in the (1,1,0)- (***a***) , (0,0,1)- (***b***) and (0,0,4)- (***c***) sections of the ice *Ih* structure. The unit cell and the place where the center of the helium molecules (R_{He} = 1.22A) is to be found are shown in the Figure.

(2) In spite of the fact that the ice *Ih* structure is very porous, its cavities are small (d = 3.54*A*, Fig. 2). They can accommodate only **He, Ne** and $\mathbf{H_2}$ molecules. For other guest molecules these cavities are too small. However, as I have already mentioned, due to tetrahedral coordination of oxygen in a water molecule and relatively variable *H*-bond lengths and angles water can build a number of framework structures close in energy content of which those are stabilized that possess the best complementary ability towards the given guest and for which most suitable external conditions are created. At present the number of polyhedral frameworks (TABLE 1) which can include guest molecules 3.8*A* to 10 *A* in size and larger exceeds 10 [4-7].

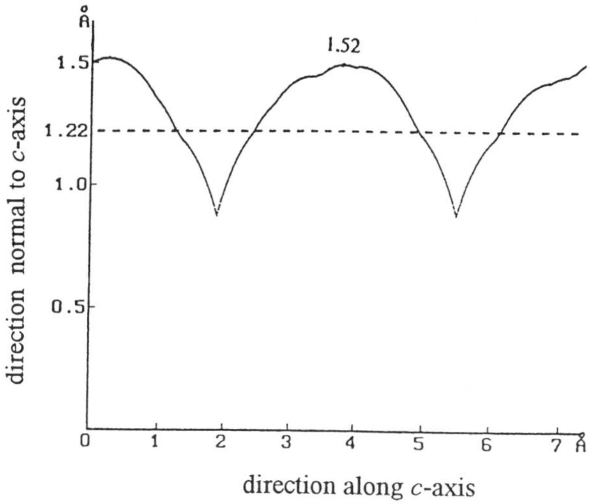

Figure 2. The maximum radius of a sphere inscribed in the channel of the ice *Ih* structure along the *c*-axis.

(3) A peculiarity of hydrates is that their frameworks have two kinds of cavities: the so called, large and small ones (Fig. 3). For a structure to be stable the large cavities would have to be practically fully occupied, while the small ones may be partially occupied or even vacant. Therefore, it would be more correct, more fundamentally to distinguish between unstable (large) and relatively stable (small) cavities. We do not change the accepted, habitual names, but will keep this in mind.

3. Complementarity and Stability

Now I will show by the example of hydrates of quaternary ammonium salts, the dependence of the stability of hydrates on how effectively the guest uses the vacant space. In this case the guest is a cation each of whose radicals occupies one of the above cavities arranged in the structure tetrahedrally (Fig. 4). The vertex shared by all four cavities is occupied by the central cation

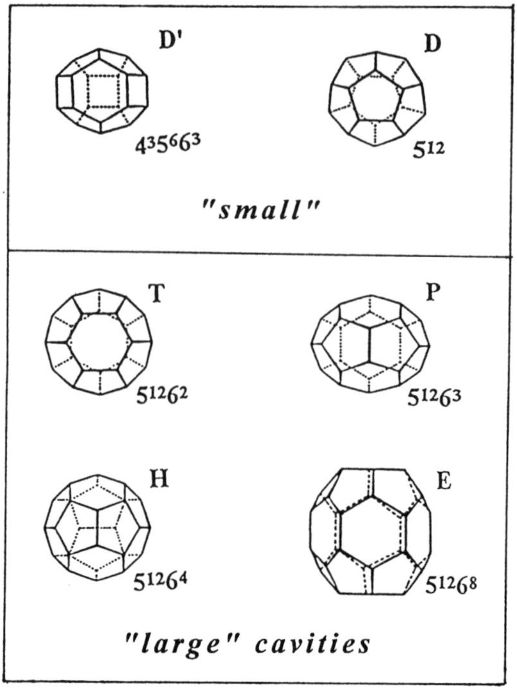

Figure 3 The most common cavities in the clathrate hydrate frameworks

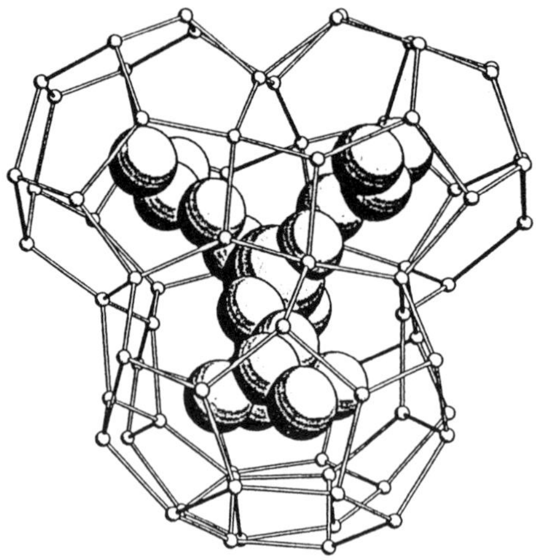

Figure 4 **i-Am$_4$NF** in the ***P'***-cavity (***P'*** is 15-hedron $4^25^96^37^1$)

TABLE 1 Structural information on idealized water clathrate frameworks [2, 4-7]

type, space group	cell parameters, A; ρ, g/cm^3	unit cell formula	h_s,*	examples of guest molecules
		I c e f r a m e w o r k s		
hexagonal, ice *Ih,* ***P6$_3$/mmc***	*a* 4.48 *c* 7.31 ρ 0.917	**2C*4H$_2$O#**	2	**H$_2$, He, Ne**
cubic, ice *Ic,* ***F$\bar{4}$3m***	*a* 6,35 ρ 0.93 (-130°)	**8C*8H$_2$O**	1	**H$_2$, He, Ne**
trigonal, ice *II* ***R$\bar{3}$***	*a* 7.78 α 113.1° ρ 1.15	**2C*12H$_2$O**	6	**H$_2$, He, Ne**
	F r a m e w o r k s c o n t a i n i n g		*D-c a v i t y*	
cubic I *CS-I* ***Pm3n***	*a* 12.0 ρ 0.796	**6T*2D*46H$_2$O**	$7\frac{2}{3}$ *a* $5\frac{3}{4}$ *b*	*cyclo*-**C$_3$H$_6$,** **CH$_4$, Xe, BF$_4$**
cubic II *CS-II* ***Fd3m***	*a* 17.1 ρ 0.812	**8H*16D*136H$_2$O**	17 *a* $5\frac{2}{3}$ *b*	**SF$_6$, THF, CCl$_4$** **Ar, Kr**
hexagonal I *HS-I* ***P6/mmm***	*a* 12.4 *c* 12.5 ρ 0.719	**2P*2T*3D*40H$_2$O**	10 *a* 38 *c*	**Br$_2$** (?) ***i*-Am$_4$NF**
hexagonal II@ *HS-II* ***P6$_3$/mmc***	*a* 12.1 *c* 19,7 ρ 0.812	**4H*8D*68H$_2$O**	17 *a* 15 *d*	**(CH$_3$)$_4$NF+** **(C$_3$H$_7$)$_4$NF**
hexagonal III *HS-III* ***P6/mmm***	*a* 12.3 *c* 10.2 ρ 0.768	**E*2D'*3D*34H$_2$O**	34*e*	*cyclo*-**CH$_3$C$_6$H$_{11}$ + H$_2$S**
tetragonal I *TS-I* ***P4$_2$/mnm***	*a* 23.5 *c* 12.3 ρ 0.757	**4P*16T*10D*172H$_2$O**	8.6 *a* 32.4*c*	**Br$_2$** **Bu$_4$N$^+$**
orthorombic *RS* ***Pbam***	*a* 23.5 *b* 19.9 *c* 12.1 ρ 0.782	**4P*4T*14D*148H$_2$O**	$12\frac{1}{3}$ *a* 34.5*c*	**Br$_2$** (?) **Bu$_3$PO**

(the continuation of TABLE 1)

type, space group	cell parameters, A; ρ, g/cm*3	unit cell formula	h_s,*	examples of guest molecules
		Frameworks without D-cavity		
tetragonal II	*a* 15.4		8.5 [a]	
TS-II	*c* 12.0	**8P*4d*68H$_2$O**[$]	7.5[c]	**(CH$_3$)$_4$NOH**
I4/mcm	ρ 0.709		32.0[c]	***i*-Am$_4$PBr**
tetragonal III	*a* 16.9		7.25[a]	
TS-III	*c* 17.1	**16P'*4Q*116H$_2$O**[$]		
I4$_1$/a	ρ 0.710		27.0[c]	***i*-Am$_4$NF**
cubic III	*a* 18.8			
CS-III	ρ 0.701	**16A*18Q*156H$_2$O**	9.75	**(CH$_3$)$_3$CHNH$_2$**
I43d				
cubic IV	*a* 7.68		6	**HPF$_6$**
CS-IV	ρ 0.793	**2B*12H$_2$O**[$]		
Im3m			5	**(CH$_3$)$_4$NOH**

* h_s is the structural hydrate number

[#] **C** is the wide part of the channel in the ice structures where the guest molecules can be accomodated

[$] **d, Q, B, P', A** is decahedron ($4^2 5^8$), 8-hedron ($4^4 5^4$), 14-hedron, 15-hedron ($4^2 5^9 6^3 7^1$) and 17-hedron, respectively, see also Figure 3.

[a] only large cavities are occupied

[b] all types of cavities are occupied

[c] see paragraph **3.**

[d] **(CH$_3$)$_4$N$^+$** is arranged in the *H*-cavity, **(C$_3$H$_7$)$_4$N$^+$** occupies the D_4-cavity, **F$^-$** is included in a hydrophilic way and this double hydrate formula is **(CH$_3$)$_4$NF*0.5(C$_3$H$_7$)$_4$NF*15H$_4$O**

[e] this structure is stable only as double hydrates

atom which displaces the water molecule and does not form any *H*-bonds with the adjacent molecules [5-8] (*hydrophobic inclusion*). The simple anion (of the **F', Cl'** type and others) with water molecules build a framework, forming *H*-bonds between them (*hydrophilic inclusion*). Figure 5 shows the position of the butyl and isoamyl radicals in the *T*-cavity. As is seen from the Figure, isoamyl uses the cavity space efficiently, which is why its hydrate, ***i*-Am$_4$NF*32H$_2$O** of tetragonal structure I melts at a temperature 4.2^o higher than does the hydrate **Bu$_4$NF*32H$_2$O** of the same structure (Fig. 6) [9]. Similar energies of the water frameworks are the reason why a slight change of conditions is sufficient for the same guest to stabilize different clathrates. It is seen from Figure 6 that in each of the two systems more than one hydrate crystallizes, and in tetraisoamylammonium system even in the range 1%

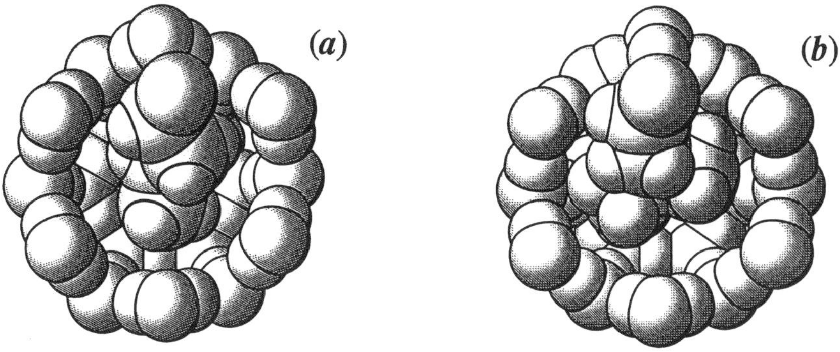

Figure 5 **Bu** *(a)* and **i-Am** *(b)* in the *T*-cavity

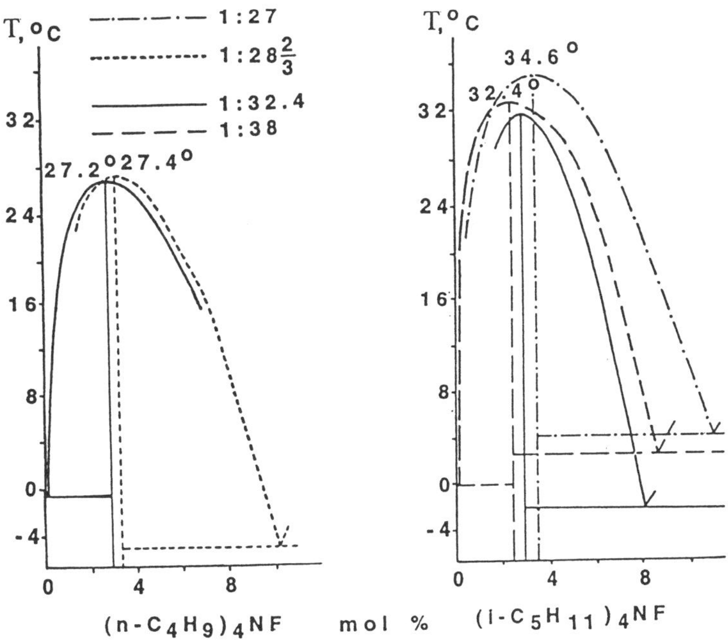

Figure 6 Phase diagram of the $\mathbf{Bu_4NF}$ — $\mathbf{H_2O}$ and $\mathbf{i\text{-}Am_4NF}$ — $\mathbf{H_2O}$ systems. The melting point of tetragonal tetraisoamylammonium fluoride hydrate 1:32.4 (continuous line) 4.2^{o} higher than that of the analogous hydrate of tetrabutylammonium fluoride.

mol three (!) hydrates of different structures are formed. This means that in water systems the guest and host subsystem are interconnected more closely than, for example, in classic hydroquinone or urea systems where the ability of the host to adapt for the guest results in some distortion of the framework while the structural pattern remains the same. In this case the situation changes so that not only the host can choose the most suitable guest, but the guest also acquires the ability to choose the best suited framework.

A vivid picture can be observed in the stability set of hydrates of tetrabutylammonium carboxylates of the same structure (Fig. 7,*a*) [10]. The stability of the hydrate increases as the anion hydrophobic part fills the vacant dodecahedral cavities and drops abruptly if its size exceeds that of the cavity. The dodecahedral (small) cavities in *TS-I* are vacant if the anion is formate, which, by displacing two water molecules with its oxygen atoms, builds one of the framework edges. Its melting point is 11.8^{o}C. The acetate-anion methyl group already occupies a part of the small cavity, additionally stabilizing the structure, and the hydrate melting point increases to 14.9^{o}. As is seen from Figure 8,*a* propionate (16.8^{o}) and butyrate (15.5^{o}) hydrates correspond to the cavity most closely. Further growth of the anion causes some distortion of the cavity and decrease of the stability of the hydrates. Besides, in this case butyrate stabilizes the large cavity quite well, forming a second hydrate within the same framework, but of different composition, 1:40. The small cavities are again vacant and the stability is lower. The picture for most stable hydrates in these series of systems is analogous, but stability maximum is removed to butyrate (Figure 7,*a*, line 1)

A similar dependence of the stability of the hydrates on the spatial complementarity of the guest and host subsystems is observed in the series of hydrates of dicarboxylates (Figs. 7,*b*, 8,*b*). However, the choice of the guest is more rigorous in this case since the correspondence between the dimensions of the hydrophobic chain and the cavity is not enough and certain conditions have to be created for the hydrophilic inclusion of the second carboxylic group to become possible.

4. Structure of Hydrates and Their Stability at High Pressures

Since in the structures of different hydrates the ratio of the small cavities to the large ones varies in the wide range ($2>Rt>0$), the packing coefficients vary considerably ($0.47<k<0.60$) if the small cavities are vacant (the guest molecule can occupy only the large cavities) and the behaviour of hydrates under pressure varies, too: from destabilization (dt/dp = -2.5 *K/kbar* for **THF*17H$_2$O**; k = 0.485) in the case of the classic cubic structure II to a considerable stabilization (Fig. 9) in the case of *CS-I* hydrates (dt/dp = 7.0 *K/kbar* for **C$_2$H$_4$O*6.7H$_2$O**; k = 0.537), and finally, of hydrates with completely occupied cavities dt/dp is as much as 10 *K/kbar* for **THF*0.5Pr$_4$NF*16H$_2$O** (k = 0.591) [5]. In the last example the large *H*-cavities are occupied by the average size THF-guest molecules and the small

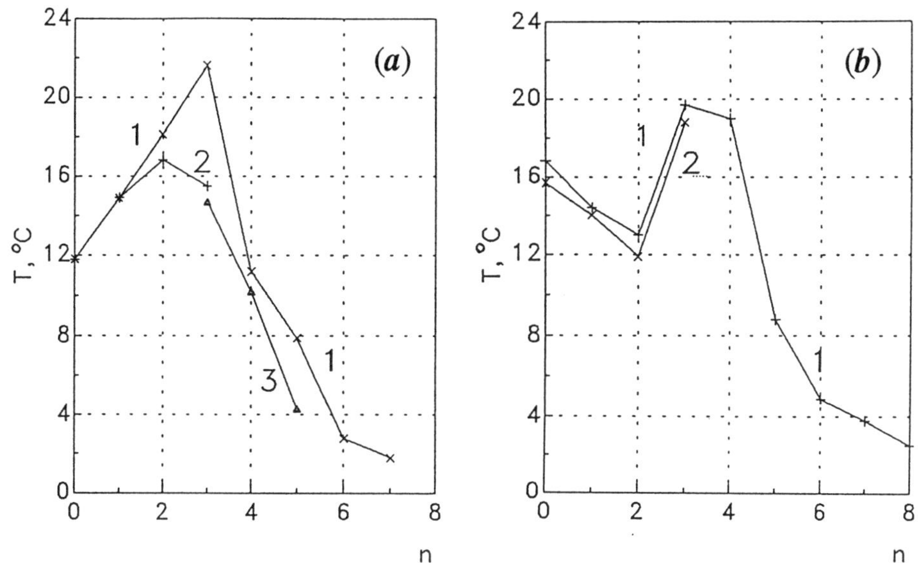

Figure 7. The dependence of melting point of the hydrates of the tetrabutylammonium carboxylates - **$(Bu_4N)H(CH_2)_nCOO)$** - (*a*) and dicarboxylates - **$(Bu_4N)_2OOC(CH_2)_nCOO)$** - (***b***) on the anion length; n - the number of carbon atoms in the hydrocarbon anion chain, 1 - the most stable hydrate, 2 - tetragonal hydrate of the composition 1:31.4 (***a***) and 1:63 (***b***), 3 - tetragonal hydrate of the composition 1:40 (dodecahedral cavities are vacant).

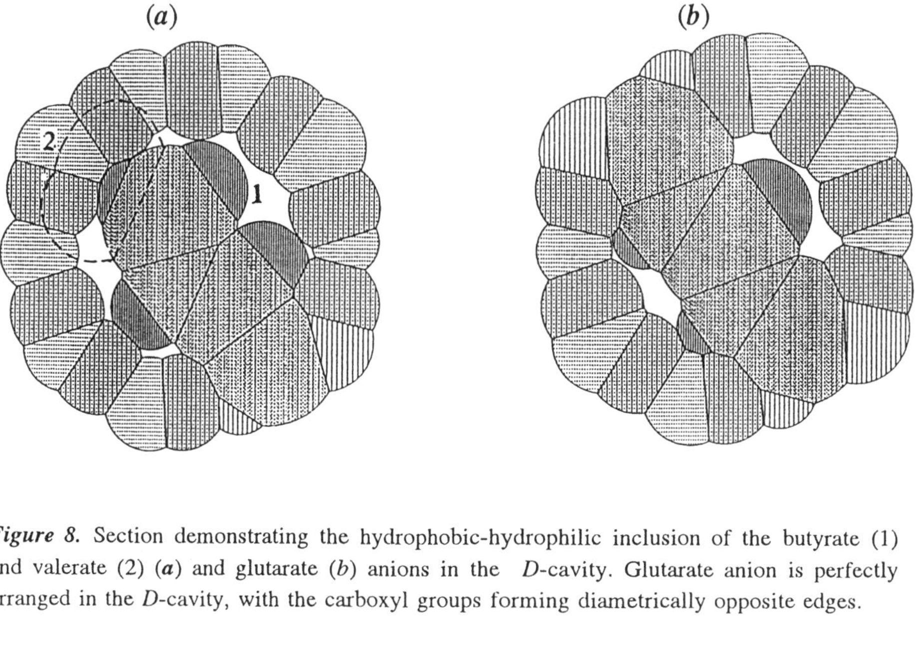

Figure 8. Section demonstrating the hydrophobic-hydrophilic inclusion of the butyrate (1) and valerate (2) (***a***) and glutarate (*b*) anions in the *D*-cavity. Glutarate anion is perfectly arranged in the *D*-cavity, with the carboxyl groups forming diametrically opposite edges.

dodecahedral ones (more exactly, the complex D_4-cavities) are occupied by the help tetrapropylammonium-cation guests (Fig. 10).

4.1 TRANSFORMATION OF HYDRATES OF THE CS-II INTO THE CS-I

As in the classic *CS-II* hydrates with stoichiometry 1:17 the *Rt*-ratio is the largest[1] (TABLE 2) for the hydrates which can exist with absolutely vacant small cavities

TABLE 2 The ratio (***Rt***) of the number of the small cavities (n_s) to large ones (n_l) in some clathrate water frameworks (see Table 1, too)

structure	CS-II	HS-II	RS	HS-I	TS-I	CS-I
Rt $= n_s / n_l$	2.0	2.0	1.17	0.75	0.5	0.33

their packing coefficients are the smallest (after the ice *Ih,* of course) and, therefore, they are destabilized by pressure. Sooner or later with pressure increase they become thermodynamically unstable and are replaced by hydrates with a stoichiometry about 1:7 (Fig. 11). There is good reason (stoichiometry about 1:7 and increasing melting (decomposition) point under pressure) to believe that the latter are hydrates of *CS-I* [11]. Recently [12] the parameters of the cubic unit cell (a = 12.08 *A*) were determined, giving support to this conclusion. The list of the studied systems in which *CS-II* hydrates are formed at the ambient pressure, their upper pressure stability limit and lower pressure stability limit for the appearing hydrates 1:7 are given in TABLE 3. In the only case with thiophene, hydrate 1:7 was not discovered because its crystallization field was overlapped by the guest crystallization field.

4.2 OPTIMUM PRESSURE RANGE AND UPPER STABILITY BOUNDARY OF CLASSIC CLATHRATE HYDRATES

The presence of the guest component makes it possible to solve quite a complicated problem of constructing water structures with moderately good packing which makes them stable under pressure [5]. Indeed, as we can see, the guest inclusion in clathrate water frameworks causes the packing coefficient to increase from 0.43 for ice *Ih* to 0.47-0.52 for *CS-II* hydrates (with the stoichiometry 1:17) via 0.53-0.58 for *CS-I* hydrates to 0.59-0.60 for hydrates with all the cavities completely

[1]In the framework of hexagonal structure III (*HS-III*) this *Rt*-ratio is equal to five, i.e. it is considerably higher than that for CS-II (*Rt* = 2), but double hydrates of this structure are known to exist only when a considerable number of the small cavities is occupied by any help-guest molecules.

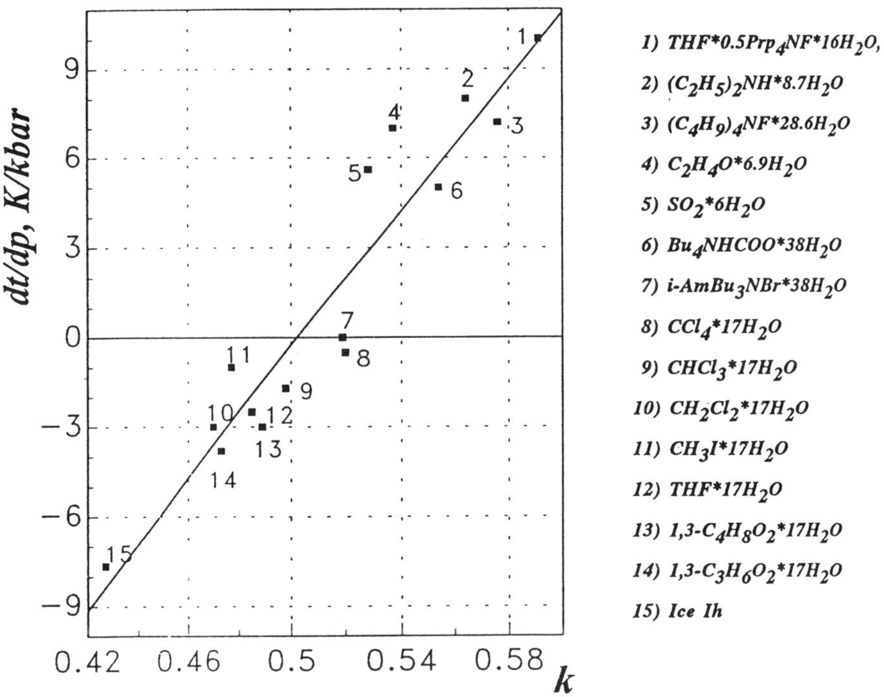

Figure 9. Packing coefficient (***k***) dependence of ***dt/dp*** of melting points of some hydrates

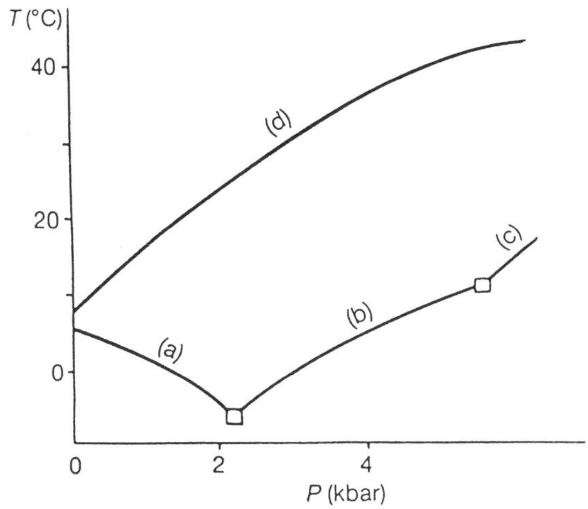

Figure 10 . Congruent melting curves of tetrahydrofuran hydrates: (a) **$THF*17H_2O$**, (b) **$THF*7H_2O$**, (c) **$THF*5H_2O$**, and (d) the double hydrate **$THF*0.5Pr_4NF*16H_2O$**.

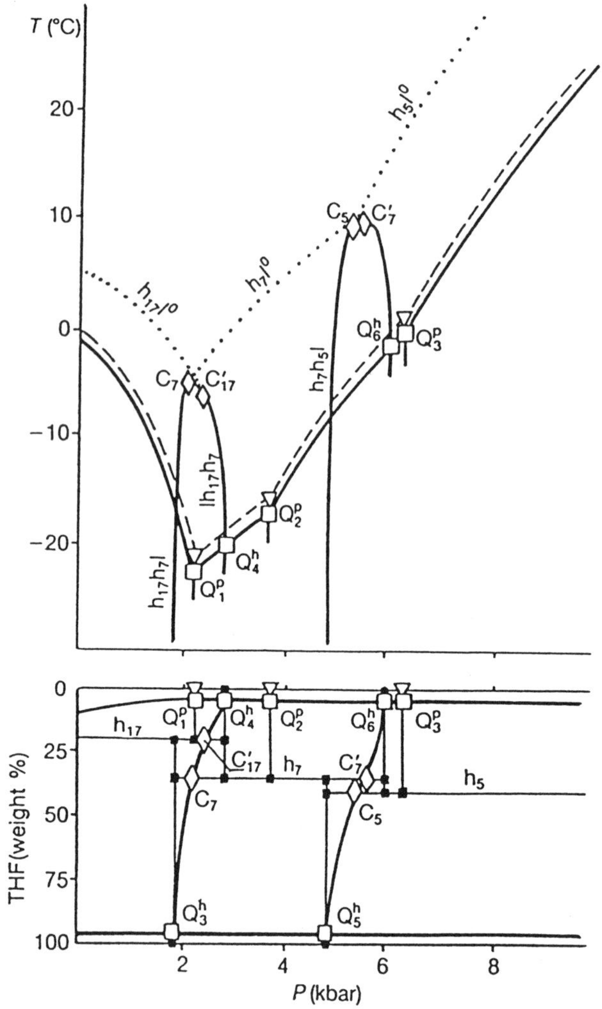

Figure 11 P,T- and P,X-projections of the water–tetrahydrofuran system phase diagram (P,X-projection does not show its subsolidus part). Solid lines show 3-phase equilibria; dashed lines show monovariant equilibria in the water system; dotted lines mark the hydrates congruent melting monovariant equilibria. □—the points of the 4-phase equilibrium (Q) in a binary system. On the P,X-projections the squares designate the compositions of the coexisting phases in the quadruple point (light squares stand for the liquid phase); ◇—the points of nonvariant equilibrium of the beginning (C) and the end (C′) of the hydrate congruent melting with increasing pressure; ∇—the points of nonvariant equilibria in unary systems.

TABLE 3 The behaviour of the hydrates of *CS-II* at high pressure

guest	m.p., °C	*dt/dp, K/kbar**	*B. kbar*#	*A, kbar*&
TMO	-9.2i@	-3.0	1.1	P<0.001
THF	4.3c	-2.5	2.8	1.8
1,3-dioxolane	-2.6c	-3.8	2.4	0.9
1,3-dioxane	-3.5c	-3.0	3.0	1.8
1,4-dioxane	-11.3i		2.5	2.0
tetrahydropyran	-0.8i	-2.0		1.2
CH_3I	4.5s	-1.0		0.5$
CH_2Cl_2	i.6s	-3.0		0.6$
$CHCl_3$	1.7s	-1.7		2.6$
CCl_4	1.4s	-0.5		0.8$
CH_2Br_2	3.4s			1.7$

* for melting point at 1 atm

upper pressure stability boundary of CS-II hydrates

& pressure of appearance of 1:7 hydrates

@ c, i and m - congruent melting point, incongruent melting point with decomposition to ice and liquid and incongruent melting point with decomposition to two liquids

$ pressure of quadruple point $l_1h_{17}h_7l_2$ where l_1, l_2 - liquids rich in water and guest, respectively, h_{17}, h_7 - hydrates **$G*17H_2O$** and **$G*7H_2O$**, respectively.

occupied (Fig. 9). As we have already mentioned that is the reason why all hydrates are more stable to pressure effect than ice *Ih* and there is a broad spectrum pressure effect on the hydrate decomposition: from destabilization to different degrees of stabilization. That is why clathrate formation is more pronounced under pressure even for the hydrates destabilized by pressure (Fig. 12). However, the complementarity of the guest molecule and cavity (even in the best case) is yet not sufficient for the formation of the closest packing with the packing coefficient about 0.74. Therefore, under certain conditions the density of clathrate structures becomes equal to that of the fluid phase (phases), which causes the destabilization of the hydrate as is the case with the hydrates **CH_4** [13] and **SO_2** [14] (Fig. 13).The previous experience allows one to conclude that the most plausible upper pressure stability limit for the most stable hydrates of the classic structures concerned will be of the order of 15-20 *kbar*. Above these pressures selfclathrate ices *VI-VIII* will be more stable [2].

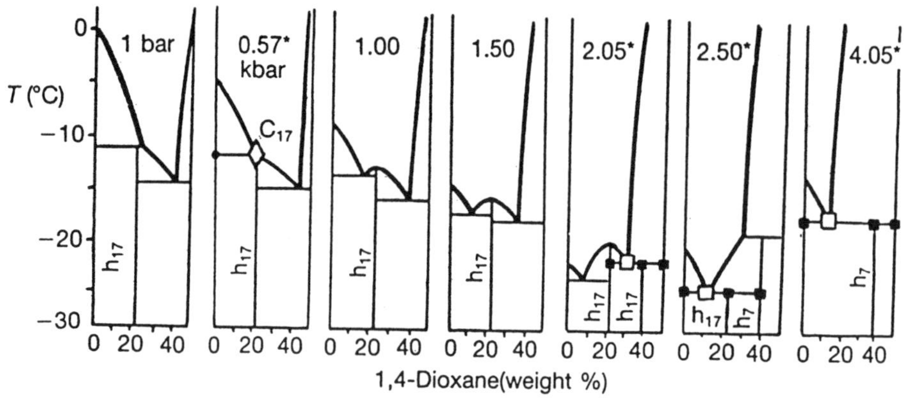

Figure 12 Isobaric sections of the water — *1,4*-dioxahe system. A specific feature of the system in question is that the hydrate melting incongruently at normal pressure, melts congruently in the range of 0.6 -2.3 *kbar.*

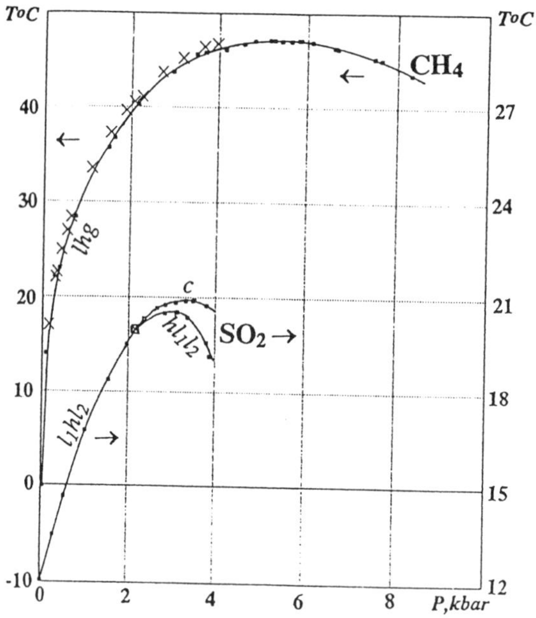

Figure 13 Decomposition points of the $\mathbf{CH_4}$ (• - [13], x - [15]) and $\mathbf{SO_2}$ [14] hydrates at high pressures, ***h*** is the hydrate, ***l*** is the liquid, ***g*** is the gas, ***c*** is the congruent melting line of the $\mathbf{SO_2}$ hydrate.

5. Solid Solutions and Hydrates on the Basis of the Frameworks of Ice *Ih,* Ice *Ic* and Ice *II*.

A completely different type of situation occurs in such small guest molecules as $\mathbf{H_2}$, **He** and, maybe, **Ne**. They can be arranged in the cavities of common ice (Fig 1, 2), ice *Ic* (Fig. 14, 16) and ice *II* (Fig. 15, 16), forming quite dense structures with the limit stoichiometry $\mathbf{G{*}2H_2O}$, $\mathbf{G{*}H_2O}$ and $\mathbf{G{*}6H_2O}$, respectively. This makes their solubility in ices *Ih* and *II* perceptible enough even at relatively low pressures [16-18], iskhoric solution biing formed

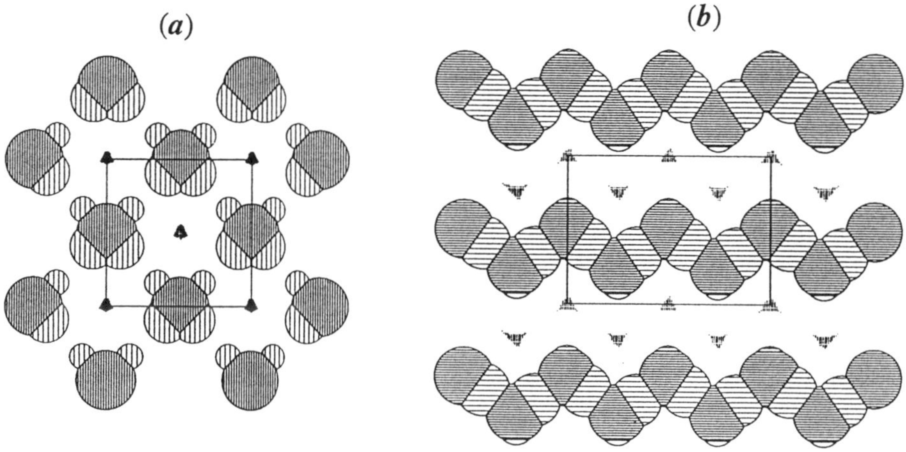

Figure 14 The dimensions and the shape of the channels in the (0,0,1)- (***a***) and (1,1,0)-sections (***b***) of the ice *Ic* structure (see Figure 1).

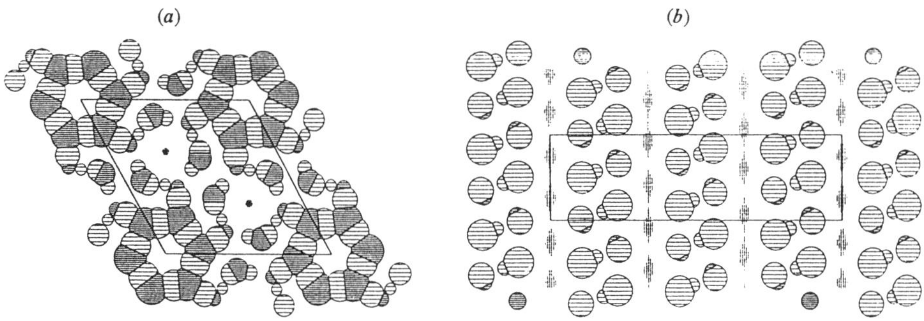

Figure 15 The dimensions and the shape of the channels in the (0,0,1)- (***a***) and (1,1,0)-sections (***b***) of the ice *II* structure (see Figure 1).

of the I and II types[2]. It was shown [20] that at very high pressures the hydrates H_2*6H_2O and H_2*H_2O are formed on the basis of the ice *II* and ice *Ic* frameworks, respectively (Fig. 17). The latter must be very stable at high pressure due to its compact structure (k = 0.714) and including of hydrogen practically does not distort the structure : a = 6.36 *A* (ice *Ic*) and a = 6.43 *A* (H_2*H_2O). On the contrary, the interpenetrating of the structures of ice *Ic* one into another (accompanied by the formation of ice VII) results in the essential stretching of the framework essentially (a=6.82 *A*). That is why at very high pressures (up to 300 *kbar* [20]) the hydrogen monohydrate is more stable phase than the densest ice modification

6. Utilization of Gas Hydrates

There are a lot of examples of commercial utilization of gas hydrates and hydrate formation processes (and they appear to have a considerable promise) : storage of gases, the freshening of water, separation and purification of substances, the drying of gases, elimination of high temperature fogs, ets. [21]. I will not continue this list but I would like to say some words about the natural gas hydrate deposits and the global effects that might be caused by them. In 1970 in Russia the discovery made by V.Vasil'ev, Yu.Makogon, A.Trebin, A.Trofimuk and N.Cherski was registered under the title "The Property of Natural Gases to Occur in the Earth Crust in a Solid State" [22]. By now huge gas hydrates deposits have been discovered both in sediment of continental stratum in Northern Latitudes and at the bottom of the World Ocean. According to the estimates available today [1] the mass of carbon in gas hydrates exceeds its content in all the other fuels, including natural gas in a free state. So, it might seem that there is enough fuel not only for people living now but also for the coming generations. However the radiative activity of methane is about 20 times greater than that of carbon dioxide; that is why its commercial recovery may cause the green-house effect. Very handsome and realistic hypotheses of the global rise in temperature and freezing are discussed, and the processes of the formation and decomposition of gas hydrates are believed to play the determining part in this models. Besides, enormous discharges of the gas may be caused by seismic activity in places where deposits of gas hydrates are found. For instance, these processes might prove to be one of the reasons for the anomaly within Bermuda Triangle [1, 23]. This is one of the examples showing that the objects of clathrate (and in a wider sense, of supramolecular) chemistry are not exotic and can essentially affect the life of the mankind, and, therefore, are worth close attention and studying.

---

[2] Iskhoric solution, in contrast to alloxenic (resulting from the substitution of one guest for another) and allokiric (the substitution of one host for another) solutions, is a variable composition phase which is due to different filling degree of the cavities [19]. Iskhoric solutions of the first type form on the basis of a stable host modification. The filling degree of the cavities (y) is within $0 < y < 1$. The second type iskhoric solutions form on the basis of a metastable host modification. Here $y^o < y < 1$, and $0 < y^o$.

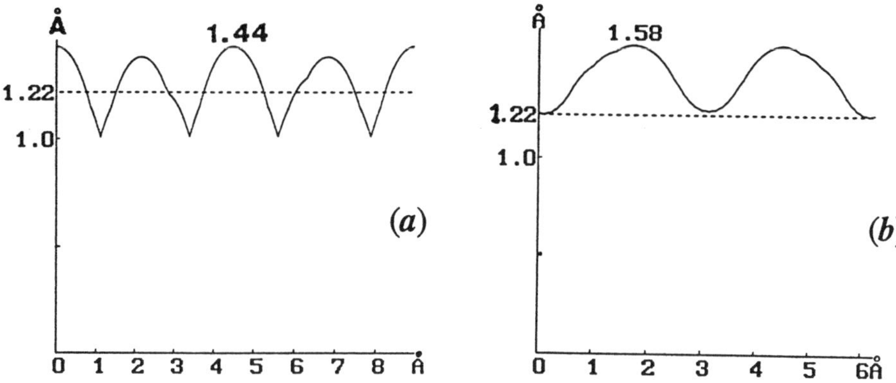

Figure 16 The maximum radius of a sphere inscribed in the channel of the ice *Ic* structure in the 1,1,0-direction *(a)* and ice *II* structure along the *c*-axis *(b)*.

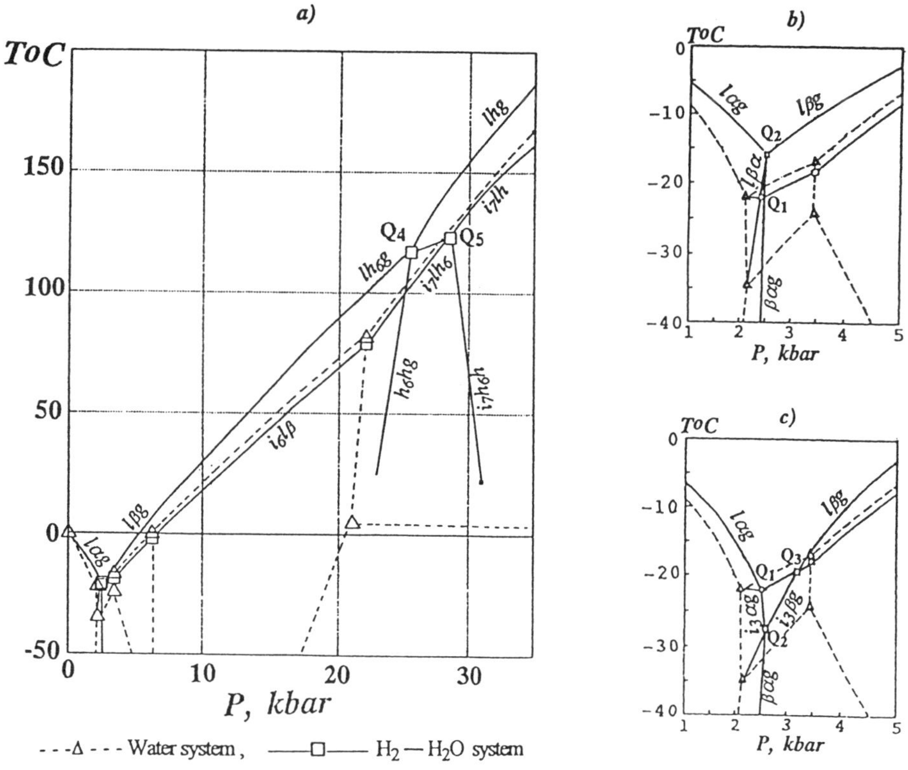

Figure 17 (*a*) P,T-Projection of the H_2 — H_2O system at high pressure (semischematic), (*b*) and (*c*) are different possible versions of the phase equilibria at average pressures, α and β are solid solutions of hydrogen in ices *Ih* and *II*, respectively, **Q** is the quadruple point.

7. Conclusion

One can see that structural information in combination with the study of phase diagrams is extremely useful. Since in the studied compounds practically no specific chemical interactions occur between the guest and the host, thermodynamic stability of the hydrates is mainly determined by spatial complementarity. Slight (often the slightest possible) changes in the spatial correspondence between the guest and the host subsystem cause small but strictly defined and amenable to the quantitative study changes in the stability of a clathrate system on the whole. The situation becomes more complicated and at the same time more diversified due to the ability of water to build a number of frameworks close in energy content results in the formation of not only various clathrates with different guests but also in the formation of various clathrates with the same guest when equilibrium conditions are changed. Different content of the vacant small cavities in the structures of different hydrates results in different pressure affects on them: from a destabilizing to a considerably stabilizing one. This (together with the fact that the ice *Ih,* being the most porous structure, is destabilized by pressure most of all) makes it possible to obtain information on clathrate formation which could not be obtained at ambient pressures. This is feature of water systems, since in the majority of the other guest - host systems the host does nor possess the properties typical of water, and a change in pressure in these cases results only in quantitative changes. In the structures of three (ices *Ih, Ic* and *II*) out of ten ice modifications known now there are cavities that can include only the smallest molecules (**H_2, He**). In the presence of the excess hydrogen solid solutions and clathrates based on the frameworks of the above ices suppress all other modifications throughout the studied temperature and pressure range (up to 300 kbar), which is due to a very good complementarity (k = 0.714). As to the classic clathrate hydrates, their guest - host complementarity is not so good (the maximum k = 0.60), and it seems that even for the most stable hydrate the upper (pressure) boundary might be evaluated to be 15 - 20 kbar

Professor Mario FARINA paid much attention to the structural-thermodynamic approach to the study of the guest - host systems. He emphasized that it is very productive and informative, and contributed much to this area of knowledge [24, 25]. This lecture is dedicated to his memory.

Acknowledgements

The author acknowledges, with thanks, the Organizing Committee of 22nd International School of Crystallography for the kind invitation and financial support. This work was possible owing to the financial support of the International Science Foundation (Grant RBR 000). The author gratefully thanks N.V.Udachina for translating this paper into English.

References

1. **Kvenvolden, K.V.**, (1994) Natural Gas Hydrate Occurrence and Issues, in *International Conference on Natural Gas Hydrates,* Eds. Sloan, E.D., Happel, J., and Hnatow, M.A., *Ann. N-Y.Academy*, **715**, 232
2. **Eizenberg, D. and Kautsman, V.**, (1975) *Struktura i svoistva vody (Structure and Properties of Water)*, Gidrometeoizdat, Leningrad.
3. **Kitaigorodski, A.I.**, (1971) *Molekuliarnye Kristally (Molecular Crystals)* Nauka, Moskva.
4. **Ripmeester, J.A., Ratcliffe, C.I., Klug, D.D., and Tse, J.S.**, (1994) Natural Gas Hydrate Occurrence and Issues, in *International Conference on Natural Gas Hydrates,* Eds. Sloan, E.D., Happel, J., and Hnatow, M.A., *Ann. N-Y.Academy*, **715**, 161
5. **Dyadin, Yu.A., Bondaryuk, I.V., Zhurko, F.V.**, (1991) Clathrate Hydrates at High Pressures, *Inclusion Compounds*, Eds. Atwood, J.L., Davies, J.E.D. and MacNicol, D.D., Oxford University Press, Oxford, 5, 213
6. **Jeffrey G.A. and McMullan,** (1967) The Clathrate Hydrates, *Progr.Inorgan.Chem.* **8,** 43.
7. **Jeffrey G.A.**, (1984) Hydrate Inclusion Compounds, *Inclusion Compounds*, Eds. Atwood, J.L., Davies, J.E.D. and MacNicol, D.D., Academic Press, London, **1**, 135
8. **Lipkowski, J., Suwinska, K., Rodionova T.V., Udachin, K.A., and Yu.A.Dyadin,** (1994) Phase and X-Ray Study of Clathrate Formation in the Tetraisoamylammonium Fluoride — Water System, *J.Inclus.Phenom.* **17,** 137
9. **Dyadin, Yu.A., Terekhova, I.S., Polianskaia, T.M. and Aladko, L.S.,** (1976) Clathrate Hydrates of Tetrabutylammoniun Fluoride and Oxalate, *Zh.Stuktur.Khim.,***17,** 655; *Ch.A.* **86:**10947d (1977).
10. **Dyadin, Yu.A., Gaponenko, L.A., Aladko, L.S., and Bogatyryova, S.V.,** (1984) Clathrate Hydrates of Tetrabutylammonium Carboxylates and Dicarboxylates, *J.Inclusion Phenom* **2,** 259
11. **Dyadin, Yu.A., Kuznetsov, P.N., Yakovlev I.I. and Pyrinova, A.V.,** (1973) The Water — Tetrahydrofuran System in the Crystallyzation Field under High Pressures up to 9 kbar, *DAN USSR,* **208,** 103; *Ch.A.* **78:**102558r (1973)
12. **Zakrzewski, M., Klug, D.D, and Ripmeester, J.A,** (1994) On the Pressure-induced Phase Transformation in the Structure II Clathrate Hydrate of Tetrahydrofuran, *J.Inclusion Phenom.* **17,** 237
13. **Dyadin, Yu.A. and Aladko, E.Ya.** unpublished data.
14. **Berkum, J.G.van ang Diepen, G.A.,** (1979) Phase equilibria in SO_2+H_2O: the sulfur dioxide gas hydrate, two liquid phases, and the gas phase in temperature range 273 - 400 *K* and an pressures up to 400 MPa, *J.Chem.Thermodynamics,* **11,** 317
15. **Marschall, D.R., Saito, S.H. and Kobayashi, R.**, (1964) Hydrates at High Pressures: I. Methane—Water, Argon—Water and Nitrogen—Water Systems, *AIChEJ,* , 22.
16. **Arnold, G.P., Wenzel, R.G., Rabideau, S.W., Nereson, N.G., and Bowman, L.,** (1971) Neutron Diffraction Study of the Ice Polymorphs under Helium Pressure, *J.Chem.Phys.,* **55,** 589
17. **Londono, D., Kuhs, W.F., and Finney, J.L.,** (1988) Inclathration of helium in ice II : the first helium hydrate, *Nature,* **332,** 141. (see also *J.Chem.Phys.,* **97,** 547 (1992)).
18. **Dyadin, Yu.A., Aladko, E.Ya.,.Udachin, K.A., and Tkacz, M.,** (1994) The Solubility of Helium and Hydrogen in Ice Ih at High Pressures, *Polish J.Chem.,* **68,** 343
19. **Dyadin, Yu.A.,** (1995) On the Stoichiometry of Clathrates, *Supramolecular Chemistry,* (in press)
20. **Vos, W., Finger, L., Hemley, R., and Ho-kwang Mao,** (1993) Novel H_2 — H_2O Clathrates at High Pressures *Phys.Rev.Let.*, **17,** 3150
21. **Makogon, Yu.F.,** (1974) *Gidraty Prirodnykh Gasov (Natural Gas Hydrates)*, "Nedra", Moscow.
22. **Vasil'ev, V.G., Makogon, Yu.F., Trebin, A.F., Trofimuk, A.A., and Cherski, N.V.,** (1970) The Property of Natural Gases to Occur in the Earth Crust in a Solid State,*"Otkrytia, izobretenia i tovarnye znaki", (Discoveries, Inventions and Comercial Licence Signs),* N 10, 4
23. **Collett, T.S.,** (1994) Permafrost-Associated Gas Hydrate Accumulations, in *International Conference on Natural Gas Hydrates,* Eds. Sloan, E.D., Happel, J., and Hnatow, M.A., *Ann. N-Y.Academy*, **715**, 247
24. **Farina, M.,** (1994) Phase Diagrams of Crystalline Adducts Containing Two Volatile Components, *J.Inclusion Phenom.*, **14,** 101
25. **Farina, M.,** (1984) Inclusion Compounds of Perhydrotriphenylene, *Inclusion Compounds*, Eds. J.L.Atwood, J.E.D.Davies and D.D.MacNicol, Academic Press, London, N-Y., **2,** 69.

CYCLODEXTRINS AND FRAGMENTS OF STARCH AND CELLULOSE: CRYSTAL STRUCTURES, SELF-ASSEMBLY AND HYDROGEN BONDING

W. SAENGER, W. HINRICHS, K. GESSLER, TH. STEINER
Institut für Kristallographie, Freie Universität Berlin,
Takustr. 6, D-14195 Berlin, Germany

This article reviews our structural studies on (1) the cyclodextrin family, on (2) maltohexaose, which is a fragment of amylose and a linear analog of the cyclodextrins, and on (3) β-D-cellotetraose, which is a fragment of cellulose (Fig. 1). In all these oligosaccharides, the main structural motif is D-glucopyranose, which is linked to its neighbours with α(1-4) bonds in the cyclodextrins and in amylose, and by β(1-4) bonds in cellulose. The study on β-D-cellotetraose has, in particular, shown that with the oligosaccharides, we are now in a similar position as with the oligonucleotides. With both, it is possible to obtain HPLC-pure material in sufficient quantities so that single crystals can be grown and, should they not be of the size required for conventional X-ray diffraction studies, synchrotron radiation can be employed to collect the diffraction data. If the packing of the molecules in the single crystals is comparable to that observed by fiber diffraction in the corresponding polymers, the oligosaccharides can be extended by computer modeling to "infinite" structures which serve as good models for the re-interpretation of the fiber data, should there be discrepancies in the models.

G. Tsoucaris et al. (eds.), Crystallography of Supramolecular Compounds, 243–263.

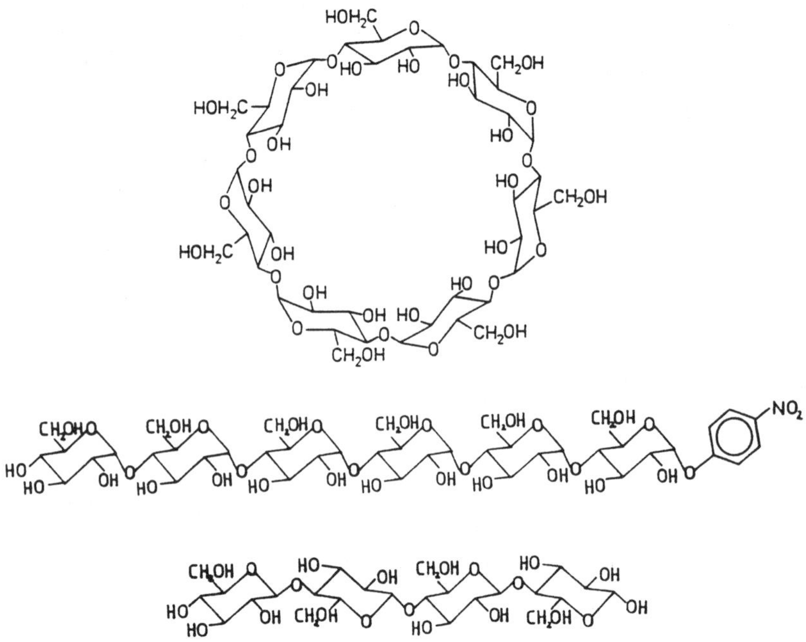

Figure 1. Chemical structures of (*top*) β-cyclodextrin (β-CD); (*center*) p-nitrophenyl-α-D-maltohexaoside; (*bottom*) β-D-cellotetraose

1. Structural Features of Cyclodextrins and Linear Oligosaccharides of the Types Amylose and Cellulose

1.1. CYCLODEXTRINS

The cyclodextrins are a family of four cyclic oligosaccharides consisting of 6 to 9 D-glucopyranoses, called α, β, γ, δ-cyclodextrin [1,2]. The glucoses adopt the $^4C_1$ chair conformation and there are only two reported structures of methylated cyclodextrins, where one glucose is in another conformation [3a, b]. The shape of the annular cyclodextrins is roughly toroidal with the narrower rim lined with O6-hydroxyl groups, and the wider one with O2 and

O3 hydroxyl groups. Consequently, the cyclodextrins are hydrophilic at the outside, whereas their central cavity is hydrophobic due to the presence of the C3-H and C5-H groups and of the ether-like O4 atoms.

If dissolved in water, the cyclodextrins are not "empty", but they contain water molecules in their cavities. This water is easily displaced by "guest" molecules, which enter into the cavity to form inclusion complexes. Depending on the size of the guest molecules, 1:1, 1:2 or 3:2 guest:cyclodextrin ratios are observed. Dissociation constants as well as other thermodynamic and kinetic parameters can be measured by conventional spectroscopic and relaxation methods. Complex formation is, in general, associated with dissociation constants in the order of 10^{-2} to 10^{-4} mol/l. In essence, the only requirement for complex formation appears to be the spatial fitting of the substrate (guest) into the cyclodextrin cavity. The character of the substrate may range from hydrophobic (noble gases, paraffins) to hydrophilic (carboxylic acids, salts), and does not markedly influence the thermodynamic parameters. For these, ΔH is largely negative and ΔS can be either positive or negative. Several effects such as hydrophobic and van der Waals interactions, hydrogen bonding, expulsion of "activated" water are contributing to complex formation. The inclusion properties have made cyclodextrins very good models to study non-covalent interactions which are so important in biological systems. In fact, the accumulation of hydroxyl groups at the surface of the cyclodextrins provide them with catalytic properties, so that cyclodextrins have been used successfully as enzyme models [1,2].

The annular structure of the cyclodextrins is stabilized by intramolecular interglucose hydrogen bonding O2···O3'. These hydrogen bonds form a ring, which was found broken in only three cases: (i) crystalline α-cyclodextrin · $6H_2O$ (Fig. 2), where the cavity is distorted by rotation of one of the glucoses out of the ring formed by the other five, leading to reduction of the cavity volume so that only two water molecules are enclosed [4], (ii) crystalline α-CD · benzaldehyde ·$6H_2O$ [5] and the related αCD · hydroquinone · $6H_2O$ [6], where the α-CD molecules are elliptically distorted to adjust to the shape of the guest, so that one O2···O3 bond is broken, and (iii) the per-2,3,6-trimethylated cyclodextrins, where all the hydroxyl groups are chemically modified so that no hydrogen bonding is possible. In crystal structures of α-, β- and γ-cyclodextrin, water of hydration molecules and O-H groups are frequently disordered, and we observe dynamical reorientation of the flip-flop type, which was first described for the crystal structure of β-cyclodextrin · $12H_2O$ (*vide infra*).

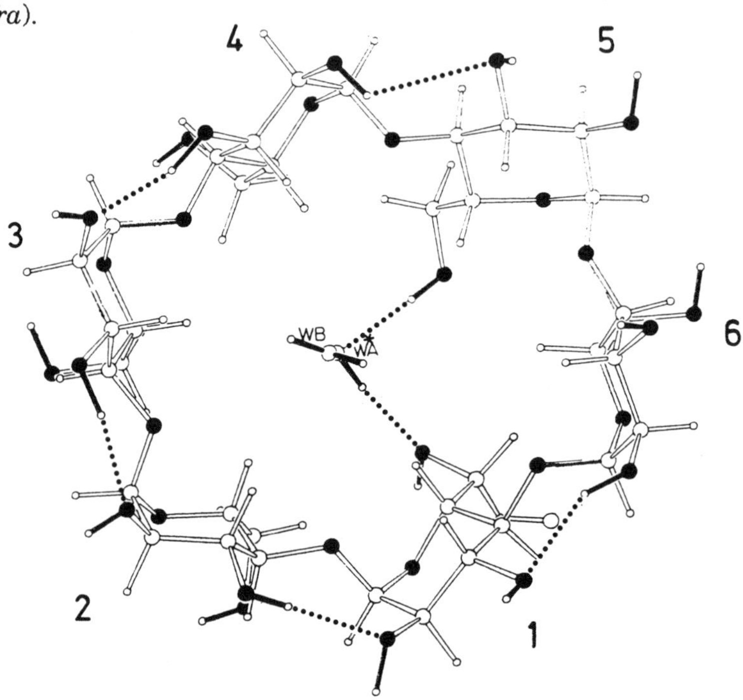

Figure 2. Molecular structure of α-Cyclodextrin (α-CD) in the complex with $6H_2O$. Water molecules WA and WB are enclosed in the α-CD cavity. Hydrogen bonds are drawn as dotted lines, glucoses are numbered 1 to 6; there is no intramolecular, interglucose hydrogen bond between glucoses 5 and 6.

1.2. MALTOHEXAOSE

If the cyclodextrins are cleaved, we obtain linear oligosaccharides of the amylose type. These are readily available up to seven glucoses long and one of them, with a protected reducing end, p-nitrophenyl-maltohexaoside, was crystallized in the presence of $Ba(I_3)_2$ [7,8]. In this crystal structure, which contains 27 water of hydration molecules, the oligosaccharide adopts the shape of a wound ribbon, with all the O2, O3 hydroxyl groups on one side and all the O6 hydroxyl groups on the other (Fig. 3). Two of these ribbons are oriented antiparallel and form a left-handed helical structure, which is stabilized by intra- and intermolecular hydrogen bonding between the O2 and O3 hydroxyl groups. The helix has a central cavity, in which two I_3^- molecules are enclosed in near-linear arrangement. The p-nitrophenyl groups are on both ends of this short double helix, and they interact in the crystal lattice (*vide infra*). Of importance for the stabilization of this crystal structure is also the coordination of the Ba-cation, which involves four glucose units of four different molecules.

1.3. CELLOTETRAOSE

In the crystal structure of β-D-cellotetraose the structural organization of the individual D-glucopyranose units in the molecules is very different from that in cyclodextrins and in the maltohexaoside [9]. This is because in β(1-4) linked D-glucopyranoses, alternating units have to be rotated 180° since otherwise O2 and O3' groups of adjacent units would interfere sterically; as a result, a straight, ruler-like structure is formed. All the glucoses are in a comparable $^4C_1$ chair conformation with the C6-O6 bonds in *gauche, trans* orientation. The molecular structure is again stabilized by intramolecular,

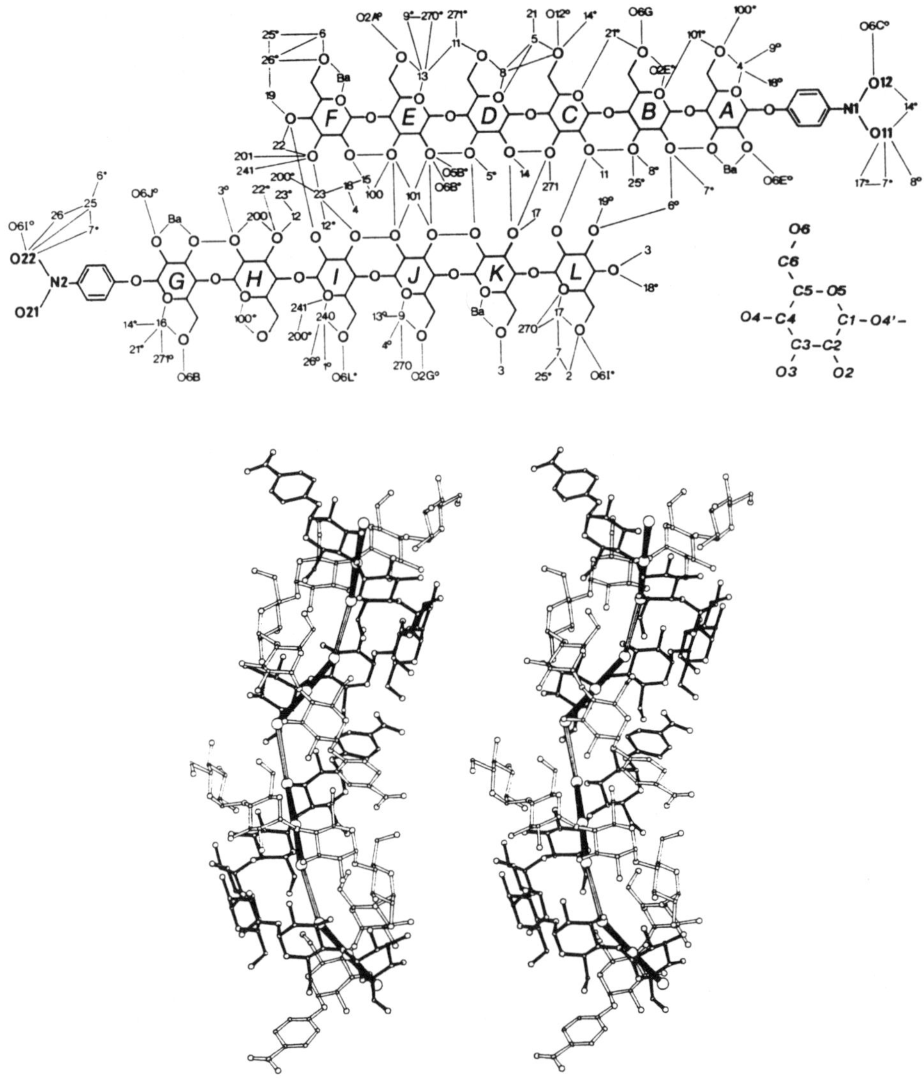

Figure 3. The structure of $(p\text{-nitrophenyl-}\alpha\text{-D-maltohexaoside})_2 \cdot Ba(I_3)_2 \cdot 27H_2O$. (*Top*) schematic picture of hydrogen bonding interactions (thin lines) within and between maltohexaose OH-groups and water molecules (indicated by numbers). (*Bottom*) Stereo view of the double helix showing two asymmetric units. The stacking interaction between p-nitrophenyl groups is at the center right; Ba-cations, water molecules and hydrogen bonds are not drawn.

interglucose hydrogen bonding, which is now of the three-center type with O3-H as donor and O5', O6' as acceptors (Fig. 4).

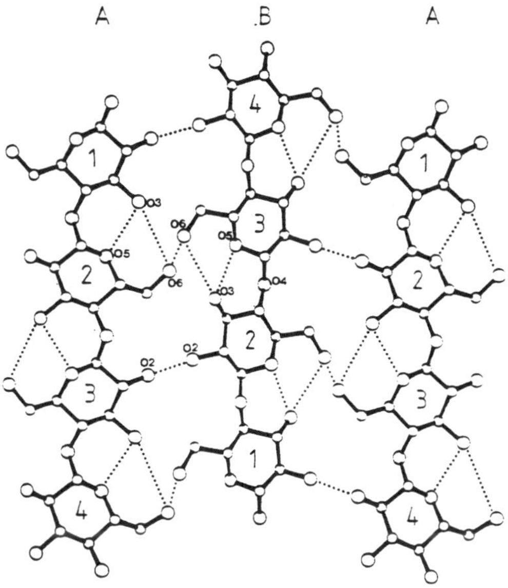

Figure 4. Section of the crystal structure of β-D-cellotetraose. Molecules A and B are antiparallel, intra- and intermolecular hydrogen bonds are indicated by dotted lines.

2. Self-Organization of Cyclic and Linear Ologosaccharides in the Crystal Lattice

The three oligosaccharides pack differently in the crystal lattices. For the cyclodextrins, there is a strong dependence between crystal packing and the inclusion of certain guest molecules. The p-nitrophenyl-maltohexaoside double helices pack as "infinite" double helices enclosing "infinite" polytriiodide, and β-D-cellotetraose packs similar as observed for microcrystalline cellulose II. The packing patterns are described below in more detail:

2.1. CYCLODEXTRINS

If the cyclodextrins are crystallized from aqueous solutions without the addition of guest molecules, they form cage-type complexes, in which the cyclodextrin molecules are arranged in herringbone fashion such that the cavity of one molecule is closed on both ends by adjacent molecules. This cage-type pattern is also observed if inclusion complexes between α-cyclodextrin and small molecular guests like I_2, methanol, propanol, acetic acid, butyric acid are formed, whereas long molecular guests or ionic guests give rise to channel-type structures [10], Fig. 5. In these channels, the cyclodextrins are stacked like coins in a roll. In most cases, the stacking is either of the type head-to-head or of the type head-to-tail. More complicated stakking sequences occur only occasionally.

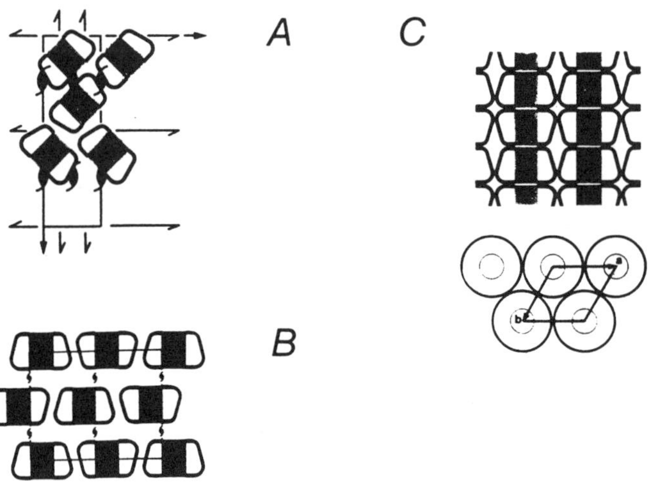

Figure 5. Packing patterns of cyclodextrins. (A) herringbone-type cages; (B) brick-type cages; (C) channel-type. The cavities are indicated by shading.

The columns are always stabilized by intermolecular hydrogen bonding between the 02, 03 and 06 hydroxyl groups. The channel-type arrangement predominates in the β- and γ-cyclodextrins, with certain modifications in the packing of β-cyclodextrin, where the molecules frequently form head-to-head dimers with basket-like cage which accommodates up to three guest molecules. These dimers are either stacked as true channels, or they are displaced along the channel axis so that the channel, in the extreme case, consists of well-aligned cages.

If α-cyclodextrin is complexed with aromatic guest molecules like p-iodoaniline, it adopts an elliptical form because the aromatic ring is wedged into the cavity. These complexes often crystallize in a cage-type form in which the molecules are arranged in planes, and these planes are displaced relative to each other so that a brick-type pattern results [10].

2.2. MALTOHEXAOSE

The packing pattern of the p-nitrophenyl-maltohexaoside is comparable to the channel-type structures of the cyclodextrins (Fig. 3 bottom). The reason is probably that the triiodide molecules interact with each other to form poly-triiodide in a slightly zigzagged form, and the left-handed double helices of the antiparallel arranged p-nitrophenyl-maltohexaoside molecules wind around this poly-triiodide. Each of the double helices contains one p-nitrophenyl moiety at each end, and they are arranged such that these aromatic moieties are stacked at a distance of 3.4 Å, thereby forming and stabilizing an "infinite" double helix. Crystallographically, the molecules in this double helix are related by a crystallographic twofold screw axis (space group $P2_12_12_1$), i.e. each of the short double helices is rotated with respect to its neighbors by 180°.

2.3. CELLOTETRAOSE

In the crystal structure of β-D-cellotetraose, two ruler-shaped molecules (A and B) are accommodated in the triclinic unit cell, space group P1. The two molecules are oriented antiparallel and have slightly different conformations of the glucopyranose residues and about the glycosidic bonds but the intramolecular, interglucose hydrogen bonding is virtually the same. The molecules are arranged in sheets which are stabilized by intermolecular hydrogen bonding, and there are also hydrogen bonds between the sheets so that the crystal structure is stabilized by a three dimensional network of hydrogen bonds. Since the ruler-like molecules are arranged head-to-tail, only slight rearrangement is necessary to produce a "polymer" structure. It corresponds closely to the structure of cellulose II which was proposed earlier on the basis of fiber X-ray diffraction data coupled with computer aided molecular modelling [11,12] (Fig. 6). There are, however, discrepancies concerning differences in the orientation of the C6-O6 bonds in the fiber structure (*gauche,trans* in one molecule and *trans,gauche* in the other), which also gives rise to different intra- and intermolecular hydrogen bonding. Since ^{13}C CPMAS NMR spectra have shown that all the C6-O6 bonds in cellulose II have the same conformation [13,14] as found in the single crystal study of β-D-cellotetraose, this is a strong argument for the model developed by us and suggests that the fiber X-ray diffraction data should be re-interpreted.

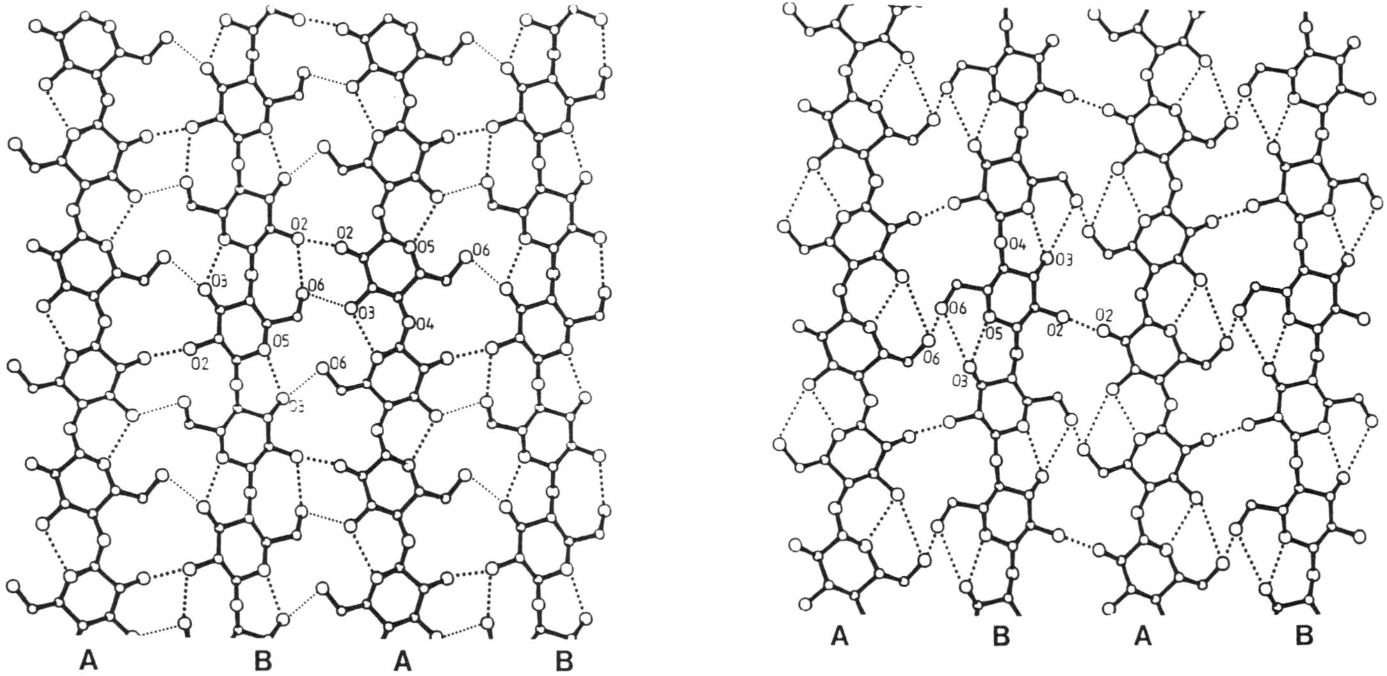

Figure 6. Crystal structures of polymer cellulose II derived on the basis of (*left*) fiber X-ray diffraction patterns and (*right*) the crystal structure of β-D-cellotetraose. Note differences in C6-O6 orientation and in hydrogen bonding schemes. Molecules A and B are in antiparallel orientation.

3. Cyclodextrin Hydrates as a Source of New Insight into Hydrogen Bonding

If cyclodextrins are crystallized without guest molecules, they form cage type complexes, and water molecules are included in the cavities and also located between the cyclodextrin molecules. These cyclodextrin hydrates grow in exceptionally large crystals which are suitable for neutron diffraction work. This permits to locate H-atons as precisely as C and O, and to unambiguously define hydrogen bonds. We have performed a series of seven neutron diffraction studies on cyclodextrin complexes, which supply a wealth of accurate structural data of hydrogen bonds. The individual studies are listed in the following table, and the most important results are summarized below.

Structure	Temperature	Reference
α-CD $\cdot$ 6 H_2O	Room Temp.	15
α-CD $\cdot$ (Benzyl-alcohol)$_2$ $\cdot$ 6 H_2O	Room Temp.	16
β-CD $\cdot$ 11 D_2O	Room Temp.	17
β-CD $\cdot$ 11.6 D_2O	120 K	18
β-CD $\cdot$ Ethanol $\cdot$ 8 D_2O	15 K	19
β-CD $\cdot$ Ethanol $\cdot$ 8 D_2O	Room Temp.	20
γ-CD $\sim$ 16 D_2O	110 K	21

3.1. COOPERATIVITY OF O-H···O HYDROGEN BONDS

Because the O-H group is simultaneously a hydrogen bond donor and acceptor, it can be involved in hydrogen bonded chains O-H···O-H···O-H. In these arrangements, the O-H groups mutually increase their polarisation, thereby strengthening the individual hydrogen bonds by approximatly 25 % as compared to isolated hydrogen bonds O-H···O. This is called the 'cooperative effect' [22,23]. A large part of our knowledge on cooperative hydrogen bond networks was derived from the cyclodextrin neutron crystal structures. The cooperative arrangements form linear chains or they close up as cycles. Depending on the hydrogen bond orientation, the arrangements are called 'homodromic' (**1**, all hydrogen bonds run in the same direction), 'antidromic' (**2**, one double donor and one double acceptor) and 'heterodromic' (**3**, neither **1** nor **2**) [24].

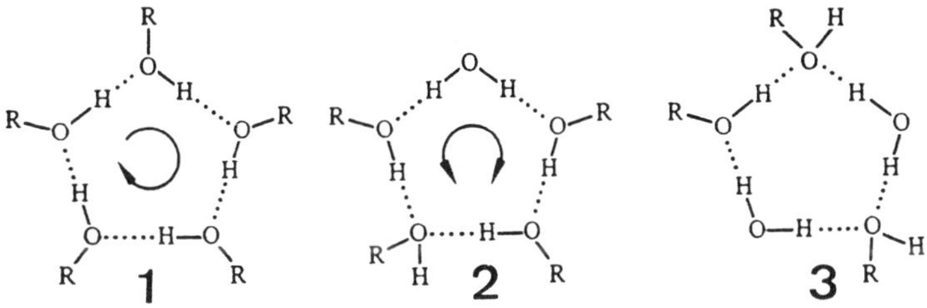

Within such arrangements, the hydrogen bond geometries are very flexible. They are limited, however, by the shortest possible H···H approach of 2.05 Å [25].

3.2. THREE-CENTER O-H···O HYDROGEN BONDS

If an O-H 'donor' forms hydrogens with two 'acceptor' O-atoms simultaneously, this is called a 'three-center hydrogen bond' (because H is bonded to three O-atoms) [23]

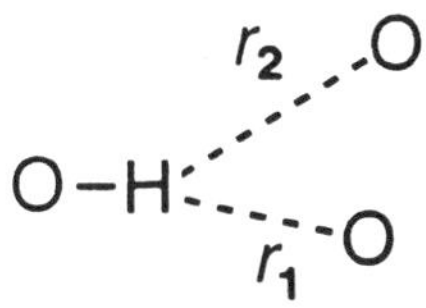

Normally, the two H···O separations are unequal, with the shorter one called the 'major component' and the longer called the 'minor component' of the three-center bond. The three-center bonds are an integral part of hydrogen bonding in carbohydrates [26], and in cyclodextrin crystal structures, they make up the exceptionally large fraction of ca. 50 % of all hydrogen bonds. Typical examples of three-center hydrogen bonds are shown in Figure 7, which represents the hydrogen bonding in β-CD · ethanol · $8D_2O$ at 15 K [19]. Despite their abundance, direct structural evidence for the strength of minor hydrogen bond components could be given only recently from cyclodextrin low-temperature neutron diffraction data: it is well known that in O-H···O hydrogen bonds, the covalent O-H bond is significantly lengthened [23]. In three-center bonds, the formation of a minor component with an H···O distance < 2.4 Å lengthens the O-H bond by ~ 0.01 Å in addition to the lengthening due to the major component; this clearly shows the structural significance of three-center bonds [27,28].

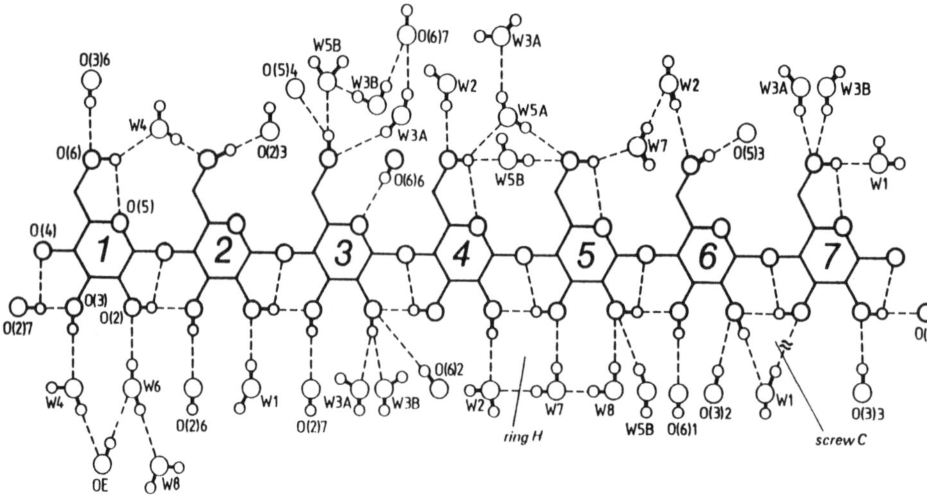

Figure 7. Hydrogen bonding around the β-CD molecule in βCD-ethanol-octahydrate at 15K [19].

3.3. FLIP-FLOP DYNAMICS OF HYDROGEN BONDS

In the neutron diffraction study of β-CD · 11 H_2O at room temperature [17], 35 of the total 53 O-H···O hydrogen bonds showed two half occupied hydrogen positions between the oxygen atoms, O-(1/2H)···(1/2H)-O. This was interpreted as a dynamic equilibrium between two states of the form

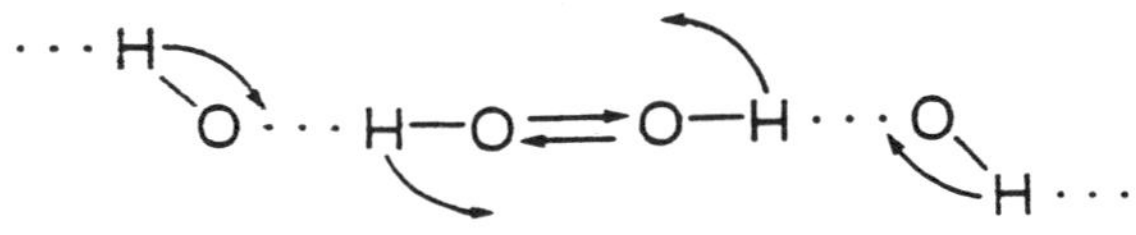

This kind of *flip-flop* hydrogen bonding occurs, for example, between hydroxyl groups, but water molecules are also involved. Many of these flip-flops are interconnected to larger systems. The same kind of orientational disorder was also observed in β-CD · ethanol · 8H_2O [20], and α-CD · (benzyl alcohol)$_2$ 6H_2O [16], and we suppose that it is a property of many more CD-complexes.

The flip-flop disorder disappears upon cooling [18], indicating that it is of dynamic nature. Molecular dynamics simulations predicted dynamic disorder on a 10^{11}sec^{-1} time scale [29], and direct proof for the fast dynamic nature was given by subsequent incoherent neutron scattering [30] and solid state NMR [31] studies. These complementary experimental studies show that the jump rates of the individual hydrogen bonds, which have different surroundings, cover several decades, extending from the NMR time scale ~ 10^6 sec^{-1} to motions as fast as 2 x 10^{11}sec^{-1}.

3.4. C-H···O HYDROGEN BONDS

The weak C-H···O hydrogen bonds are very common in organic crystal structures. With typical bond energies 1-2 kcal/mol, they are much weaker than O-H···O interactions (typically 4-5 kcal/mol), but strong enough to have significant influence on molecular conformation and on crystal packing [32]. Based on precise structural data on hydrophilic molecules enclosed in the hydrophobic cyclodextrin cavities, we were for the first time able to describe the role of C-H···O interactions in the coordination of water molecules. If water molecules cannot arrange in the preferred tetrahedral O-H···O hydrogen bond coordination due to steric hindrance, they tend to resort to the weaker C-H···O_w interactions rather than leaving the multiple acceptor potential unsatisfied [33,34]. A particulaly frequently observed configuration is water accepting one O-H···O_w and an additional C-H···O_w interaction, with the latter completing a four-fold coordination.

Whereas in crystal structures of native cyclodextrins, O-H···O hydrogen bonding is the dominant motif, C-H···O interactions become a major intermolecular interaction in some methylated cyclodextrins, where no O-H donors, but many C-O-C acceptors are available [3b].

3.5. DIFFUSION OF WATER MOLECULES IN THE CRYSTAL LATTICE OF β-CYCLODEXTRIN

If crystals of β-cyclodextrin · $12H_2O$ are exposed to ambient atmosphere, they lose about three molecules of water and retain their crystal lattice, but the volume of the unit cell is somewhat reduced. This process is reversible, i.e. if a dehydrated crystal is exposed to water vapor, it absorbs about three water molecules and the original crystal structure is restored. This process is relatively fast and completed within about 15 minutes, according to the changes in unit cell constants [35].

This process must involve diffusion of water molecules through the crystal lattice. The diffusion takes place although the packing pattern is of the fish-bone type and such that no continuous channels are formed in the crystal. We assume that there must be local movement of the cyclodextrin molecules which permits the water molecules to travel through the crystal lattice. This is probably associated with the disorder of water molecules in the cavity of β-cyclodextrin and in interstices between the β-cyclodextrin molecules [17] and with the above mentioned dynamic flip-flop disorder of hydroxyl groups and water molecules.

To explain the water diffusion in β-CD hydrate, we performed H_2O/D_2O and $H_2{}^{16}O/H_2{}^{18}O$ exchange experiments [36]. If fully hydrated β-cyclodextrin · $12H_2O$ is exposed to D_2O vapor, all the O-H hydrogen atoms of water and of the hydroxyl groups are exchanged by D. The reaction is completed within one hour and involves all the O-H groups in the crystal lattice. This could be due to proton transfer from the surface of the crystal to the interior, or to the exchange of complete water molecules. To investigate this further, cyclodextrin · $12H_2{}^{16}O$ was exposed to $H_2{}^{18}O$ vapor, and the exchange of $H_2{}^{16}O$ versus $H_2{}^{18}O$ was monitored by mass spectrometry. Since the kinetics of the $H_2{}^{16}O/H_2{}^{18}O$ exchange are comparable to those of the H/D exchange, this shows unambiguously that the water molecules are exchanged as such and that proton transfer occurs only locally.

Acknowledgement:

This work was supported by Deutsche Forschungsgemeinschaft (Sa 196/12, Sa 196/16 and Sa 196/25-1), Fonds der Chemischen Industrie and Deutscher Akademischer Austauschdienst (322-inida-dr).

References

1. Szejtli, J. (1988) *Cyclodextrin Technology*, Kluwer, Dortrecht.
2. Saenger, W. (1980), Cyclodextrins in Research and Industry. *Angew. Chem. Int. Ed. Engl.* **19**, 344-362.

3a. Harata, K., Hirayama, F., Arima, H., Uekama, K. and Miyaji, T. (1992), Crystal Structure of Heptakis (2,3,6-Hi-O-methyl)-β-Cyclodextrin Complexes with m-Iodophenol and 4-Biphenylacetic Acid. Guest. Induced Conformational change of a Pyranose Ring. *J. Chem. Soc., Perkin Trans.* **2**, 1159-1166.

3b. Caira, M. R., Griffith, V. J., Nassimbeni, L. R. and van Oudtsboorn, B. J. (1994), Unusual $^1C_4$ Conformation of a Methylglucose Residue in Crystalline Permethyl-β-Cyclodextrin Monohydrate. *Chem. Soc., Perkin Trans,* **2**, 2071-2072.

4. Manor, P. C and Saenger, W. (1974), Topography of Cyclodextrin Inclusion Complexes. III. Crystal and Molecular Structure of Cyclohexaamylose Hexahydrate, the $(H_2O)_2$ Inclusion Complex. *J. Amer. Chem. Soc.* **96**, 3630-3639.
5. Harata, K., Uekama, K., Otagiri, M., Hirayama, F., Ogino, H. (1981), The Structure of the Cyclodextrin Complex. X. Crystal Structure of α-Cyclodextrin-Benzaldehyde (1:1) Complex Hexahydrate. *Bull. Chem. Soc.Jpn,* **54**, 1954-1959.
6. Steiner, Th. and Saenger, W. (1994), Crystal Structure of the Inclusion complex α-Cyclodextrin-Hydroquinone Hexahydrate. *Carbohydrate Letters*, **1**, 143-150.
7. Hinrichs, W., Büttner, G., Steifa, M., Betzel, Ch., Zabel, V.,Pfannemüller, B., Saenger, W. (1987), An Amylose Antiparallel Double Helix at Atomic Resolution. *Science* **238**, 205-208.
8. Hinrichs, W., and Saenger, W. (1990), Crystal and Molecular Structure of the Hexasaccharide Comples (p-Nitrophenyl α-maltohexaoside)2 $\cdot Ba(I_3)_2 \cdot 27\ H_2O$. *J. Amer. Chem. Soc.* **112**, 2789-2796.
9. Geßler, K., Krauß, N., Steiner, Th., Betzel, Ch., Sandmann, C. and Saenger, W. (1994), Crystal Structure of β-D-Cellotetraose Hemihydrate with Implications for the Structu-

re of Cellulose II. *Science* **266**, 1027-1029.

10. Saenger, W. (1985), Nature and Size of Included Guest Molecule Determines Architecture of Crystalline Cyclodextrin Host Matrix. *Israel J. Chem.* **25**, 43-50.

11. Kolpak, F. J. and Blackwell, J.(1976), Determination of the Structure of Cellulose II. *Macromolecules* **9**, 273-278 .

12. Stipanovic, A. J. and Sarko, A. (1976), Molecular and Crystal Structure of Regenerated Cellulose II. *Macromolecules* **9**, 851-857.

13. Isogai, A., Usuda, M., Kato, T., Urgu, T. and Atalla, R. H. (1989), Solid-State CP/MAS ^{13}C NMR Study of Cellulose Polymorphs. *Macromolecules* **22**, 3168-3172.

14. Dudley, R. L., Eyfe, C. A., Stephenson, P. J., Deslandes, Y., Hamer, G. K. and Marchessault, R. H. (1983), High-Resolution ^{13}C CP/MAS NMR Spectra of Solid Cellulose Oligomers and the Structure of Cellulose II. *J. Amer. Chem. Soc.* **105**, 2469-2472.

15. Klar, B., Hingerty, B. E. and Saenger, W. (1980), Topography of Cyclodextrin Inclusion Complexes. XII. Hydrogen Bonding in the Crystal Structure of α-Cyclodextrin Hexahydrate: The Use of a Multicounter Detector in Neutron Diffraction. *Acta Crystallodgraphica*, Sect. **B36**, 1154-1165.

16. Steiner, Th., Starikow, E. B., Zabel, V., Hingerty, B. E., Johnson, C. K. and Saenger, W. manuscript in preparation. X-ray study: Steiner, Th. and Geßler, K. (1994), Aromatic molecules included into and contacting the outer surface of cyclomaltohexaose (α-cyclodextrin): crystal structure of α-cyclodextrin-$(benzylalcohol)_2$-hexahydrate. Carbohydr. Res. **260**, 27-38.

17. Betzel, Ch., Saenger, W., Hingerty, B. E. and Brown, G. M. (1984), Circular and Flip-Flop Hydrogen Bonding in β-Cyclodextrin Undecohydrate: A Neutron Diffraction Study. *J. Am. Chem. Soc.* **106**, 7545-7557.

18. Zabel, V., Saenger, W. and Mason, S. A. (1986), Neutron Diffraction Study of the Hydrogen Bonding in β-Cyclodextrin Undecahydrate at 120K. From Dynamic Flip-Flops to Static Homodromic chains. *J. Am. Chem. Soc.* **108**, 3664-3673.

19. Steiner, Th., Mason, S. A. and Saenger, W. (1990), Cooperative O-H···O Hydrogen Bonds in β-Cyclodextrin-Ethanol · Octahydrate at 15K: A Neutron diffraction study. *J. Am. Chem. Soc.* **112**, 6184-6190.

20. Steiner, Th., Mason, S. A. and Saenger, W. (1991), Disordered Guest and Water Molecules, Three-Center and Flip-Flop O-H···O Hydrogen Bonds in Crystalline β-Cyclodextrin Ethanol Octahydrate at T = 295K: A Neutron and x-Ray Diffraction Study. *J. Am. Chem. Soc.* **113**, 5676-5687.

21. Ding, J., Steiner, Th., Zabel, V., Hingerty, B. E., Mason, S. A. and Saenger, W. (1991), Neutron Diffraction Study of the Hydrogen Bonding in Partially Deuterated γ-Cyclodextrin · 15.7 D_2O at T = 110K. *J. Am. Chem. Soc.* **113**, 8081-8089.

22. Del Bene, J. E. and Pople, J. A. (1973), Theory of molecular interactions. III. A comparison of studies of H_2O polymers using different molecular orbital sets. *J. Chem. Phys.* **58**, 3605-3608.

23. Jeffrey, G. A. and Saenger, W. (1991), *Hydrogen Bonding in Biological Structures*, Springer-Verlag. Berlin.

24. Saenger, W. (1979), Circular hydrogen bonds. *Nature (London)*, **279**, 343-344.

25. Steiner, Th. and Saenger, W. (1991), H···H van der Waals Distance in Cooperative O-H···O-H···O Hydrogen Bonds Determined From Neutron Diffraction Data. *Acta Crystallographica*, Sect. **B47**, 1022-1023. (1992), Addendum ibid, **48**, 551-552.

26. Ceccarelli, C., Jeffrey, G. A. and Taylor, R. (1981), A Survey of O-H···O hydrogen bond geometries determined by neutron diffraction. *J. Molec. Struct.* **70**, 255-271.

27. Steiner, Th. and Saenger, W. (1992), Covalent Bond Lengthening of Hydroxyl Groups Involved in Three-Center and in cooperative Hydrogen Bonds. Analysis of Low Temperature Neutron Diffraction Data. *J. Am. Chem. Soc.*, **114**, 7123-7126.

28. Steiner, Th. and Saenger, W. (1994), Lengthening of the Covalent =-H Bond in O-H···O Hydrogen Bonds Re-examined from Low Temperature Neutron Diffraction Data of Organic Molecules. *Acta Crystallographica*, **50**, 348-357.

29. Koehler, J. E. H., Saenger, W. and van Gunsteren, W. F. (1988), The flip-flop hydrogen bonding phenomenon. *Eur. Biophys. J.* **15**, 153-168.

30. Steiner, Th., Saenger, W. and Lechner, R. E. (1991), Dynamics of Orientationally Disordered Hydrogen Bonds and of Water Molecules in a Molecular Cage. A Quasielastic Neutron Scattering Study of β-Cyclodextrin · 11 H_2O. *Molec. Phys.* **72**, 1211-1232.

31. Usha, M. G. and Wittebort, R. J. (1992), Structural and Dynamical Studies of the Hydrate. Exchangeable Hydrogens, and Included Molecules in β- and γ-Cyclodextrins by Powder and Single-Crystal Deuterium Magnetic Resonance. *J. Am. Chem. Soc.* **114**, 1541-1548.

32. Desiraju, G. (1991), The C-H···O Hydrogen Bond in Crystals: What is it? *Acc. Chem. Res.* **24**, 290-296.

33. Steiner, Th. and Saenger, W. (1992), Geometry of C-H···O Hydrogen Bonds in Carbohydrate Crystal Structures. Analysis of Neutron Diffraction Data. *J. Am. Chem. Soc.* **114**, 10146-10154.

34. Steiner, Th. and Saenger, W. (1993), Role of C-H···O Hydrogen Bonds in the Coordination of Water Molecules. Analysis of Neutron Diffraction Data. *J. Am. Chem. Soc.* **115**, 4540-4547.

35. Steiner, Th. and Koellner, G. (1994), Crystalline β-Cyclodextrin Hydrate at Various Humidities: Fast, Continuous and Reversible Dehydration Studied by X-Ray Diffraction. *J. Am. Chem. Soc.* **116**, 5122-5128.

36. Steiner, Th., Moreira da Silva, A. M., Teixeira-Dias, J. J. C.., Müller, J. and Saenger, W. (1995), Rapid Water Diffusion in a Cage-Type Crystal Lattice: β-Cyclodextrin Dodecahydrate. Angew. Chem. in press.

CLATHRATION AND SOLVATION OF MOLECULES

J. LIPKOWSKI
Institute of Physical Chemistry, Polish Academy of Sciences,
Kasprzaka 44, 01 224 Warszawa, Poland

1. Introduction

Clathrates are heteromolecular associates representing a variety of supramolecular structures at the multimolecular level. Clathrates are not a rarity in chemistry. Chlorine was first isolated by Davy [1] in the form of a clathrate hydrate. Organic clathrates are as old as synthetic organic chemistry since urea was obtained by Wöhler in the form of a clathrate [2]. Chemists know that it happens quite often that products of chemical reactions are obtained in the form of "solvates" (or "adducts", etc.) in which solvent and/or other foreign molecules co-crystallize with the "main" component. The ordered arrangements of these species comprised of "main" and "solvent", or "host" and "guest" components, contain important information on intermolecular interactions. Crystalline"inclusion" or "clathrate" compounds provide unique opportunities to study model structures which are likely to occur in liquid solutions.

The meaning of *clathratus* as 'closed or protected by cross bars of trellis' was originally used by H.M. Powell to describe how SO_2 molecules are held in the crystal structure of $3C_6H_4(OH)_2 \bullet SO_2$ [3]. This definition corresponds to that of a topological compound formed by two components, H (host) and G (guest), in which the H-H interactions determine the structure; the guest playing the role of the *filler* of the spaces left empty by the host. The absence of any specific, chemical G-H interaction in clathrate inclusion compounds has been taken by Weber and Josel [4] as the most characteristic feature by which to differentiate between *complexes* and *clathrates* in their classification and nomenclature of host-guest compounds.

Presently, it is quite common practise to use the terms *clathrate*, and inclusion or addition *molecular complexes* for two- or multicomponent molecular crystals in which there is no apparent chemical interaction between the components. In other words, *clathration* is used instead of *solvation* if only there is no specific, chemical interaction between the *solvent* and *solute*.

In this paper several recent, and for the most part presently unpublished, results are presented from the joint point of view of x-ray crystallography as well as physico-chemical analysis of the intermolecular interactions between the guest (solute) species and its environment in the crystal lattice. The structures include crystalline *hydrate* structures [5,6] in which the guest component is a molecule or quarternary

G. Tsoucaris et al. (eds.), Crystallography of Supramolecular Compounds, 265–283.

alkylammonium salt, and *coordination complexes* [7] in which the host component is of the MX_2A_4 chemical formula and the guest is an organic species. The lattices are quite different from a chemical standpoint (polar and hydrophilic water molecules forming the host lattice in hydrates, and hydrophobic metalloorganic complexes in the latter example) however some common phenomena may be observed on their inclusion compound formation.

2. **Background information**

2.1. THERMODYNAMIC CONDITIONS OF CLATHRATE STABILITY

The thermodynamic condition for the stability of a clathrate phase [8] is given by:

$$\ln(1-\Theta) = \frac{-m\delta(T)}{kT}$$

where $\delta(T) = \mu_\beta - \mu_\alpha$ stands for the difference of chemical potentials of the host component in its clathrate (β) and non-clathrate (α) phase, and Θ represents the fraction of cavities in the clathrate structure which are occupied by guest molecules, and m is the host/cavity molar ratio in the β-phase. Equation (1) describes the equilibrium state in which both clathrate and non-clathrate phases of the host component can co-exist. Clearly, the equilibrium guest content (Θ) is the minimum amount of the guest required for the clathrate phase to be stable, from a thermodynamic standpoint. Within the range of composition $\Theta < y < 1$, non-stoichiometry (meant as deviations from $y = 1$) may be very significant if the $\delta(T)$ value is small. Another characteristic feature of clathrate solutions, the minimum guest content Θ is a function of temperature, is given by:

$$\frac{\partial \ln(1-\Theta)}{\partial T} = -m\frac{\partial[\delta(T)/kT]}{\partial T} = m\frac{H_{\overline{\beta}} - H_{\overline{\alpha}}}{RT^2_-}$$

where H_β-H_α stands for the difference of the molar enthalpies of the β and α phases. Equations given above refer to simple clathrate structures in which: (i) there is just one sort of cavity able to accommodate single quest molecules, (ii) absorption of the guest does not significantly influence the host crystalline lattice including its oscillation modes, and (iii) intermolecular guest-guest interactions in the clathrate phase are negligibly small. The above conditions are fulfilled in some of the simplest structures, for example argon in the β-quinol clathrate. With $\delta(T) = 343$ J/mole, m=3 and H_β-H_α =670 J/mole [9] one finds that Θ equals 0.34 at 25°C, and decreases to 0.28 at 60°C.
In descriptive terms, it is commonly assumed that:
(i) the energy required to transform a non-clathrated host into its clathrate structure needs to be small since it has to be compensated by rather weak, non-bonded quest-host interactions. This is, from experience, realized when the host substance alone does not form efficiently packed homomolecular solid phases because of either its inappropriate molecular shape, and/or hydrogen bonding pattern in the solid;

(ii) clathrates are, in principle, non-stoichiometric solids with possible deviations from stoichiometry which are strongly dependent on the host structure and, for a given clathrate, dependent on temperature as well. In many real clathrate structures more complicated patterns are obtained. For example, in clathrate hydrates, cages of more than one type are available for guest species. Thus, under equilibrium conditions there are two Θ values (each for a specific cavity type) characteristic for the system, as given below

$$-\Delta\mu^{\circ}_{w} = \frac{RT}{23}[3\ln(1-\Theta_L) + \ln(1-\Theta_S)]$$

where $\Delta\mu^{\circ}_{w}$ denotes the chemical potential of water molecules of the clathrate structure relative to ice, and Θ_L ,Θ_S stand for guest occupancies of the large and small cavities, respectively. An elegant experimental proof of different stoichiometries for the small and large cages, and of the overall non-stoichiometry of the hydrate clathrates, has been given by Ripmeester [5]. With xenon as the guest species the values are $\Theta_L = 0.98$ and $\Theta_S = 0.72$. The hydration number, as given by

$$n = 23/(3\Theta_L + \Theta_S)$$

ranges from 6.04 for CH_4 to 7.26 for C_2D_4 in structure I hydrates [5].

2.2. STRUCTURE

The rather obvious feature of clathrate inclusion, in view of the definitions given in Section 1, is the good steric fit between the guest and host components. However, such conditions cannot be fulfilled if not for some exceptional G-H pairs, and this is the reason why clathrate compounds seem to be a rarity in chemistry. On the other hand, there are numerous examples of two- or multi-component solids in which fragments of clathrate structures can easily be seen. A combination of chemically specific (ionic, covalent or hydrogen bonding) and non-specific (clathrate) interactions is characteristic for many supramolecular structures at the *meso*scopic level.

Among the large varieties of different clathrate structures there are two main types which may be defined as, respectively, "skeletal" and "packing" architectures. In the former type intermolecular hydrogen bonding leads to formation of three-dimensional host networks having sizeable voids which can accommodate foreign ("guest") species. This sort of host clathrate structure includes β-quinol and ice. The "packing" clathrate structure type refers to compounds in which host molecules, having an unfavourable shape from the point of view of dense packing, show a pronounced ability to crystallize in forms possessing intermolecular voids which are large enough to accommodate suitable guest atoms, molecules or ions. Although differences in physicochemical characteristics between the two clathrate types are of a quantitative rather than qualitative nature, the above classification seems to be useful in description of the structures. In skeletal clathrates it is both customary and convenient to represent their structures as assemblies of polyhedra constructed from, say, tetrahedral OH_4 units (water host structures, each H_2O molecule donating two and accepting two hydrogen bonds). There is a tendency in nature to build regular pentagonal dodecahedra from the tetrahedral units. The resulting spatial structure, containing some extra cages (in

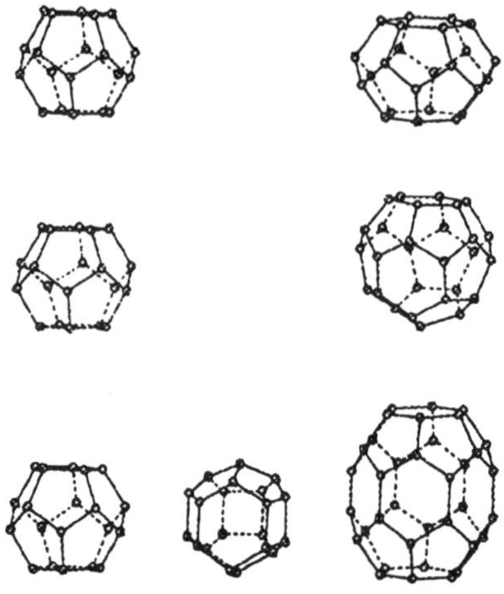

Figure 1. The cavities found in the three major types of clathrate hydrates, according to [5], from top to the bottom: structure **I**, **II** and **H**.

addition to the dodecahedra representing small cavities) is unstable without absorbed guest species filling the voids in the structure. As seen in Figure 1, the guest moieties have the choice between some cavities differing in shape and size. If the cages are too small to accommodate a guest then a combination of cavities is possible, as found by Jeffrey et al. [10], to form more sizeable voids (Figure 2).

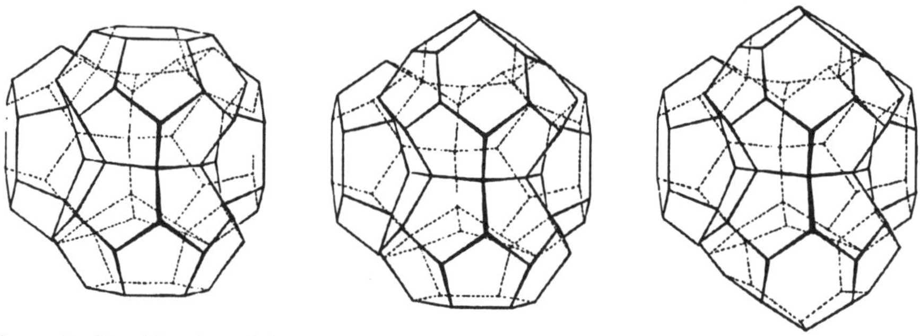

Figure 2. Combined cavities of four 14-hedra, three 14-hedra and one 15-hedron, and two 14-hedra and two 15-hedra (from left to right), according to Jeffrey [10].

No introductory description of the basic clathrate hydrate structures will be given here since there are excellent review papers on this topic in the recent literature [5,6,10,11]. Suitable examples of the packing clathrate structure type are the series of the, so called, Werner clathrates [7]. These are products of the co-crystallization of some coordination complexes with either solvent molecules or some other guest compounds. Among the host complexes there are some which display such a pronounced clathrate-forming ability that preparation of the host in a pure, non-clathrated crystalline form presents difficulties. A good example is $Ni(NCS)_2$(4-Methylpyridine)$_4$ for which solid clathrates have been known since the 1950's while the non-clathrated form was not identified untill 1974 [7]. The variety of structural types and stoichiometries of this particular host is illustrated by the data given in Table 1.

TABLE 1. Crystal Structure Types of $Ni(NCS)_2$(4-Methylpyridine)$_4$ (Host) Intermolecular Inclusion Compounds

Structure type	*example of a guest compound*	*guest/host ratio/*	*Space group*
Cage	m-Bromonitrobenzene	0.67 : 1	R3
Organic zeolite	Benzene	1 : 1	$I4_1/a$
Layered structure	Naphthalene	2 : 1	C2/c
	1-Methylnaphthalene	2 : 1	$P2_1/c$
	2-Methylnaphthalene	2 : 1	P1
	Bromobenzene	2 : 1	$Pna2_1$

Illustrations of the structures listed in Table 1 are given in Figures 3 through 5. In all

these structures their physicochemical behaviour is strongly dependent upon the size and shape of the guest molecule(s) present in the voids of the host crystalline lattice, as well as the formally assigned structure type based on the packing of the host molecules. The cage structure of the m-bromonitrobenzene clathrate might behave like a channel-type structure if the guest molecules would have appropriate size and shape. The same holds for organic zeolite in which some guest species are held firmly as in a cage structure while the other guest species can easily diffuse in and out. As already

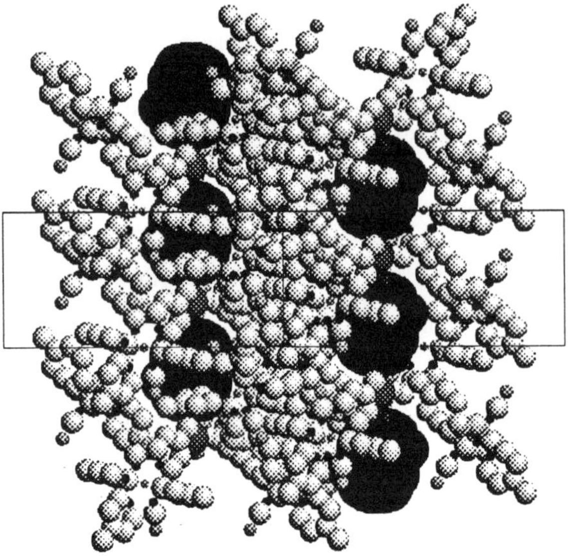

Figure 3. A packing diagram showing channel-type arrangement of cavities (black fields) in the structure of the host $Mg(NCS)_2(4\text{-Methylpyridine})_4$ [12].

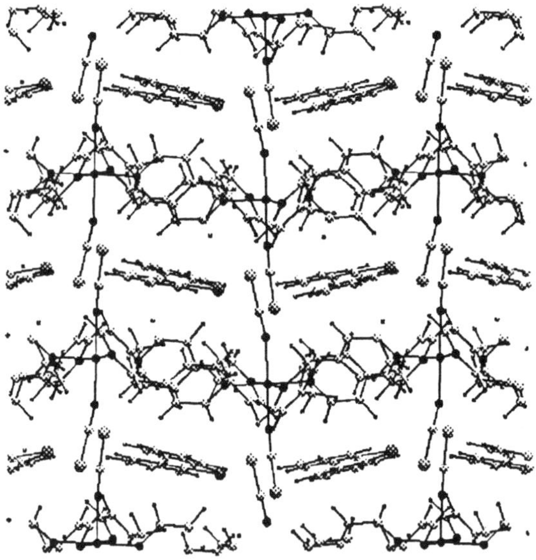

Figure 4. A layered packing of the $Ni(NCS)_2(4\text{-Methylpyridine})_4 \bullet 2(\text{chlorobenzene})$ clathrate [13].

mentioned, the porous structure of the organic zeolite host displays a pronounced contraction-dilation ability according to the amount of guest material absorbed in its intracrystalline pores. The minimum volume of an *organic zeolite* unit cell is attained when no guest is absorbed. Such *empty* clathrate structures, as well as some partially filled lattices, although unstable thermodynamically, show significant kinetic stability which can be studied experimentally. The organic zeolite structure of this host complex can vary its molar volume (as given in [7]) from 472 cm^3 (empty structure) to 547 cm^3 (m-xylene clathrate). This difference is greater than the lattice volume expansion during the $\alpha \rightarrow \beta°$ phase transformation (431 to 472 cm^3 ; $\beta°$ denoting the *empty* phase). Host lattice dilation may cause significant effects on guest diffusion kinetics in the solid and on the guest-lattice interaction energy.

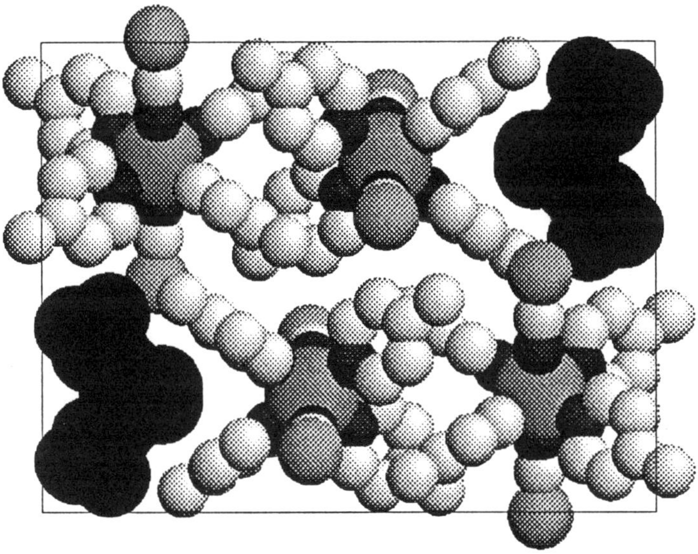

Figure 5. The *organic zeolite* crystal structure of the $Ni(NCS)_2(4\text{-Methylpyridine})_4$ host complex viewed along **b** [14]. Half width of the unit cell content is shown for the sake of clarity; space occupied by guest molecules shown in black.

2.3. INTERMOLECULAR INTERACTIONS

When analyzing the experimentally derived thermochemical data on clathration from the point of view of the guest-lattice intermolecular interactions in clathrate phases it is convenient to take the overall clathration process as a sum of the following three components:

a) the host lattice transformation $\alpha \rightarrow \beta$ (corresponding to the enthalpy change $\Delta H_{\alpha \rightarrow \beta°}$),

b) guest evaporation or sublimation (ΔH_{evap}), and

c) sorption of the gaseous guest in the empty β-host (ΔH_{sorp}).

Then, the overall heat effect measured upon clathration is the sum

$$\Delta H_{clath} = \Delta H_{evap} + \Delta H_{\alpha\to\beta} + \Delta H_{sorp}$$

However, as shown experimentally [15], the host lattice is not as rigid as assumed in the simple thermodynamic treatment of clathrate behaviour, and some additional terms thus have to be considered in order to quantitatively describe the measured enthalpies of clathration. There are, at least, two significant contributions to the overall heat of clathration. First, the host lattice may be sensitive to the sort and amount of the absorbed guest; the difference between the empty clathrate phase and the stoichiometric inclusion being, in some cases, even higher than the lattice expansion which occurs at phase transformation $\alpha \to \beta$ [15]. Thus, the enthalpy of dilation of the β-lattice of the host (ΔH_{dil}) must be considered as an additional component in the equation above. This effect is not only different for various guest components, but is dependent on the final concentration of the guest as well. In other words, the enthalpy of clathration of a given guest compound is a function of its concentration in the solid phase, and the resulting expression for the enthalpy of clathration is

$$\Delta H_{clath} = \Delta H_{evap} + \Delta H_{\alpha\to\beta^\circ} + \Delta H_{dil} + \Delta H_{sorp}$$

where $\Delta H_{\alpha\to\beta^\circ}$ denotes the enthalpy change associated with the $\alpha\to\beta^\circ$ phase transformation, and ΔH_{dil} represents the heat effect which comes from the dilation of the host lattice. Since the host lattice expands or contracts its cages respective to the amount of the absorbed guest (averaged in the crystal), one might expect the enthalpy of sorption to be dependent on the clathrate's composition. This is indeed the case as shown experimentally in [15]. Based on thermochemical investigations of clathration enthalpy of three-component mixtures of o-, m- and p-xylenes by $Ni(NCS)_2(4\text{-}MePy)_4$ it has been found that the partial molar enthalpies of clathration (H_{clath}) of each of the isomers depends upon the concentration of the given species in the clathrate (y), and the increment is always positive, (i.e. the more molecules of an isomer are present in the clathrate the higher partial molar enthalpy of clathration of that given isomer). The experimentally derived functions are given below:

$$(H_{clath})_o = 2.41 + 2.17\, y_o$$

$$(H_{clath})_m = 3.20 + 1.35\, y_m$$

$$(H_{clath})_p = 6.01 + 1.50\, y_p$$

where indexes *o*, *m* and *p* stand for ortho, meta and para-xylene, respectively, and the enthalpies are given in the kcal mol^{-1} units. The overall enthalpy of clathration of a given mixture of the isomers is then given by

$$\Delta H_{clath} = (H_{clath})_o\, y_o + (H_{clath})_m\, y_m + (H_{clath})_p\, y_p$$

Since the dependence is primarily dependent on the amount of the given component and <u>not</u> on the lattice volume, one may assume that the absorbed guest molecule "knows" whether the adjacent cage is filled or empty since the energy of its interaction with the lattice is "sensitive" to the degree of filling of the cavities. Molecular modelling calculations [16] have shown that there are no significant guest-guest interactions in the experimentally studied examples. Thus, the mechanism of the dependence seems to function *via* guest-lattice rather than guest-guest interactions. Put in other words, the host structure adapts itself in order to accommodate a guest molecule in such a way that the next absorbed molecule of the same kind will find the structure more suitable for absorption, and so on. It may be considered a sort of *induced fit* behaviour in solid inclusion compounds. This would seem reasonable assuming that an absorbed molecule can "feel" not only if its adjacent neighbouring cavities are filled or empty, and also "recognizes" if the cavities contain similar or different guest species. This being so, one would expect a dependence of the partial molar enthalpy of clathration of a guest on the qualitative composition of the enclathrated mixture, as given above.

One may expect larger enthalpy effects when, instead of isomeric guest molecules, a mixture of species which are dissimilar in chemical and geometrical parameters is used for clathration. Such is indeed the case, as was observed in the examples of simultaneous clathration of 4-methylpyridine and a solvent (methanol or acetone). The enthalpy of clathration of 4-methylpyridine (the partial molar value) varies from 2.8 kcal mol^{-1} (in ternary Mepy/acetone clathrate) [17] to 4.8 kcal mol^{-1} (in MePy clathrate) [18].

3. Recent Developments

3.1. WERNER CLATHRATES

From the general point of view of the, so-called, *clathratogenic ability* of Werner complexes, the most characteristic development seems to be the generalization of inclusion compound formation. Initially, in the 1950s, two main conditions were stressed as determining the success of clathrate-formation: (1) the central ion should, at least preferably, be an element from the transition *d*-series, and (2) the pyridine (or similar ligand) should have a substituent (or substituents) which prevents the complex molecules from close packing in the crystal lattice.The former condition for clathrate formation contains an implicit suggestion that some guest-host charge-transfer interactions may be important from the point of view of the clathrates' stability. On the other hand, the latter condition regarding clathrate formation stresses the role of steric interactions. If the host complex is able to form close packed mono-component crystal structures by itself, i.e. without any guest, then it would display little, if any, tendency to form heterocomponent solids. Based on the most recent experimental data it may be concluded that neither of the above two conditions is valid in a general, qualitative sense. As seen in Figure 6, magnesium (II) complexes are rather potent

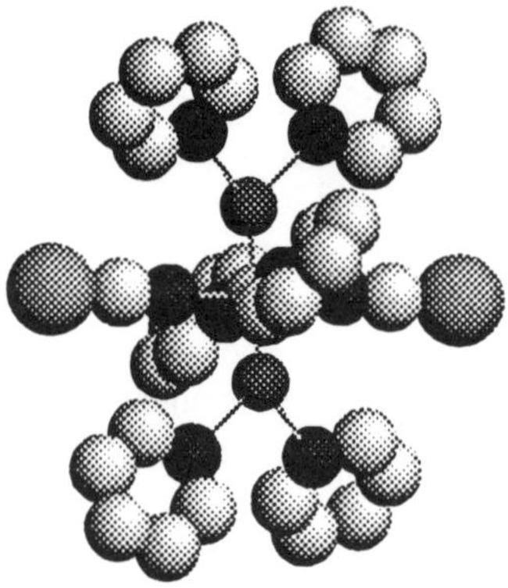

Figure 6. Molecule of the Mg(NCS)2(4-Methylpyridine)2(H2O)2•4(4-Methylpyridine) complex [20].

clathrate hosts and, even more convincing, these compounds are isostructural to some formerly known solvates of calcium complexes. Furthermore, a series of pyridine complexes (containing simple, i.e. unsubstituted, pyridine ligands) prepared in the form of solvates [21] of different compositions and crystal structures, shows that complicated ligands are not necessary for clathrate-formation. In particular, the $MX_2(py)_4$ •2(py) intermolecular complexes are well known to coordination chemists under the formula $MX_2(py)_6$. The recently determined crystal structure of $Ni(NO_3)_2(py)_4$•2(py) shown in Figure 7 may serve as a good illustration of the above statement. This is noteworthy that this is just one of a long series of either isostructural or just similar crystal structures of hexa-pyridinate complexes.

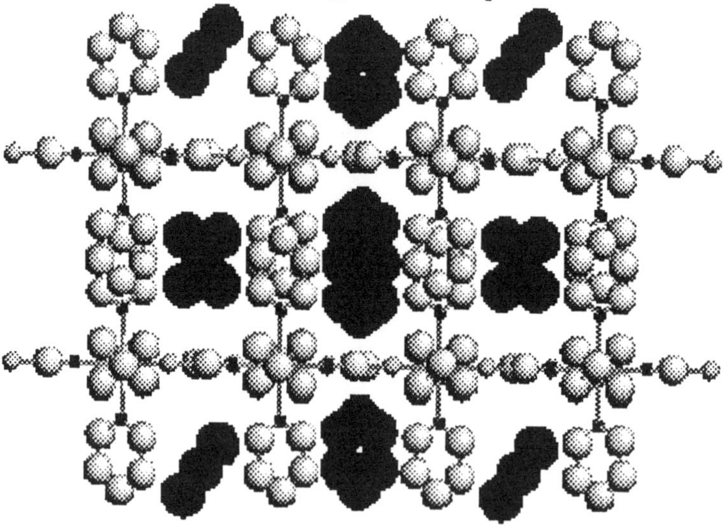

Figure 7. Molecular packing in the Ni(NCS)2(pyridine)4•2(pyridine) solvate [21]; guest molecules contour emphasized in black colour, host atoms in arbitrary radii.

The conditions for a complex to have 'clathratogenic' ability, mentioned previously, seem to preserve, in part, their validity in a quantitative sense. Thus some Werner complexes show a more pronounced clathratogenic ability than others, more or less following the above rules. In the Authors own opinion it seems that, from the two conditions mentioned, the steric factors are of much more significance. Summarizing recent observations one may note that a clathrate-forming ability appears as a quite general physical property of coordination complexes. The so-called *Werner clathrates* or *Schaeffer clathrates* formed by MX_2A_4 hosts are very convenient model systems to study in detail the physico-chemical characteristics of the products formed, their diverse structures, properties, etc. However, virtually any complex molecule may, under suitable conditions, crystallize in the form of a *solvate* which, depending on its composition and structure, may behave either as a *clathrate*, *intercalate,* or *organic zeolite*. The common characteristic between them is that there is no chemical bond between the host and guest and, as a rule, the product is a non-stoichiometric solid. It seems that a very similar conclusion emerges from the recent papers by Nassimbeni and coworkers [23], following comprehensive structural, thermodynamic and kinetic studies on a long series of Werner clathrates.

It is worth mentioning here that the formation of solid inclusion compounds of Werner complexes may help significantly in understanding the molecular structure of the complexes. It is rather well known that in the solution phase of coordination complexes different forms of the complexes may co-exist at an equilibrium, like *cis-trans* coordination, or *tetrahedral* versus *octahedral* geometry. Crystallization of the complex as a pure, mono-component solid leads, in general, to precipitation of the most stable form which can be submitted to x-ray structure studies while the other forms can only exist in solution. Clathrates, being a type of "solid solution", expand the possibilities with which to explore the molecular structures of some of the other forms as well. There are some fascinating examples in the recent literature to prove this. One, which was reported by Nassimbeni et al. [24], shows that the $Ni(NCS)_2$(α-phenylethylamine)$_4$ complex, when crystallized in a pure form, is found to have a *cis*-configuration of anionic, isothiocyanate ligands. In contrast, when co-crystallizing to form a clathrate solid (when the guest was *s*-buthylbenzene), the *trans*-structure of the complex is more stable. Dyadin and coworkers call this phenomenon *"contact stabilization"* of the host molecular structure [25]. Another effect belonging to this category is preparation of some Zn^{2+} , Cd^{2+} and Cu^{2+} octahedral MX_2A_4 complexes in the form of solid inclusion compounds while the pure compounds are thermodynamically unstable as monocomponent solids, as shown in Figure 8 with the example of $Zn(NCS)_2$(4-Methylpyridine)$_4$. The stable monocomponent solid in the system is $Zn(NCS)_2$(4-Methylpyridine)$_2$ in which Zn^{2+} is tetrahedrally coordinated [27] but the octahedral complex could be investigated by x-ray diffraction of its inclusion compounds. Two different conformers of the complex could be found in this way. Also, the so-called, *distortion isomers* [28] of $Cu(NCS)_2$(4-Methylpyridine)$_4$ could be studied in the solid phase due to different possible packing modes. The complex, being unstable (similar to the above mentioned Zn octahedral complex), is able to co-

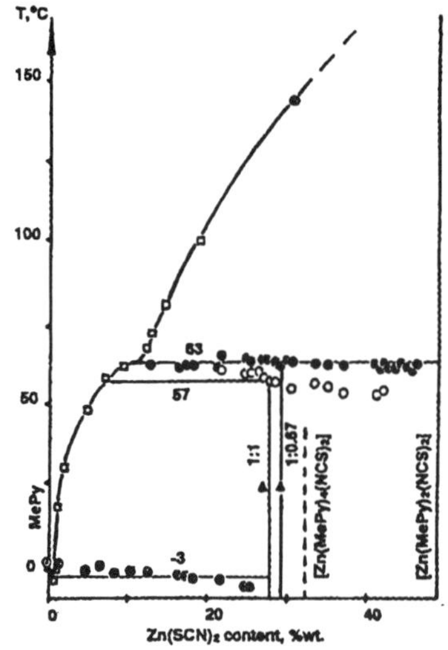

Figure 8. The phase diagram of the $[Zn(MePy)_2(NCS)_2]$-MePy system. The circles stand for DTA (circles for different types refer to different processes), the squares for solubility technique and the triangles for analytical data (reprinted from [27]). [⊗ The eutetic effects $MePy_s$ +liquid phase + 1:1; o The peritectic effects phase +1:1 +1:0.67; • The peritectic effects phase +1:0.67 + $[Zn(MePy)_2(NCS)_2]$; o The metastable effects O Liquidus from DTA data.]

crystallize with 4-Methylpyridine. At temperatures below 17° C the stable solid has the $Cu(NCS)_2$(4-Methylpyridine)$_4$ •2(4-Methylpyridine) composition and exhibits a blue colour. Above 17°C the stable form is $Cu(NCS)_2$(4-Methylpyridine)$_4$•0.67(4-Methylpyridine) and has a green colour. These colour differences are associated to the structural differences of the coordination sphere of the Cu^{2+} ion [29], as illustrated in Figure 9.

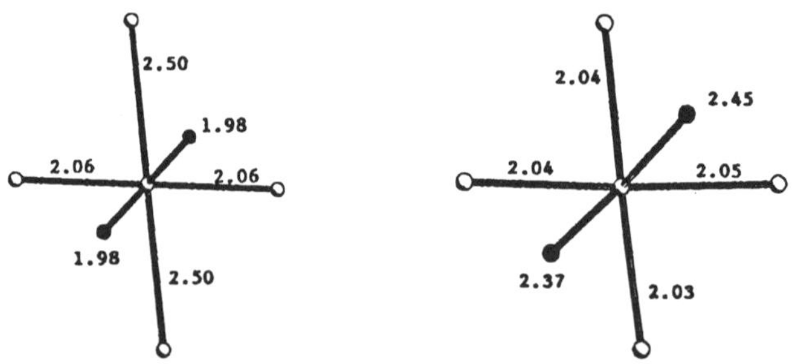

Figure 9. Comparison of the geometry of the first coordination sphere of the host $[Cu(MePy)_4(NCS)_2]$ molecule in the green (a) and blue (b) clathrates. The open circles represent the N_{Mepy} donor atoms while dashed circles are for the isothiocyanate N_{CS} atoms.

3.2. HYDRATES

Recent developments in the field of clathrate hydrates will be illustrated here by using the results of x-ray single crystal diffraction analysis of some new alkylammonium fluoride hydrates, the hydration patterns of crown ethers, and some of their molecular complexes. The respective phase studies of these hydrates are supposed to be presented as a separate chapter of this volume.

3.2.1. *Hydrates of Alkylammonium Fluorides*

The series begins from the tetraisoamylammonium fluoride hydrate (1 : 27) [30] which is shown in Figure 10. The host framework is formed from water molecules and fluoride anions, and is thus anionic. The guest tetraisoamylammonium cation occupies

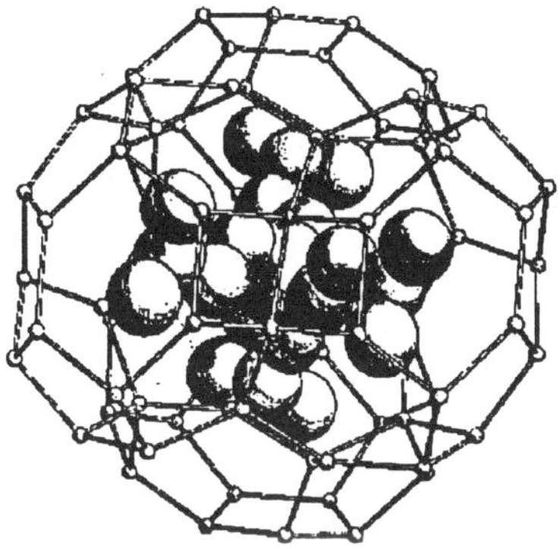

Figure 10. The structure of tetraisoamylammonium fluoride hydrate: large cavity with the guest cation. For the sake of clarity: hydrogen atoms omitted, guest atoms represented as shaded balls.

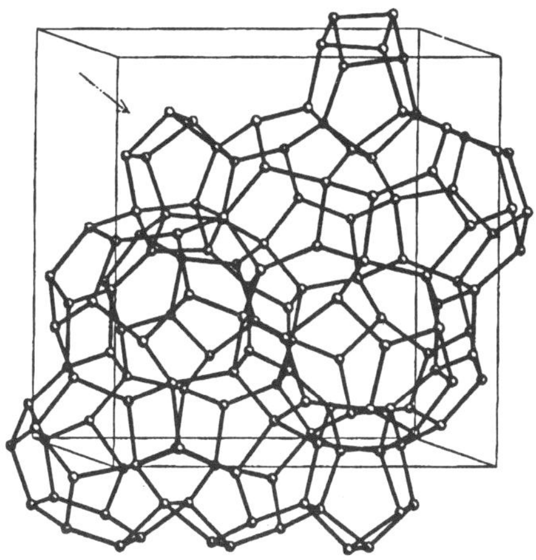

Figure 11. Packing of host polyhedra in the tetraisoamylammonium fluoride hydrate [30] showing small cavities (indicated by arrow).

a cavity of rather complex structure which also shows, when compared to previously known structures of comparable cavity size, significant deviations from the tetrahedral geometry due to the intermolecular bonding pattern of water. The extreme case of that is the small cavity shown in Figure 11 in which square planar faces are observed. This

is also, to our knowledge, the smallest cavity ever found formed by host water molecules in the solid hydrates.
The next example shown in Figure 12 represents a somewhat more open structure in which guest cations (tetraethylammonium) are arranged along open channels in the water framework.

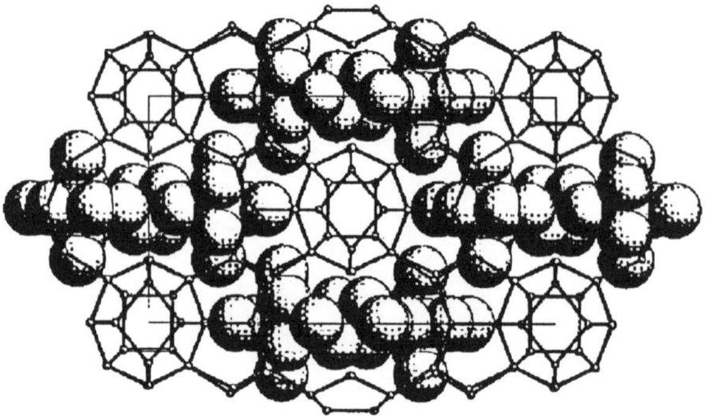

Figure 12. The structure of tetraethylammonium fluoride hydrate (1 : 10.5 mol/mol) [31].

The tetra-n-propylammonium fluoride hydrate [32] shows a layered structure (Figure 13) in which the three-dimensional intermolecular bonding between water molecules is preserved only partially, i.e. within the layer formed by H_2O molecules.

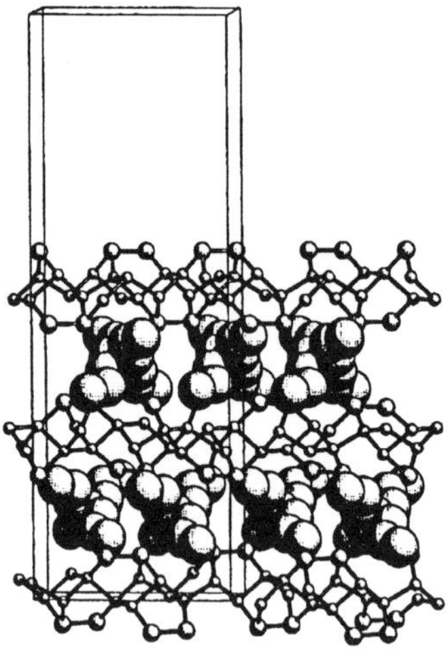

Figure 13. The layered structure of tetra(n-propyl)ammonium fluoride hydrate [32].

In Figure 14, in which the structure of the tetraethylammonium fluoride pentahydrate is displayed, jonly a two-dimensional pattern of the host water is observed [31].

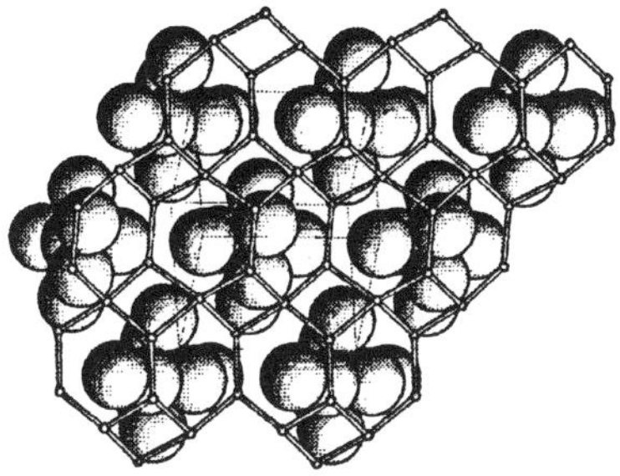

Figure 14. A projection of layered hydrate structure. Guest: tetraethylammonium fluoride.

When going from the cage structures as in Figure 10 to the layered ones as illustrated in Figures 13 and 14, fewer and fewer intermolecular water-water hydrogen bonds are seen and, at last, one can hardly identify any "host" or"guest" component in the structure.

By inspection of the structures given in these Figures, and of the phase diagrams of the systems studied (*cf.* the chapter by Dyadin et al.), it may be concluded that the type of structure of the host-guest associates is strongly dependent upon the composition of the solution phase from which the solid crystallizes. If, for example, a layered structure is obtained, for which less host material is necessary when compared to the cage type, then at higher guest concentrations in solution the crystallization of layered compounds is observed.

3.2.2. *Hydrates of Crown Ethers and Their Complexes*

In a recently studied series of hydrate structures of the crown-type molecular receptors it has been found that some supramolecular patterns can also be found in

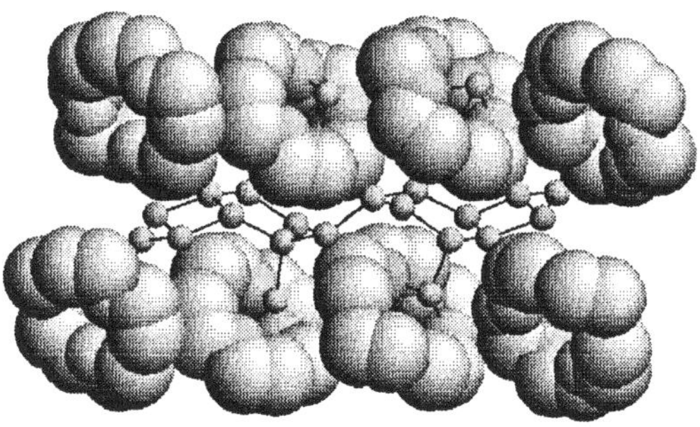

Figure 15. Figure 15. The structure of 18-crown-6 hexahydrate [33]

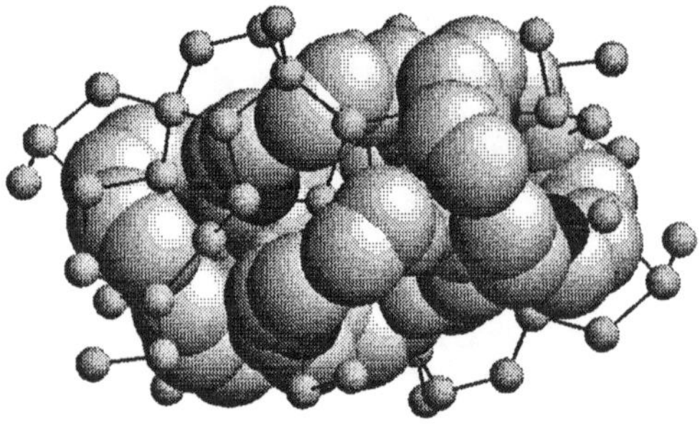

Figure 16. The structure of 1,10-diaza-18-crown-6 nonahydrate [33].

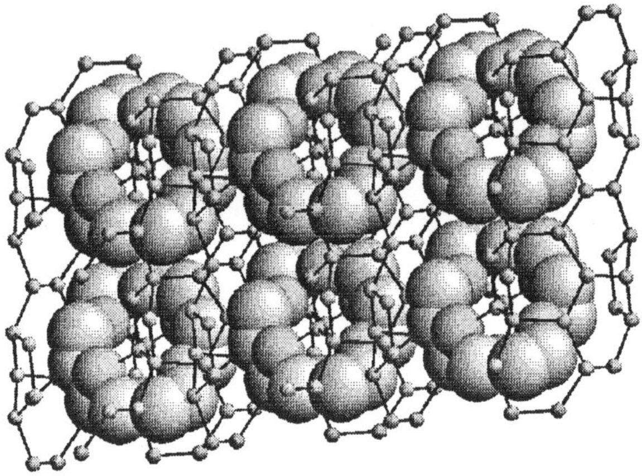

Figure 17. The structure of 18-crown-6 dodecahydrate [33].

these compounds. Although there is no contribution of ionic guest-lattice interactions, as is the case of alkylammonium salt hydrates, in the crown/water structures there is more guest-host interaction than clathration forces alone. As may clearly be seen by inspecting Figures 15 through 17, there is some hydrogen bonding between water and the crown species. Nevertheless, there are many structural features common to both the alkylammonium hydrates and the crown/water complexes in the solid phases studied. Perhaps the most convincing illustration of that hypothesis arises from comparison of the supramolecular architecture of water molecules in the alkylammonium cage hydrate (Figure 10) to that found in the crown : water solid (Figure 16).

When the internal crown ether receptor site is blocked by complexation of another species, (e.g. by an aminoacid or peptide moiety), then water-crown hydrogen bonding is no longer feasible. Still high hydration of the complexes is possible as has been shown by Lipkowski et al. [34]. The structure of the glycylglycine/water/18-crown-6 ternary compound, which is illustrated in Figure 18, shows two distinct types of the water-complex interaction. One is 'typical' hydrogen bonding between water molecules and the hydrophilic end groups of the peptide (i.e. the carboxylic end group, the peptide existing in the zwitterionic form in the structure). The other is *hydrophobic hydration*, i.e. clathration of the hydrophobic part of the complex by the supramolecular assembly of water molecules.

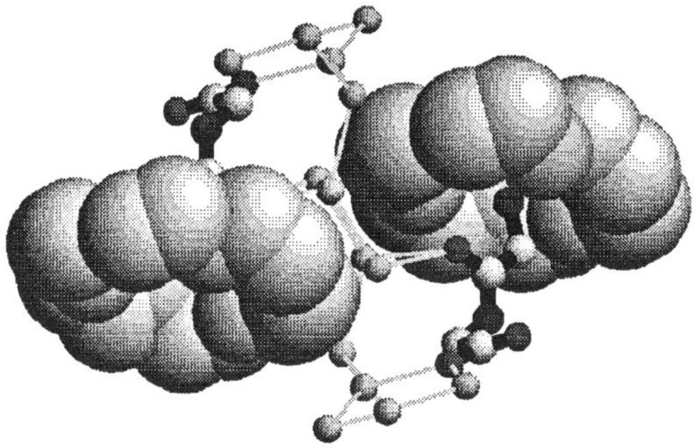

Figure 18. The structure of glycylglycine/water/18-crown-6 ternary compound.

4. **Discussion**

As demonstrated above, hydrate structure type and stoichiometry is not restricted to the particular example of alkylammonium hydrate salts. Nor is structure type limited to a given guest species fitting well into host cavities. The host may adapt its lattice in order to accommodate different guest species and, analogous to Werner clathrates, the clathratogenic ability term has its significance in a quantitative, rather than qualitative, sense. The extensive experimental studies performed recently by Dyadin et al. and Ripmeester and coworkers, seem to confirm the above conclusion. If there is a possibility to form more or less hydrated solids by changing the experimental conditions for crystallisation then, in some sense, the series of crystal structures given above demonstrates a variable "degree of clathration". In other words, the variable structural contributions of clathration, as for the particular kinds of intermolecular interaction, influence the overall stability of the crystalline products. For this reason it is not always easy to unambiguously distinguish between "inclusion" and "addition" structures. It seems more important to consider clathration as a sort of collective interaction at the multimolecular level. Thus, clathration may be assumed as the distinguishing factor contributing to complete pattern of solvent-solute interactions.
Based on the analogy of the structures obtained irrespective of very significant differences in qualitative chemical compositions of the crystalline products, one may suggest that x-ray structures of clathrates (or solvates etc.) may be considered as snapshots of some of the aggregation modes of the guest/host or solute/solvent molecular systems in the liquid or solution state. This concept applied to the chemistry of liquid solutions is rather old (older even than the use of x-ray diffraction to study the structure of solids !). Now, with the modern experimental techniques for structure investigations well established, it would be interesting to examine whether x-ray crystallographic work may be considered a relevant experimental technique to model heteromolecular associates which are likely to occur in solution phases. The illustrations reported above are limited to 'static' pictures only, and certainly do not

represent the complete structural information derived from x-ray diffraction. It should be noted here that in most of the inclusion compounds described in this paper, rather interesting and valuable data on molecular dynamics are contained in the x-ray structures. These topics are not reported here since the results are preliminary and the analysis of some molecular disordering has only been completed for a few of the structures given above.

Acknowledgement: The author acknowledges, with thanks, the assistance of Mr. Paweł Staszewski in preparation of illustrations to this chapter.

5. References

1. Davy Sir H. (1811), *Philos. Trans. R. Soc.* (London), 101, 30.
2. Wöhler F. (1849), *Ann.Chem.Liebigs*, 69, 297.
3. Powell H.M. 1948, *J. Chem. Soc.*, 612.
4. Weber E. and Josel H.-P. (1983), A proposal for the classification and nomenclature of host-guest type compounds, *J. Inclusion Phenomena* **1**, 79-85.
5. Ripmeester J.A. and Ratcliffe C.I. (1991) Solid state NMR studies of inclusion compounds, in J.L. Atwood, J.E.D. Davies and D.D. MacNicol (eds.) *Inclusion Compounds*, Oxford University Press, vol.5, chapter 2 pp. 37-89, and references therein.
6. Dyadin Yu.A., Bondaryuk I.V. and Zhurko F.V. (1991) Clathrate hydrates at high pressures, in J.L. Atwood, J.E.D. Davies and D.D. MacNicol (eds.) *Inclusion Compounds*, Oxford University Press, vol.5, chapter 7 pp. 213-275.
7. Lipkowski J. (1984), Inclusion compounds formed by Werner MX_2A_4 coordination complexes, in J.L. Atwood, J.E.D. Davies and D.D. MacNicol (eds.), *Inclusion Compounds*, Academic Press, London, vol. 1, ch. 3, pp. 59-103.
8. van der Waals J.H. and Platteeuw J.C. (1959), Clathrate solutions, *Adv. Chem. Phys.*, **2**, 1-57.
9. Powell H.M. (1964) Clathrate compounds, in L. Mandelcorn (ed.), *Non-stoichiometric Compounds*, Academic Press, NY-London, ch. 7, pp. 438-490, and references therein.
10. Jeffrey G.A. (1984), Hydrate inclusion compounds, in J.L. Atwood, J.E.D. Davies and D.D. MacNicol (eds.) *Inclusion Compounds*, Academic Press, London, vol. 1, ch. 5, pp. 135-190, and references therein.
11. Davidson D.W. (1973), Clathrate hydrates, in F. Franks (ed.) *Water: A Comprehensive Treatise*, Plenum Press, New York, vol. 2, ch. 3, pp. 115-203.
12. Lipkowski J. and Soldatov D.V. (1994), Two clathrates of bis(isothiocyanato)tetrakis(4-methylpyridine)magnesium(II) as host with 4-methylpyridine as a guest, *J. Inclusion Phenomena and Molecular Recognition in Chem.* **18**, 317-329.
13. Lipkowski J. and Soldatov D.V. (1995), The layered structure of chlorobenzene (guest) clathrate with $Ni(NCS)_2$(4-Methylpyridine)$_4$ (host), *J. Inclusion Phenomena and Molecular Recognition in Chem.*, in preparation.
14. Lipkowski J., Suwińska K., Andreetti G.D. and Stadnicka K. (1981), Clathrate Inclusion Compounds of bis(isothiocyanato)tetrakis(4-methylpyridine)nickel(II). II. Arrangement of guest molecules in the β-$Ni(NCS)_2$(4-Methylpyridine)$_4$ porous crystals, *J. Mol. Structure*, **75**, 101-112.
15. Lipkowski J., Starzewski P. and Zielenkiewicz W. (1982), Thermochemical studies on clathration phenomena. Part III. Competitive clathration of xylene isomers by the $Ni(NCS)_2$(4-Methylpyridine)$_4$ complex, *Polish J. Chem.*, **56**, 349-358.
16. Starzewski P. (1982), Calorimetric study of clathration phenomena (*Pol.*), PhD thesis, Warszawa.
17. Lipkowski J., Zielenkiewicz A., Hatt J. and Zielenkiewicz W. (1989), Calorimetric and thermogravimetric study of competitive clathration in the β-$Ni(NCS)_2$(4-MePy)$_4$ host, *J. Inclusion Phenomena and Molecular Recognition in Chem.* **7**, 519-527.
18. Chajn J., Lipkowski J. and Zielenkiewicz W. (1977), Thermochemical studies on clathration phenomena. Part I. Enthalpy of clathration of isomeric xylenes by the $Ni(NCS)_2$(4-Methylpyridine)$_4$ complex, *Roczniki Chem.* **51**, 1431-1442.

19. Lipkowski J. and Soldatov D. (1993), Crystal and molecular structure of bis(isothiocyanato)tetrakis(pyridine)magnesium(II), *J. Coord. Chem.*, **28**, 265-269.
20. Lipkowski J. and Soldatov D. (1993), Crystal structure of bis(isothiocyanato)diaquabis(pyridine)magnesium(II) solvate with pyridine 1 : 4, *Supramolecular Chem.* **3**, 43-46.
21. Soldatov D. V., Lipkowski J. and Gratchev E.V.(1995), Crystal structure of bis(nitrato)tetrakis(pyridine)nickel(II) clathrate with pyridine as the guest, $[Nipy_4(NO_3)_2]\bullet 2py$ (*Russ.*), *Zh. Strukt. Khim.*, , in press.
22. Dyadin Yu.A., Soldatov D.V., Lipkowski J. and Logvinenko V.A. (1995), Contact stabilization of host complex molecules during clathrate formation: the pyridine - zinc nitrate and the pyridine - cadmium nitrate systems, *J. Coord. Chem.*, in press.
23. Lavelle L. and Nasimbeni L.R. (1993), Studies on Werner clathrates. Part 13. Selective criteria in Werner clathrates. Six crystal structures with bis(isothiocyanato)tetra(4-vinylpyridine)nickel(II) as host, *J. Inclusion Phenomena and Molecular Recognition in Chem.*, **16**, 25-54.
24. Nassimbeni L.R., Niven M.L. and Zemke K.J. (1985), Di-isothiocyanatotetrakis(α-phenylethylamine)nickel(II): Structures of host and clathrate complexes, *J. Chem. Soc., Chem. Commun.*, 1788-1791.
25. Dyadin Yu.A. and Kislykh N.V. (1991), A contact stabilization of host molecules in the clathrate formation, *Mendeleev Commun.*, **4**, 134-138.
26. Lipkowski J. (1990), Structure of bis(isothiocyanato)bis(4-Methylpyridine)zinc(II), *J. Coord. Chem.* **22**, 153-158.
27. Lipkowski J., Soldatov D.V., Kislykh N.V., Pervukhina N.V. and Dyadin Yu.A. (1994), Phase and structural study of clathrate formation in the $[Zn(MePy)_2(NCS)_2]$-MePy system (MePy = 4-Methylpyridine), *J. Inclusion Phenomena and Molecular Recognition in Chem.* **17**, 305-316.
28. Gazo J., Boca R., Jona E., Kabesova M., Macaskova L., Sima J., Pelikan P. and Vallach F. (1982), Equatorial-axial interactions in complexes as a manifestation of mutual influence of ligands, *Coord. Chem. Rev.*, **43**, 87-131.
29. Lipkowski J., Dyadin Yu.A. and Kislykh N.V. (1995), Crystal structure of the $[Cu(MePy)_4((NCS)_2]\bullet 2MePy$ clathrate, *J. Inclusion Phenomena and Molecular Recognition in Chem.*, in preparation.
30. Lipkowski J., Suwińska K., Rodionova T.V., Udachin K.A. and Dyadin Yu.A. (1994), Phase and x-ray study of clathrate formation in the tetraisoamylammonium fluoride-water system, *J. Inclusion Phenomena and Molecular Recognition in Chem.* **17**, 137-148.
31. Udachin K.A. and Lipkowski J., *unpublished data.*
32. Lipkowski J., Luboradzki R., Udachin K. and Dyadin Yu. (1992), A layered clathrate hydrate structure of tetrapropylammonium fluoride, *J. Inclusion Phenomena and Molecular Recognition in Chem.* **13**, 295.
33. Udachin K.A. and Lipkowski J. (1995), Hydration patterns of some crown ethers, *Supramolecular Chem.*, in preparation.
34. Lipkowski J., Kulikov O.V. and Zielenkiewicz W. (1992) Complexation of peptides with crown ethers. Structure and thermal behaviour of ternary compounds: 18-crown-6/glycylglycine/water and 18-crown-6/ L-α-alanyl-L-α-alanine/water, *Supramolecular Chem.* **1**, 73-79.

INCLUSION COMPOUNDS: RELATING STRUCTURE TO KINETICS AND THERMODYNAMICS

L R NASSIMBENI
Chemistry Department, University of Cape Town
Rondebosch 7700, South Africa

Inclusion compounds, made up of host and guest components, are interesting systems to study by a variety of physico-chemical techniques, because many of their macro-properties can be interpreted in terms of their molecular structures. In particular, the strengths of the intermolecular interactions and the topology of host-guest assembly allow us to explain such properties as their thermal stabilities, their kinetics of formation and the mechanism accompanying their decomposition.

Standard techniques which can be usefully employed in the study of inclusion compounds are shown in Figure 1.

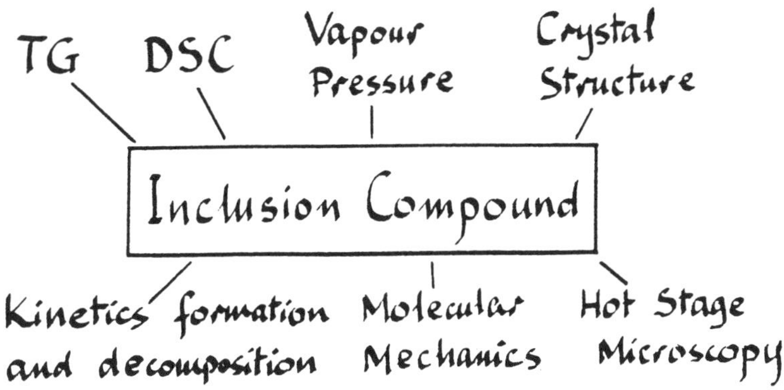

Figure 1. Techniques employed to study inclusion compounds

This paper will briefly review these techniques and illustrate their potential in the study of crystalline inclusion compounds.

G. Tsoucaris et al. (eds.), Crystallography of Supramolecular Compounds, 285–305.

The process which is under study is illustrated in Figure 2 [1].

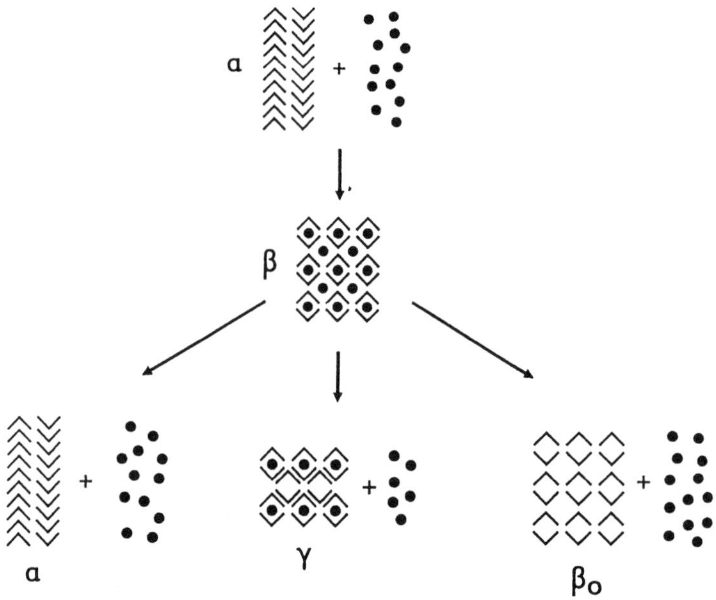

Figure 2. Schematics of the formation and decomposition of an inclusion compound

When a suitable host compound H is dissolved in a volatile guest G or is exposed to its vapour, it forms an inclusion compound HG_n(β-phase). The latter may decompose in one of several ways. It may lose the guest and the host may revert to its non-porous α-phase:

$$HG_n(\beta) \rightarrow H(\alpha) + nG\uparrow \tag{1}$$

Alternatively, the host may lose the guest but retain its structural integrity, displaying the empty cage or β_o structure:

$$HG_n(\beta) \rightarrow H(\beta_o) + nG\uparrow \tag{2}$$

Thirdly, it may lose only part of the guest and form a new γ phase:

$$HG_n(\beta) \rightarrow HG_m(\gamma) + (n\text{-}m)G\uparrow \;,\; n>m \tag{3}$$

Thermal Gravimetry, TG and Differential Scanning Calorimetry, DSC, are suitable techniques for monitoring the decomposition of the inclusion compound (β phase).

Figure 3 gives a schematic representation of the thermal analysis of an inclusion compound which decomposes in two steps, forming a stable intermediate. The corresponding DSC illustrates the two endotherms occurring due to guest loss and the last endotherm which results from the melting of the host compound.

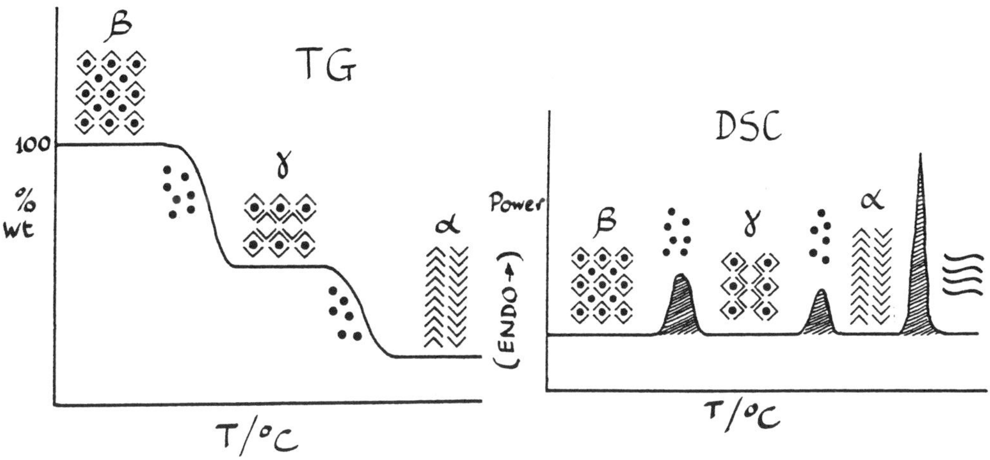

Figure 3. TG and DSC of the decomposition of an inclusion compound forming a stable intermediate

In principle the DSC curves yield the onset temperature for the guest release, T_{on}, the host melting point, T_m, and the corresponding enthalpy changes ΔH_G and ΔH_m.

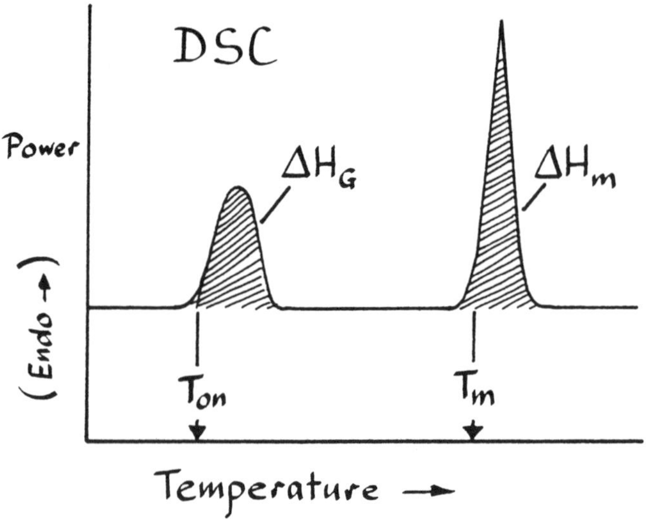

Figure 4. Parameters measured by DSC

While the temperature at the onset of a given endotherm is generally reproducible to ± 0.5°C, this is not true of the enthalpy changes. The poor precision in the enthalpy measurements by DSC is due to a number of experimental factors such as the crystallite size, the heating rate, the flow rate of the purging gas and the geometry of the calorimeter. The crystallite size distribution is often an intractable problem, in that the inclusion compounds are often too labile to withstand crushing and sieving. We have therefore devised an apparatus which allows us to measure the vapour pressure of a volatile guest at various temperatures.

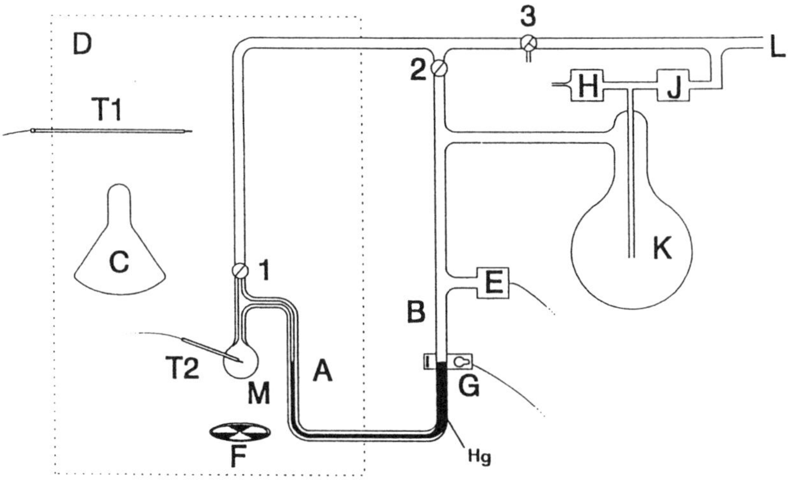

A	sample limb	H	solenoid valve 1
B	transducer limb	J	solenoid valve 2
C	500W lamp	K	buffer flask
D	oven	L	to vacuum pump
E	pressure transducer	M	sample flask
F	circulating fan	T1	thermocouple 1
G	optical source and detector unit	T2	thermocouple 2

Figure 5. Vapour pressure apparatus

The apparatus is shown diagrammatically in Figure 5, and consists of a transducer which converts pressure to a voltage signal which is amplified and fed to a computer. The temperature of the oven and of the specimen are monitored by thermocouples. We used an indirect method of measuring the pressure by utilising a U-tube containing mercury. This design has the advantage of protecting the transducer from corrosive vapours and keeping it outside the oven, thus avoiding the need for temperature compensation. The apparatus is fully automated. As the vapour pressure rises in the flask, the mercury rise is detected optically and this automatically opens a solenoid valve which lets in air from the atmosphere and balances the pressure. During a cooling cycle the system works in reverse by opening another valve to a vacuum pump. A typical result for the vapour pressure of Bis-(9,9′-dihydroxy-9,9′ difluorene).0.5 ethanol is shown in Figure 6, and the semilogarithmic plot yielded values of $\Delta H° = 17.9$ kJ mol^{-1} and $\Delta S° = 50.1$ JK^{-1} mol^{-1} [2].

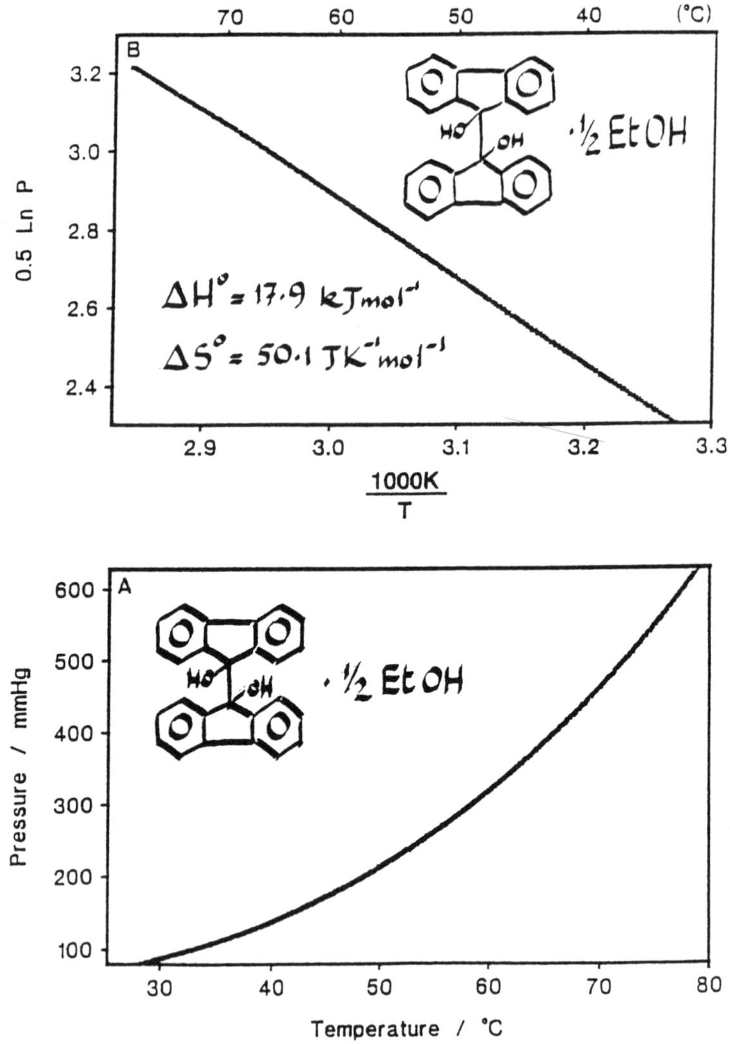

Figure 6. Vapour pressure versus temperature curve for difluorene-ethanol compound

An interesting example of applying these techniques is the inclusion compound formed between 2,2′-bis(9-hydroxy-9-fluorenyl)biphenyl and diethyl ether [3]. The structure of the compound is shown in Figure 7, and is characterised by an O–H....O hydrogen bond between host and guest.

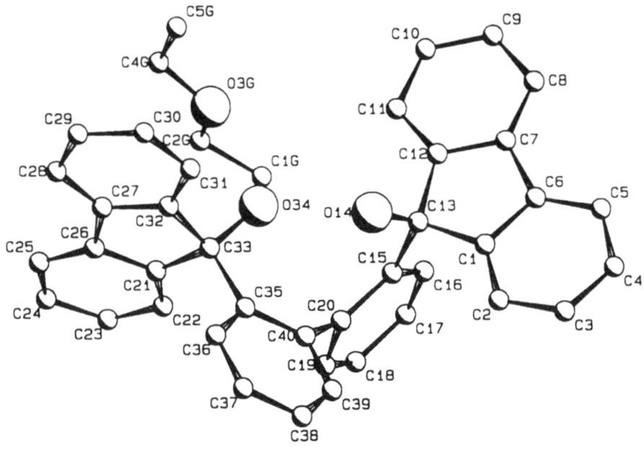

Perspective view of the host

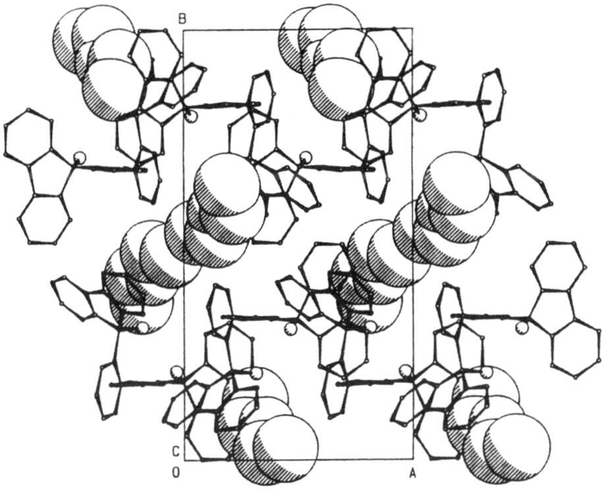

Figure 7. **Structure of fluorenylbiphenyl-ether compound**

The TG - DSC curves are shown in Figure 8, in which the first endotherm A, due to the guest release, has T_{on} = 125.8°C, which is 91.3°C <u>above</u> the normal boiling point of diethyl ether. The endotherm B and exotherm C are attributed to phase changes in the desolvated host, while endotherm D is due to the final melting of the host.

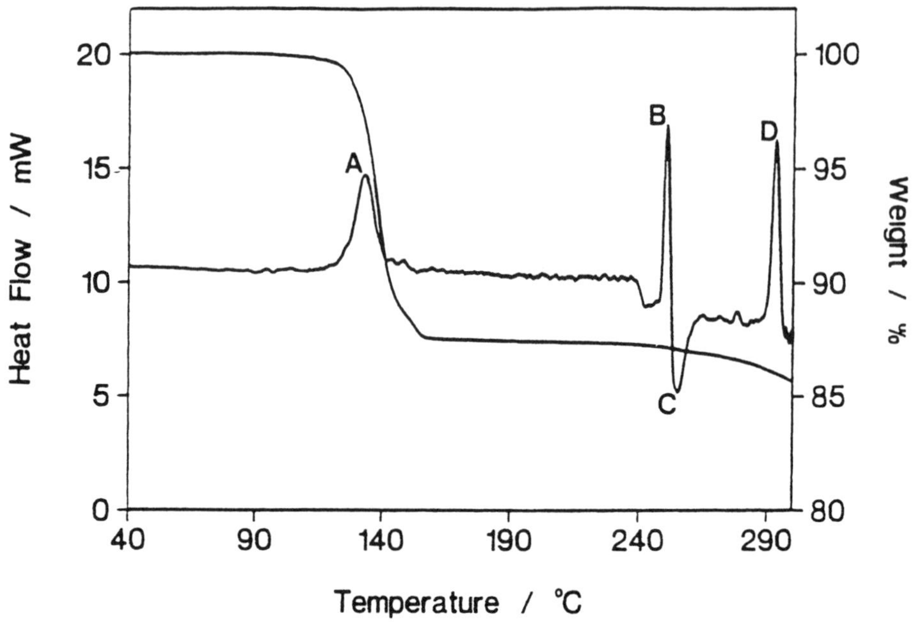

Figure 8. TG and DSC of fluorenylbiphenyl.ether compound

We measured the vapour pressure of this inclusion compound between 25°C and 85°C and the resulting semilogarithmic plot shown in Figure 9 displays a distinct discontinuity which occurs at 56°C and the two slopes correspond to ΔH values of 24.7 and 46.0 kJ mol^{-1}.

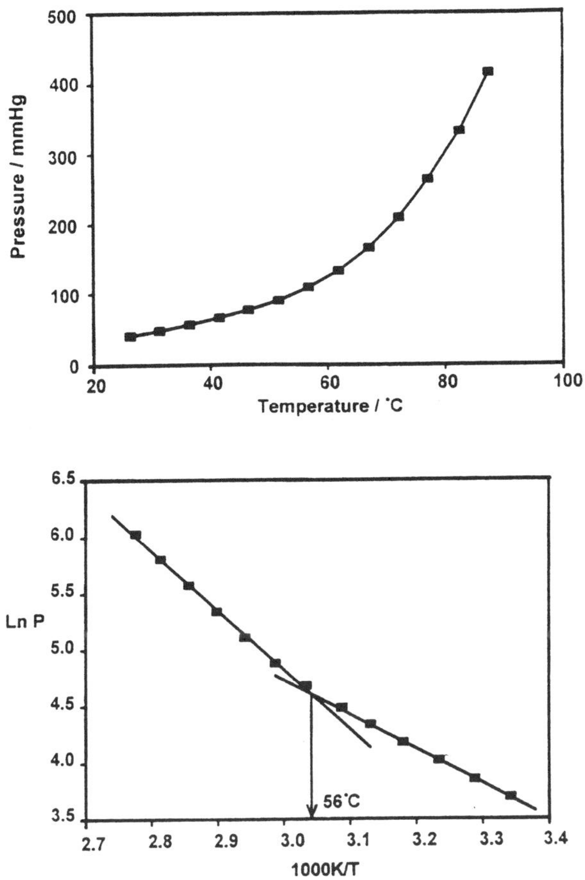

Figure 9. Vapour pressure versus temperature plots for bifluorenylbiphenyl ether compound

There are cases where a combination of thermal analysis and X-ray crystallography are not sufficient to characterise an inclusion compound unambiguously. Recently we synthesised the inclusion compound between 2,2′-bis(diphenylhydroxymethyl)-1,1′-biphenyl (BDHMB) and wet benzene. The final stoichiometry was host.H_2O.$^1/_6$benzene and the compound crystallised in the space group R3. However the thermal analysis, shown in Figure 10 was unclear, in that the TG showed a single step decomposition and the DSC did not allow differentiation between the water and benzene guest molecules.

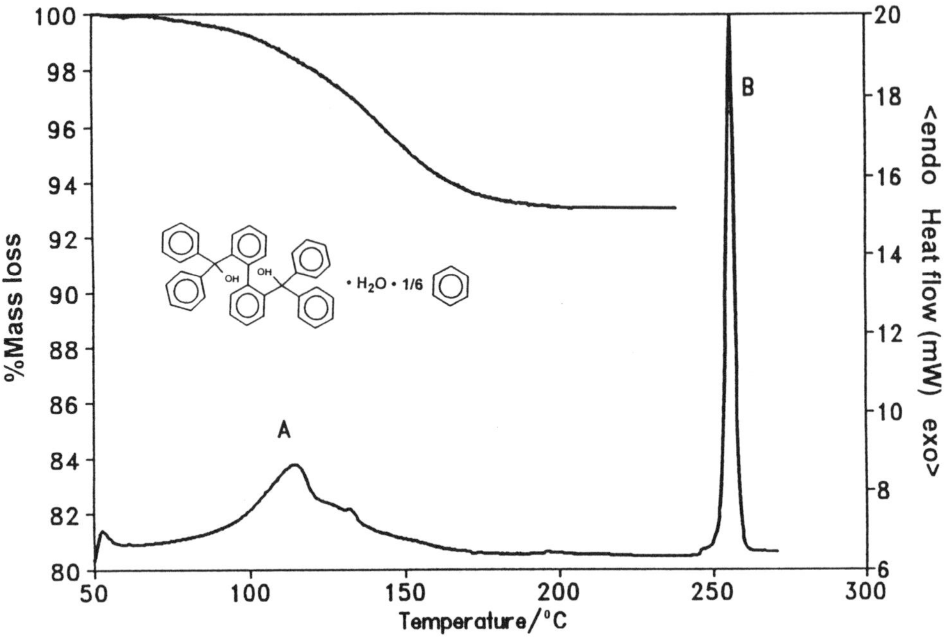

Figure 10. TG and DSC of BDHMB.H_2O.$^1/_6$ C_6H_6

The difference electron density map, phased on the host atoms, yielded a single peak at 1.4Å from the centre of inversion, which the symmetry multiplied into a regular hexagon of a benzene ring. However we were reluctant to accept this as unequivocal evidence of the presence of benzene and we therefore measured the ^1H NMR spectrum of the pure host and compared it to that of the inclusion compound, shown in Figure 11. The latter showed an extra sharp resonance at 7.36 ppm, assignable to benzene, and this was confirmed by spiking the pure host solution with a small quantity of benzene.

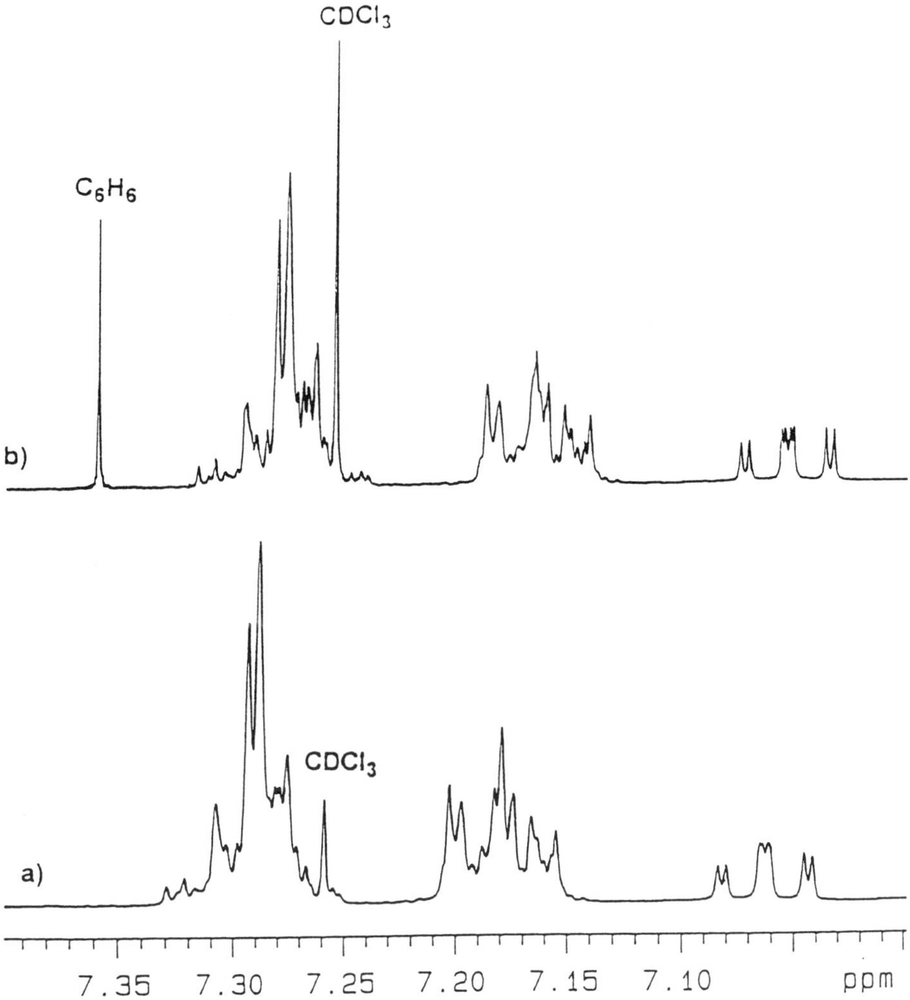

Figure 11. ^{1}H NMR spectra of BDHMB and its benzene inclusion compound

The final structure is shown in Figure 12, and is an interesting example of a high symmetry inclusion compound exhibiting double recognition. The host molecules are locked in a hexagon by a ring of 18 hydrogen bonds, resulting in an hexagonal hydrophobic cavity which encompasses the benzene guest [4].

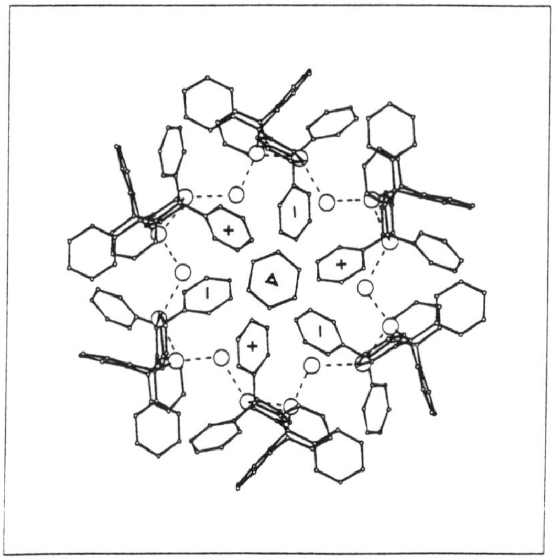

Figure I2. Structure of BDHMB.$H_2O.^1/_6$ C_6H_6

KINETICS OF DECOMPOSITION

Isothermal Methods

The theory of decomposition of solids is well established and has been succintly reviewed by Brown [5], and thermal decomposition data can be obtained by TG at selected temperatures. We have elucidated the structures of two inclusion compounds of 9,10-derivatives of *trans*-9,10-dihydroxy-9,10-dihydroanthracene with benzene as guest and have analysed their kinetics of desolvation [6]. The p-tert-butylphenyl derivative (DDTBDA) has a host:guest ratio 1:3 and desolvates in a single deceleratory step which follows first-order kinetics [$-\ln(1-\alpha) = kt$]. The di-α-naphthyl derivative (DDNDA) has a host:guest ratio of 1:1 and the α versus time curves are of truncated symoidal shape which are best described by Prout-Tompkins model [$\ln\{\alpha/(1-\alpha)\} = kt$]. Isothermal runs, conducted over a range of temperatures, yielded the Arrhenius plots and activation energies of 87 kJ for DDTBDA (compound 1) and 116 kJ for DDNDA (compound 2), and are shown on Figure 13. The topology of the structures shows that the guests in compound 1 lie in intersecting channels and are relatively mobile, while the benzene guests in compound 2 are located in cavities formed by the host network, and cannot escape without disruption of the crystal structure. this justifies the higher activation energy obtained for compound 2.

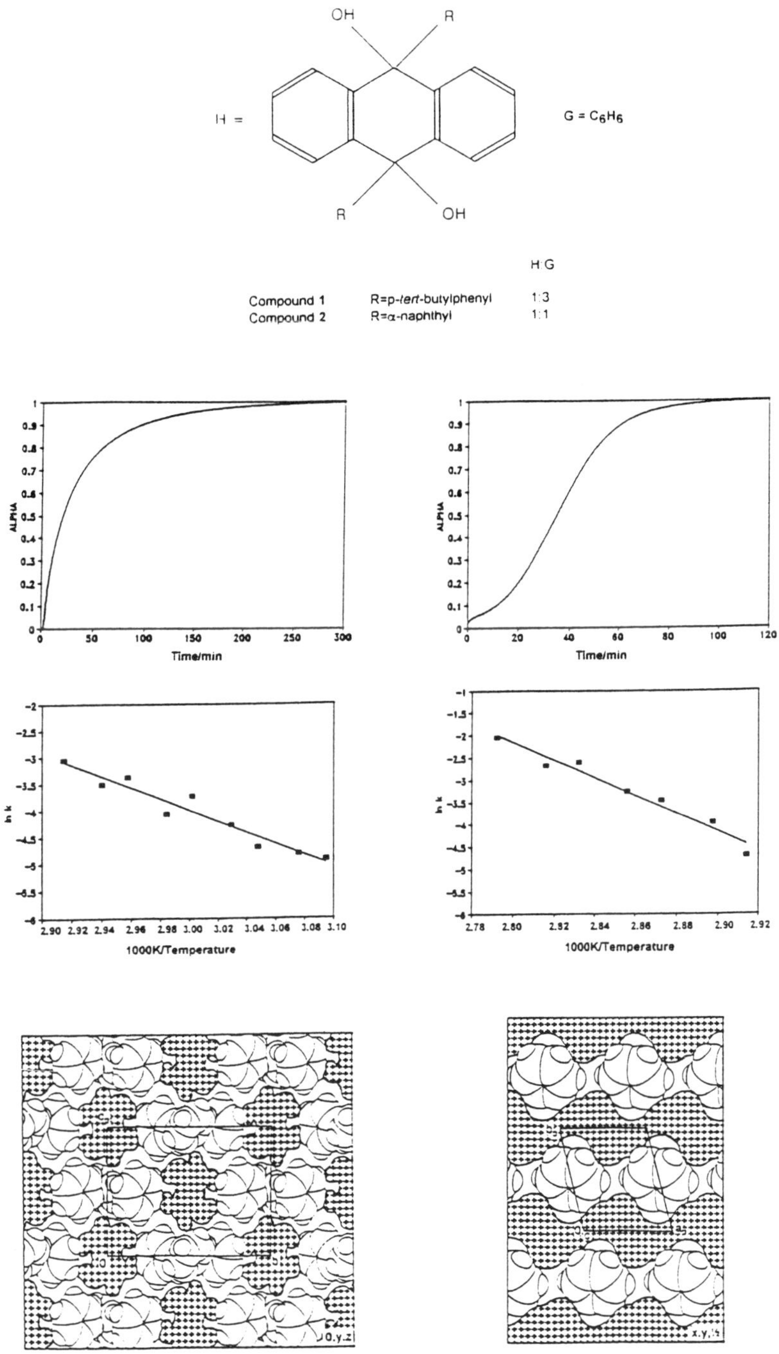

Figure 13. Kinetics and structures of dihydroanthracene derivatives with benzene

Non-isothermal Methods

Isothermal methods of analysis yield reliable kinetic parameters but are generally laborious and time consuming. Differential methods, which are based on weight loss versus temperature at several heating rates, are quicker, but less accurate, and thus suffer from magnification of experimental scatter. Nevertheless if precaution is taken with particle size distribution of the solid inclusion compounds, non-isothermal methods can be made to yield reproducible results of the energy of activation of the decomposition reaction. We have used the method of Flynn and Wall [7] in which the rate of decomposition is given by

$$d\alpha/dT = (A/\beta)f(\alpha)e^{-Ea/RT}$$

where α is the extent of reaction, T is the absolute temperature, A = pre-exponential factor of the Arrhenius equation, β is the constant heating rate, Ea is the activation energy and $f(\alpha)$ is a function of the extent of reaction (weight loss). The method requires one to record decomposition curves at various heating rates and to plot the log β versus T^{-1} values at selected extents of decomposition. We have used this method to analyse the decomposition of the inclusion compound of *trans*-9,10-dihydroxy-9,10-diphenyl-9,10-dihydroanthracene (DDDA) with acetonitrile [8]. The results are shown in Figure 14, and the accompanying semilogarithmic plot yielded activation energies varying from 83 − 115 kJ mol^{-1}. Similar methods have been used to obtain activation energies of the decompositions of Hofmann clathrates with a variety of aromatic guests [9].

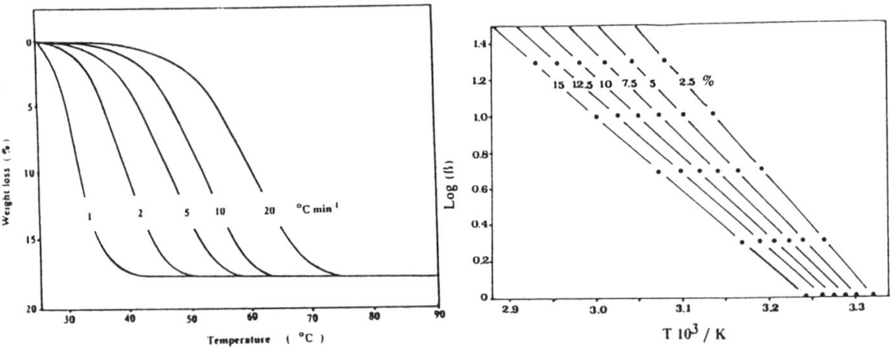

Figure 14. Thermal decomposition of DDDA with CH_3CN at variable heating rates

KINETICS OF ENCLATHRATION

The rates of formation of inclusion compounds are generally more difficult to analyse than the corresponding desolvation process. We have concentrated on gas-solid reactions, and have constructed an automated magnetic suspension balance for this purpose, which is shown in Figure 15. The apparatus consists of a reaction vessel connected to a vacuum pump, a conventional balance and a feedback control circuit and microcomputer. The system works by magnetic levitation and allows the sample in the pan to be exposed to a gaseous guest at pre-chosen pressure and temperature [10].

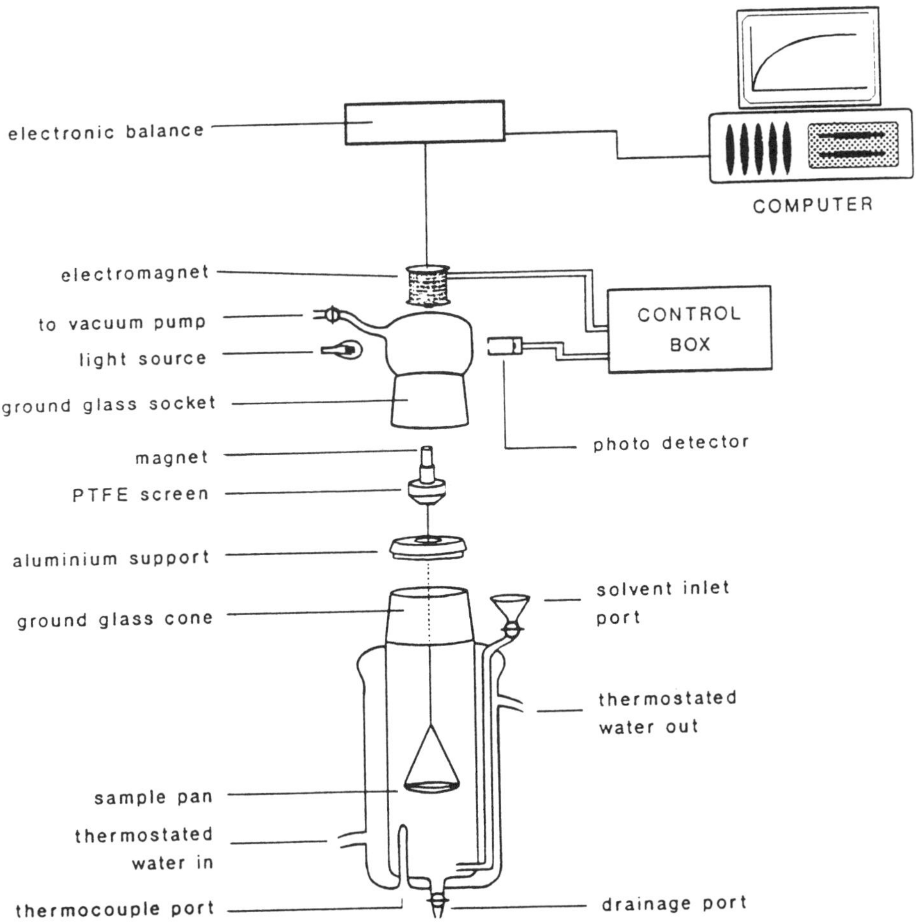

Figure 15. Schematic diagram showing the principal features of the balance system

Using this apparatus we have studied the reaction between *trans*-9,10-dihydroxy-9,10-diphenyl-9,10-dihydroanthracene (DDDA) and acetone vapour [11]. The interesting feature of this reaction is that at a given temperature it requires a threshold pressure Po of the vapour in order to proceed, and in addition, the velocity of reaction decreases with increasing temperature, thus exhibiting anti-Arrhenius behaviour.

We have explained the experimental results in terms of the rate law $\ln(1-\alpha) = k_f[(P-P_o)/Po]t$ where α is the extent of reaction and k_f is the forward rate constants at pressure P. Carrying out the reaction at various temperatures yielded an activation energy of 26.8 kJ.

A plot of the observed rate constants k_{obs}, versus guest vapour pressure at three temperatures (a) 284 K, (b) 293 K and (c) 303 K is shown in Figure 16.

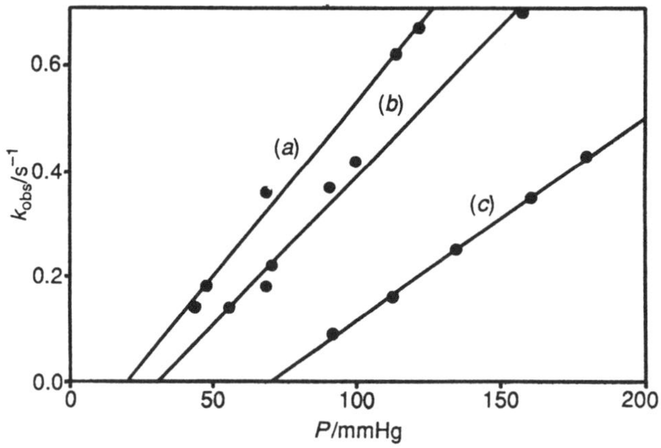

Figure 16. k_{obs} versus P for DDDA.acetone reaction

The above apparatus requires approximately 300 mg of host compound in order to obtain results of high precision. This is not always convenient, and we have therefore constructed a Quartz Micro Balance (QMB) which allows us to carry out the kinetics of inclusion and desolvation with very small quantities of host compound, as low as 1 μg. The principle depends on the variation in frequency of a quartz crystal, typically a flat cylinder 1 cm in diameter and 1 mm thick, which has gold electrodes attached to the circular faces. The crystal is dipped into a strong solution containing the host compound and allowed to dry under vacuum, resulting in a thin layer of host compound evenly distributed on the surfaces of the quartz crystal. This in then exposed to a vapour of the guest, and the mass gain is measured by the change in frequency of the crystal. We are currently assembling an apparatus which will contain four such quartz crystals and connecting it to a microcomputer. This will allow us to measure the kinetics of inclusion and desolvation of three host compounds for a given guest simultaneously, at a given temperature and pressure. This technique has been employed in the study of bulky hydroxyl host compounds with volatile alcohols [12].

SELECTIVE ENCLATHRATION

The process of guest inclusion can be exploited in the separation of close isomers, and this procedure has considerable industrial interest. We have used the host 1,1-Bis-(4-hydroxylphenyl)cyclohexane to separate the isomers of phenylene diamine [13]. The experiments are carried out by setting up a series of crystallisations of the host in mixtures of the two isomers in different proportions, and the mole fractions of the guests obtained from the crystals are compared with those of the starting solutions. The results of these competition experiments are shown in Figure 17, together with details of the crystal structures of the para- and the ortho-isomers of phenylène diamine. The para-isomer is stabilised by four host:guest hydrogen bonds while the ortho-isomer only has two. This renders the para-isomer more stable, and this is reflected in the shape of the competition curve. An important aspect of this host-guest reaction, is that it can be carried out with the host in water suspension, making this process industrially attractive.

SOLID – SOLID REACTIONS

The formation of inclusion compounds by direct reaction of solid host with solid guest is an intriguing process, because the precise mechanism of such reactions is open to question. We have monitored the solid-solid reaction occurring between 1,1,6,6-tetraphenyl-hexa-2,4-diyne-1,6-diol and benzophenone, and followed its kinetics using infra-red spectrophotometry [14]. Recently we have studied the solid-solid reaction between sulphadimidine (SD) and salicylic acid by monitoring the X-ray powder patterns which were recorded at various intervals during manual grinding of the two solids.

The kinetics could be followed by monitoring the decay of the (120) peak of the sulphadimidine and the growth of the (041) peak of the complex. The peak intensities follow first order kinetics at room temperature, with a half life

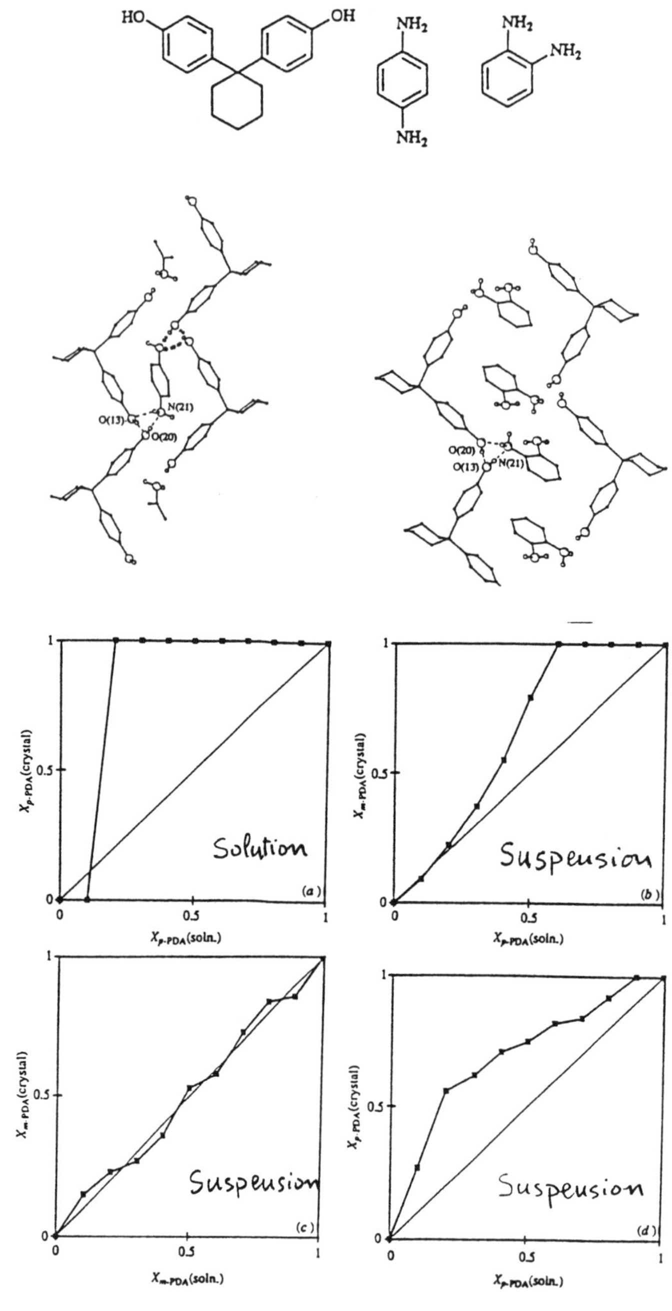

Figure 17. **Structure and selectivity of phenylene diamine compound**

of 6.3 mins. One of the problems is that the temperature of the reaction is difficult to measure, and localised hot spots probably occur in the grinding mixture. The results of this reaction are shown in Figure 18.

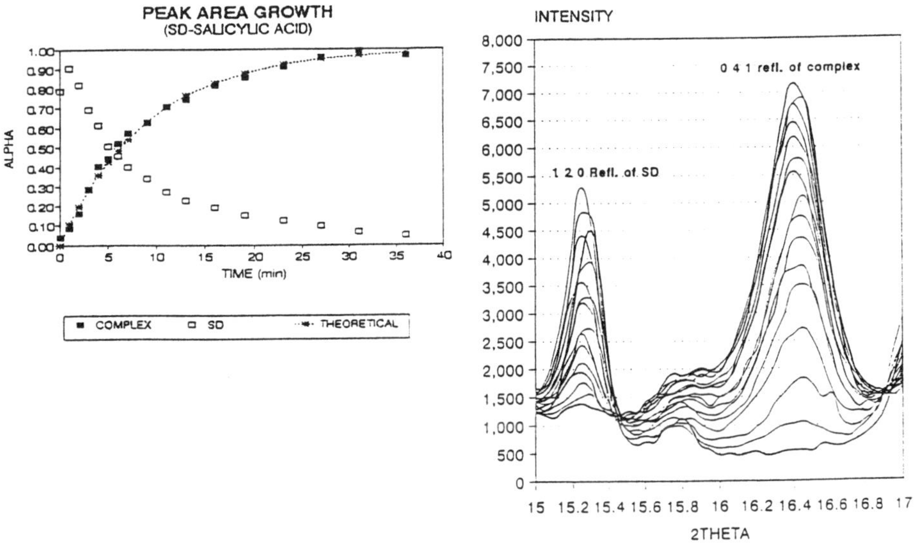

Figure 18. Kinetics of solid-solid reaction

Conclusion

The application of various physico-chemical techniques to inclusion compounds is an exciting field of study. It allows us to establish accurate thermodyanmic parameters relating to their stability and to their modes of formation and decomposition. We are currently carrying out molecular mechanics calculations in order to attempt correlations between crystal energies and measured thermodynamic quantities. This will lead to a greater understanding of the macroproperties of inclusion compounds in relation to their crystal and molecular structure.

References

1. Bond, D.R., Caira, M.R., Harvey, G.A. and Nassimbeni, L.R. (1990) Complexation with diol host compounds. 5. Structures and thermal analyses of inclusion compounds of *trans*-9,10-dihydroxy-9,10-diphenyl-9,10-dihydroanthracene with 2-butanone, 4-vinylpyridine, 4-methylpyridine and 2-methylpyridine, *Acta Crystallographica* **B46**, 771-780.
2. Barbour, L.J., Caira, M.R., Nassimbeni, L.R., Wierig, A. and Weber, E. (1995) Complexation with diol host compounds. 17. Structure and thermal analysis of 9,9′-dihydroxy-9,9′-bifluorene with ethanol, 1-butanol and pyridine, *Supramolecular Chemistry,* in press.
3. Caira, M.R., Nassimbeni, L.R., Winder, N., Weber, E. and Wierig, A. (1994) Complexation with diol host compounds. 16. Structure and thermal stability of 2,2′-bis(9-hydroxy-9-fluorenyl)biphenyl diethyl ether, *Supramolecular Chemistry*, **4**, 135-138.
4. Caira, M.R., Coetzee, A., Koch, K.R., Nassimbeni, L.R. and Toda, F. (1995) High-symmetry inclusion compound with mixed guests, *J. Chem. Soc. Perkin Transactions* 2, submitted.
5. Brown, M.E. (1988) *Introduction to thermal analysis*, Chapman and Hall, London.
6. Barbour, L.J., Caira, M.R., Coetzee, A. and Nassimbeni, L.R. (1995) Kinetics of desolvation of inclusion compounds of 9,10-derivatives of *trans*-9,10-dihydroxy-9,10-dihydroanthracene with benzene, *J. Chem. Soc. Perkin Transactions* 2, submitted.
7. Flynn, J.H. and Wall, L.A. (1966) A quick, direct method for the determination of activation energy from thermogravimetric data, *Polymer Letters*, **4**, 323-328.
8. Caira, M.R., Nassimbeni, L.R., Schubert, W-D. and Toda, F. (1992) Complexation with hydroxy hosts. 9. Structures and thermal analysis of inclusion compounds of trans-9,10-dihydroxy-9,10-diphenyl-9,10-dihydroanthracene with acetonitrile and 3-hydroxyproprionitrile, *Thermochimica Acta*, **206**, 265-271.
9. Kitazawa, T., Mizushima, Y., Shiraha, J., Taguchi, R., Katoli, A., Hasegawa, T., Nishikiori, S.I. and Iwamoto, T. (1991) Thermoanalytical investigation of guest-release process in aromatic-guest clathrates of three-dimensional metal complex hosts, *J. Incl. Phenom. and Mol. Recognition in Chem.*, **10**, 29-37.

10. Barbour, L.J., Achleitner, K. and Greene, J. (1992) A system for studying gas-solid reaction kinetics in controlled atmospheres, *Thermochimica Acta*, **205**, 171-177.
11. Barbour, L.J., Caira, M.R. and Nassimbeni, L.R. (1993) Kinetics of inclusion, *J. Chem. Soc. Perkin Transactions* 2, 2321-2322.
12. Ehlen, A., Wimmer, C., Weber, E. and Bargon, J. (1993) Organic clathrate-forming compounds as highly selective sensor coatings for gravimetric detection of solvent vapours, *Agnew. Chem. Int. Ed. Engl.*, **32**, 110-112.
13. Caira, M.R., Horne, A., Nassimbeni, L.R., Okuda, K. and Toda, F. (1995) Selective inclusion of phenylenediamine isomers by 1,1-bis(4-hydroxyphenyl)cyclohexane, *J. Chem. Soc. Perkin Transactions* 2, in press.
14. Bond, D.R., Johnson, L., Nassimbeni, L.R. and Toda, F. (1991) Complexation with diol host compounds. 6. Structure and dynamics of enclathration by 1,1,6,6-tetraphenyl-hexa-2,4-diyne-1,6-diol, *J. Solid State Chem.*, **92**, 68-79.

THE SELF-ASSEMBLY OF GUANOSINE DERIVATIVES AND FOLIC ACID

GIOVANNI GOTTARELLI and GIAN PIERO SPADA
Università di Bologna, Dipartimento di Chimica Organica "A. Mangini", Via S. Donato 15, 40127 Bologna (Italy)

PAOLO MARIANI
Università di Ancona, Istituto di Scienze Fisiche Via Ranieri, 65 - 60131 Ancona (Italy)

1. Introduction

It has been known for several years that highly concentrated DNA in water forms liquid crystalline mesophases [1]. The phases have been identified as cholesteric, at lower concentration, and hexagonal, at higher concentration [2] (see Figure 1).

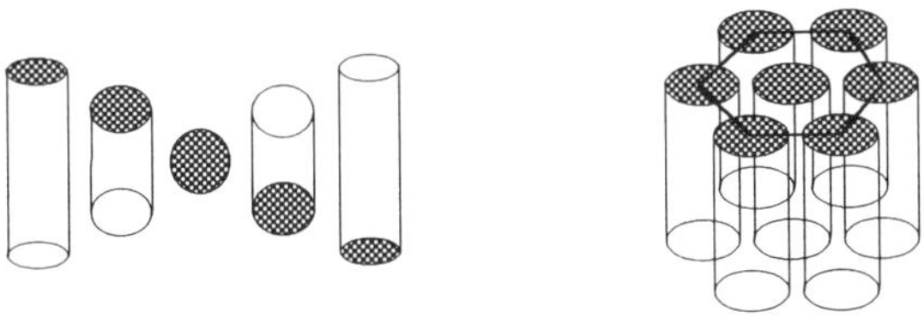

(a) *(b)*

Figure 1. Cholesteric *(a)* and hexagonal *(b)* arrangements of elongated objects, *e.g.* DNA.

Relatively short fragments of DNA (*ca.* 100 b.p.) form cholesterics which can be aligned by magnetic fields; the cholesteric axis is aligned parallel to the field (the individual double helices are instead perpendicular to the field) to give either fingerprint or planar textures (see Figure 2).

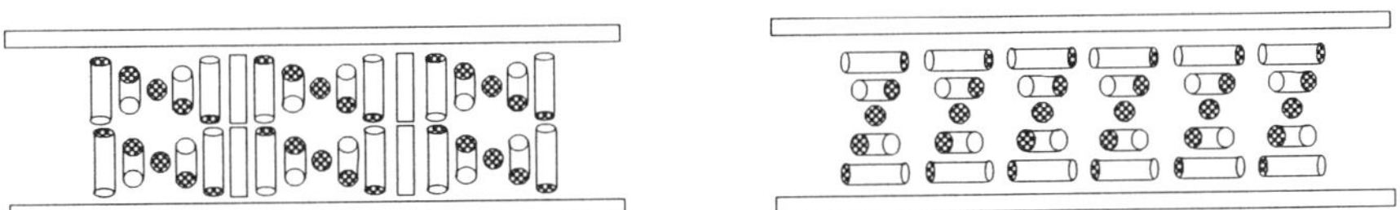

(a) *(b)*

Figure 2. Fingerprint *(a)* and planar *(b)* orientation of a cholesteric relative to the cell surface.

G. Tsoucaris et al. (eds.), Crystallography of Supramolecular Compounds, 307–330.

In the planar texture, the cholesteric helix axis is perpendicular to the cell windows; this texture is very useful for spectroscopic measurements as it shows little or no birefringence. In the fingerprint texture instead, the helix axis is parallel to the cell windows and a typical striped pattern appears; from the distance between the stripes, the cholesteric pitch can be measured [3].
The first example of mesomorphic behaviour in the guanine family concerns the dimer d(GpG) **1** [4]. Subsequently also alkaline folates **2** were reported to give liquid crystalline phases [5].

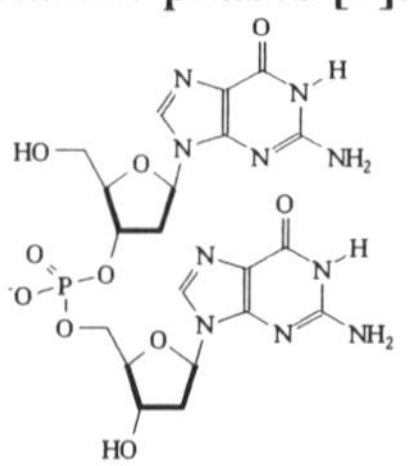

(1)

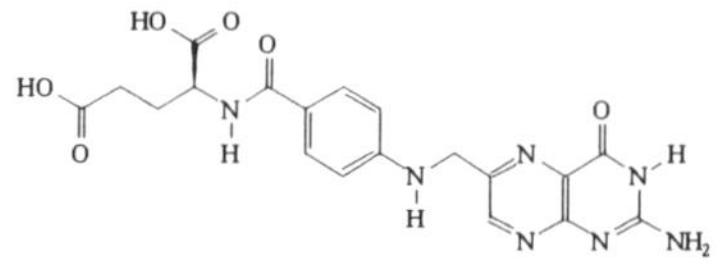

(2)

Guanine and pterine are the heterocyclic bases present in the two molecules; both bases possess a particular sequence of groups which act as donors or acceptors of hydrogen bonds. These groups are of fundamental importance in the self-recognition and self-assembly processes.

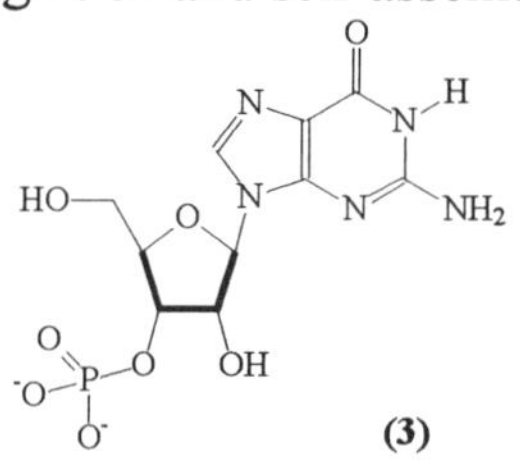

(3)

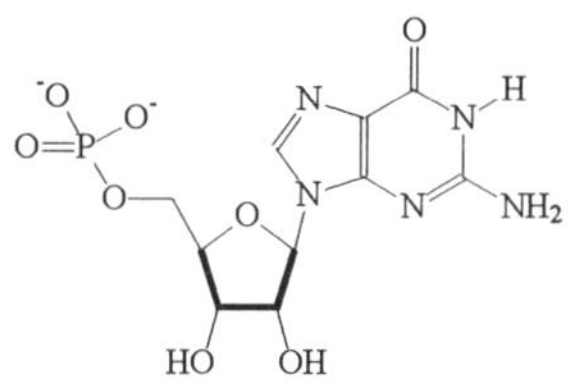

(4)

The ability of guanosine derivatives to form gels has been known for more than eighty years since the work of Bang [6]. An X-ray study of the fibre obtained from a gel of guanosine 3'-phosphate **3** allowed Gellert *et al.* [7] to propose the tetrameric arrangement depicted in Figure 3 (G-quartet) in which the four molecules are connected by a special scheme of hydrogen bonds (Hoogsteen), as a consequence of the sequence of donor and acceptor groups present in the base.

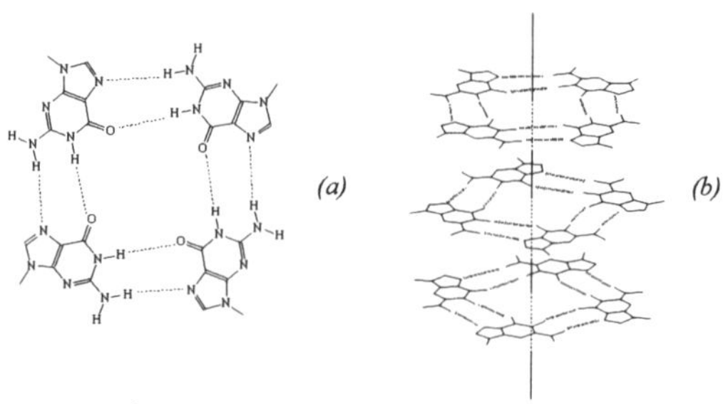

Figure 3. The tetrameric arrangement, G-quartet *(a)*, of the guanine moieties and the piling up *(b)* of three contiguous quartets.

In the aggregate, the guanosine tetramers are piled on top of one another, with sugar protruding at the periphery. However, the tetramers do not stack in

register, but are rotated with respect to each other to give an overall helical structure. In the case of guanosine 5'-phosphate **4**, a continuous helix seems instead to be formed; however, this change from the planar tetramers requires only small distortions of hydrogen bonds. If the fibre is obtained in the presence of an excess of NaCl, the planar tetramers seem to form again [8].
The same tetrameric motif was found also in the self-structure formed by polyguanylic acid by two independent fibre X-ray investigations [9,10]: the assembled species is a four-stranded, right-handed helix with parallel sugar phosphate filaments held together by the familiar G-quartets.
An extensive investigation on gels formed by several guanosine derivatives was carried out by Guschlbauer and coworkers [11]; the presence of the tetramers was assumed also in gels formed at very low concentration.
The X-ray study of the liquid crystalline phases, particularly of the more ordered hexagonal one, has given very useful information on the structure of the assembled columnar species and is of fundamental importance for the study of the self-assembly process in isotropic solution.

2. X-Ray Diffraction Structural Analysis

X-Ray diffraction analysis is a powerful and well-established technique for the determination of mesomorphic structures [12-14].
The first step in the structure analysis of a mesomorphic system is to construct the phase diagrams, *i.e.* to characterize the different phases and to determine their range of existence as a function of the usual variables, temperature and concentration. From the experimental standpoint, the techniques used for the study of lyotropic phases are those currently used in crystallography laboratories. However, the X-ray cameras must fulfill a few requirements, as the composition of the sample, especially the water content, must be kept under control and the experiments must be carried out at controlled temperature. Moreover, the X-ray diagrams must be fully explored, as the dimensions of the structure elements can be fairly large (from several hundred Å to about 10 Å) and at the same time, information on the short-range molecular organization (from about 10 Å to few Å) must be obtained. The conventional X-ray cameras generally do not satisfy all the above conditions. Most of the experiments described in this paper were therefore carried out with Guinier-type cameras of special design (see for example ref. [15]), operating *in vacuo* or in air. In all the reported experiments, the samples were held in a vacuum-tight cylindrical cell provided with thin mica windows and the sample holder was kept at constant temperature. For intensity measurements, the samples were rotated continuously during the exposure, in order to reduce spottiness arising from possible macroscopic monodomains.

The data of X-ray diffraction experiments from lyotropic samples can be analysed considering a convenient operational separation. In fact, the long-range organization of the structure elements gives rise to the reflections observed in the so-called low-angle region (scattering angle 2θ ranging from 10' to 10°), which specifies the crystalline lattice and the symmetry of the structure, while the short-range molecular organization gives rise to usually one or two reflections in the so-called high-angle region (centred around 2θ angles of about 22°) [12]. If the experiments are carried out on unoriented samples, only the spacings s (s being defined as (2 sinθ)/λ, where λ is the X-ray wavelength) and the intensities of the reflections are known. The problem is to index the low-angle reflections and to determine the symmetry of the lattice and the structure of the phase, discriminating the reflections that are systematically absent, because of symmetry, from those whose intensity happens to be weak for other reasons. As usual in crystallography, the symmetry of the lattice is determined by finding one equation, among those which define the spacings of the reflections for the different symmetry systems [12,16], with which all the observed spacings agree. Once the symmetry is found, the dimensions of the unit cell can be calculated. It is evident that the indexing operation must consider the possible presence of different phases coexisting in equilibrium at the same concentration: this problem is tackled by carefully exploring the phase diagram. It should be observed that the resolution of the low-angle scattering data is intrinsically low (typically in the order of 15-10 Å), as a consequence of the liquid crystalline order. Therefore, the number of observed reflections could be too small to identify the structure unambiguously. From the unit cell parameters, considering the chemical composition of the sample and on the basis of some fundamental assumptions (namely that the lyotropic molecules segregate into regions from which water is excluded and the interface between the water and the hydrophobic residues is covered by the hydrophilic groups), the shape and the dimensions of the structure elements can be obtained [12].

The final step in the structure determination is the calculation of the electron density maps and the comparison of the results with chemically independent information. In order to solve the crystallographic phase problem, several direct and indirect methods have been proposed and applied in the case of lyotropic structures [17,18]. However, some unusual constraints are encountered in these systems: due to the delicate thermodynamical phase equilibrium, the use of the isomorphous replacement technique is for example precluded. On the other hand, the modelling of the structure elements is not usually easy. In the experiments described in this paper, the phase problem has been solved by using a direct approch, based on the Shannon theorem [19]: the so-called swelling series experiment (see for example refs. [20,21]).

3. X-Ray and Neutron Small Angle Scattering Investigations of the Isotropic Phases

X-Ray (SAXS) or Neutron (SANS) Small Angle Scattering are methods for analysing the structure of colloid and macromolecular systems in solution [22-25]. A sharp and well-collimated incident beam penetrates the solution under investigation, and the radiation scattered to very small angles is measured as a function of the scattering vector Q, defined as $(4\pi \sin \theta) / \lambda$. A well-developed theory permits the derivation of several geometrical and structural parameters of the particles in solution from the scattering curve [22,23,25]. In particular, in the case of solutions of mono-disperse particles with concentration sufficiently small for there to be no interference between waves scattered by different particles, the Guinier approximation applies: for small values of Q, the scattering intensity follows an exponential course. In logarithmic form this reads as:

$$\ln I(Q) = \ln I(Q)_{Q=0} - (R_g^2 Q^2 / 3) \qquad (1)$$

Therefore, a plot of ln I(Q) versus Q^2, the so-called Guinier plot, can furnish the radius of gyration of the scattering particles (R_g, *i.e.* the square root of the mean square distance of all individual scattering centres of the particle from the centre of gravity of the particle) and the intensity scattered at zero angle, $I(Q)_{Q=0}$, which depends on the concentration and volume of the scattering particles and on the contrast (differences in electron densities or in scattering length densities between the solvent and the scattering particles in the case of X-ray and neutron scattering experiments, respectively). Likewise, in the case of rod-like particles, linear behaviour is expected in the small Q-range in a Guinier plot of the cross-section factor ln(I(Q)Q) versus Q^2, according to:

$$\ln [I(Q)\, Q] = \ln [I(Q)\, Q]_{Q=0} - (R_c^2 Q^2 / 2) \qquad (2)$$

From this plot, the cross-sectional intensity scattered at Q=0 (which depends on the mass per unit length) and the radius of gyration of the cross-section R_c can be determined. A simple relationship between R_g and R_c allows the determination of the height h of the rod-like particle [25]:

$$h^2 = 12\,(R_g^2 - R_c^2) \qquad (3)$$

The radius of gyration R_g can be given explicitly in terms of the geometrical parameters of the scattering particles in a few simple cases [22,24].

In the series of guanosine and analogous derivatives discussed in this paper, the structural properties of the scattering particles existing in the isotropic phases were investigated both by X-ray and neutron scattering experiments [26-28]. For neutron experiments, in order to improve statistics, D_2O was used as a solvent. Guinier plots ln I(Q) versus Q^2 and ln [I(Q) Q] versus Q^2 were used to evidence the presence of cylindrical particles and to determine their dimensions. According to the structure of the cholesteric and hexagonal phases, the experimental data were fitted by using a model of rod-shaped aggregates formed by a stacking of guanosine or folate tetramers. As a result, the number of piled tetramers could be obtained.

4. The Formation of Columnar Liquid Crystals by Deoxyguanosine Derivatives and Alkaline Folates.

2'-Deoxyguanylyl-(3'-5')-2'-deoxyguanosine **1** in water forms LC phases at relatively low concentration (2.5% w/w). Microscopic textures are indicative of the presence of a cholesteric and a hexagonal phase at lower and higher concentration respectively. The cholesteric phase can easily be aligned with a magnetic field to give a planar or a fingerprint texture without unwinding the cholesteric helix which is oriented parallel to the applied field.
This behaviour indicates that the "objects" composing the phase have negative diamagnetic anisotropy. The overall characteristics of the LC phases, including the magnetic behaviour, are very similar to those displayed by DNA with an average length of around 100 base pairs. This similarity, together with what was known about the structure of the gels and the four-stranded helix of poly(G), already give indications about a probable structure of the aggregates [4].
X-Ray experiments give evidence on the columnar nature of the different phases observed as a function of concentration and temperature. According to the diffraction profiles observed for thermotropic columnar LCs [14], a narrow band centred in the high-angle region at about s = $(3.4 \text{ Å})^{-1}$ is indicative of a phase made of columnar structure elements; this reflection is in fact related to the nature of the intracolumnar order (see Figure 4). In our systems this is observed in both the cholesteric and hexagonal phases [29].

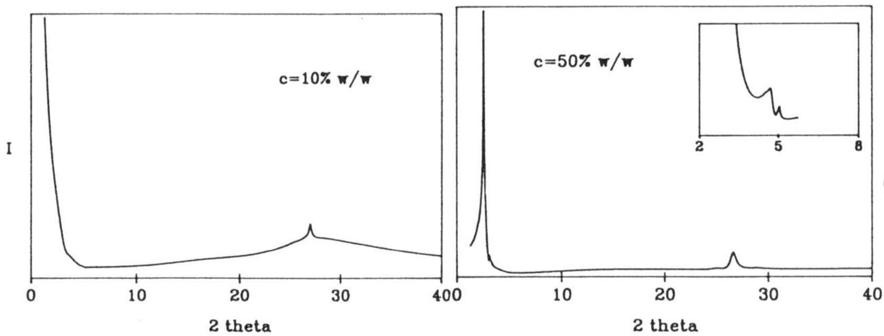

(a) *(b)*

Figure 4. Typical X-ray diffraction patterns of cholesteric *(a)* and hexagonal *(b)* phases formed by **1**.

The low angle ($s < (10\ \text{Å})^{-1}$) diffraction profile gives information on the arrangement of the columns. When the sample, observed with the optical microscope, shows the cholesteric texture, the low-angle X-ray diffraction region is characterized by only a diffuse region whose position strongly depends on the water content. This profile confirms that we are dealing with a cholesteric phase characterized by the presence of columns; the band observed refers to the mean distance between neighbouring cylinders, which changes as a function of concentration.

At higher concentrations, in correspondence to the hexagonal texture detected by optical microscopy, the low angle profiles are characterized by narrow reflections (from 3 to 5 Bragg peaks), which indicate that the phase is highly ordered (see Figure 4). The peak positions are in the order 1, $\sqrt{3}$, $\sqrt{4}$, ... and could be indexed considering a 2-D hexagonal lattice of p6m symmetry (space group No.17 of the International Tables of Crystallography [16]). For this symmetry group, the equation which defines the spacing of the reflection is in fact [12,16]

$$s_{hk} = (h^2 + k^2 - hk)^{1/2} / a \tag{4}$$

where a is the unit cell dimension and h, k are the Miller indices of the reflections. This phase is formed by parallel columnar aggregates packed in a hexagonal array, each aggregate being surrounded by six others at a distance equal to the unit cell dimension.

At constant temperature a changes continuously as a function of the sample concentration. As is usual for chromonic phases [30] (see also ref. [12] for the corresponding behaviour of lyotropic phases), the distance between the rods increases as the water content increases. Assuming that water does not penetrate the columns and they have a circular section and infinite length, from the unit cell dimension the radius could be determined according to

$$R = (\sigma\, c_v / \pi N)^{1/2} \tag{5}$$

where σ is the area of the 2-D hexagonal unit cell ($\sigma = a^2 \sqrt{3} / 2$), N the number of columns per unit cell ($N = 1$) and c_v the d(GpG) volume concentration [31]. At room temperature, the results indicate a roughly constant radius of *ca.* 12.8 Å, in good agreement with that of the G-quartet. Similar results were obtained also for the other derivatives investigated.

From the intensity of the low-angle diffraction peaks and by "swelling" experiments, the 2-D electron density map has been determined. In particular, the phase problem has been solved considering the continuous changes of the observed intensity as a function of the water content (see ref. [20,24]). Within the same phase, one can in fact expect that at all concentrations the electron distribution of the cylinders will remain fairly constant and that the increased water content will increase only the separation between the columns. Considering that in these conditions the structure factors sample a unique continuous curve, the amplitudes of the reflections observed change continuously as a function of the s values of the reflections. From the position of the zeros, the signs of the corresponding structure factors can be derived. In this way, an electron density map can be calculated (Figure 5).

(a)

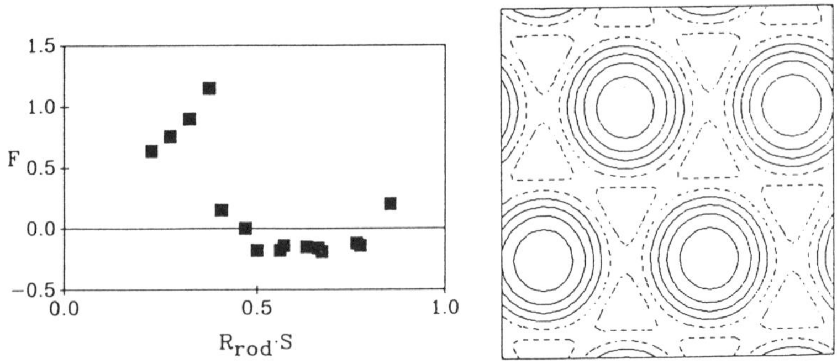

(b)

Figure 5. Determination of the signs of the structure factors *(a)* in the hexagonal phase and its electron density distribution *(b)*.

The map shows clearly the electron-dense columnar projections, roughly circular; the rods appear surrounded by a smooth region of lower electron density, which is associated with water. The radius of the columnar cross-sections directly measured from the map is in very good agreement with that deduced from equation (5).

From all these data, a model for the "objects" which are the phase building blocks can be deduced: they are chiral columnar aggregates composed of a stacked array of planar or quasi-planar tetramers; the tetramers are formed by Hoogsteen-bonded guanine moieties. The tetrameric planes are stacked on top of one another, but not in register; they are rotated at a positive angle one with respect to the other to give a right-handed helix. As reported for poly(G), the sugar-phosphate backbones protrude at the periphery of the structure.

In the cholesteric phase, which exists in the low concentration part of the phase diagram, the columns are relatively free to move and their surface chirality generates the cholesteric helical structure. In the hexagonal phase, which occurs at higher concentrations, the free space available is much less and the columns are forced to adopt a more compact hexagonal packing in which the weak chiral forces are overpowered by the severe space limitations. In all cases, no long-range column-to-column correlation of the tetramer position exists as the rods may freely translate in a direction perpendicular to the two-dimensional hexagonal cell. This interpretation was confirmed by X-ray diffraction experiments on magnetically oriented samples [32].

Other guanosine derivatives which show the formation of columnar LCs, all characterized by the presence of the tetrameric motif, are compounds **5-13**. The critical concentrations for phase transitions are reported in Table 1.

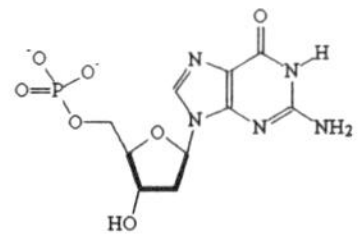

(5)

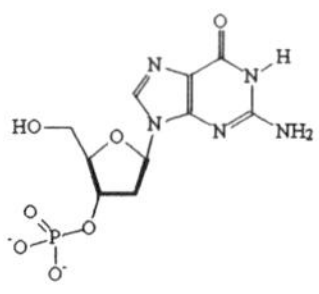

(6)

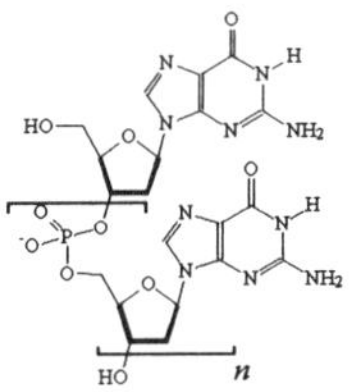

n=2 (7)

n=3 (8)

n=5 (9)

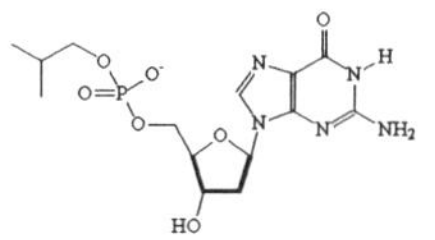

(10)

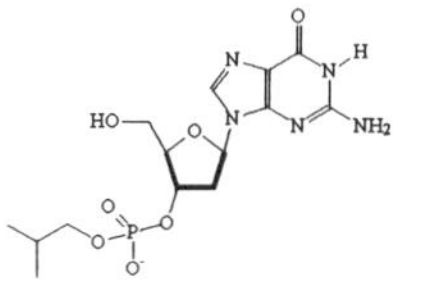

(11)

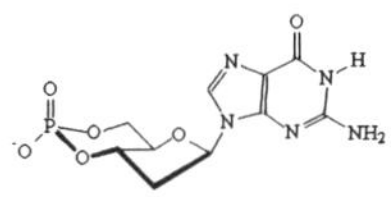

(12)

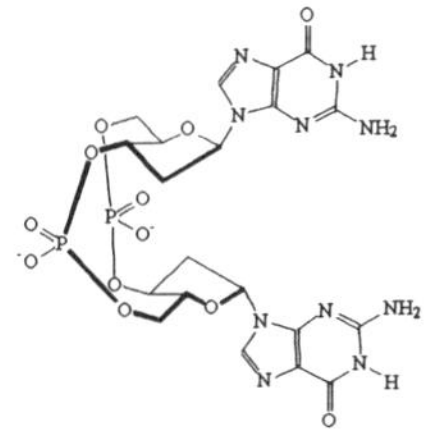

(13)

In the homologous series d(GpG) **1**, d(GpGpG) **7**, d(GpGpGpG) **8** and d(GpGpGpGpGpG) **9**, which possess only internucleoside phosphate groups, the critical concentration at which the cholesteric phase appears seems to follow a definite trend related to the ratio of negative charges/guanine units. This ratio is indicative of both the electrostatic repulsive interaction and the hydrophilic/hydrophobic balance, the smaller the ratio, the easier the formation of the cholesteric phase [33].

Table 1. The phase sequences and the transition concentrations for the guanosine derivatives investigated.

Compound		Counterion	Phase sequence						
(**1**)	d(GpG)	Na^+	I	2.5%	N*		18%		H
		K^+	I	1.5%	N*		15%		H
(**7**)	d(GpGpG)	Na^+	I	8%	N*		25%		H
(**8**)	d(GpGpGpG)	Na^+	I	13%	N*		*ca.* 35%		H
(**9**)	d(GpGpGpGpGpG)	Na^+	I	18-20%	N*		35%		H
(**5**)	d(pG)	NH_4^+	I	30%	N*		40%		H
(**6**)	d(Gp)	NH_4^+	I	5%	N*	35% H_b	58% Sq	68%	H
(**10**)	d(*i*-Bu-pG)	NH_4^+	I						
(**11**)	d(Gp-*i*-Bu)	NH_4^+	I	25%	N*	38% H_b	52% Sq	70%	H
(**12**)	d(cGp)	NH_4^+	I	6.5%	N*		16%		H
(**13**)	d(cGpGp)	NH_4^+	I			40%			H
(**14**)	d(GpGpApGpG)	Na^+	I	12-20%	N*		35-40%		H

Also d(GpGpApGpG) **14** forms a cholesteric and a hexagonal phase similar to the others. The critical concentrations are not very different from those of d(GpGpGpG) and d(GpGpGpGpGpG), indicating that the presence of adenine in the central position does not inhibit the self-assembly process [34].

For the other derivatives, this ratio seems of minor importance: emblematic is the example of d(pG) **5** and its mono butyl ester **10**, where the less hydrophilic molecule does not show LC phases; steric factors which affect the assembly process seem to be predominant. Also the difference in the critical concentration for the formation of the cholesteric phases of d(pG) **5** (30% w/w) and d(Gp) **6** (5% w/w) (this molecule and its isobutyl ester **11** show also a complex polymorphism with two hexagonal and one square phase [35]) is impressive and leads to similar conclusions. On the other hand, also the gel formation [11] and the fibre structure [7,8] of the two isomers are rather different: as reported in the introduction, the former molecule seems to form a continuous helix and follows a one-step assembly process; the latter seems to follows a stepwise aggregation and gives a fibre with strictly planar tetramers [11].

It should be noticed that the cyclic monomer phosphate **12** gives both the cholesteric and hexagonal phases, therefore the columns are formed even in the absence of a free OH in the 3' or 5' position. Hydrogen bonds are therefore not essential for promoting the vertical stacking of the tetramers. Stacking interactions and binding effects from ions seem to give dominant contributions to the formation of the columns.

Alkaline folates form, in pure water, only a hexagonal phase [5]. The overall characteristics of this phase are of the type described for the guanosine family; in particular, the diameter of the cylinders, calculated from diffraction experiments is *ca.* 30 Å and corresponds to a pterine tetramer (Figure 6).

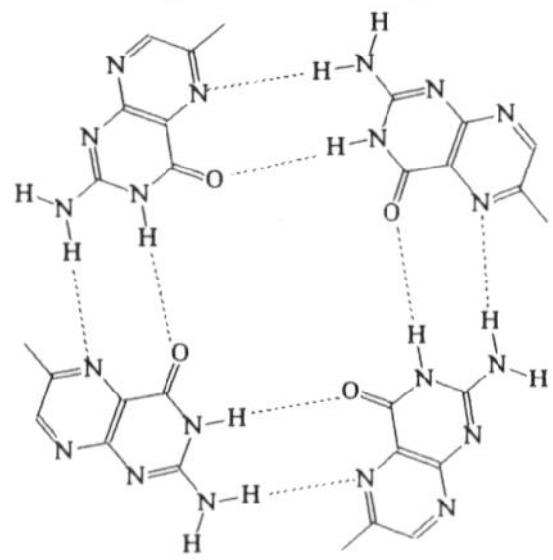

Figure 6. Tetrameric arrangement of the pterine moiety.

The critical concentration is rather high (35% w/w). However, in the presence of NaCl 1M, at *ca.* 27%, the solution becomes cholesteric. This cholesteric phase can be aligned with a magnetic fields to give planar or fingerprint textures, just like the majority of guanosine derivatives; this enables the determination of the cholesteric pitch, which is rather short (*ca.* 14 μm). All the data indicate that we are again dealing with a chiral columnar aggregate with pterine tetramers perpendicular to the long axis of the rod; the chiral glutamic residue protrudes from the column and gives the surface chirality which is related to the short pitch observed [36].

It can be noticed that for folate in pure water, a stable cholesteric phase cannot be detected; a stable cholesteric phase exists instead in the presence of an excess of Na^+ ions. Furthermore, in the case of d(GpG), the critical concentrations change in passing from the sodium (I 2.5% N* 18% H) to the potassium salt (I 1.5% N* 15% H). Also for d(pG), the critical concentration diminishes remarkably in passing from pure water to a 4M KCl solution [37].

In order to explain these and other experimental data, one has to understand the process of growth of the columns and their stability as a function of the concentration and ions present both in isotropic solution and in the mesomorphic state.

5. The Self-Assembly Process in Isotropic Water Solution

Small Angle Neutron Scattering (SANS) is a powerful technique for studying the shape and dimentions of assembled species in isotropic solution.
SANS experiments were carried out in pure D_2O for d(pG), d(GpG), d(GpGpG), d(GpGpGpG) and d(GpGpGpGpGpG) at 1 and 1.5% w/w concentrations [26]; except for d(pG), all the compounds revealed the presence of aggregates with a cylindrical shape. Modelling the cylinders with the radius of the guanosine tetramer, as deduced from X-ray work on the hexagonal phases, their length can be extracted. Furthermore, from the analysis of the scattered intensity at zero angle, one can also deduce the number of scattering particles per unit volume and hence the percent of aggregation. The results [26] are reported in Table 2.

Table 2. The number (N) of scattering particles per unit volume (10^{-17} cm^3), the percent aggregation and the length (L) of the columnar aggregates formed by homoguanylic derivatives at 1 and 1.5% w/w concentration as deduced from SANS experiments.

		(1) d(GpG)	**(7)** d(GpGpG)	**(8)** d(GpGpGpG)	**(9)** d(GpGpGpGpGpG)
c=1%	L (Å)±3.0	69.2	62.1	62.9	59.7
	Aggregation (%)	43.6	36.6	33.1	35.8
	N	1.10	1.00	0.89	0.99
c=1.5%	L (Å)±3.0	71.2	69.1		
	Aggregation (%)	50.2	39.0		
	N	1.84	1.43		

The length of the aggregates decreases very slowly in passing from the dimer to the hexamer and the same seems to happen also for the aggregation percent. At 1.5% concentration, the aggregation increases *ca.* 10%. All the data reported refer to freshly prepared solutions not treated thermically. In certain cases, the thermal history of the sample plays an important role in determining the dimensions of the aggregates.
The effect of thermal cycles and ions has been studied in some detail by a strictly related technique, Small Angle X-ray Scattering (SAXS) [27]; the geometrical parameters obtained by measurements carried out on 1% solutions in water with and without added electrolytes are reported in Table 3.

The different behaviour of the Na and K salts of d(GpG) is remarkable: at 20°C, both salts show observable cylinders with comparable length; at 60°C (above the melting temperature as determined by CD), in both cases no aggregates could be detected; after cooling again to 20°C, only the K salt again shows cylindrical aggregates, which are, incidentally, slightly longer than for the freshly prepared sample, while the Na salt does not show detectable aggregates.

The effect of an added excess of ions is also remarkable: in the presence of an excess of NaCl or KCl, at 60°C no melting is observed and, after cooling, slightly longer aggregates are formed.

Table 3. The length of the columnar aggregates formed by d(GpG) **1** at 1% w/w concentration as deduced from SAXS experiments (n.d.: not detectable).

		T	L(Å)
In water	Na salt	20°C	70±15
		60°C	n.d.
		20°C after cooling	n.d.
	K salt	20°C	76.7±6.9
		60°C	n.d.
		20°C after cooling	94±11
In KCl 1 M	Na salt	20°C	96.4±7.2
		60°C	95±10
		20°C after cooling	107±10
	K salt	20°C	110.8±8.8
		60°C	111.5±4.7
		20°C after cooling	119.1±6.4

The self-assembly process of all these chiral molecules can easily and conveniently be followed by CD spectroscopy. The spectra of the isolated monomeric species can be observed in diluted solutions without added salts or, in more concentrated solutions at higher temperature; they are usually drastically different from those of the assembled species. For assembled species, exciton double signed bands are often observed in correspondence to a single absorption band [38]; the spectra become similar to those of biological polymers (DNA, proteins). The technique is very useful also for measuring melting temperatures and can be more sensitive and selective than ordinary absorption measurements.

In the case of guanosine derivatives, several advantages are available in interpreting CD spectra: first, the CD spectra of the four-stranded helices of poly(G) and poly(dG) are known and also the structure of this unusual helix, which is similar to our columnar aggregates, is known in detail from fibre X-ray work on poly(G): the helix is right-handed and the G-quartets are stacked perpendicularly to the helix axis [9,10]. Secondly, the spectroscopic properties of guanine have been extensively investigated. Consequently, it is possible to calculate the CD spectra of poly(G) and similar molecules with a relatively simple exciton treatment [39,40], having a reliable model with which to check the calculations.

From this approach, the chirality of the columnar aggregates could be deduced: in the low-energy part of the spectrum, the guanine chromophore displays two well-characterized electronic transitions corresponding to an absorption maximum at *ca.* 250 nm and to a shoulder at *ca.* 280 nm. In the four-stranded helix of poly(G), the transition at *ca.* 250 nm gives rise to a non-symmetric exciton couplet with a stronger positive component at *ca.* 260 nm and a weaker negative band at ca. 240 nm. The second transition appears as a shoulder of the positive 260 nm band.

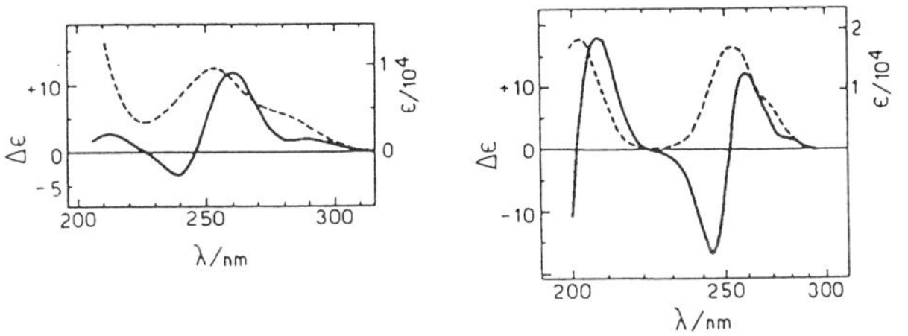

Figure 7. The experimental (left) and simulated (right) absorption (dashed line) and CD (full line) spectra of poly(G).

It follows that, whenever the columnar deoxyguanosine aggregates give similar spectra, it is possible to assign a right-handed absolute stereochemistry. The correlation has both empirical and theoretical validity.

From the sign of the CD couplet centred at *ca.* 250 nm, the absolute handedness reported in Table 4 can be deduced.

Table 4. The columnar helicity (as deduced by the exciton CD corresponding to the guanine chromophore; - and + signs refer to left and right handedness, respectively) and the cholesteric helicity and pitch for guanosine and pterine derivatives.

Compound		Columnar helicity	p (μm)	Cholesteric helicity
d(Gp)	**(6)**	(-)		*M*
d(Gp-*i*-Bu)	**(11)**	(-)		*M*
d(pG)	**(5)**	(-)	-31	*M*
d(GpG)	**(1)**	(+)	-70	*M*
d(GpGpG)	**(7)**	(+)	+140	*P*
d(GpGpGpG)	**(8)**	(+)	+100	*P*
d(GpGpGpGpGpG)	**(9)**	(+)	+43	*P*
Sodium Folate + NaCl	**(2)**		-14	*M*

Other information on the geometry of the columns can be deduced. In derivatives d(pG), d(GpGpG), d(GpGpGpG), d(GpGpGpGpGpG) [33] and d(cGp), the band shapes are similar to poly(G) , the CD spectrum is dominated by the 260 nm band which is positive for d(GpGpG), d(GpGpGpG) and d(GpGpGpGpGpG), and negative for d(pG) and d(cGp). However, d(GpG) displays a positive 260 nm band in the presence of K^+ or Sr^{2+} ions or a positive 290 nm band in the presence of Rb^+ or Cs^+; with Na^+, both 260 and 290 nm bands are observed. The 260 or 290 nm bands have been discussed also for the assembled species of G-rich oligonucleotides connected to telomere repeat sequences [41].

From the stereochemical point of view, the presence of the 260 nm band seems to be associated to a rather regular structure, similar to that of poly(G), with the tetrameric planes perpendicular to the helix axis, while a dominant 290 nm band indicates consistent variations with respect to this situation [27,39,40].

The finding that K^+ ions allow the passage from a distorted to a regular structure is related to the size of this ion, which is ideal for positioning itself between two adjacent tetrameric planes, coordinating all eight sorrounding oxygen atoms [8].

For d(GpG), indication of the stability of the aggregates in terms of melting temperatures (T_m) were obtained by a detailed variable temperature CD investigation [27]. The stabilizing effect of the different ions, as deduced from melting temperatures, are in the order: Sr ≈ K > Rb > Na > Cs.

The results are in excellent agreement with those from SAXS reported above: the conditions which favour a lengthening of the columns are the same as those which increase T_ms; in particular, the effect of K^+ ions is remarkable. We have therefore parallel behaviour of stability and length of the aggregates, both factors playing an important role in the LC formation.

In the case of folates, in pure D_2O no scattering aggregates can be detected by SANS even at a relatively high concentration (4% w/w) [28]. At the same concentration and with an excess of Na^+ or K^+ ions instead, the presence of scattering aggregates can be detected. The aggregates are cylindrical and, assuming a diameter of *ca.* 30 Å, as deduced from X-ray work on the hexagonal phase [5,36], the length of the cylinders can be calculated: the aggregates are composed of a stack of *ca.* 9 disks [28].

As data on the pterine electronic transitions are not reported in the literature, CD spectroscopy can be used only empirically. In particular, no information on the handedness of the columnar aggregates can be deduced. CD spectra indicate very clearly the presence of aggregates also at lower concentrations (beginning at 10^{-4} M), for which no SANS signals can be measured; this points to the formation of aggregates of smaller dimensions, too small to be detected by the scattering technique. As will be reported in the next section, the folate column length varies considerably with the concentration.

For the assembled folates, rather different spectra are obtained in pure water and in the presence of an excess of Na^+ and K^+ ions. The spectra in pure water

display single melting to give the spectrum of the free folate. The spectra obtained in the presence of salts give a clear double melting.
These results, together with NOESY and NOE data, were interpreted considering the existence of two assembled forms: one (with added ions) is more compact, and has a stronger tetramer-to-tetramer interaction; the other has a different stereochemistry of the aromatic core region and a weaker tetramer-to-tetramer interaction. In both cases, the glutamic chains are rather mobile [28].
These data in isotropic solutions are in full agreement with those obtained in the mesomorphic state and reported in the next section.
In the case of folates, as for guanine derivatives, the polymorphism observed is related to the growth of the columnar aggregates; the factors which promote the growing process lower the transition concentrations and stabilize in particular the cholesteric phases.

6. Column Length and Growth in the Mesomorphic State

The structural properties of the columns can be derived from the variation of the hexagonal cell dimension a, and (tentatively) of the main distance between the cylinders in the cholesteric phase, as a function of the volume concentration c_v. Further information can be obtained by the analysis of shape and position of the high angle 3.3 Å peak.
Assuming that in the hexagonal phase the columns are infinitely long and have a circular section with radius R, the relation between the cross-section area of the rod and the two-dimensional hexagonal unit cell surface is [12]

$$\pi R^2 = (\sqrt{3} / 2)\, a^2\, c_v \qquad (6)$$

In the columnar system discussed, the structure of the disk-shaped G-quartets is not expected to change as a function of concentration [42,43], so R can be considered constant. For infinite cylinders therefore, a will change with the volume concentration as

$$a = (2\, \pi\, R^2 / \sqrt{3})^{1/2}\, c_v^{-1/2} \qquad (7)$$

This means that the rods move apart only laterally as dilution proceeds.
If the rod length is finite, the cylinders pack in the two-dimensional hexagonal cell again with parameter a, but also with an average vertical distance C between the cylinder centres. If L is the length of the cylinders, one obtains

$$L\, \pi\, R^2 = C\, (\sqrt{3} / 2)\, a^2\, c_v \qquad (8)$$

therefore the variation of a with c_v depends on the variation of the ratio L/C, *i.e.* the fraction of water in the C direction. In particular, the relationship $a \propto c_v^{-1/3}$ can be demonstrated when the interparticle distances decrease uniformly in the three dimensions as a function of concentration (*i.e.* when $L/C = 2R/a$ in the hexagonal domain) [36,37,43,44].

The two situations can be interpreted in terms of length and flexibility [45,46,47] of the aggregates: a $c_v^{-1/2}$ dependence may be indicative of infinite (with respect to the lateral distance) cylinders, while a $c_v^{-1/3}$ dependence may be interpreted in terms of finite cylinders.
The columnar stacking of the teramers can be studied in more detail by the analysis of the high-angle (s = (3.4 Å)$^{-1}$) peak. By fitting a gaussian and a locally linear background to every peak, one can have an accurate value of the position of the peak, from which the distance between the stacked G-quartets is obtained, and of its width at half maximum, which is correlated to the average number of stacked tetramers in one column.

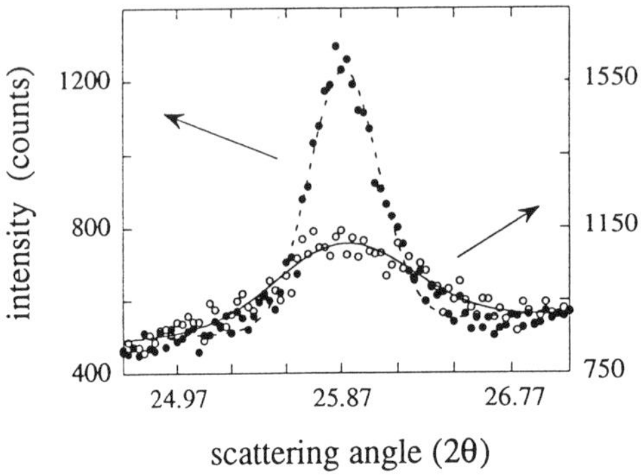

Figure 8. High angle diffraction peak (s = (3.4Å)$^{-1}$) observed at c = 50% (closed circles) and 23% (open circles) for d(pG) in KCl 4 M.

To obtain this value, numerical simulations of the scattering profile of the columnar aggregates must be performed: the tetramers were approximated by homogeneous disks with a thickness of 2.35 Å [36,37].The scattering amplitude of n of these disks, stacked one above the other at a distance of 3.4 Å, was summed up, squared and averaged over all orientations. This procedure is correct as long as there is no 3-dimensional correlation between the positions of the tetramers in different columnar aggregates. The average number of stacked discs in one column was determined by comparing the calculated and measured peak width; the average correlation length was obtained by multiplying n by the experimental stacking distance. Using these approaches and considering the results for the self-assembly in the isotropic solutions (preceding section), the pictures reported in Figure 9 can be obtained.
In the case of d(GpG), the dependence is of the -1/3 type in the cholesteric phase and indicates finite cylinders, while in the hexagonal phase infinite cylinders seem to be present. The cross-over between finite and infinite objects is near the N* to H transition [32]. The description which can be deduced is that the aggregate length grows rather quickly with concentration in the

isotropic and in the cholesteric phase. In the hexagonal phase, they are so long that their length cannot be compared to the lateral distance.

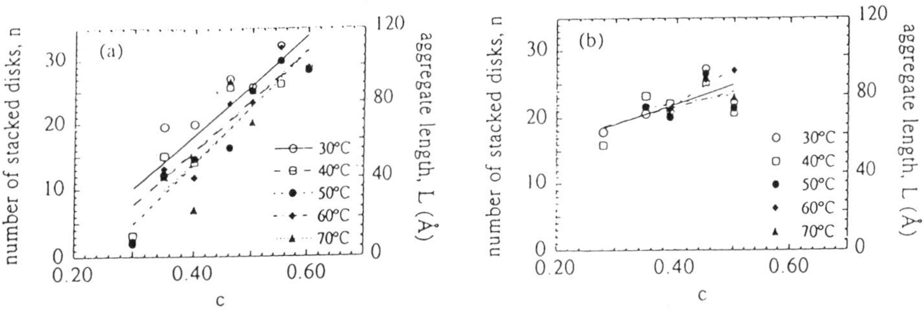

Figure 9. Number of stacked disks (and the aggregate length) as a function of folate concentration (w/w) for different temperature for disodium folate in water (a) and in NaCl 1 M (b).

In the case of derivative d(pG) instead, the columns seem to be finite also in the hexagonal phase and the length growth seems to be slow [37]. The fact that SANS is unable to detect aggregates at 1% concentration is in agreement with this result. For stereochemical and, to a lesser extent also for hydrophilicity reasons, the assembly process of d(pG) seems to be difficult. Accordingly, the critical concentration at which the N* phase appears is much higher than for d(GpG).

Also for folates with and without added salts, the -1/3 exponent indicates the formation of finite cylinders in the hexagonal phase [36]. In pure water, the distance between the disks increases as a function of the water content and temperature. The average length of the aggregates increases continuously, as a function of concentration, from 5 piled disks at the I to H transition to *ca.* 30 at a 60% concentration. In NaCl instead, the stacking distance remains almost constant, as does the size of the aggregates: also at low concentration one has *ca.* 20 piled disks.

The stabilisation operated by Na^+ ions (and, to a lesser extent, by K^+), seems similar to that described above for K^+ in the guanine system.

This behaviour is in full agreement with the results obtained by SANS in isotropic solution: at 4% concentration, no aggregates could be detected in pure water, while, in NaCl, the data indicate the presence of aggregates composed of *ca.* 9 disks [28].

The formation of a stable cholesteric mesophase only in the presence of NaCl is related to the fact that only in this condition are the aggregates long enough at relatively low concentration (*ca.* 27% w/w). Without added salts one needs a higher concentration to obtain aggregates long enough to give a mesophase, and, at these concentrations, there is no space available for the formation of the cholesteric helix; the more compact H phase is therefore observed.

This picture is in good agreement with the theoretical prediction of phase diagrams for assembling lyotropics obtained recently by Taylor and Herzfeld [45].The calculated phase diagrams display a nematic between the isotropic and the hexagonal phases only when the aggregation is strong; for weak aggregation the direct I to H transition is predicted.

7. Stereochemistry of the Chiral Packing in the Cholesteric Mesophases

Data on the handedness and pitch of the cholesteric columnar mesophases can be obtained by using the magnetic alignment described above.The determination of the pitch values is carried out by measuring the spacing between the lines of the fingerprint textures. The determination of the cholesteric handedness is carried out by CD spectroscopy on samples with planar alignment (in order to avoid artefacts connected to strong birefringence of the sample) [48]. As reported in ref. [48] a positive CD in the absorption region indicates a right-handed cholesteric, a negative CD a left-handed cholesteric. The handedness was confirmed by recording the CD spectra in the presence of ethidium bromide which intercalates into four-stranded oligoguanylates [33]. The handedness and pitch values are reported in Table 4.

With the exception of d(GpG), in all the deoxyguanosine oligomers investigated, right-handed columns generate right-handed cholesterics.

A model for the packing of hard helical screws, which recognize their reciprocal chirality and spontaneously generate an overall helical structure, is reported in Figure 10.

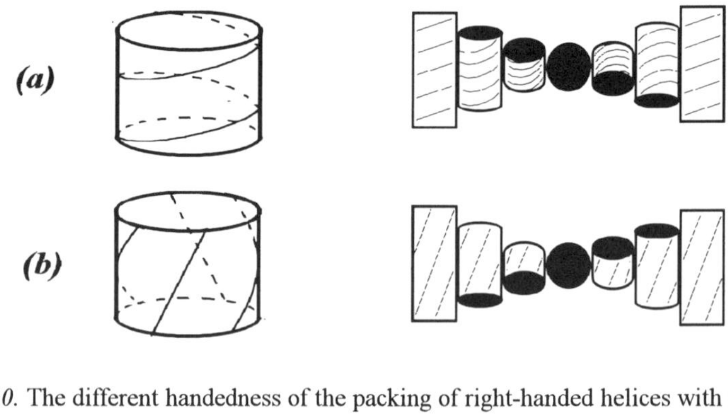

Figure 10. The different handedness of the packing of right-handed helices with different pitch-to-diameter ratios.

The model gives a simplified picture of the interaction, by hard-core repulsion only, of chiral columns in the cholesteric phase. In the model of Figure 10, right-handed screws pack to give a right-handed superstructure. However, the model is sensitive not only to the handedness of the screws, but also to the ratio pitch/diameter of the individual objects: for a given handedness of the screws (*e.g.* right-handed) one can obtain a definite handedness of the superstructure (*e.g.* right-handed) only if $p/d < \pi$; if $p/d > \pi$ an opposite-handed superstructure is generated [48].

It is certainly an oversimplification to compare the cholesteric phases given by the guanine derivatives to the model of Figure 10 in which neither charges nor dispersive interactions nor excluded volume effects exist. Furthermore, the geometry of the four-stranded aggregates is complex and it is impossible to evaluate precisely the external diameter of the cylinders. Nevertheless, this model, which has a critical value of the p/d ratio of the individual screws around which there is an inversion of the superhelical handedness, has some utility: it gives a qualitative indication of the effect that a process of winding or unwinding the helical aggregates might have on the cholesteric handedness.

If we consider the guanine columnar assembled species as screws (the fact that they are four-stranded helices does not affect qualitatively the results), all the screws but that derived from d(GpG) behave as if their p/d ratio is smaller than the critical value. For d(GpG) the helical system seems to be unwound to give p/d larger than the critical value.

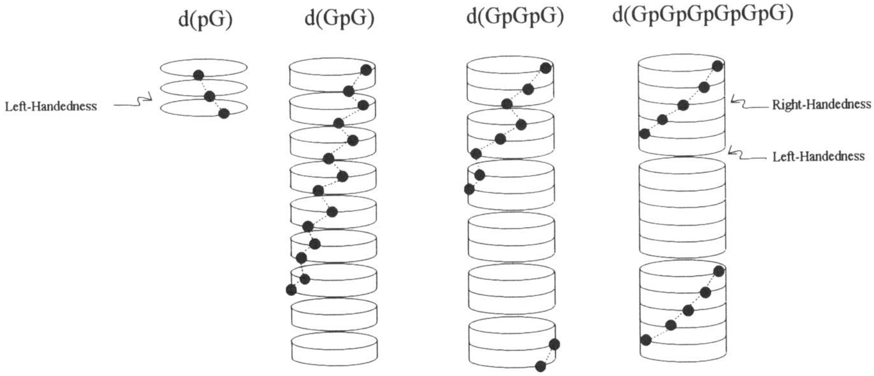

Figure 11. A representation of the chiral columnar aggregates formed by oligoguanylate derivatives (the filled circles represent the sugar units; for clarity, only one sugar per tetramer is shown).

In the homologous series d(GpG), d(GpGpG), d(GpGpGpG) and d(GpGpGpGpGpG), in which only internucleoside phosphate groups are present, the different characteristics of the cholesterics and also the opposite sign displayed by d(GpG) can be understood: in order to construct an aggregate composed of 18 disks (which for example is the average length observed by SANS in isotropic solutions of *ca.* 1% concentration), in the case of d(GpGpGpGpGpG), three discrete tetrameric aggregates have to be connected by two sets of similar non-covalent interactions (stacking forces and probably also hydrogen bonds between the free 3' and 5' OH groups) (Figure 11). In the case of d(GpG) instead, nine discrete aggregates have to be connected by seven sets of non-covalent interactions.

It is known from X-ray work and CD data on poly(G) and poly(dG) that the four-stranded structures are right-handed, and the same applies also to the aggregates formed by homoguanylic oligomers (even if in the case of d(GpG), in the absence of added K^+, the CD intensity is much weaker). The phosphodiester linkages between the tetrameric planes is therefore connected to a right-handed structure. The contribution of non-covalent interactions to the helical handedness is not necessarily of the same sign. Let us assume that they are of the opposite sign, or maybe of the same sign but with a smaller angle of rotation between the G-quartets; in both cases. they cause unwinding of the four-stranded helices.

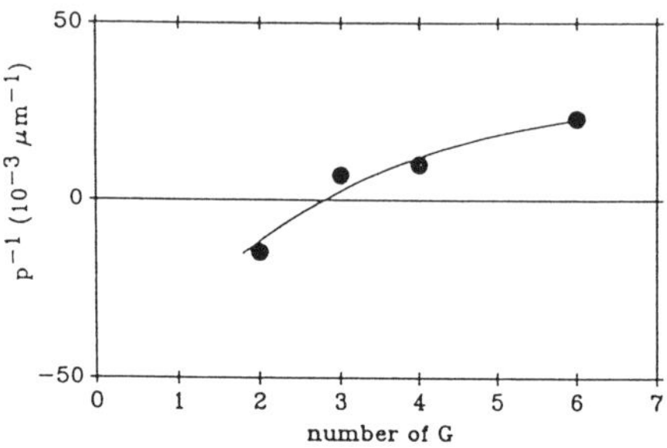

Figure 12. Reciprocal pitch of the cholesterics obtained from oligoguanylates (+ and - refer to left and right-handed cholesterics, respectively)

The unwinding effect is obviously greater for d(GpG) > d(GpGpG) > d(GpGpGpG) > d(GpGpGpGpGpG). In the case of d(GpG), it is not surprising that we have an inversion of the sense of packing as a result of the increased p/d ratio, which could be greater than the critical value.

The variation of the reciprocal pitch of the cholesterics with the number of guanosine residues is in full agrement with this hypothesis: the value p^{-1} (which is a measure of the twist between the chiral columns in the cholesteric phase) varies smoothly with the number of residues as predicted from the screw model. The assumption that the columns are formed by packing of discrete objects, for which we have some evidence from CD spectra for d(GpG) and higher oligomers (in the case of d(GpG), also a continuous aggregation mechanism could be operative in some cases), is not a necessary prerequisite and a similar reasoning could apply also to a continuous assembly model.

With regard to this point, the high-resolution crystal structure of the hexanucleotide d(TpGpGpGpGpT) in the presence of Na^+ ions, reported very recently [49], is particularly instructive. The usual G-quartets are present and connect four molecules of the oligonucleotide with parallel strands. In the crystal, two of these aggregates are connected coaxially by a sodium ion in a head-to-head way to give a cylindrical structure formed by G-quartets. The terminal thymine molecules protrude from the cylindrical, structure which is similar to a right-handed four-stranded helix.

Acknowledgement

We thank Consiglio Nazionale delle Ricerche (Italy) and Ministero dell'Università e della Ricerca Scientifica e Tecnologica (Italy) for financial support.

References

1. Robinson, C. (1961) Liquid crystalline structure in polypeptide solutions, *Tetrahedron* **13**, 219-234.
2. Livolant, F., Levelut, A. M., Doucet, J., and Benoit, J. P. (1989) The highly concentrated liquid-crystalline phase of DNA is columnar hexagonal, *Nature* **339**, 724-726; and references therein.
3. Spada, G. P., Brigidi, P., and Gottarelli, G. (1988) The determination of the handedness of cholesteric superhelices formed by DNA fragments, *J. Chem. Soc., Chem. Commun.*, 953-954.
4. Spada, G. P., Carcuro, A., Colonna, F. P., Garbesi, A., and Gottarelli, G. (1988) Lyomesophases formed by the dinucleoside phosphate d(GpG), *Liq. Cryst.* **3**, 651-654.
5. Bonazzi, S., De Morais, M. M., Gottarelli, G., Mariani, P., and Spada, G. P. (1993) Self-assembly and liquid crystal formation of folic acid salts, *Angew. Chem., Int. Ed. Engl.* **32**, 248-250.
6. Bang, I. (1910) Guanylic acid, *Biochem. Z.* **26**, 293-311.
7. Gellert, M., Lipsett, M. N., and Davies, D. R. (1962) Helix formation by guanylic acid, *Proc. Natl. Acad. Sci. USA* **48**, 2013-2018.
8. Saenger, W. (1984) *Principles of Nucleic Acid Structures*, Springer, New York, pp. 315-320; and references therein.
9. Arnott, S., Chandrasekaran, R., and Martilla, C. M. (1974) Structures for polyinosinic acid and polyguanylic acid, *Biochem. J.* **141**, 537-543.
10. Zimmermann, S. B., Cohen, G. H., and Davies, D. R. (1975) X-ray fiber diffraction and model building study of polyguanylic acid and polyinosinic acid, *J. Mol. Biol.* **92**, 181-192.
11. Guschlbauer, W., Chantot, J. F., and Thiele, D. (1990) Four-stranded nucleic acid structures 25 years later: from guanosine gels to telomer DNA, *J. Biomolec. Struct. & Dynam.* **8**, 491-511; and references therein.
12. Luzzati, V. (1968) X-ray diffraction studies of lipid-water systems, in D. Chapman (ed.) *Biological Membranes*, Academic Press, London, pp. 71-123.
13. Mariani, P., Rustichelli, F., and Torquati, G. (1991) Structure of mesophases: X-ray diffraction, in R. Bartolino, L. Fronzoni and F. Simoni (eds.), *Physics of Liquid Crystals*, Gordon and Breach, New York, pp. 3-44.

14. Weber, P., Guillon, D., and Skoulios, A. (1991) Hexagonal columnar mesophases from phtalocyanine. Upright and tilted intracolumnar molecular stacking, herringbone and rotationally distorted columnar packing, *Liq. Cryst.* **9**, 369-382.
15. Tardieu, A. (1972) Etude Cristallographique de systemes lipides-eau, Ph. D. Thesis, Université Paris Sud.
16. (1952) International Tables for X-ray Crystallography, The Kynoch Press, Birmingham.
17. Luzzati, V., Gulik-Krzywicki, T., and Tardieu, A. (1968) Polymorphism of lecithins, *Nature* **218**, 1031-1034.
18. Luzzati, V., Mariani, P., and Delacroix, H. (1991) X-ray crystallography at macromolecular resolution: a solution of the phase problem, *Makromol. Chem. Macromol. Symp.* **15**, 1-17.
19. Sayre, D. (1952) Some implications of a theorem due to Shannon, *Acta Cryst.* **5**, 843-843
20. Franks, N. P. (1976) Structural analysis of hydrated egg lecithin and cholesterol bilayers. I. X-ray diffraction, *J. Mol. Biol.* **100**, 345-358.
21. Gulik, A., Luzzati, V., De Rosa, M., and Gambacorta, A. (1985) Structure and polymorphism of bipolar isopranyl ether lipids from Archaebacteria, *J. Mol. Biol.* **182**, 131-149.
22. Guinier A. and Fournet, G. (1955) *Small angle scattering of X-rays*, Wiley, New York.
23. Jacrot, B. (1976) The study of biological structures by neutron scattering from solution, *Rep. Prog. Phys.* **39**, 911-953.
24. Chen, S. H. and Lin, T. L. (1987) Colloidal solution, in R. Celotta and J. Levine (eds.) *Methods in Experimental Physics*, Vol.23; D. L. Price and K. Skold (eds.), *Neutron Scattering*, Part B, Academic Press, New York, pp. 489-543.
25. (1982) Glatter, O. and Kratky, O. (eds.) *Small Angle X-ray scattering*, Academic Press, New York.
26. Carughi, F., Ceretti, M., and Mariani, P. (1992) Structural organization of guanosine derivatives in dilute solutions: small angle neutron scattering analysis, *Eur. Biophys. J.* **21**, 155-161.
27. Bonazzi, S., Gottarelli, G., Spada, G. P., Mariani, P., Romanzetti, S., Garbesi, A., and La Monaca, A. (1995) A study of the self-assembly of 2'-deoxyguanylyl-(3'-5')-2'-deoxyguanosine, d(GpG), by SAXS and CD, *Gazz. Chim. (It.)*, in press.
28. Carsughi, F., Di Nicola, G., Gottarelli, G., Mariani, P., Mezzina, E., Sabatucci, A., and Spada, G. P. (1995) The self-recognition and self-assembly of folic acid salts in isotropic water solution, *J. Am. Chem. Soc.*, submitted.
29. Mariani, P., Mazabard, C., Garbesi, A., and Spada, G. P. (1989) A study of the structure of the lyomesophases formed by the dinucleoside phosphate d(GpG). An approach by X-ray diffraction and optical microscopy, *j. Am. Chem. Soc.* **111**, 6369-6373.
30. Perahia, D., Luz, Z., Wachtel, E. J., and Zimmermann, H. (1987) NMR and X-ray diffraction of the 7,7'-disodiumchromoglycate-water lyomesophases, *Liq. Cryst.* **2**, 473-489.
31. Iball, J., Morgan, C. H., and Wilson, H. R. (1963) Fibres of guanine nucleosides and nucleotides, *Nature* **150**, 688-689.

32. Amaral, L. Q., Itri, R., Mariani, P. and Micheletto, R. (1992) Structural study of the aggregates formed by the dinucleoside phosphate G2 in aqueous solution, *Liq. Cryst.* **12**, 913-919.

33. Bonazzi, S., Capobianco, M., De Morais, M. M., Garbesi, A., Gottarelli, G., Mariani, P., Ponzi Bossi, M. G., Spada, G. P., and Tondelli, L. (1991) Four-stranded aggregates of oligodeoxyguanylates forming lyotropic liquid crystals: a study by circular dichroism, optical microscopy, and X-ray diffraction, *J. Am. Chem. Soc.* **113**, 5809-5816.

34. Bonazzi, S., Garbesi, A., Gottarelli, G., Spada, G. P., and Mariani, P.; manuscript in preparation.

35. Mariani, P., De Morais, M. M., Gottarelli, G., Spada, G. P., Delacriox, H., and Tondelli, L. (1993) Structural analysis of the lyotropic polymorphism of four-stranded aggregates of 2'-deoxyguanosine 3'-monophosphate derivatives, *Liq. Cryst.* **15**, 757-778.

36. Ciuchi, F., Di Nicola, G., Franz, H., Gottarelli, G., Mariani, P., Ponzi Bossi, M. G., Spada, G. P. (1994) Self-recognition and self-assembly of folic acid salts: columnar liquid crystalline polymorphism and the column growth process, *J. Am. Chem. Soc.* **116**, 7064-7071.

37. Franz, H., Ciuchi, F., Di Nicola, G., De Morais, M. M., and Mariani, P. (1994) Unusual lyotropic polymorphism of deoxyguanosine-5'-monophosphate: X-ray diffraction analysis of the correlation between self-assembling and phase behaviour, *Phys. Rev. E* **50**, 395-402.

38. Mason, S.F. (1982) *Molecular Optical Activity and the Chiral Discrimination*, Cambridge University Press, Cambridge.

39. Cech, C.L. and Tinoco,I., Jr. (1976) Circular dichroism calculations for polyinosinic acid in proposed multi-strand geometries, *Nucl. Acid Res.* **3**, 399-404.

40. Gottarelli, G., Palmieri, P., and Spada, G. P. (1990) On the exciton optical activity of the four-stranded helix of poly(G), *Gazz. Chim. (It.)* **120**, 101-107.

41. Balagurumoorthy, P., Brahmachary, S. K., Mohanty, D., Bansal, M., and Sasisekharan, V. (1992) Hairpin and parallel quartet structures for telomeric sequences, *Nucl. Acid Res.* **29**, 4061-4067.

42. Zimmerman, S.B. (1976) X-ray study by fiber diffraction methods of a self-aggregate of guanosine-5'-phosphate with the same helical parameters as poly(rG), *J. Mol. Biol.* **106**, 663-672.

43. Amaral, L. Q., Gulik, A., Itri, R., and Mariani, P. (1992) Micellar hexagonal phases in lyotropic liquid crystals, *Phys. Rev. A* **46**, 3548-3550.

44. Mariani, P. and Amaral, L. Q. (1994) Micellar growth in hexagonal phases of lipid systems, *Phys. Rev. E* **50**, 1678-1681.

45. Taylor, M.P. and Herzfeld, J. (1991) Shape anisotropy and orderd phases in reversibly assembling lyotropic systems, *Phys. Rev. A* **43**, 1892-1905.

46. Hentschke, R. and Herzfeld, J. (1991) Isotropic, nematic, and columnar ordering in systems of persistent flexible hard rods, *Phys. Rev. A* **44**, 1148-1155.

47. Hentschke, R. (1991) Effect of persistent flexibility on the isotropic, nematic and columnar ordering in self-assembling systems, *Liq. Cryst.* **10**, 691-702.

48. Gottarelli, G., Spada, G. P. (1994) Application of CD to the study of some cholesteric mesophases, in K. Nakanishi, N. Berova, and R. W. Woody (eds.), *Circular Dichroism, Principles and Applications*", VCH, New York, pp. 105-119.

49. Laughlan, G., Murchie, A. I. H., Norman, D. G., Moore, M. H., Moody, P. E. C., Lilley, D. M., Luisi, B. (1994) The high-resolution crystal structure of a parallel-stranded guanine tetraplex, *Science* **265**, 520-524.

TAILOR-MADE AUXILIARIES FOR THE CONTROL OF NUCLEATION, GROWTH AND DISSOLUTION OF CRYSTALS.[1]

M. Lahav and L. Leiserowitz
Department of Materials & Interfaces
Weizmann Institute of Science, Rehovot 76100, Israel.

1. Introduction

Although crystals, and their precipitation as a process, play an important role in pure and applied science, our ability to design crystals with desired properties, and our control over crystallization, is still at a rudimentary stage. Indeed our ignorance of the overall process of crystal growth is brought to the fore by a comparison with the controlled crystal growth provided in biomineralization. Organisms can mold crystal of specific morphologies, sizes and orientations in the form of composites of minerals and organic materials with characteristics vastly different from those of their inorganic counterparts.[4] Nevertheless the technological advances in controlling the microstructure of inorganic materials on the subnanometer scale have led to tremendous success in the preparation of ceramics, synthetic layered microstructures and semiconducting materials. Paradoxically, advances in the field of organic materials have not been so striking although the Langmuir-Blodgett technique for the preparation of films holds great promise as pyroelectric, piezoelectric and frequency doubling devices, etc. The tread that binds these diverse topics involves molecular interactions at the interface. Many important studies on crystal growth of a kinetic and thermodynamic nature, sometimes taking into account molecular interactions, have been reported.[5] In recent years a stereochemical approach has been adopted involving growth and dissolution of organic and inorganic crystals in the presence of "tailor-made" auxiliaries, which interact with specific crystal faces and may be crystal growth inhibitors or nucleation promotors.[2,6] With the assistance of molecular auxiliaries it is possible to examine a variety of processes. Crystals may be engineered with desired morphologies,[7,8] a technique which, paradoxically could reduce concentration of impurities in growing crystals. Preferred crystallization of a desired phase

G. Tsoucaris et al. (eds.), Crystallography of Supramolecular Compounds, 331–345.

may be induced for the resolution of optical isomers,[9,10] or the precipitation of metastable polar crystalline polymorphs.[11,12] The absolute configuration of chiral molecules or polar crystals can be assigned.[13] With occlusion of the additive into the crystal, the symmetry of the crystal may be reduced and so make a "centrosymmetric" crystal display frequency doubling with laser light (i.e. second harmonic generation).[14] Molecular crystals can be etched at desired faces on dissolution[15] and the effect of solvent on crystal growth may be elucidated.[16] It has also proven possible to engineer surfaces such as Langmuir monolayers for epitaxial crystal growth,[17-18] and to acquire information relevant to the mechanism of the growth of crystals in biological environments. Furthermore, the epitaxy suggested that uncompressed Langmuir monolayers form two-dimensional (2-D) crystalline aggregates on the water surface, as was quantitatively demonstrated by grazing incidence X-ray diffraction.[19] The knowledge acquired from such diffraction experiments led to experiments involving the use of alcohol monolayers for induced nucleation of ice.[20,21] More recently, it became possible to demonstrate directly by diffraction and atomic force microscopy, the formation of 3-D crystalline clusters of bolaamphiphiles and alkanes at the air-liquid interface, and to model in a direct mode, the role played by "tailor-made" additives on the early stages of crystal nucleation.[22,23,24] Here we present a birds eye view of the different above mentioned, yet connected, topics.

2. Morphological Crystal Engineering

The habit of a crystal is defined by its relative rates of growth in different directions; the faster the growth in a given direction the smaller the face developed perpendicular thereto. When growth is inhibited in a direction perpendicular to a given face, the area of this face is expected to increase relative to the areas of other faces of the same crystal. Differences in the relative surface areas of the various faces can therefore be directly correlated to the inhibition in different growth directions. Dramatic morphological changes observed during growth of organic crystals in the presence of additives revealed a high degree of specificity in the interaction of the foreign material with the different structured surfaces of the crystalline matrix. Indeed, from a systematic study on a large variety of organic compounds crystallized in the presence of additives of molecular structure similar to the structures of the corresponding substrate molecules labelled "tailor-made" additives (or auxiliaries), we deduced a stereochemical correlation between the structures of the affected surfaces and the molecular structure of the inhibitor. We could infer that the "tailor-made" additive is adsorbed on the growing crystals, but only at certain surfaces and then with the part of the adsorbate that differs from that of the substrate emerging from the crystal. Once this mechanism was established, it became possible to exploit it to systematically modify the morphology of crystals by tailoring additives that bind at preselected faces and thus inhibit growth in a predictable manner.[2,7,8] This approach has also been applied for the direct assignment of the absolute configuration of chiral molecules and polar crystals.[13] The method is simple and useful particularly for pedagogical purposes.

3. Reduction of Symmetry of Solid Solutions

During crystallization of a substrate in the presence of the tailor-made additives, a small amount of the latter is occluded into the bulk of the growing crystal. We deduced, and confirmed experimentally, that this occlusion occurs only through specific faces and may indeed occur only at specific sites on the faces, following rules dictated by crystal symmetry and molecular orientation. The symmetry rules arise from the fact that in general the 2-D symmetry of a crystal face is lower than the 3-D symmetry in the crystal bulk, thus the additives may in general be adsorbed and eventually occluded only via a subset of the different surface sites. Consequences of this selectivity are an anisotropic distribution of the additive inside the crystal and a reduction in its symmetry.[13] This implies that the so-called single crystals of solid solutions may be composed of segments of crystals of different local symmetries coherently compounded together, the symmetry of each dictated by the surface through which the guest molecules have been occluded. Such lowering of symmetry has been demonstrated in several systems by both macroscopic and microscopic methods. These include, enantiomeric separation of tailor-made chiral additives in "centrosymmetric" host crystals,[25] second harmonic generation in "centrosymmetric" crystals,[14] topochemical solid-state reactions,[26] and X-ray and neutron diffraction.[26,27] McBride has demonstrated independently the reduction in symmetry of some peroxides by applying the method of optical birefringence.[28]

4. "Tailor-made" Etchants for Organic Crystals

Growth and dissolution under conditions close to equilibrium may be considered as reverse processes that can be interchanged by altering the degree of saturation of the solution. Consequently, an inhibitor of growth of a given crystal face must, in principle, affect the rate of dissolution of that same crystal face as well. Since crystal dissolution starts at sites of emerging dislocations, the effect of addition of designed additives manifests itself macroscopically by causing stereospecific etching at the same faces that are affected upon growth. This correlation is general and was successfully applied to a large variety of molecular crystals.[15] In a manner analogous to the crystal growth process, the tailor-made etchants may also be used for assignment of absolute configuration of chiral molecules and polar crystals.

5. Effect of Solvent on Crystal Growth and Morphology

Solvent has a strong influence on the habit of crystalline materials, however the role played by solvent-surface interactions in enhancing or inhibiting crystal growth is still unclear. To date there have been two different approaches to clarify this point. In one theory, proposed by Bennema[29,30] and others, favorable interactions between solute and solvent on specific faces leads to reduced interfacial tension, causing a transition from a smooth to a rough interface and a concomitant faster surface growth. Alternatively, it has been proposed that preferential adsorption at specific faces will inhibit their growth as removal of a bound solvent poses an additional energy barrier for continued growth.[31-33] Our studies on the role played by tailor-made additives are in keeping with the latter approach. However, solvent-surface interactions cannot be as clearly analyzed as those between the tailor-made additives and the crystal.

In order to help clarify this problem the following experimental approaches were adopted. Use was made of crystalline solvates where the solvent of crystallization plays the dual role of solvent and crystal solute. Such crystals were grown in the presence of tailor-made solvent, which is a slightly modified version of the solvate solvent. It was found that when crystalline hydrates were grown from aqueous solution in the added presence of methanol (the tailor-made solvent) that the change in morphology was interpretable[34] in a manner akin to the effect of a tailor-made inhibitor. In another approach use was made of a rare family of plate-like crystals composed of amphiphilic molecules packed in layers stacked head-to-tail in a polar arrangement so that the polar plate-like crystals are hydrophobic on one face and hydrophilic on the opposite face (Chain-like amphiphilic molecules normally pack head-to-head and tail-to-tail in the crystalline state). In CH_3OH solution the polar plate-like crystals were found to grow almost four times faster at the hydrophobic side than at the hydrophilic side.[35] We also found that H_2O and other polar solvents wet the hydrophilic face more strongly than the hydrophobic face.[36] All these results are in keeping with the theory that strong solvent-surface interaction inhibit crystal growth.

An entirely different approach involved strong selective adsorption of solvents at a subset of molecular surface sites, of type *A*, and repulsion of solvent at the remaining set of surface sites, of type *B*, on the crystal face. It was found that such a crystal surface can lead to fast growth, as was shown by experiments on the crystals of (R,S) alanine and the γ-form of glycine in aqueous solution.[34] Thus if solvent is strongly bound to a subset of surface sites and repelled (or weakly bound) at the remaining set of surface sites a different set of rules for growth may apply, more in line with the theory proposed by Bennema.[29]

6. Control of Polymorphism

Any crystalline process must cross the nucleation barrier. It is generally assumed that, as the molecules start to associate in supersaturated solutions, they form embryos with structures resembling those of the to-be-grown crystals. For polymorphous systems embryos of all the observed phases may be formed. Each type of embryo would resemble structurally the crystal into which it will eventually develop. If this hypothesis is correct, one may use the structural information on a mature crystal to design inhibitors of one particular crystalline phase. The overall consequence of this stereospecific inhibition process can be that the unaffected phase, if less stable, will grow by kinetic control.

For simplicity we consider two distinct and simple cases where the logic of the above arguments apply. In one case the molecules are chiral racemic and which crystallize as a conglomerate, namely yield two enantiomorphous, equienergetic, crystalline phases. Kinetic resolution of the two conglomerate forms may be achieved with the assistance of tailor-made chiral resolved inhibitors which would, in solution, bind enantioselectively and stereoselectively to only one of the enantiomeric crystalline phases and so prevent its eventual growth. This principle has been demonstrated for many systems.[9,10]

Another case involved the preferred precipitation of a crystal with a polar axis at the expense of the dimorph which is centrosymmetric and so nonpolar. The concept is that in crystals with a polar axis all molecules are aligned in the same direction vis-a-vis the polar axis. In centrosymmetric crystals neighboring molecules are arranged in an antiparallel manner. Thus an appropriate tailor-made additive will inhibit growth of the centrosymmetric form at the opposite ends of the crystal, and so prevent its appearance, but will inhibit growth of the polar form only at one end of the crystal. This concept has been demonstrated in two reported systems.[11,12]

7 "Tailor-Made" Surfaces for Promotion of Crystal Nucleation

Crystal nucleation is generally a heterogeneous process. This means that the activation barrier for nucleation is lowered by interaction with foreign surfaces. This process can happen at different levels of specificity, ranging from non specific adsorption to epitaxial growth. One can therefore envisage induced nucleation of desired crystalline structures, even with a specific crystal orientation, by designing appropriate nucleation promotors which match the structure of the crystal on a specific plane. Langmuir monolayers, composed of water-insoluble amphiphilic molecules at the air-water interface, proved to be a very expedient nucleating system for the design of a synthetic surface; being provided by the polar head groups at the amphiphile in contact with the water. It has proven possible to induce, by virtue of a proposed structural match, oriented crystallization of glycine ($^{+}H_3NCH_2CO_2^{-}$) under α-amino acid monolayers[17] such as palmitoyl lysine ($C_{15}H_{31}CONHC_4H_8CHNH_3^{+}CO_2^{-}$). Several crystals other than glycine have been nucleated at the water surface via Langmuir monolayers by virtue of either a (partial structural fit or an electrostatic attraction). These include *p*-hydroxybenzoic acid,[37] silver propionate,[38] NaCl.[39] Mann and coworkers crystallized the vaterite and calcite form of $CaCO_3$ as part of model studies on biological mineralization.[40] Fendler and coworkers applied this method for the growth of nanoparticles of semiconducting CdS and PbSe.[41]

8. Crystalline Self-Assembly of Amphiphilic Monolayers on the Water Surface

Evidence for the formation of self-organized 2-D crystalline aggregates (crystallites) composed of water-insoluble amphiphilic molecules, at the air-water interface, has until recently, been indirect; it has generally been assumed that the degree of lateral ordering in the uncompressed state is very low, if at all existent. The newly developed method of grazing incidnce X-ray diffraction (GIXD) is a powerful tool for exploring this field[43] since it allows for the detection of 2-D crystallites formed by the amphiphilic molecules and for the determination of their size and structure, almost on the atomic level. Using GIXD, the 2-D crystallinity of uncompressed monolayers was shown to be a wide-spread phenomenon. It was demonstrated for long chain fluorocarbon α-amino acids at room temperature[19] and for hydrocarbon chain alcohols,[44] amides, acids and acids bound to divalent ions[45] such as Cd^{2+}, at temperatures below 10°C. The GIXD spectrum of the Cd^{2+} fatty acid monolayer gave direct evidence of an ordered Cd^{2+} counterionic layer bound to the acid. This result is in agreement with induced crystallization of an inorganic salt at the Langmuir monolayer interface, from which it was deduced that the counterion is bound to the head group of the monolayer in an ordered arrangement.[17,19]

Amphiphilic monolayers on the water surface are an ideal vehicle for the study of the effect of solvent on (2-D and 3-D) crystal growth and dissolution. 2-D crystals have an advantage, over

3-D crystals, because solvent molecules, present in the aqueous subphase, come into contact only with the polar head groups of the monolayer on a well-characterized interface. Moreover, GIXD offers the possibility to monitor in real time, changes in crystallinity. We have been able, by GIXD, to demonstrate that solvent can retard, or inhibit, the growth or induce dissolution of monolayer crystallites. Scanning force and cryoelectron microscopies were also applied, for monitoring the dynamics of nano-clusters.

9. Induced Nucleation of Ice.

Pure water can be supercooled to temperatures lower than -20°C. Therefore, the induction or inhibition of the freezing point of ice, in particular through the role of auxiliaries, has far-reaching ramifications for the living and the non-living world. Promotion of ice nucleation, has been exploited in the induced precipitation of rain by silver iodide seeded in clouds.[46] On the other hand, it can result in wide scale damage to non-coniferous plants in temperate climes by frost bacteria[47-49] which nucleate ice at temperatures as high as -3°C. An intriguing example of inhibition involves the activity of antifreeze proteins in the blood serum of fish in polar seas; these proteins lower the freezing point by no more than 2°C, but this small difference is vital for their survival.[50-52]

Ice nucleation has been widely studied, yet the role played by auxiliaries involved in the promotion of ice nucleation, is still unclear. Several mechanisms have been proposed such as a lattice or structural match between the substrate and the nucleated ice crystal[53] or an electric field present in the substrate which orients the water molecules to induce ice nucleation.[54] To unravel the complexity of this process, we designed experiments where such mechanisms can be studied. There is a close epitaxial fit between the *ab* cell axes of hexagonal ice ($a = b = 4.5$Å, $\gamma = 120°$) and of the 2-D cell of the *compressed* monolayer of the aliphatic long chain alcohol $C_{21}H_{43}OH$ ($a = b$ 4.5Å, $\gamma = 113°$) at just above 0°C as determined from GIXD measurements by Dutta and coworkers.[55] We found that uncompressed monolayers of aliphatic long chain alcohols were very efficient ice nucleators,[20,21] in contrast to water-soluble alcohols which are antifreeze agents. By the application of a cryoelectron microscopic methodology it became possible to demonstrate directly epitaxial-like growth of hexagonal ice underneath the monolayer, with the (0001) face attached to the monolayer. The induced freezing point of ice for $C_nH_{2n+1}OH$ with n odd reached an asymptotic temperature just below 0°C for an upper value of $n = 31$; for n even the freezing point reached a plateau of -8°C for n in the upper range of 22 to 30. GIXD measurements on *uncompressed* monolayers of alcohols $n = 30,31$ on water demonstrated formation of ordered clusters[24,56] at a temperature below 8°C (see section 8) with a close lattice match to ice. It was possible to extract the overall packing arrangements of the alcohol monolayers, and to correlate qualitatively the ice nucleation behaviour with chain length and odd-even parity.

GIXD analysis of the 2-D alcohol clusters on water has demonstrated, with chain shortening, the amount of crystalline material is reduced, and the molecular motion of the hydrocarbon chains becomes more pronounced. Indeed a rotator phase was proposed for $C_nH_{2n+1}OH$ (n = 15-13), in keeping with their poor ice nucleating activity.[56]

The GIXD data of the $C_{31}H_{63}OH$ monolayer on water as the latter was cooled to freezing just below 0°C, indicated not only that the monolayer maintained its crystalline integrity, but that the average domain size of many of the induced ice crystals in their *ab* plane was as low as 25Å. Thus if we assume that the "critical" ice nucleus is a hemisphere in shape and of diameter 25Å, than the number of H_2O molecules in such a cluster is in the range of ~100.[57]

The GIXD pattern of the C_{31} and C_{30} alcohols are very similar to each other, indicating very similar packing arrangement. The molecular chains are tilted by an angle of about 8° from the normal to the monolayer plane. Thus we may deduce from the ice nucleating results that the hydrocarbon chain structures for n = 30 and 31 are similar and so the orientation of their -OH groups with respect to the water surface must be different. In order to support this model, the series $C_nH_{2n+1}X\ C_mH_{2m}OH$ ($X = CO_2$, CONH) were synthesized, and their ice nucleating behavior examined. In these amphiphilic systems the polar carboxy ester or the amide group fixes the tilt angle of the hydrocarbon chains. Therefore, the orientation of the -OH groups on the water surface depends only on m, i.e. the even or odd number of methylene groups between the ester or the amide to the OH groups, but not on the odd or even value of ***n***. The two-dimensional crystal structures of the two esters (n = 19; m = 9 and m = 10) which induce ice nucleation at -5°C and -8°C respectively) on water were determined. According to the GIXD data, their structures are indistinguishable. The orientation of the molecular chains in these two systems are the same following to atom-atom potential energy calculations, which then yielding the two different orientations of their hydroxy groups. These studies demonstrated that a stringent structural fit is required between the 2-D net of the alcohol OH groups, which involves also the orientation of the -OH and its lone pair system, and the *ab* layer of hexagonal ice for efficient freezing of water to take place.

10. Self-organization into 3-D dimensional multilayer crystals at the air/water interface.

In the examples discussed above, we considered the self-organization of amphiphilic molecules into crystalline monolayers. Recently we demonstrated, for a variety of systems, the formation of 3-D crystalline clusters at the air/liquid interface. For example, bola-amphiphilic molecules (α,ω alkane diols, or a,ω alkane dicarboxylic acids) were found to self-aggregate into three-dimensional crystallites upon spreading on water.[22-24] The number of layers formed was found to depend on parameters such as the conditions of preparation, percentage of coverage, chain length. The thickness of these crystallites could be measured by scanning force microscopy (SFM) after their deposition on smooth mica surfaces. In these systems it was also possible to demonstrate the role played by "tailor-made" additives on the multilayer formation. For example the multilayer growth of $HO(CH_2)_{22}OH$ was inhibited upon addition of 10% of $C_{21}H_{43}CO_2H$ or $C_{24}H_{49}OH$, resulting in the formation of monolayers and some residual bilayer. GIXD pattern was modeled in a form of a bilayer with monoclinic "space group" symmetry $P2_1$ a.

More recently it could be demonstrated that the alkanes (*n* = 23,24,27,28,29,32,36 and 50) when deposited on water self-aggregate into crystalline films. The longer chain alkanes C_{32}, C_{36} and C_{50} form monolayers in which the chains are aligned normal to the water surface. The packing arrangements of the multilayers for *n* = 23-29 were found to resemble the structures of their macroscopic three dimensional crystals. Analysis of the GIXD patterns, demonstrated that the number of layers increases with chain shortening; for *n* =29 a bilayer was obtained, whereas for *n* = 23 crystallites with more than 20 layers were found. Addition of "tailor-made" growth inhibitors to the spreading solutions of $C_{23}$and C_{24}, in the form of alcohols slightly longer than the alkanes led to bilayer formation.[58]

The structure of the bilayers and multilayers of the alkanes, are identical to the structures of their 3-D corresponding macroscopic crystals. This suggests that the nucleus assumes the structure of the mature crystal at a very early stage.

11. Outlook

We have summarized some ideas and results regarding the various stages of crystal growth, dissolution, and nucleation. These processes are of importance in providing us with possibilities for the construction of new materials. The advent of analytical tools, such as synchrotron X-radiation, the scanning tunneling microscope, and the atomic force microscope should provide new insights for the understanding of these processes.

References

1. This is an expanded abstract of two lectures presented at the school. For further details the reader should allude to two comprehensive recent reviews.[2,3]

2. I. Weissbuch, R. Popovitz-Biro, M. Lahav and L. Leiserowitz, (Lead Article) *Acta Cryst.* **B51**, 115-148 (1995).

3. I. Weissbuch, R. Popovitz-Biro, L. Leiserowitz and M. Lahav, In: 100 Years of the Lock-and-Key Principle, Ed. J.P. Behr, John Wiley & Sons, Chapter 6, 1994, pp. 173-246.

4. S. Weiner, *CRC Crit. Rev. Biochem.* **20**, 365 (1986).

5. H.E. Buckley, *Crystal Growth*, Wiley, New York (1951); A.F. Wells, *Phil. Mag.* **37**, 605 (1946); R. Kern, *Bull. Soc. Fr. Miner. Crystallogr.* **76**, 397 (1953); B.M. Smythe, *Austr. J. Chem.* **20**, 466 (1986).

6. L. Addadi, Z. Berkovitzh-Yellin, I. Weissbuch, J. van Mil, L.J.W. Shimon, M. Lahav, L. Leiserowitz, *Angew. Chem.* **24**, 466 (1986).

7. L. Addadi, Z. Berkovitz-Yellin, N. Domb, E. Gati, M. Lahav, L. Leiserowitz, *Nature* **26**, 21 (1982).

8. Z. Berkovitz-Yellin, J. van Mil, L. Addadi, M. Idelson, M. Lahav and L. Leiserowitz, *J. Am. Chem. Soc.* **107**, 3111 (1985).

9. L. Addadi, S. Weinstein, E. Gati, I. Weissbuch, M. Lahav, *J. Am. Chem. Soc.* **104**, 4610 (1982).

10. D. Zbaida, I. Weissbuch, E. Shavit-Gati, L. Addadi, L. Leiserowitz and M. Lahav, *Reactive Polymers* **6**, 241-253 (1987).

11. E. Staab, L. Addadi, L. Leiserowitz, M. Lahav, *Adv. Mat.* **2**, 40 (1990).

12. I. Weissbuch, L. Leiserowitz and M. Lahav, *Adv. Mat.* **6**, 953-956 (1994).

13. L. Addadi, Z. Berkovitz-Yellin, I. Weissbuch, M. Lahav and L. Leiserowitz, in: *Topics in Stereochemistry*, Wiley, New York, Vol. **16**, 1-85 (1986).

14. I. Weissbuch, M. Lahav, L. Leiserowitz, G.R. Meredith, H. Vanherzeele, *Chemistry of Materials*, **1**, 114 (1989).

15. L.J.W. Shimon, M. Lahav, L. Leiserowitz, *J. Am. Chem. Soc.* **107**, 3375 (1985); L.J.W. Shimon, M. Lahav, L. Leiserowitz, *Nouv. J. Chim.* **10**, 723 (1986).

16. F.C. Wireko, L.J.W. Shimon, Z. Berkovitz-Yellin, M. Lahav and L. Leiserowitz, *J. Phys. Chem.* **91**, 471 (1987).

17. E.M. Landau, S. Grayer-Wolf, M. Levanon, L. Leiserowitz, M. Lahav and J. Sagiv, *J. Am. Chem. Soc.* **111**, 1436 (1989).

18. E.M. Landau, S. Grayer-Wolf, J. Sagiv, M. Deutsch, K. Kjaer, J. Als-Nielsen, L. Leiserowitz and M. Lahav, *Pure & Appl. Chem.* **61**, 673 (1989).

19. D. Jacquemain, S. Grayer-Wolf, F. Leveiller, M. Lahav, L. Leiserowitz, M. Deutsch, K. Kjaer and J. Als-Nielsen, *J. Am. Chem. Soc.* **112**, 7726 (1990).

20. M. Gavish, R. Popovitz-Biro, M. Lahav and L. Leiserowitz, *Science* **250**, 973 (1990).

21. R. Popovitz-Biro, J.L. Wang, J. Majewski, E. Shavit, L. Leiserowitz and M. Lahav, *J. Am. Chem. Soc.* **116**, 1179 (1994).

22. R. Popovitz-Biro, J. Majewski, L. Margulis, S. Cohen, L. Leiserowitz and M. Lahav, *J. Phys. Chem.* **98**, 4970 (1994).

23. R. Popovitz-Biro, J. Majewski, L. Margulis, S. Cohen, L. Leiserowitz and M. Lahav, *Adv. Mater.* **6**, 956 (1994).

24. J. Majewski, R. Edgar, R. Popovitz-Biro, K. Kjaer, W.G. Bouwman, J. Als-Nielsen, M. Lahav and L. Leiserowitz, *Angew. Chem. Int. Ed. Engl.* **34**, 649 (1995).

25. I. Weissbuch, L. Addadi, Z. Berkowitz-Yellin, E. Gati, S. Weinstein, M. Lahav and L. Leiserowitz, *J. Am. Chem. Soc.***105**, 6615 (1983).

26. M. Vaida, L.J.W. Shimon, Y. Weisinger-Lewin, F. Frolow, M. Lahav, L. Leiserowitz and R.K. McMullan, *Science* **241**, 1475 (1988).

27. Y. Weisinger-Lewin, F. Frolow, R.K. McMullan, T.F. Koetzle, M. Lahav and L. Leiserowitz, *J. Am. Chem. Soc.* **111**, 1035 (1989).

28. J.M. McBride, S.B. Bertman, *Angew. Chem. Int. Ed. Engl.* **28**, 330 (1989); *Ibid*, **28**, 1377 (1989).

29. P. Bennema, G. Gilmer, in: *Crystal Growth, An Introduction.* Hartman, P. Ed., North Holland, Amsterdam, 274 (1973); M. Elwenspoek, P. Bennema and J.P. van der Eerden, *J. Cryst. Growth* **83**, 297 (1987).

30. J.R. Bourne, R.J. Davey, *J. Cryst. Growth* **36**, 278 (1976).

31. A.F. Wells, *Phil. Mag.* **37**, 184 (1946); A.F. Wells, *Discuss. Faraday Soc.* **5**, 197 (1949).

32. Z. Berkovitz-Yellin, *J. Am. Chem. Soc.* **107**, 8239 (1985); Z. Berkovitch-Yellin, I. Weissbuch and L. Leiserowitz, in: *Morphology and Growth Units of Crystals*, pp. 247-278, published by TERRAPUS, Tokyo, 1989.

33. S.N. Black, R.J. Davey, T.D. McLean, *Mol. Cryst. Liq. Cryst. Inc. Nonlin. Opt.* **161**, 283 (1989).

34. L.J.W. Shimon, M. Vaida, L. Addadi, M. Lahav and L. Leiserowitz, *J. Am. Chem. Soc.***112**, 6215 (1990).

35. J.L. Wang, M. Lashav and L. Leiserowitz, *J. Phys. Chem.* **96**, 15 (1992).

36. J.L. Wang, M. Lahav, L. Leiserowitz, *Angew. Chem. Int. Ed.* **30**, 696 (1991).

37. I. Weissbuch, G. Berkovitz, M. Lahav and L. Leiserowitz, *J. Am. Chem. Soc.* **112**, 5874 (1990).

38. I. Weissbuch, J. Majewski, K. Kjaer, J. Als-Nielsen, M. Lahav and L. Leiserowitz, *J. Phys. Chem.* **97**, 12848 (1993).

39. E.M. Landau, R. Popovitz-Biro, M. Levanon, L. Leiserowitz and M. Lahav, *Mol. Cryst. Liq. Cryst.* **134**, 323 (1986).

40. S. Mann, B.R. Heywood, S. Rajam and J.D. Birchall, *Nature* **334**, 692 (1988); S. Mann, *Nature* **332**, 119 (1988).

41. X,K, Zhao, J. Yang, L.D. McCornick and J.H. Fendler, *J. Phys. Chem.* ***96***, 9933 (1992).

42. S. Grayer-Wolf, L. Leiserowitz, M. Lahav, M. Deutsch, K. Kjaer and J. Als-Nielsen, *Nature* **328**, 63 (1987).

43. For two recent reviews see: J. Als-Nielsen, D. Jacquemain, K. Kjaer, F. Leveiller, M. Lahav and L. Leiserowitz, *Physics Reports*, **246**, 251 (1994); D. Jacquemain, S. Grayer-Wolf, F. Leveiller, M. Deutsch, K. Kjaer, J. Als-Nielsen, M. Lahav and L. Leiserowitz, *Angew. Chem.* **31**, 130-152 (1992).

44. D. Jacquemain, F. Leveiller, S. Weinbach, M. Lahav, L. Leiserowitz, K. Kjaer and J. Als-Nielsen, *J. Am. Chem. Soc.* **113**, 7684-7691 (1991).

45. F. Leveiller, D. Jacquemain, M. Lahav, L. Leiserowitz, M. Deutsch, K. Kjaer and Als-Nielsen, *Science* **252**, 1532 (1991).

46. B. Vonnegut, *J. Appl. Phys.* **18**, 593 (1947); M.L. Corrin, J.A. Nielsen, *J. Phys. Chem.* **72**, 643 (1968).

47. R.C. Schnell, C. Vali, *Nature* **263**, 163 (1972).

48. L.R. Maki, E.L. Galyon, M. Chang Chien, D.R. Colwell, *Appl. Microbiol.* **28**, 456 (1974).

49. S.E. Lindow, D.C. Arny, C.D. Upper, *Phytopathology* **68**, 523 (1978).

50. A.L. Devries, *Philos. Trans. R. Soc. London,* Ser. **B304**, 575 (1984).

51. R.E. Fenney, T.S. Burcham, Y. Yeh, *Ann. Rev. Biophys. Biophys. Chem.* **15**, 59 (1986).

52. D.S.C. Yang, M. Sax, A. Chakrabartty, C.L. Hew, *Nature* **333**, 232 (1988).

53. N.Y. Fletcher, *The Chemical Physics of Ice*. Cambridge Uni. Press, Cambridge, pp. 73-103 (1970).

54. H.R. Prappacher, *Z. Angew. Math. Phys.* **14**, 590 (1963).

55. S.W. Barton, A. Goudot, O. Bouloussa, F. Rondelez, B. Lin, F. Novak, A. Acero, S.A. Rice, *J. Chem. Phys.* **96**, 1343 (1992).

56. J. Majewski, R. Popovitz-Biro, W.G. Bouwman, K. Kjaer, J. Als-Nielsen, M. Lahav and L. Leiserowitz, *Chem. Eur. J.* **1**, 304 (1995).

57. J. Majewski, R. Popovitz-Biro, K. Kjaer, J. Als-Nielsen, M. Lahav and L. Leiserowitz, *J. Phys. Chem.* ***98***, 4087 (1994).

58. S. Weinbach, I. Weissbuch, K. Kjaer, W. Bouwman, J. Als-Nielsen, M. Lahav and L. Leiserowitz, *Adv. Mat.* **7**, 857 (1995).

SYNTHETIC RECEPTORS : A MODULAR APPROACH TO LARGE STRUCTURES

I. HIGLER, W. VERBOOM and D. N. REINHOUDT
Laboratory of Organic Chemistry
University of Twente
P.O. Box 217
7500 AE Enschede
The Netherlands

ABSTRACT. Combination of medium-sized building blocks comprises a new strategy for the synthesis of artificial receptor molecules. These building blocks should have relatively rigid structures to which functional groups can be attached, such as calix[4]- and -[6]arenes, resorcin[4]arenes, porphyrins, and cyclodextrins. Combination of a β-cyclodextrin with a calix[4]arene moiety resulted in water soluble receptor molecules. Large multifunctional receptors have been prepared by the covalent combination of two calix[4]arenes and one Zn-tetraarylporphyrin. The controlled conformation of *p-tert*-butylcalix[6]arenes which are alternately substituted with methoxy groups, was used to synthesize a cryptocalix[6]arene consisting of a calix[6]arene and a cyclotriveratrylene unit. A new type of diastereoisomerism (*carceroisomerism*) was found during the synthesis of carceplexes constructed from one calix[4]arene and one resorcin[4]arene. In a (2+2)-mode of coupling, the combination of calix[4]arenes and resorcin[4]arenes gave a large rigid receptor molecule of nanosize dimensions *(holand)*. The combination of two calix[4]arenes with one resorcin[4]arene resulted in three different compounds (*endo-endo, endo-exo*, and *exo-exo*) with large hydrophobic surfaces to which steroids can be complexed.

G. Tsoucaris et al. (eds.), Crystallography of Supramolecular Compounds, 347–368.

1. Introduction

Nature constructs receptors by the combination of a limited number of building blocks. Amino acids are combined to proteins, nucleosides to DNA and RNA, and monosaccharides to carbohydrates. Systematic variation of the monomer sequence gives access to an almost infinite number of difficult receptors. This is achieved at the expense of a high molecular weight which might be even larger when the biologically active species is composed of several sub-units that are organized by non-covalent bonds.

Most synthetic receptors have been prepared by *de novo* synthesis using modern synthetic methodologies, which allow an almost unlimited variation. This strategy focuses on the complementarity of functional groups of receptor and guest. The drawback is that for each individual guest molecule a new synthetic pathway has to be developed.

In recent years we have developed a new strategy for the synthesis of artificial receptors which is a compromise between the two extremes described above. We use medium-sized, relatively rigid molecules to which functional groups can be attached. Calix[4]- and -[6]arenes, resorcin[4]arenes, and cyclodextrins have the required structures for such building blocks.

In addition to coupling via covalent bonds these building blocks can be connected by non-covalent bonds to give *objects* with molecular weights of more than 5000 D.

In this article we will illustrate our new strategy with a few examples of the synthesis of receptor molecules by the covalent linkage of medium-sized, relatively rigid molecules to which functional groups can be attached. Examples of the combination of calix[4]arenes with macrocycles and building blocks such as β-cyclodextrins, porphyrins, and resorcin[4]arenes are presented.

2. Combination of Calix[4]arenes with Macrocycles

Calix[4]arenes have been combined with polyethylene glycols, terphenyls, metallosalophenes and metallosalenes to give calixcrown ethers, calixspherands, anion, and ditopic receptors, respectively.

2.1. CALIXCROWN ETHERS

Calixcrown ethers can be synthesized by the selective bridging of *p-tert*-butylcalix[4]arene with glycol derivatives. The obtained 1,3-dihydroxy-*p-tert*-butylcalixcrown ethers can be rigidified by subsequent alkylation of the two remaining phenolic groups with substituents larger than C_2H_5; the resulting compounds can be formed as (a mixture of) different stereoisomers. All ligands are selective towards potassium cations with a surprisingly high K^+/Na^+ selectivity, as high as 1.18 x 10^4 for **1** (Chart 1). The partial cone isomer is in all cases the preferred conformation for binding of cations and showed the highest free energy of complexation [1].

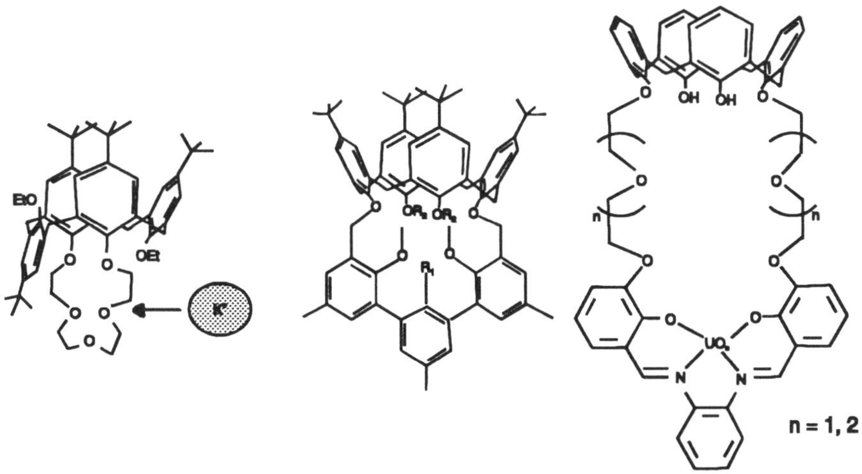

(1) **(2)** $R_1 = CH_2CH_3$, $R_2 = CH_3$ **(4)**

(3) $R_1 = CH(CH_3)_2$, $R_2 = CH_3$

Chart 1

2.2. CALIXSPHERANDS

Calixspherands can be synthesized by reaction of *p-tert*-butylcalix[4]arenes with terphenyl derivatives and subsequent alkylation of the remaining hydroxyl groups (Chart 1). Appropriately substituted calixspherands form kinetically stable complexes with Na^+, K^+, and Rb^+, the complexes being present in the partial cone conformation. The kinetic stability was determined both by ^1H NMR spectroscopy and a method based on the exchange of radioactive rubidium or sodium in the complexes for nonradioactive sodium in different solvents. Both methods showed that the kinetic stability of the different complexes is strongly increased when the size of the group on the central aromatic ring of the *m*-terphenyl is increased. The latter showed that the rate of decomplexation is the rate-limiting step in the exchange of rubidium in the complex for sodium present in solution. In $CDCl_3$ saturated with D_2O, the Rb^+ complexes exhibit half-lives of decomplexation, of 139 h and 180 d, for $[\mathbf{2}\cdot Rb]^+$ and $[\mathbf{3}\cdot Rb]^+$, respectively. The complexes with Na^+ and K^+ have half-lives of decomplexation of several years [2].

The cavity of the calixspherands is also complementary to Ag^+. Although the calixspherands lack the soft donor ligands (sulfur and nitrogen) which are preferred for the complexation of the soft silver cation, the calixspherands form kinetically stable complexes with Ag^+, in $CDCl_3$ saturated with D_2O. The half-lives of decomplexation are 51 h ($[\mathbf{2}\cdot Ag]^+$) and 131 h ($[\mathbf{3}\cdot Ag]^+$) [3].

2.3. CALIXSALOPHEN CROWN ETHERS

Macrocycles that contain a uranyl containing salen or salophen unit are known as complexing agents for urea or other neutral molecules that contain a nucleophilic group [4]. Calixsalophen crown ether (**4**, Chart 1) was obtained by macrocyclization of a dialdehyde functionalized calix[4]arene and a suitable diamine. The resulting receptor molecules are highly lipophilic and consequently useful as carriers for urea in supported liquid membranes. Furthermore, they have a phenolic group on either side of the crown ether ring, which can in principle be used for the incorporation of functional groups, either as additional binding sites or for catalyzing reactions of a complexed substrate [5].

2.4. CALIX[4]ARENE BASED BIFUNCTIONAL RECEPTOR

By combination of anionic and cationic binding sites in one molecule, bifunctional receptors can be obtained which are capable to complex simultaneously anionic and cationic species. Calix[4]arenes containing four preorganized esters at the lower rim complex alkali metal cations with a high selectivity for Na^+ [6]. Neutral metalloclefts and metallomacrocycles containing both an immobilized Lewis acidic UO_2-center and amido C(O)NH groups as additional binding sites, are excellent receptors for anions with a high selectivity for dihydrogen phosphate $H_2PO_4^-$ [7].

Using calix[4]arene as a molecular platform, both four ester groups and a uranyl containing salen moiety were attached to one calix[4]arene to give a neutral calix[4]arene-based bifunctional receptor which contains both anionic and cationic sites, see **5** in Chart 2.

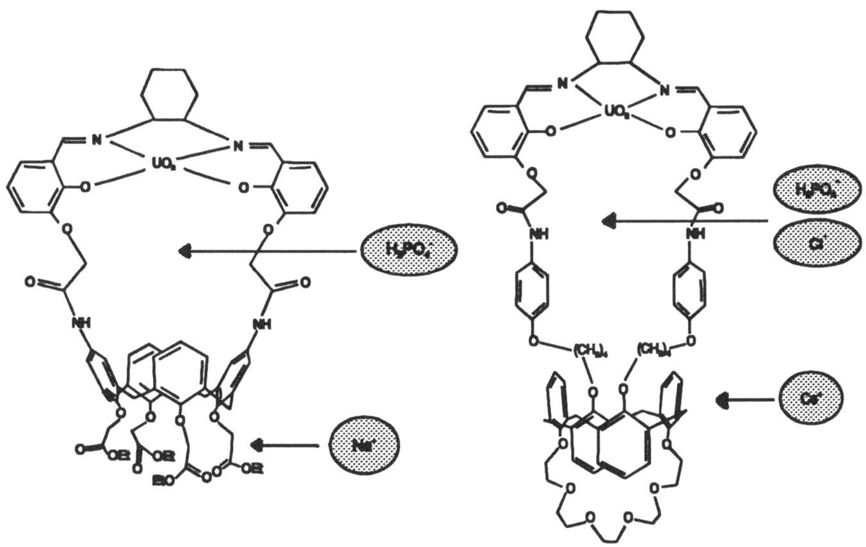

(5) **(6)**

Chart 2

A study of the binding ability of this receptor showed that it selectively binds dihydrogen phosphate $H_2PO_4^-$; only weak binding was observed for the divalent HPO_4^{2-} anion. From ^1H NMR dilution experiments, an association constant of 3.5×10^2 M^{-1} in DMSO-d_6 was calculated . In the positive FAB mass spectrum of the 1:1 complex of the ditopic receptor with NaH_2PO_4 an intense peak corresponding to the [host + Na^+]$^+$ was observed, while the corresponding negative FAB mass spectrum of the same sample yielded an intense peak for [host + $H_2PO_4^-$]$^-$. This proves the complexation of both cation and anion in one bifunctional receptor molecule [8].

Our strategy to combine the anion and the cation binding sites in one receptor molecule was also the basis of a neutral bifunctional receptor which can be used for carrier-assisted cotransport in which the anion and cation of a hydrophilic salt are bound and transported simultaneously through a membrane. As in the previous example, the combination of a Lewis acid UO_2-center and amido C(O)NH moieties was used as receptor site for dihydrogen phosphate ($H_2PO_4^-$) or chloride anions. The calix[4]arene crown-6 (1,3-alternate) fragment was used for the selective complexation of cesium cations (**6**, Chart 2). The transport of CsCl by the monofunctional carriers (anion and cation) exhibits low flux values (0.07-0.42 x 10^{-7} mol m^{-2} s^{-1}). This can be expected because when only one of the ionic species is complexed, the uncomplexed chloride cannot sufficiently penetrate the lipophilic membrane. For the bifunctional carrier with CsCl a significant flux (1.20 x 10^{-7} mol m^{-2} s^{-1}) was observed. This carrier showed a surprisingly low flux (0.89 x 10^{-7} mol m^{-2} s^{-1}) for $CsNO_3$, although NO_3^- is more lipophilic than Cl^-. This proves that *both* anion and cation binding sites are involved in the complexation and that the presence of only an anion or a cation binding site in the receptor molecule is not sufficient for effective transport of a *hydrophilic* salt such as CsCl. This also suggests a preference of hydrophilic CsCl over lipophilic $CsNO_3$ [9].

3. Combination of Calix[4]arenes with Other Building Blocks

More recently we have investigated combinations of β-cyclodextrins and calix[4]arenes, of porphyrins and calix[4]arenes, of cyclotriveratrylene and calix[6]arenes and of resorcin[4]arenes and calix[4]arenes.

3.1. CALIX[4]ARENES COMBINED WITH β-CYCLODEXTRINS

Cyclodextrins are known as a unique group of naturally occurring cyclic D-glucose oligomers, capable of the complexation of hydrophobic guest molecules in aqueous solution by predominantly hydrophobic interactions [10].

Modification of β-cyclodextrins with a calix[4]arene moiety at the secondary face resulted in receptor molecules which are very effective hosts for the fluorescent dyes 1-anilino-8-naphthalenesulphonate (ANS, **(7)** in Chart 3) and 2-*p*-toluidino-6-naphthalenesulphonate (TNS, **(8)** in Chart 3) with association constants much higher than reported thus far in the literature for cyclodextrins capped at the primary or secondary hydroxyl face. The strongly increased binding capacity is attributed to the additional shielding of the guest by the upper rim of the calix[4]arene which can cover the secondary hydroxyl face of the cyclodextrin cavity [11].

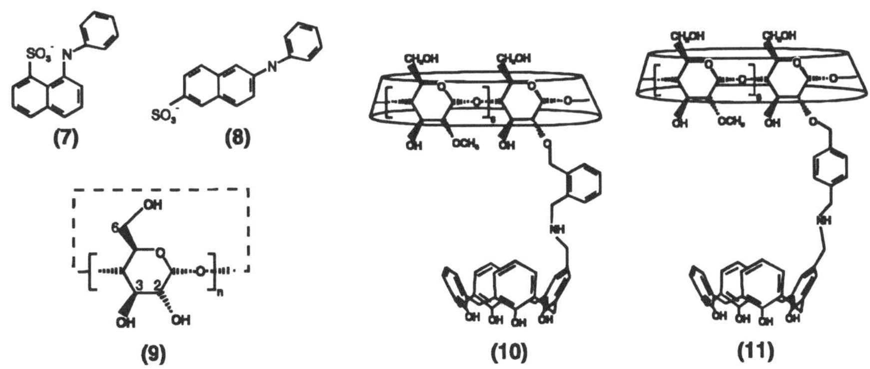

Chart 3

For the combination of calix[4]arenes with cyclodextrins it is necessary to monofunctionalize the cyclodextrins at the secondary face. Two methods have been reported for this monofunctionalization. In the first method a functional group is introduced at the C(3) position (see **9** in Chart 3) of the cyclodextrin by first tosylation of a secondary hydroxyl group of a silylated cyclodextrin and subsequent nucleophilic ring opening of the intermediate cyclodextrin mannoepoxide. The second method comprises the direct and selective introduction of functional groups at the secondary hydroxyl face of silylated cyclodextrins via monoalkylation of the C(2)-oxyanion, leaving the configuration of all glucose units in the cyclodextrin unaffected. The second method was also followed in the synthesis of a secondary face tethered cyclodextrin dimer and for the synthesis of two novel cyclodextrin-calix[4]arene host molecules. For the latter compounds, silylated β-cyclodextrins, functionalized with an *ortho*- or *para*-aminoxylyl spacer, were coupled to a calix[4]arene monoformyl derivative by reductive amination. After desilylation of the cyclodextrin residues two water soluble cyclodextrin-calix[4]arene receptors (**10** and **11** in Chart 3) were obtained of which the opening of the hydrophilic cavity at the secondary hydroxyl face is flexibly shielded by the aryl residues of the calix[4]arene moiety [12].

3.2. CALIX[4]ARENES COMBINED WITH PORPHYRINS

Large multifunctional receptors for molecular recognition have been prepared by covalent combination of two calix[4]arenes and one Zn-tetraarylporphyrin. Zn-porphyrins are being used in biomimetic chemistry for recognition of biologically important molecules such as nucleotides and amino acids.

Covalent combination of porphyrin and calix[4]arene building blocks may lead to receptors with multifunctional complexing properties. Two biscalix[4]arene-Zn-tetraarylporphyrins were prepared by the condensation of pyrrole with 1,3-diformylcalix[4]arenes. The aldehyde moieties were immobilized either at the *lower* or the *upper* rim of the calix[4]arene platform, (**12** and **13** (M=Zn) in Chart 4). These receptors possess an active metallocenter which is able to complex anions and neutral species, and two lipophilic calix[4]arene moieties in a large preorganized cavity [13].

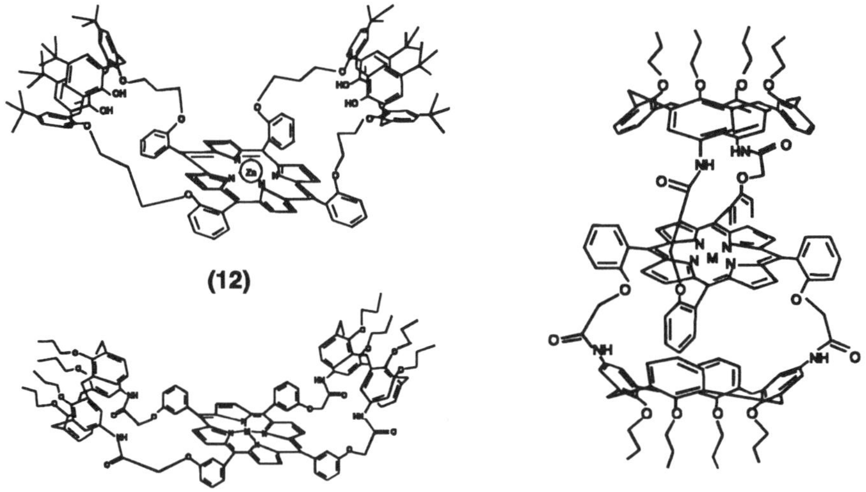

(12) (13) (14)

Chart 4

The active metalloporphyrin core in biscalix[4]arene-Zn-porphyrin **14** (M=Zn) is covalently incorporated into a hydrophobic and rigid *egg-shaped* biscalix[4]arene cavity. The difference with the biscalix[4]arene-Zn-tetraarylporphyrins **12** and **13** is that in **14** the calix[4]arenes are coupled via distal positions, and in the former cases via the proximal positions of the porphyrin. For this reason the core of biscalix[4]arene-Zn-porphyrin **14** provides a more effective shielding and encapsulation of a substrate [14].

3.3. CRYPTOCALIX[6]ARENES

Since calix[6]arenes are even more flexible than calix[4]arenes, capping would provide a method to obtain conformationally more defined calix[6]arene derivatives.

In a detailed study of the conformational properties and dynamic behavior of calix[6]arenes, we found that *p-tert*-butylcalix[6]arenes which are alternately substituted with methoxy groups and bulky substituents predominantly exist in a flattened cone conformation [15]. The *syn* arrangement of the bulky substituents in space provides an ideal stereochemistry to cap the lower rim of calix[6]arene via an intramolecular cyclotrimerization. Veratryl alcohol can be easily derivatized and cyclotrimerized to cyclotriveratrylenes in reasonable to good yields [16].

Cryptocalix[6]arenes **15** (Chart 5) containing both a calix[6]arene and a cyclotriveratrylene unit, were prepared via three-point capping of calix[6]arenes to give compounds with an increased rigidity and a well-defined conformational and dynamic behavior.

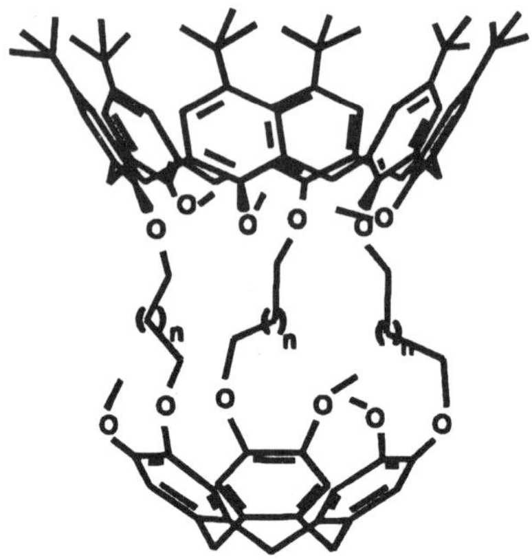

(15)

Chart 5

The ^{1}H NMR spectra of these cryptocalix[6]arenes clearly prove the C_3 symmetry of these molecules. Low temperature ^{1}H NMR spectra showed the existence of a minor conformer beside the major C_3 conformer, whose resonances coalesce upon raising the temperature. In the minor conformer one anisole moiety has rotated in such a way that its *tert*-butyl group is located in the cavity [17].

4. Combination of Calix[4]arenes with Resorcin[4]arenes

Coupling reactions of calix[4]arenes and modified resorcin[4]arenes have also been investigated. Direct combination of *upper rim* tetrafunctionalized calix[4]arenes (with alkoxy groups at the *lower rim*) with tetrafunctionalized resorcin[4]arenes is not possible because of the flexibility of the calix[4]arenes. Reaction of mono(chloroacetamido)calix[4]arene with a tetrahydroxycavitand gave the 1:1 coupled product (calix-resorcinarene **16** in Chart 6) in 61% yield. Combination of an *upper rim* 1,3-difunctionalized calix[4]arene with a tetrahydroxycavitand afforded predominantly the 2:1 calix-resorcinarene **17** (Chart 6) in 47% yield [18].

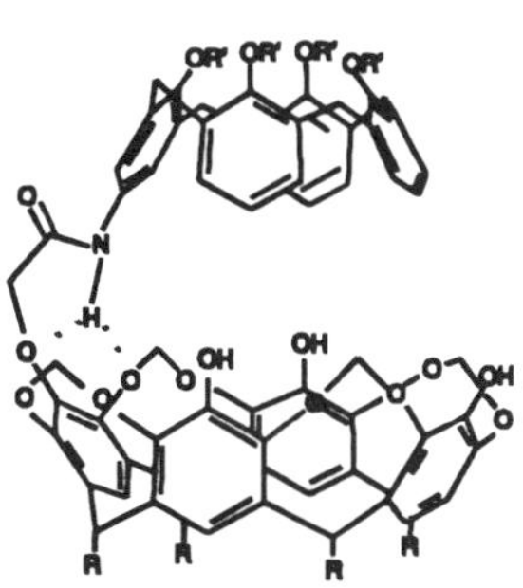

(16)

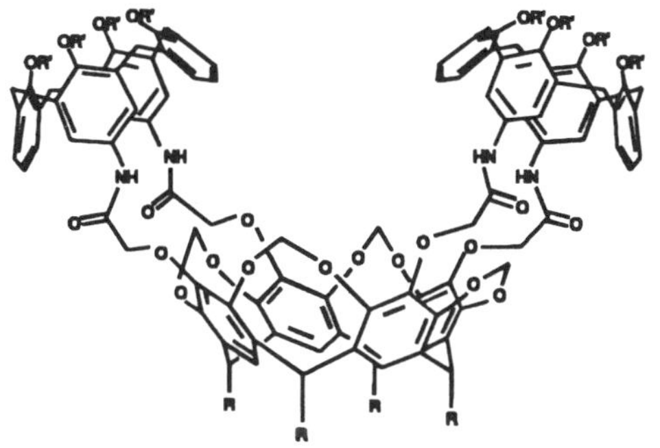

(17)

Chart 6

The combination of tetrahydroxycavitand and *upper rim* 1,2-difunctionalized calix[4]arenes is possible in different ratios depending on the reaction conditions.

4.1. CALIX[4]ARENE BASED CARCEPLEXES

Carcerands are characterized by their ability to trap guest molecules in their interior space [19]. During synthesis they form carceplexes by capturing guest molecules from the medium that cannot leave the cavity without the rupture of at least one covalent bond of the host. Carcerands are usually formed by combination of two resorcinarene-based cavitands; these carcerands have D_{4h} symmetry. Different orientations of the guest molecule inside such cavities do not lead to diastereomeric structures due to the overall symmetry of the carceplex. However, a carceplex constructed from both a calix[4]arene and a resorcin[4]arene has C_{4v} symmetry which could lead to diastereomeric structures caused by different orientations of the guest [20].

Stepwise 1:1 coupling of the respective building blocks gave calix[4]arene-based carceplexes which possess a non-symmetrical cavity that is completely closed. In this way the calix[4]arene-based carceplexes with *N,N*-dimethylformamide (DMF), 1-methyl-2-pyrrolidinone (NMP) **(18)**, and *N,N*-dimethylacetamide (DMA) as guests were isolated (Chart 7).

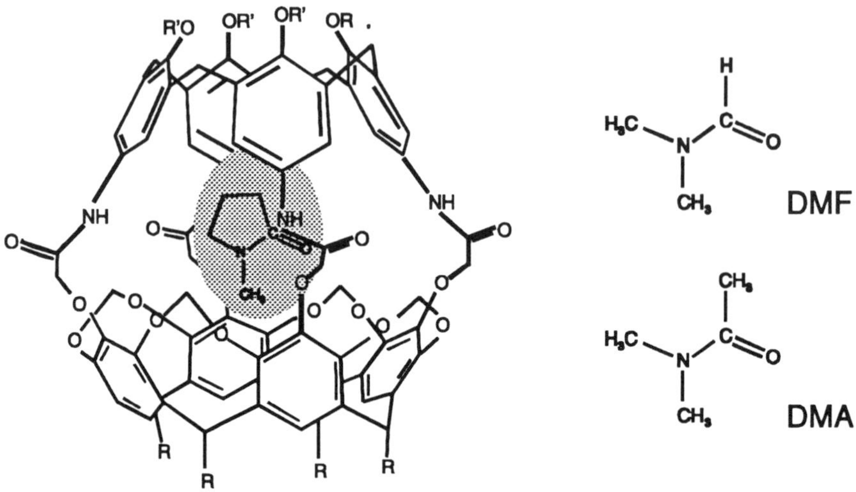

(18) carceplex • NMP

Chart 7

These carceplexes contain one guest molecule inside the cavity and they can be present as two different diastereomers (*carceroisomerism*). NMR and computational studies have revealed a good relation between experimental and calculated activation energies of the isomerization process [21].

4.2. HOLANDS

In a (2+2)-mode of coupling, the calix[4]arene and resorcin[4]arene building blocks give a large, rigid receptor molecule. This molecule **19**, a holand, can be synthesized starting either from the *endo* 1:1 isomer **20** or the *endo-endo* 2:1 isomer **21** (see Chart 8).

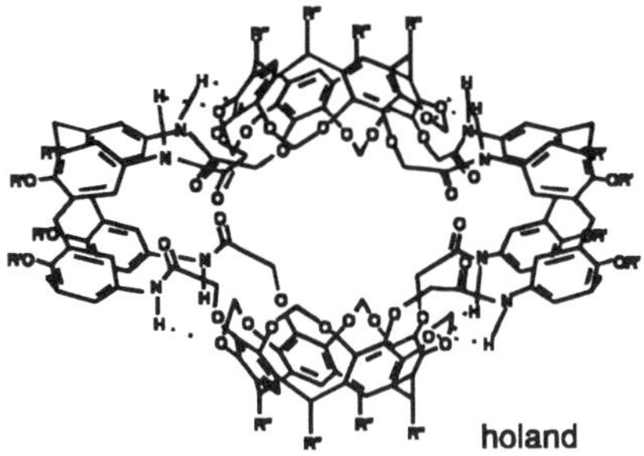

holand

(19)

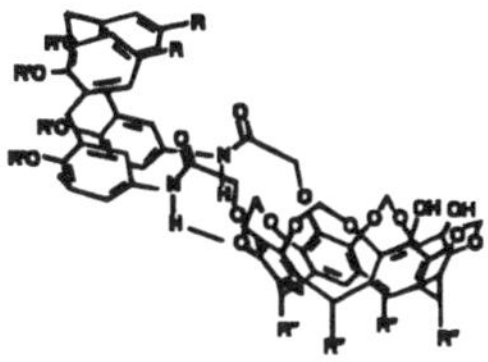

1:1 endo

(20)

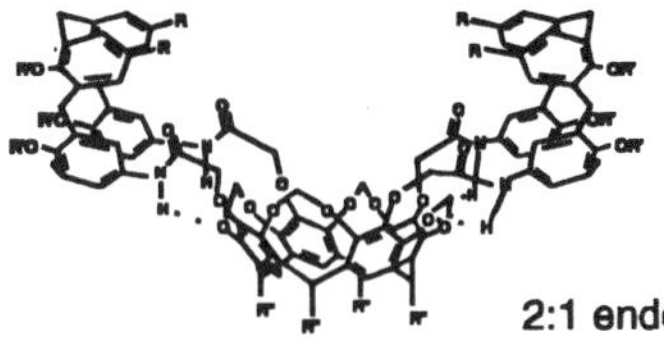

2:1 endo-endo

(21)

Chart 8

The formation of the 2:2 structure was proven by FAB mass spectrometry; its ^1H NMR spectrum reflects the high degree of symmetry as expected for the holand. For its size, the molecule is extremely rigid: the calix[4]arene and cavitand moieties, which are intrinsically rigid, are connected by highly preorganized spacers. The holand contains a cavity of nanosize dimensions. According to CPK models, the axes are 1.5 and 2.0 nm and the molecule has a calculated internal volume of approximately 1.0 nm^3 (1000 $Å^3$). During a 50 ps dynamics simulation (at 300 K in a $CHCl_3$ box with 46 Å edges) of the energy-minimized structure of this molecule, four chloroform molecules entered the cavity and stayed there during the entire simulation without changing the shape of the cavity, supporting the rigid structure of the holand [20].

4.3. SYNTHETIC RECEPTORS THAT COMPLEX STEROIDS

The chemistry of synthetic receptors for a variety of cations, anions, and small neutral molecules is well known, but the synthesis of receptor molecules for larger guest molecules is relatively unexplored. The holand, with a large and rigid cavity of nanosize dimensions, was studied for suitable guest molecules using the computer simulation program DOCK. This systematic search revealed an excellent fit for a number of steroids, in particular prednisolone-21-acetate **(22)** (see Chart 9).

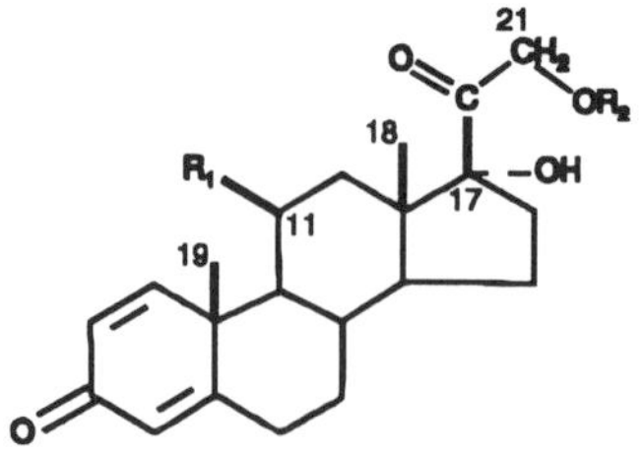

	R_1	R_2
(22) prednisolone acetate	-OH	$-C(O)CH_3$
(23) prednisolone	-OH	-H
(24) prednisone acetate	=O	$-C(O)CH_3$

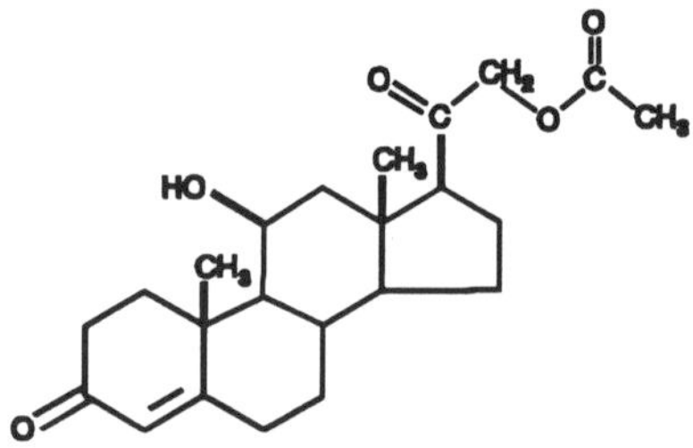

(25) Corticosterone acetate

Chart 9

The complexation of steroids in *apolar* solvents has hardly been studied, in contrast with that in polar solvents. The complexation of steroids in aqueous solution by water soluble azacyclophanes [22] or cyclodextrins [23] is well known and the stability and selectivity is mainly governed by hydrophobic interactions.

Coupling of two calix[4]arenes with one resorcin[4]arene gives three different isomers: *endo-endo* **21**, *endo-exo* **26**, and *exo-exo* **27** (see Charts 8 and 10). These molecules have a large hydrophobic surface (which is in the case of **21** a part of the cavity of the holand) to which steroids can be complexed [24].

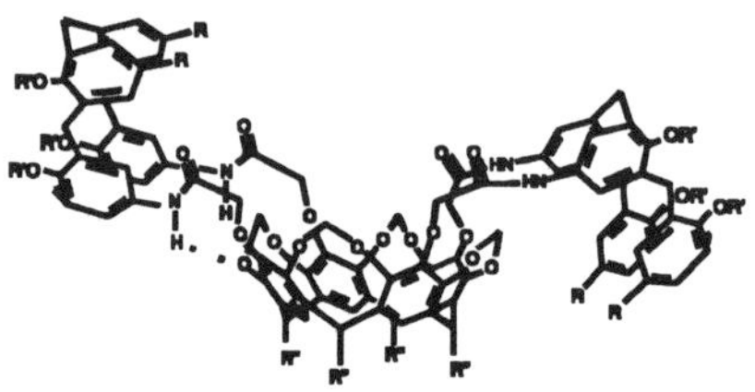

(26) endo-exo

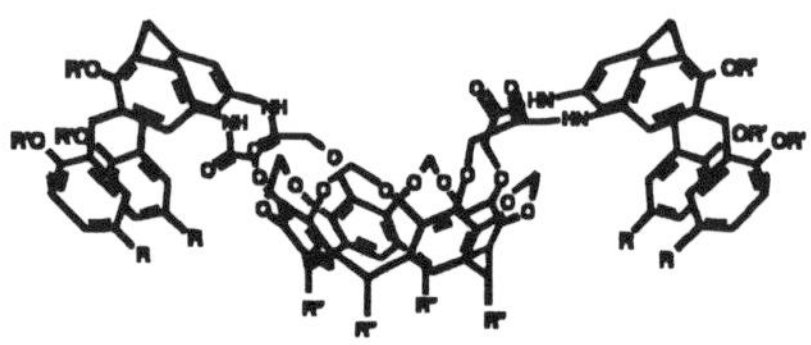

(27) exo-exo

Chart 10

Upon the addition of prednisolone-21-acetate **(22)** to solutions of these receptor molecules (**21**, **26**, or **27**, with R=H) in $CDCl_3$ at 25 °C, several host proton signals, *viz.* the amide resonances, split into two signals of equal intensity. The observed splitting is larger for the *exo* amide protons (*ca.* 0.4 ppm) than for the *endo* amide protons (*ca.* 0.1 ppm). The signals of prednisolone-21-acetate **(22)** show considerable shifts. The singlet for the acetyl methyl group at δ 2.0 shifts and becomes much broader, even when the host is present in low concentrations (ratio 1:10). The AB quartet of the C-21 methylene group shifts upfield by *ca.* 0.2 ppm and also the singlets for both methyl groups (C-18 and C-19) show upfield shifts. The observed splitting of the amide proton signals of the receptor molecules upon complexation of prednisolone-21-acetate **(22)** is due to the chirality of the guest molecule. As a result of the complexation, enantiotopically related protons in the free host become diastereotopic in the chiral complex. However, in the case of the *endo-endo* conformer and the *exo-exo* conformer, which both exhibit C_{2v} symmetry, the homotopic amide protons give rise to only one signal as a result of fast exchange between the free host and the complex on the ^{1}H NMR chemical shift timescale.

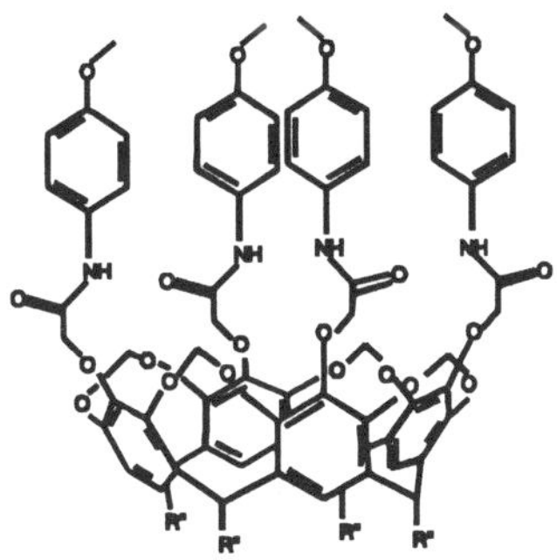

(28)

Chart 11

Prednisolone-21-acetate **(22)** is complexed by receptors **21**, **26**, and **27** with association constants of 4.3 - 8.3 x 10^{2} M^{-1} in $CDCl_3$ at room temperature. The relatively small difference in association constant between *endo-endo* **21** (R=H) and *exo-exo* **27** (R=H) suggests that the upper rim cavity of the calix[4]arene fragments contributes little to the binding. In order to prove that the calix[4]arene fragments in these receptor molecules play a crucial role in the complexation of prednisolone-21-acetate **(22)**, a cavitand carrying four *para*-methoxyphenyl-aminocarbonyl-methoxy substituents **28**, was synthesized (see Chart 11).

Neither the signals of the host nor the guest signals shifted upon the addition of up to 10 equivalents of prednisolone-21-acetate **(22)** to a solution of host in $CDCl_3$. This unambiguously proves that complexation of prednisolone-21-acetate **(22)** by the diastereoisomeric 2:1 products is not simply a result of the presence of four amide moieties at the upper rim of a cavitand, but that preorganization of the amide spacers due to the presence of the calix[4]arene fragments is at least partially responsible for the observed complexation.

By comparison with other steroids it appeared that at least three functional groups are involved in complexation. In order to determine which functionalities in the guest favor complexation by these receptors, the related corticosteroids prednisolone **(23)**, prednisone-21-acetate **(24)** and corticosterone-21-acetate **(25)** were also studied. Addition of prednisolone **(23)** to a solution of any of the 2:1 products in $CDCl_3$ did not give rise to any significant shift of guest or host protons. With prednisone-21-acetate **(24)**, having a keto function at C-11 instead of a hydroxyl group, the splitting of the amide proton signals was absent in case of the *endo-endo* and strongly diminished in the cases of the *endo-exo* and *exo-exo* isomers ($\Delta\delta < 0.1$ for the *exo* amide protons). Significant shifts of guest proton signals were not observed for any of the three diastereoisomers. Very similar results were obtained with corticosterone-21-acetate **(25)** a steroid that lacks the hydroxyl group at C-17.

These results show that at least three functional groups in prednisolone-21-acetate **(22)** are involved in complexation. First of all, the acetoxy group at C-21 that can interact both *via* CH-π interactions of the slightly acidic acetyl group with the aromatic rings of the cavitand, and *via* hydrogen bonding interactions of the carbonyl group with the amide protons. Secondly, the hydroxyl groups at C-11 and C-17 seem also involved in the complexation, most probably via hydrogen bonding.

Addition of prednisolone-21-acetate **(22)** to a solution of holand **19** in $CDCl_3$ did not give significant shifts or splitting of signals in the ^1H NMR spectrum of both host and guest. Probably the extreme rigidity of the host prevents the molecule from accommodating the structural deformations necessary for complexation.

4.4. COMBINATION OF CALIX[4]ARENES WITH PARTLY BRIDGED RESORCIN[4]ARENES

Calix[4]arenes can also be combined with partly bridged resorcin[4]arenes. Combination of *upper rim* 1,3-difunctionalized calix[4]arenes with A,C-di-bridged **29** and tri-bridged resorcin[4]arenes **30** (Chart 12) yields both 1:1 and 2:1 calix-resorcin[4]arenes [25].

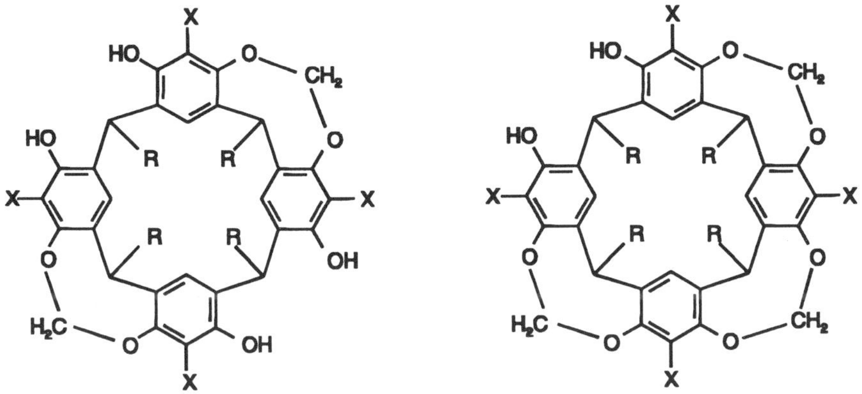

(29) **(30)**

Chart 12

Partly bridged resorcin[4]arenes possess free phenolic hydroxyl groups which render them suitable candidates for combination with *upper rim* functionalized calix[4]arenes for the synthesis of receptor molecules with large hydrophobic surfaces. The simplest way to connect partly bridged resorcin[4]arenes to *upper rim* functionalized calix[4]arenes is to react the phenolic hydroxyl groups of the resorcin[4]arene *via* a nucleophilic substitution reaction with electrophilic sites at the *upper rim* of the calix[4]arene. However, the reactivity of electrophilic groups in electron-rich aromatic rings like those in cavitands and calix[4]arenes is rather low. For this reason we decided to introduce nucleophilic groups at the *upper rim* of calix[4]arenes, *viz.* hydroxyl or amino groups. In order to connect the two buiding blocks, a spacer containing two electrophilic end groups, like $-CH_2C(O)-$, is required, which can easily be incorporated using reagents like $BrCH_2C(O)OMe$ or $ClCH_2C(O)Cl$.

Tri-bridged resorcin[4]arene derivatives **30** were combined with several 1,3-difunctionalized calix[4]arenes. The coupling of a diacid dichloride resorcin[4]arene with 1,3-diaminocalix[4]arene afforded the 1:1 calix-resorcin[4]arene **31** in 46% yield, which was comparable to the coupling reaction of dihydroxyresorcin[4]arene with 1,3-bis(chloroacetamido)calix[4]arene. The reaction of a diacid dichloride resorcin[4]arene with 1,3-dihydroxycalix[4]arene gave a 1:1 calix-resorcin[4]arene in only 13% yield, which can be attributed to the enhanced flexibility of the 1,3-dihydroxycalix[4]arene. The yield could be increased to 48% when using Cs_2CO_3 as a base (cesium-template effect). The 1:1 calix-resorcin[4]arene mainly exists as one conformer, but at low temperatures a second conformer could be observed, which is only present in small amounts (<5%). Both conformers equilibrate via rotation about one of the C(arene)-N bonds, with an activation free energy ($\Delta G^{\neq}$) of 13 kcal•mol^{-1}.

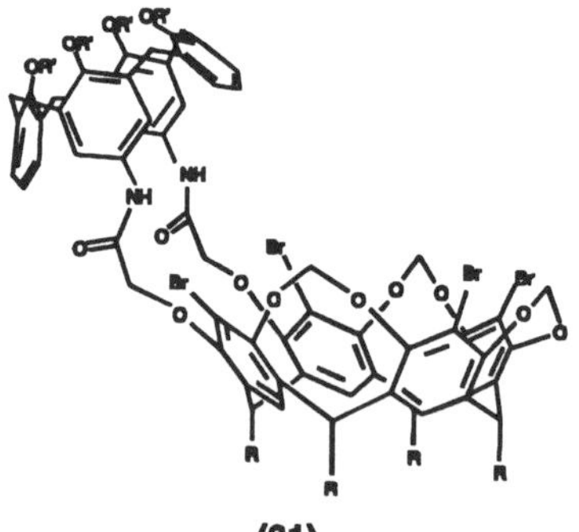

(31)

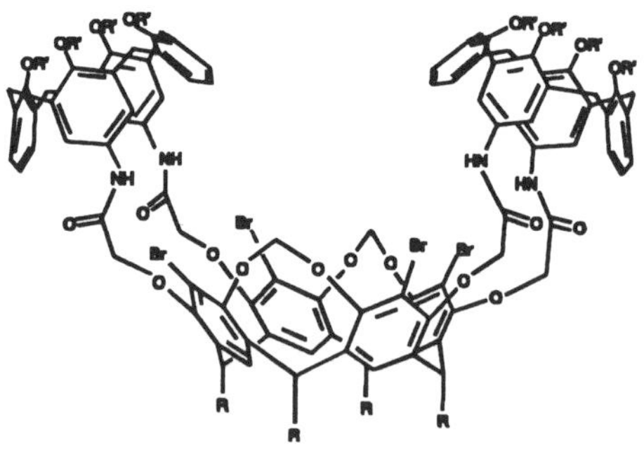

(32)

Chart 13

The 2:1 calix-resorcin[4]arenes (**32** in Chart 13) were synthesized in 20% ($R=(CH_2)_2Ph$, $R'=(CH_2)_2OCH_2CH_3$) and 53% ($R=CH_3$, $R'=(CH_2)_2CH_3$) yield, in the same way as the 1:1 calix-resorcin[4]arene starting from the tetraacid tetrachlorides A,C-dibridged resorcin[4]arenes **29**. These are present in at least three different conformations of comparable energy. To rigidify the structure of these 2:1 calix-resorcin[4]arenes several 1,3-diaminocalix[4]arenes, functionalized at the residual aromatic rings with carboxylic ester, nitro, and cyano substituents, were synthesized [26].

With these calix[4]arene derivatives, three new functionalized 2:1 calix-resorcinarenes were synthesized. Although the yields are somewhat lower than that of the unsubstituted 2:1 calix-resorcin[4]arene, it can be concluded that the presence of the carboxylic ester, nitro, or cyano groups does not severely hinder the formation of 2:1 calix-resorcin[4]arenes, or alter the structure of these host molecules to a considerable extent.

5. Conclusions

A new strategy for the synthesis of artificial receptor molecules was developed by the covalent linkage of medium-sized, relatively rigid molecules to which functional groups can be attached. The validity of the concept has been demonstrated with several examples.

6. Acknowledgements

The authors are indebted to their colleagues, whose names appear in the references, for their contribution included in this paper. Part of this work has been performed in collaboration with the University of Parma (Professors Pochini and Ungaro). Financial support from the Technology Foundation (STW), Technical Science Branch of the Netherlands Organization for Scientific Research (NWO), and the EEC Science Program is gratefully acknowledged.

7. References

1. a) Ghidini, E.; Ugozzoli, F.; Ungaro, R.; Harkema, S.; Abu El-Fadl, A.; Reinhoudt, D. N. (1990) Complexation of alkali metal cations by conformationally rigid, stereoisomeric calix[4]arene crown ethers: a quantitative evaluation of preorganization, *J. Am. Chem. Soc.* **112**, 6979-6985; b) Verboom, W.; Rudkevich, D. M.; Reinhoudt, D. N. (1994) Molecular recognition by artificial receptors, *Pure Appl. Chem.* **66**, 679-686.
2. a) Iwema Bakker, W. I.; Haas, M.; Khoo-Beattie, C.; Ostaszewksi, R.; Franken, S. M.; Den Hertog, Jr., H. J.; Verboom, W.; De Zeeuw, D.; Harkema, S.; Reinhoudt, D. N. (1994) Kinetically stable complexes of alkali cations with calixspherands: an evaluation of shielding, *J. Am. Chem. Soc.* **116**, 123-133; b) Iwema Bakker, W. I.; Haas, M.; Den Hertog, Jr., H. J.; Verboom, W.; De Zeeuw, D.; Bruins, A. P.; Reinhoudt, D. N. (1994) Functionalized calixspherands: synthesis and peptide coupling, *J. Org. Chem.* **59**, 972-976.
3. Iwema Bakker, W. I.; Verboom, W.; Reinhoudt, D. N. (1994) Kinetically stable silver complexes of calixspherands, *J. Chem. Soc., Chem. Commun.* 71-72.
4. For a few examples, see: a) Van Doorn, A. R.; Schaafstra, R.; Bos, M.; Harkema, S.; Van Eerden, J.; Verboom, W.; Reinhoudt, D. N. (1991) Molecular recognition of polar neutral molecules by metallomacrocycles: synthesis, ^{1}H NMR spectroscopy, X-ray structure, electrochemistry, and ab initio calculations, *J. Org. Chem.* **56**, 6083-6094; b) Reichwein, A. M., Verboom, W.; Reinhoudt, D. N. (1993) Binaphthyl metallomacrocycles for complexation of neutral molecules, *Recl. Trav. Chim. Pays-Bas* **112**, 358-366; c) Reichwein, A. M.; Verboom, W.; Reinhoudt, D. N. (1993) Functionalized binaphthyl salophen crown ethers as models for the enzyme urease, *Recl. Trav. Chim. Pays-Bas* **112**, 595-607.
5. Reichwein, A. M.; Verboom, W.; Harkema, S.; Spek, A. L.; Reinhoudt, D. N. (1994) Calix salophen crown ethers as receptors for neutral molecules, *J. Chem. Soc., Perkin Trans. 2*, 1167-1172.
6. Arnaud-Neu, F.; Collins, E. M.; Deasy, M.; Ferguson, G.; Harris, S. J.; Kaitner, B.; Lough, A. J.; McKervey, M. A.; Marques, E.; Ruhl, B. L.; Schwing-Weill, M. J.; Seward, E. M. (1989) Synthesis, X-ray crystal structures, and cation binding properties of alkyl calixaryl esters and ketones, a new family of macrocyclic molecular receptors, *J. Am. Chem. Soc.* **111**, 8681-8691.
7. a) Rudkevich, D. M.; Stauthamer, W. P. R. V.; Verboom, W.; Engbersen, J. F. J.; Harkema, S.; Reinhoudt, D. N. (1992) UO_2-salenes: neutral receptors for anions with a high selectivity for dihydrogen phosphate, *J. Am. Chem. Soc.* **114**, 9671-9673; b) Rudkevich, D. M.; Verboom, W.; Brzozka, Z.; Palys, M. J.; Stauthamer, W. P. R. V.; Van Hummel, G. J.; Franken, S. M.; Harkema, S.; Engbersen, J. F. J.; Reinhoudt, D. N. (1994) Functionalized UO_2-salenes: neutral receptors for anions, *J. Am. Chem. Soc.* **116**, 4341-

4351.

8. a) Rudkevich, D. M.; Verboom, W.; Reinhoudt, D. N. (1994) Calix[4]arene salenes: a bifunctional receptor for NaH_2PO_4, *J. Org. Chem.* **59**, 3683-3686; b) Visser, H. C.; Rudkevich, D. M.; Verboom, W.; De Jong, F.; Reinhoudt, D. N. (1994) Anion carrier mediated membrane transport of phosphate: selectivity of $H_2PO_4^-$ over Cl^-, *J. Am. Chem. Soc.* **116**, 11554-11555.
9. Rudkevich, D. M.; Mercer-Chalmers, J. D.; Verboom, W.; Ungaro, R.; De Jong, F.; Reinhoudt, D. N. *J. Am. Chem. Soc.* in press.
10. Szejtli, J. (1988) *Cyclodextrin technology*, Kluwer Academic Publishers, Dordrecht.
11. Van Dienst, E.; Snellink, B. H. M.; Von Piekartz, I.; Engbersen, J. F. J.; Reinhoudt, D. N. *J. Chem. Soc., Chem. Commun.*, in press.
12. Van Dienst, E.; Snellink, B. H. M.; Von Piekartz, I.; Grote Gansey, M. H. B.; Venema, F.; Feiters, M. C.; Nolte, R. J. M.; Engbersen, J. F. J.; Reinhoudt, D. N. submitted for publication.
13. Rudkevich, D. M.; Verboom, W.; Reinhoudt, D. N. (1994) Biscalix[4]arene-Zn-tetraarylporphyrins, *Tetrahedron Lett.* **35**, 7131-7134.
14. Rudkevich, D. M.; Verboom, W.; Reinhoudt, D. N. submitted for publication.
15. Van Duynhoven, J. P. M.; Janssen, R. G.; Verboom, W.; Franken, S. M.; Casnati, A.; Pochini, A.; Ungaro, R.; de Mendoza, J.; Nieto, P. M.; Prados, P.; Reinhoudt, D. N. (1994) Control of calix[6]arene conformations by self-inclusion of 1,3,5-tri-*O*-alkyl substituents: synthesis and NMR studies, *J. Am. Chem. Soc.* **116**, 5814-5822.
16. a) Collet, A. (1987) Cyclotriveratrylenes and cryptophanes, *Tetrahedron* **43**, 5725-5759; b) Collet, A.; Dutasta, J.-P.; Lozach, B. (1990) Design, synthesis, and properties of macrocyclic receptors for tetrahedral substrates, *Bull. Soc. Chim. Belg.* **99**, 617-633.
17. Janssen, R. G.; Verboom, W.; Van Duynhoven, J. P. M.; Van Velzen, E. J. J.; Reinhoudt, D. N. (1994) Cryptocalix[6]arenes; molecules with a large cavity, *Tetrahedron Lett.* **35**, 6555-6558.
18. Timmerman, P.; Nierop, K. G. A.; Brinks, E. A.; Verboom, W.; Van Veggel, F. C. J. M.; Van Hoorn, W. P.; Reinhoudt, D. N. (1995) Hydrophobic concave surfaces and cavities by combination of calix[4]arenes and resorcin[4]arenes, *Chem. Eur. J.* **1**, 134-145.
19. Cram, D. J. (1994) *Container molecules and their guests*; Monographs in supramolecular chemistry, Vol. 4; Stoddart, J. F., Ed.; Royal Society of Chemistry, Cambridge.
20. Timmerman, P.; Verboom, W.; Van Veggel, F. C. J. M.; Van Hoorn, W. P.; Reinhoudt, D. N. (1994) An organic molecule with a rigid cavity of nanosize dimensions, *Angew. Chem. Int. Ed. Engl.* **33**, 1292-1295.
21. Timmerman, P.; Verboom, W.; Van Veggel, F. C. J. M.; Van Duynhoven, J. P. M.; Reinhoudt, D. N. (1994) A novel type of stereoisomerism in calix[4]arene-based carceplexes, *Angew. Chem. Int. Ed. Engl.* **33**, 2345-2348.

22. a) Diederich, F. (1990) *Cyclophanes*; Monographs in supramolecular chemistry, Vol. 2; Stoddart, J. F., Ed.; Royal Society of Chemistry, Cambridge; b) Peterson, B. R.; Diederich, F. (1994) Dissolution of cholesterol in water by a synthetic receptor, *Angew. Chem. Int. Ed. Engl.* **33**, 1625-1628; c) Peterson, B. R.; Mordasini-Denti, T.; Diederich, F. (1995) Cavity depth and width effects on cyclophane-steroid recognition: molecular complexation of cholesterol and progesterone in aqueous solution, *Chem. Biol.* **2**, 139-146.
23. Hamada, F.; Kondo, Y.; Ito, R.; Suzuki, I.; Osa, T.; Ueno, A. (1993) Dansyl-modified γ-cyclodextrin as a fluorescent sensor for molecular recognition, *J. Incl. Phenom. Mol. Rec* **15**, 273-279.
24. Timmerman, P.; Brinks, E. A.; Verboom, W.; Reinhoudt, D. N. (1995) Synthetic receptors with preorganized cavities that complex prednisolone-21-acetate, *J. Chem. Soc., Chem. Commun.* 417-418.
25. Timmerman, P.; Boerrigter, H.; Verboom, W.; Reinhoudt, D. N. (1995) Synthesis of potential receptor molecules by combination of upper-rim-functionalized calix[4]arenes and partly bridged resorcinarenes, *Recl. Trav. Chim. Pays-Bas*, **114**, 103-111.
26. Timmerman, P.; Verboom, W.; Reinhoudt, D. N.; Arduini, A.; Grandi, S.; Sicuri, A. R.; Pochini, A.; Ungaro, R. (1994) Novel routes for the synthesis of upper rim amino and methoxycarbonyl functionalized calix[4]arenes carrying other types of functional groups, *Synthesis* 185-189.

STRUCTURAL MODELS OF BIOLOGICAL SIGNIFICANCE FROM SUPRAMOLECULAR SYSTEMS

Jerry L. Atwood
Department of Chemistry
University of Missouri-Columbia
Columbia, MO 65211

Small molecule crystallographers have been enormously successful in attacking problems of fundamental significance to chemistry and physics. The results of these endeavours have played a critical role in the development of almost all areas of chemistry, not just those directly related to structure and bonding. Great detail is known about the metrical parameters of more than 100,000 molecules subjected to single crystal X-ray diffraction study. In sharp contrast, structural studies of molecules of biological dimensions by means of X-ray diffraction techniques have been few indeed. The results of protein crystallography have, however, had a very great impact on the development of molecular biology and related fields. This is so, even though the metrical parameters of many proteins are not clearly revealed by the diffraction studies. Often, features of interest in protein structures are not revealed at all by single crystal studies. Of course, we all recognize the source of the difficulty in structurally characterizing molecules from the realm of biology: the molecules are very large and do not generally crystallize well (if they crystallize at all).

One of the major thrusts of supramolecular crystallography is to bridge the gap between small molecule and very large molecule crystallography. We seek to build up large assemblies from component parts, assemblies which will exhibit features of interest with respect to proteins. We search for the molecule or assembly of intermediate size: large enough to be relevant to biological systems but small enough to reveal details of structure and bonding.

1. O-H..., N-H..., and C-H...π Aromatic Hydrogen Bonding

The first example concerns the role of water which is exposed to "hydrophobic" molecular regions. Proteins, during the normal course of synthesis and activity, must expose hydrophobic surfaces to water. Indeed, proteins in the crystalline state are most certainly enshrouded by a shell of water molecules. Supramolecular complexes have revealed the action of water with a range of hydrophobic molecular entities. The O-H...π aromatic hydrogen bonding interaction of water, exemplified in Figure 1, is one such example [1]. Indeed, the subject of O-H...π hydrogen bonding has received considerable recent attention. Jorgensen and Severance used Monte Carlo methods to predict the O...centroid separation to be 3.11 Å in C_6H_6...H_2O with one hydrogen atom of the water molecule directed toward the center of the aromatic ring [2]. Our group then found experimental evidence for such aromatic π hydrogen bonding to water in Na_4[p-

G. Tsoucaris et al. (eds.), Crystallography of Supramolecular Compounds, 369–379.

sulfonatocalix[4]arene]·13.5 H_2O and several related complexes [1]. O-H...π hydrogen bonding has also been found in alcohols [3-5] and in a silanol [6].

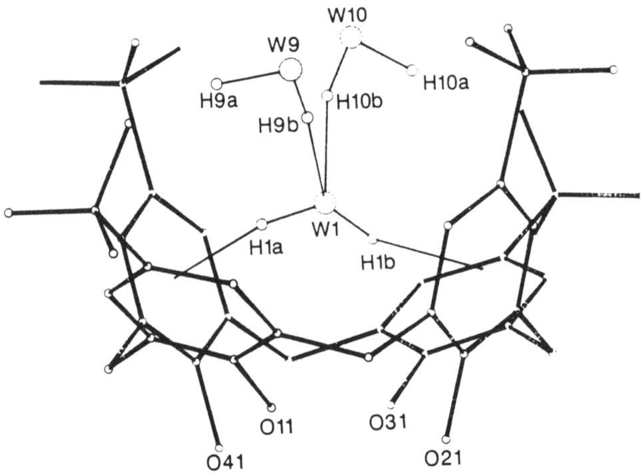

Figure 1. Environment of the water molecule W1, which displays π hydrogen bonding to two aromatic rings.

C-H...π hydrogen bonding has been found in two host-guest complexes [7] and in several other systems [8,9]. Hanton, Hunter, and Purvis have added an example of intramolecular N-H...π hydrogen bonding [10]. General crystallographic evidence for N-H...π and O-H...π hydrogen bonding has been recently discussed [11].

From these first studies of this important electrostatic interaction, it is seen that the X...centroid distances are of the order of

C...centroid = 3.5 Å,
N...centroid = 3.3 Å,
O...centroid = 3.1 Å,

as would be expected on the basis of atomic size. This is further supported by a set of *ab initio* calculations of benzene/water and benzene/ammonia. Here it is reported that the expected N...centroid distance is 0.26 Å greater than that for O...centroid [12]. However, we have now found, in a host-guest system an example of N-H...π hydrogen bonding; an abnormally short N...aromatic distance is revealed.

In the solid state, $[HNC_5H_4C_5H_4NH^{2+}]_2$[p-sulfonatocalix[4]arene]·7 H_2O displays two types of 4,4'-bipyridinium cations: one is intercalated into the bilayer structure [13] and one is partially contained in the calix[4]arene cavity. The latter one is oriented such that the $N\text{-}H^+$ portion is directed at the center of one of the aromatic rings of the calix[4]arene, Figure 2. The N...centroid distance is 3.06 Å. With the assumption of a normal N-H bond length, this leads to an H...centroid distance of 2.01 Å, much shorter than the H...centroid separation of 2.42 Å [14] found by Hanton, Hunter, and Purvis [10]. Indeed, the N...centroid distance is even shorter than the O..centroid distances, both observed [1] and calculated [2]. Further studies of aromatic π hydrogen bonding are in progress.

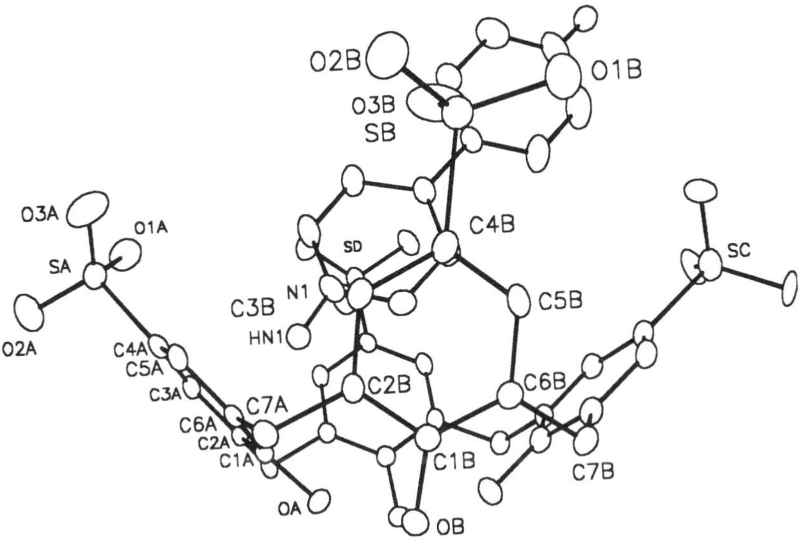

Figure 2. Structure of $[HNC_5H_4C_5H_4NH^{2+}]_2$[p-sulfonatocalix[4]arene]·7 H_2O, illustrating the N-H...aromatic π hydrogen bonding.

Supramolecular complexes in which methyl groups face layers of water molecules provide examples of C-H...O(water) interactions. Such studies are of benefit for at least two reasons: they provide proof that such interactions can exist and they afford metrical parameters which can be interpreted in terms of strengths of bonding. A model system for this interaction is found in the structure of a protonated aminomethylcalix[4]arene, [p-(dimethylaminomethyl)calix[4]arene].$(HCl)_2$·9 H_2O [15]. Here, layers of methyl groups face onto a layer of water molecules (approximately four water molecules thick). Unfortunately, the protons have not been located with the present data, but it is clear that diligent effort will reveal the nature of the C-H...O(water) interaction.

Carbonic anhydrase provides an example of the importance of ordered water in a biological system. The crystal structure of the bio-molecule [16] shows that there is an ordered arrangement of eight water molecules in a hydrophobic pocket near the site of catalytic action. The synergism of synthesis and crystallography must be noted. The existence of the hydrophobic pocket with eight water molecules provides the synthetic supramolecular chemist with the goal of constructing such a pocket from small building blocks.

2. Enzyme Models

Cyclodextrins have played an important role in the search for metalloenzyme models [17,18]. Tabushi has noted the need for a "double recognition" which involves both metal coordination recognition and hydrophobic recognition [19]. Two approaches have evolved. In one, the focus has been on the catalysis, while in the other emphasis has been placed on structural models. The second-sphere coordination of transition metals by cyclodextrins has been an outgrowth of the structural model approach [20-32]. Indeed, Stoddart has prepared complexes in which the metal is hydrogen bonded via primary-sphere ammine ligands to α-cyclodextrin, with the organic ligands held within the cavity [20,21]. We have previously reported a complex, $(H_3O^+)[La(ONC_5H_5)_2(H_2O)_6$-(calix[4]arene sulfonate)]$^-\cdot 6.5\ H_2O$, in which a metal ion is directly bonded to the upper rim of a calix[4]arene [33] (**1**) and a primary-sphere ligand is inserted in the cavity (**2**), effecting a double recognition of the ligand. The complex, shown in Figure 3, features a 9-coordinate lanthanum(III) ion. One of the nine ligands in [calix[4]arene sulfonate]$^{4-}$ which is bonded to the La^{3+} ion through a sulfonate oxygen atom. A pyridine N-oxide ligand is bonded to the metal ion and a portion of the hydrophobic aromatic ring is inserted into the hydrophobic calixarene cavity. The depth of penetration may be measured by the distance of the centroid of the pyridine ring from the plane of the calixarene -CH_2- carbon atoms. The value in this complex, 4.3 Å, may be compared with 4.0 Å for $C_5H_5NH^+$ in Na_4(pyridinium)[calix[4]arene sulfonate]$\cdot 8\ H_2O$ [34]. Thus, in the metalloenzyme analogy the pyridine N-oxide is recognized by both the metal coordination and the hydrophobic complexation.

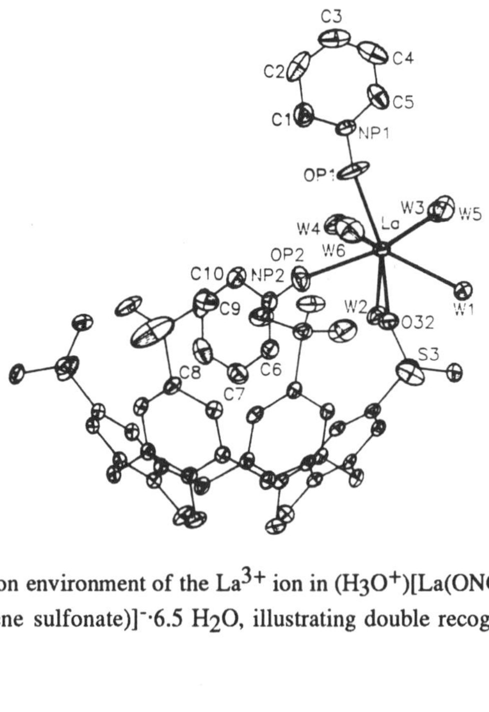

Figure 3. Coordination environment of the La^{3+} ion in $(H_3O^+)[La(ONC_5H_5)_2(H_2O)_6$-(calix[4]arene sulfonate)]$^-\cdot 6.5\ H_2O$, illustrating double recognition of the pyridine N-oxide.

The cyrstal structure is composed of alternating hydrophobic calixarene bilayers and hydrophilic layers, spanned by the metal ion complex, Figure 4. The second pyridine N-oxide ligand is intercalated into the calixarene bilayer [13].

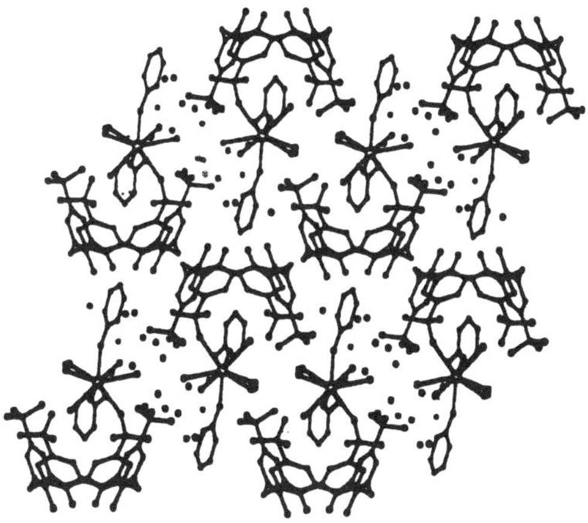

Figure 4. Crystal packing for $(H_3O^+)[La(ONC_5H_5)_2(H_2O)_6(\text{calix[4]arene sulfonate})]^- \cdot 6.5\ H_2O$, showing the bilayer arrangement of the calix[4]arenes. One metal coordinated pyridine N-oxide ligand is intercalated into the calix[4]arene bilayer.

3. Flexible versus Rigid Hosts

Supramolecular chemistry debates the merits of flexible versus rigid host/guest systems. Supramolecular crystallography can provide information about the features of both systems. Indeed, it can often provide the only convincing evidence of weak interactions. In Figure 5 is demonstrated a flexible supramolecular system which is capable of acting as a molecular Venus Flytrap [35]. A summarizing view of this study was presented in *New Scientist* [36] .

On the left portion of Figure 5 a guest molecule is shown in the calixarene cavity. This molecule seems to be unreacted, excess 1-(2-hydroxyethyl)piperazine that has been methylated at the position of the secondary nitrogen atom. Although the hydroxyethyl chain of this molecule is badly disordered, and thus is not shown in Figure 5, the remainder of the atoms are better defined. Here, again, one must make use of other physical methods to provide more evidence, even with the X-ray crystal structure in-hand. The ^1H and ^{13}C NMR spectra are also consistent with the presence of two different hydroxyethylpiperazine moieties. As Figure 5a reveals, the arms of the calix[4]arene are spread wide to provide a shallow cavity.

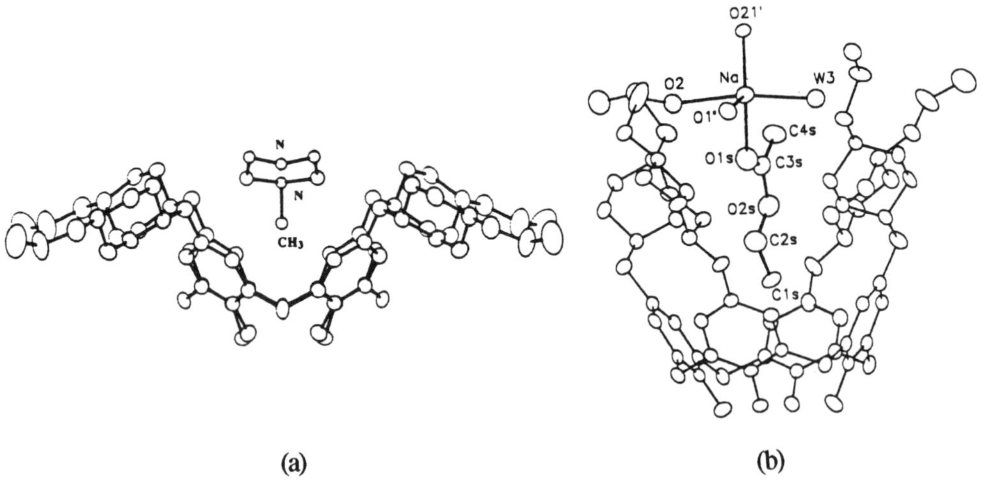

Figure 5. (a) Structure of p-[4-(2-hydroxyethyl)piperazinomethyl]calix[4]arene with the guest. The $-CH_2CH_2OH$ portion of the guest molecule was badly disordered and is not shown; (b) Structure of Na{p-[4-(2-hydroxyethyl)piperazinomethyl]calix[4]arene}.-$CH_3COOCH_2CH_3$.

Preparation of the calixarene and subsequent neutralization with 10% Na_2CO_3 solution to pH 9 leads to the removal of one of the phenolic protons resulting in the complex shown on the right of Figure 5. After crystallization from ethyl acetate, X-ray diffraction quality crystals were obtained. The structure shows that in this situation, the arms are directed up from the rim of the calixarene to produce a cavity so deep that the ethyl acetate guest molecule is completely enshrouded.

Attractive routes to the synthesis of deep cavity calixarenes are now available [37-39], and several interesting species have been crystallographically characterized [39].

4. Anion Binding

Recent work within this group has been directed towards the synthesis of calixarene-based bowl-lile host molecules in which a catalytically or redox active metal center is either embedded within, or is in close proximity to the hydrophobic cavity of a calixarene. In order to achieve this goal without the interference of solvent or other potential guest molecules the possibility of direct covalent attachment of transition metals to the calixarene aromatic rings was investigated.

Shinkai and co-workers have reported the synthesis of neutral π-chromium tricarbonyl complexes of **1c** (R = H, Y = n-Pr) in which the metal binds to the outer face of the calixarene, and have demonstrated their usefulness in chemical modification of the host molecules [40]. However, these air-sensitive metal complexes are only formed under forcing conditions (130°C) in conjunction with a very limited range of especially rigid, unhindered calix[4]arenes and are unlikely to display properties such as water solubility and the ability to accommodate polar guests, which are important in mimicking conditions *in vivo* [17,18,33,41].

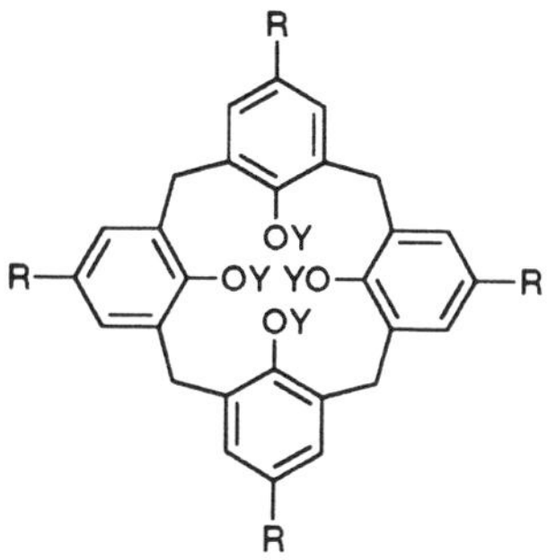

1a R = Y = H
1b R = Bu^t, Y = H
1c R = H, Y = Pr^n

Accordingly, we have synthesised a wide variety of more robust, cationic model compounds which are stable to both air and moisture and are likely to entrap a wide range of guest molecules within the hydrophobic cavity.

Treatment of the rhodium complex **2a** with Ag[BF_4] in acetone followed by refluxing with calix[4]arene **1a** in neat CF_3CO_2H resulted in the formation of the bimetallic complex [{Rh(η^5-C_5Me_5)}$_2$(η^6:η^6-$C_{28}H_{24}O_4$)][BF_4]$_4$ **3a** in 95% yield, even in the presence of excess [Rh(η^5-C_5Me_5)(acetone)$_3$]$^{2+}$. The formulation of **3a** was confirmed by a FAB mass spectrum which clearly indicated a molecular cation peak at *m/z* 898 as well as a smaller signal at *m/z* 985 corresponding to the cation in (**3a**) associated with one tetrafluoroborate anion.

Previous reports have noted that the related Cp*rhodium p-phenol complexes [Rh(η^5-C_5Me_5)(η^6-PhOH)][PF_6]$_2$ tend to deprotonate, forming [{Rh(η^5-C_5Me_5)}$_2$(η^6-PhO-H...O(η^6-)Ph)][PF_6]$_3$. Under the conditions employed in the present work, however, the calixarene oxygen atoms are fully protonated due to the presence of CF_3CO_2H (a signal due to the OH protons was located in the ^1H NMR spectrum of **3a** as an extremely broad resonance at *ca.* d 5.3 ppm, while the infrared spectrum demonstrated a broad band assigned to ν(OH) at 3420 cm^{-1}). The broadness and high field chemical shift of the OH NMR signal suggests that the hydroxylic protons in **3a** may be relatively acidic and only loosely associated with the cationic complex. Treatment of **3a** with excess Na_2CO_3 in acetone resulted in the loss of two protons to form the dicationic complex [{Rh(η^5-C_5Me_5)}$_2$(η^6:η^6-$C_{28}H_{22}O_4$)][BF_4]$_2$ **4a**.

Interestingly, if **1a** is refluxed in acetone with [Rh(η^5-C_5Me_5)(acetone)$_3$]$^{2+}$ in the presence of only a few drops of CF_3CO_2H a yellow precipitate of a further compound, **5a**, is obtained. More strikingly, the majority of resonances in the spectrum of **5a** were split into pairs (*e.g.* for the Cp* rings singlets at δ 2.16 and 2.15 ppm were observed) suggesting an asymmetric complex in which there is a slight difference between the two coordinated rings. This data, in conjunction with the less acidic conditions employed in the synthesis of **5a** leads us to formulate the complex as the singly deprotonated trication $[\{Rh(\eta^5\text{-}C_5Me_5)\}_2(\eta^6\text{:}\eta^6\text{-}C_{28}H_{23}O_4)][BF_4]_3$.

Reaction of the iridium complex **2b** with **1a** in acetone/CF_3CO_2H results in the formation of the iridium analogue of **3a**, $[\{Ir(\eta^5\text{-}C_5Me_5)\}_2(\eta^6\text{:}\eta^6\text{-}C_{28}H_{24}O_4)][BF_4]_4$ **3b** in 56% yield. Interestingly however, reaction of **2b** (pre-treated with $Ag[BF_4]$) with **1a** in neat CF_3CO_2H resulted in the surprising isolation of the tetrametallic species $[\{Ir(\eta^5\text{-}C_5Me_5)\}_4(\eta^6\text{:}\eta^6\text{:}\eta^6\text{:}\eta^6\text{-}C_{28}H_{24}O_4)][BF_4]_8$ **6b** in 76% yield, even in the presence of a 1 : 1 mole ratio of **2b** to **1a**. Calix[4]arene is extremely poorly soluble in CF_3CO_2H and a significant quantity of the undissolved macrocycle was recovered at the end of the reaction suggesting that the primary metalation step is rate limiting under these conditions. The FAB mass spectrum of **6b** exhibited peaks corresponding to calix[4]arene coordinated to one, two, three and four Cp*Ir fragments whilst the symmetrical nature of the product was clearly indicated by its ^1H NMR spectrum which exhibited a single set of resonances for the calixarene ligand at similar chemical shift to those assigned to the coordinated faces of **3b**. The observation of the characteristic AB pattern for the methylinic bridges indicates a cone conformation [41].

Reaction of complexes **2** in neat CF_3CO_2H with the more sterically hindered *p*-t-butylcalix[4]arene **1b** again led, in almost quantitative yield, to the bimetallic complexes $[\{M(\eta^5\text{-}C_5Me_5)\}_2(\eta^6\text{:}\eta^6\text{-}C_{44}H_{56}O_4)][BF_4]_4$ (M = Rh **7a**, Ir **7b**), related to **3a** and **3b**. Tetrametallic species were not formed even in the presence of excess metal complex, presumably because of the highly sterically hindered nature of the *p*-t-butylcalixarenes. Complexes **7** were characterised as described above and exist solely in the cone conformation.

Consistent with the high acidity observed for **3**, slow crystallization of complexes **7** from NO_2CH_3 / Et_2O gave two new species, **8a** and **8b**, exhibiting ^1H NMR spectra similar to that of **5a** suggesting them to be the tricationic complexes $[\{M(\eta^5\text{-}C_5Me_5)\}_2(\eta^6\text{:}\eta^6\text{-}C_{44}H_{56}O_4)][BF_4]_3$. In confirmation the iridium complex **8b** as a diethyl ether/nitromethane solvate was characterised by a single crystal X-ray structure determination, Figure 6. The structure of **8b**.Et_2O.NO_2Me, obtained from crystals grown from slow vapour diffusion of diethyl ether into a nitromethane solution of the complex, is an excellent illustration of the type of host-guest interaction which make this class of compounds of interest. The calix[4]arene host adopts the cone conformation with the two iridium centers situated upon opposite sides of the macrocycle. Deeply embedded within the hydrophobic cavity is a molecule of diethyl ether. Calculated hydrogen atom positions suggest that the -CH_3 moiety C(1)S is orientated such as to form weak C-H...p hydrogen bonds with the unmetalated aromatic rings (B, D) whille avoiding the partially positively charged coordinated decks. The hydrogen bonding network at the calixarene oxygen atoms is asymmetric with short contacts from Ob to Oa and Oc of 2.52 and 2.64 Å, respectively, whilst longer bonds are formed to Od; Od...Oa, Oc 2.71 and 2.76 Å. These distances are shorter than those observed for the longest

O...O contact in the sodium salt of the *p*-t-butylcalix[4]arene monoanion (2.92 Å) where the oxygen atoms are not bridged by a proton, whilst the distances between hydrogen bonded oxygen atoms in the latter structure are significantly less than in **8b** (2.43 - 2.58 Å). This evidence suggests that crystallographic disorder results in either Oa or Oc being deprotonated without any large energetic preference between the two possibilities. Consistent with an oxocyclohexadienyl description for the coordinated decks of the calixarene ligand, the C(1)-O distance is markedly shorter for rings A and C, 1.30 Å av., than the unmetalated rings B and D, 1.37 Å av.

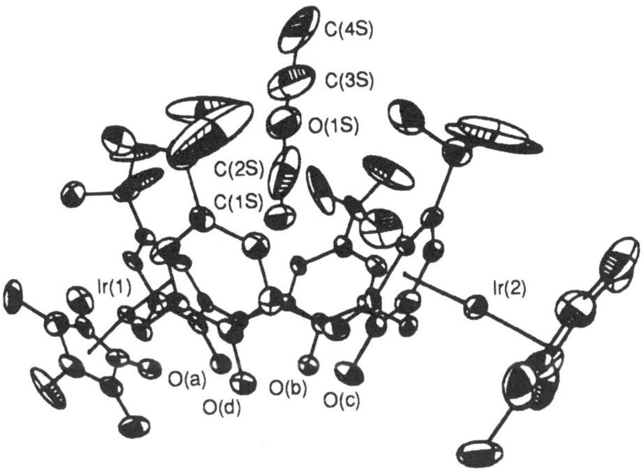

Figure 6. Structure of the cation in **8b**.$Et_2O.NO_2Me$, with the included molecule of Et_2O.

As this chemistry develops it is clear that close attention must be paid to complementary spectroscopic techniques in order to sort out the disorder, as well as to ascetain the speciation.

References

1. Atwood, J. L., Hamada, F., Robinson, K. D.;, Orr, G. W. and Vincent, R. L. (1991) *Nature* **349,** 683.

2. Jorgensen, W. L. and Severance, D. L. (1990) *J. Am. Chem. Soc.* **112**, 4768.

3. Hardy, A. D. U. and MacNicol, D. D. (1976) *J. Chem. Soc., Perkin Trans. 2* , 1140.

4. Schweizer, W. B., Dunitz, J. D., Pfund, R. A.;, Tombo, G. M. R. and Ganter, C. (1981) *Helv. Chim. Acta* **64,** 2738.

5. Rzepa, H. S., Webb, M. L., Slawin, A. M. Z. and Williams, D. J. (1991) *J. Chem. Soc., Chem. Commun.*, 1136.

6. Al-Juaid, S. S., Al-Nasr, A. K. A., Eaborn, C. and Hitchcock, P. B. (1991) *J. Chem. Soc., Chem. Commun.*, 1482.

7. Atwood, J. L., Bott, S. G., Jones, C. and Raston, C. L. (1992) *J. Chem. Soc., Chem. Commun.* , 1349; Steed, J. W., Juneja, R. K., Burkhalter, R. S. and Atwood, J. L., submitted for publication.

8. Hamor, T. A., Jennings, W. B., Proctor, L. D., Tolley, M. S., Boyd, D. R. and Mullan, T. (1990) *J. Chem. Soc., Perkin Trans. 2* , 25.

9. Grossel, M. C., Cheetham, A. K., Hope, D. A. O., Lam, K. P. and Perkins, M. J. (1979) *Tet. Lett.*, 1351.

10. Hanton, L. R., Hunter, C. A. and Purvis, D. H. (1992) *J. Chem. Soc., Chem. Commun.*, 1134. The geometrical parameters given in this contribution afford N...centroid = 3.21 Å, with the assumption that the hydrogen atom lies directly over the centroid of the aromatic ring (a difference of up to 0.15 Å is noted between the H...plane and H...centroid distances, but further details are not given).

11. Viswamitra, M. A., Radhakrishnan, R., Bandekar, J. and Desiraju, G. R. (1993) *J. Am. Chem. Soc.* **115**, 4868.

12. Bredas, J. L. and Street, G. B. (1989) *J. Chem. Phys.* **90**, 7291. However, it must be noted that these calculations over estimate the O...centroid distance by almost 0.3 Å: O...centroid = 3.40 Å; N...centroid = 3.66 Å.

13. (a) Atwood, J. L., Orr, G. W., Hamada, F., Vincent. R. L., Bott, S. G. and Robinson, K. D. (1991) *J. Am. Chem. Soc.* **113**, 2760; (b) Atwood, J. L., Orr, G. W., Hamada, F., Bott, S. G. and Robinson, K. D. (1992) *Supramol. Chem.* **1**, 15; (c) Coleman, A. W., Bott, S. G., Morley, S. D., Means, C. M., Robinson, K. D., Zhang, H. and Atwood, J. L. (1988) *Angew. Chem. Int. Ed. Engl.* **27**, 1361.

14. Estimated from the parameters given in ref 10. The results of our investigation may also be compared directly to those in ref 10: H...plane(aromatic) = 2.42 versus 1.98 Å, respectively; H...C(aromatic) = 2.72 - 2.87 versus 2.22 - 2.62 Å, respectively.

15. Atwood, J. L., Robinson, K. D., Orr, G. W., unpublished results.

16. Eriksson, E. A., Jones, T. A. and Liljas, A. (1986) in *Zinc Enzymes* (Bertini, I., Luchinat, C., Maret, W. and Zeppezauer, M., eds.), Birkhauser; Boston, p. 317; Eriksson, A. E., Jones, A. T. and Liljas, A. (1989)*Proteins* **4**, 274.;Eriksson, A. E., Kylsten, P. M., Jones, T. A. and Liljas, A. (1989) *Proteins* **4**, 283.

17. Tabushi, I. (1984) in Inclusion Compounds, Atwood, J.L., Davies, J. E. D. and MacNicol, D. D. (Eds), Academic Press, London, pp. 445-471.

18. Breslow, R. (1984) in Inclusion Compounds, Atwood, J.L., Davies, J. E. D. and MacNicol, D. D. (Eds), Academic Press, London, pp. 473-508.

19. Tabushi, I., Shimizu, N., Sugimoto, T., Shiozuka, M. and Yamamura, K. (1977) *J. Am. Chem. Soc.* **99**, 7100.

20. Alston, D. R., Slawin, A. M. Z., Stoddart, J. F. and Williams, D. J. (1985) *Angew. Chem. Int. Ed. Engl.* **24**, 786.

21. Alston, D. R., Slawin, A. M. Z., Stoddart, J. F. and Williams, D. J. (1985) *J. Chem. Soc., Chem. Commun.* , 1602.

22. Colquhoun, H. M., Stoddart, J. F. and Williams, D. J. (1988) *Angew. Chem. Int. Ed. Engl.* **27,** 1986.

23. Alston, D. R., Slawin, A. M. Z., Stoddart, J. F., Williams, D. J. and Zarzycki, R. (1988) *Angew. Chem. Int. Ed. Engl.* **27,** 1184.

24. Ashton, P. R., Stoddart, J. F. and Zarzycki, R. (1988) *Tetrahedron Lett.* **29,** 2103.

25. Alston, D. R., Ashton, P. R., Lilley, T. H., Stoddart, J. F., Zarzycki, R., Slawin, A M. Z. and Williams, D. J. (1989) *Carbohydrate Research* **192**, 259.

26. Harada, A., Hu, Y., Yamamoto, S. and Takahashi, S. (1988) *J. Chem. Soc., Dalton Trans.*, 729.

27. Hoh, T., Harada, A. and Takahashi, S. (1989) *Mem. Inst. Sci. Ind. Res.* **46,** 37.

28. Harada, A., Shimada, M. and Takahashi, S. (1989) *Chem. Lett.*, 275.

29. Harada, A., Saeki, K. and Takahashi, S. (1989) *Organometallics* **8**, 730.

30. Harada, A., Yamamoto, S. and Takahashi, S. (1989) *Organometallics* **8**, 2560.

31. Kobayshi, N. and Opallo, M. (1990) *J. Chem. Soc., Chem. Commun.*, 477.

32. Klingert, B. and Rihs, G. (1990) *Organometallics* **9**, 1135.

33. Atwood, J. L., Orr, G. W., Robinson, K. D. and Hamada, F. (1993) *Supramol. Chem.* **2**, 309.

34. Atwood, J. L., Orr, G. W., Hamada, F., Bott, S. G. and Robinson, K. D. (1992) *Supramol. Chem.* **1**, 15.

35. Atwood, J. L.. Orr, G. W., Bott, S. G. and Robinson, K. D. (1993) *Angew. Chem. Int. Ed. Engl.* **32**, 1093.

36. Bradley, D. (1993) *New Scientist*, September 25 Issue, p. 14.

37. Gutsche, C. D. and No K. H. (1982) *J. Org. Chem.* **47**, 2708; No, K. H. and Gutsche, C. D. (1982) *J. Org. Chem.* **47**, 2713.

38. Arduini, A., Pochini, A., Rizzi, A., Sicuri, A. R. and Ungaro, R. (1990) *Tetrahedron Lett.* **31**, 4653.

39. Juneja, R. K., Robinson, K. D., Johnson, C. P. and Atwood, J. L. (1993) *J. Am. Chem. Soc.* **115**, 3818.

40. Gutsche, C. D. (1989) *Calixarenes*, Royal Society of Chemistry, Cambridge.

41. Iki, H., Kikuchi, T. and Shinkai, S. (1992) *J. Chem. Soc., Perkin Trans. I*, 669; Kikuchi, T., Iki, H.,Tsuzuki, H. and Shinkai, S. (1993) *Supramol. Chem.* **1,** 103.

HOW DNA IS RECOGNIZED BY PROTEINS

W. SAENGER

Institut für Kristallographie, Freie Universität Berlin

Takustr. 6, D-14195 Berlin, Germany

In this short article, I briefly summarize the recognition of specific sequences on the DNA double helix by different structural motifs in proteins [1-3], which can be considered as interactions between two supra- or better supermolecules. The protein-DNA recognition is of importance in a biological processes which are related with gene expression. In the chromosomes, there are a large number of individual genes which cannot all be expressed simultaneously into the proteins which they are encoding, but they have to be expressed at those times at which the proteins are needed for the proper functioning of the organisms. This requires regulation of their expression, which occurs mainly at the transcriptional level, i.e. the copying of DNA into the complementary messenger RNA sequence.

This transcription is performed by RNA polymerases, large miltisubunit enzymes, which bind to a promoter sequence on the DNA and then move along the DNA double helix and synthesize messenger RNA. The binding to the promoter is regulated by DNA operator sequences which can be free, thereby permitting transcription, or blocked by proteins called repressors. Most of the repressors occur as homodimers with a twofold axis of symmetry, and they recognize DNA operator sequences which themselves have a twofold symmetric arrangement of nucleotides, called palindromes. Besides the repressors, there are a number of other proteins (transcription factors) which are involved in the regulation of RNA polymerase action. Several of these transcription factors have been analysed structurally by X-ray crystallography or NMR-spectroscopy without and with specific DNA bound to them so that we have already a rather good view and understanding of the specific interactions between these proteins and the DNA double helix.

G. Tsoucaris et al. (eds.), Crystallography of Supramolecular Compounds, 381–394.

The DNA double helix is composed of two Watson-Crick complementary strands which are oriented antiparallel and wound in a right-handed helix, with the Watson-Crick basepairs A-T and G-C in the center and perpendicular to the helix axis, and the deoxyribose-phosphate backbone is at the periphery of this double helix [4]. Since the attachment of the sugar units at the Watson-Crick basepairs is not diametrically opposite but on one side, there is a shorter separation between the glycosidic links, called minor groove, and a larger separation on the opposite side, called major groove. As Fig. 1 shows, the major groove side contains most of the functional groups of the basepairs which can be recognized by hydrogen bonding, whereas the minor groove side is less dominant. Since the DNA double helix exhibits a wide geometrical indentation at the major groove side, this is the preferred site for docking proteins [5]. In essence, the principal interactions are between amino acid side chains in protein α-helices and the major groove sides of the base pairs.

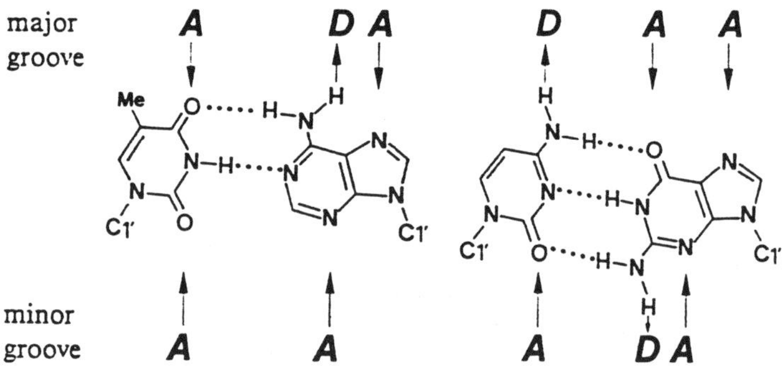

Fig. 1. Hydrogen-bonding interactions of base pairs A-T, T-A (*left*); G-C, C-G (*right*) indicated by *D* (donor) and *A* (acceptor). The discrimination between different base pairs occurs predominantly in the major (*wide*) groove, and not in the minor (*small*) groove, where only *G* can be discriminated from the other bases due to the donor function of its 2-amino group. The attachment of the deoxyribose is at C1' · After [5].

The four main classes of DNA binding proteins

The DNA binding proteins that have been studied structurally so far can be grouped into four classes depending on their respective structural motifs with which they recognize the DNA double helix, Fig. 2. These classes are:

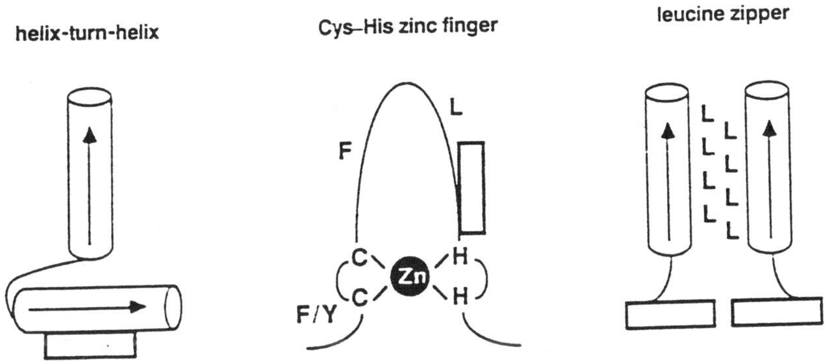

Fig. 2. Schematic diagram of the three DNA binding motifs. Capital letters indicate conserved amino acid residues, DNA double helices given by rectangular boxes. Taken from [1].

<u>Helix-turn-helix motif</u> This is a very characteristic motif in which two α-helices are joined by a short turn [6]. The N-terminal α-helix is the supporting helix and the C-terminal α-helix is the DNA recognition helix. It inserts in the major groove of the DNA double helix in different geometries, depending on the protein. There are a number of prokaryotic regulatory proteins with helix-turn-helix motifs like the repressors

lac, lambda, 434, 434cro, lambda cro, tryptophane, tetracycline, the multifunctional factor for inversion stimulation (FIS), and catabolite activator protein (CAP). In the eukaryotic regime, there are the homeodomains, which also carry helix-turn-helix motifs [7].

Zinç finger proteins constitute another large class of DNA binding proteins [8]. They contain a zinc atom which is coordinated by a total of 4 histidine N-atoms and cysteine S-groups in varying combinations. They all occur in eukaryotes as transcription factors and as parts of the nuclear receptors (glucocortikoid receptor and oestrogen receptor).

The leucine zipper is a structural motif in which two long α-helices form a coiled-coil which carry, at their N-termini, shorter α-helices with basic amino acids. The latter bind to DNA major grooves, thereby providing specific interaction. The leucine zippers are abundant and occur in eukaryotic systems [9].

Antiparallel ß-pleated sheet in a protein interacting with the major groove of DNA is less abundant and has been observed thus far in the MetJ repressor [10] and Arc protein.

In the following, I will comment on these different DNA recognizing motifs. Since my own group has been working on tetracycline repressor and factor for inversion stimulation (FIS), they will be treated in more detail.

Helix-turn-helix motifs

If the helix-turn-helix motifs of the structurally and functionally different proteins are superimposed, they have almost identical structure [6]. The reason is that there are several conserved amino acids that stabilize the helix-turn-helix motif through mutual interactions, thereby giving this motif its three dimensional shape. An illustration of the

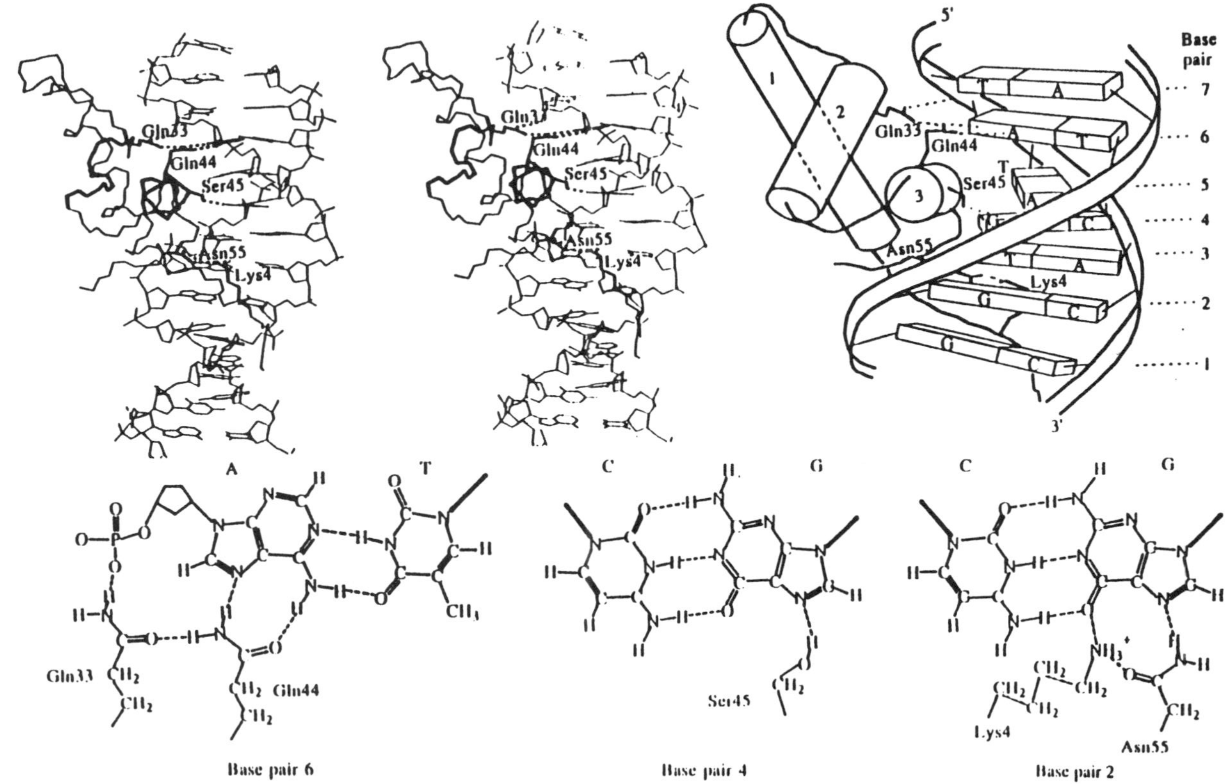

Fig. 3. Structure of the 434 repressor-operator complex. (Top left) stereo diagram, (top right) schematic drawing. (Bottom) Specrfic-interactions between amino acid side chains and base pairs 2,4,6. Taken from [12].

specific interaction between the helix-turn-helix motif and the DNA double helix is shown in Fig. 3. All the helix-turn-helix motifs that have been studied so far in complexes with their specific DNA sequences show different interactions to the extent that in tryptophane repressor, there is no direct specific contact between protein and DNA, but all the interactions are mediated by water molecules [11]

The multifunctional protein FIS is involved in the activation of transcription and in gene inversion and recombination, and recognizes a number of different DNA sites with very loose consensus sequence. FIS has been isolated from *Escherichia coli* as a homodimer, with 98 amino acids in the monomer. In the crystal structure, the 26 N-terminal amino acids are so disordered that they cannot be seen in the electron density map, and the rest of the polypeptide chain is folded into four α-helices A, B, C, D [13]. A, A' and B, B' of the two monomers interact to form a four-helix-bundle that stabilizes the dimer. α-Helices C, D and C', D' form two helix-turn-helix motifs, in which the DNA recognizing α-helices are separated by 29 Å. This is 5 Å shorter than the corresponding separation of adjacent major groves in B-DNA, 34 Å, in which these α-helices insert for binding. As a consequence, DNA has to bend ~90° when interacting with FIS, and the bending occurs at two pyrimidine/purine steps, Fig. 4. They are

Fig. 4. Complex between FIS and DNA, modeled on the basis of the crystal structure of FIS [13]. Note bending of DNA double helix which occurs at the two conserved pyrimidine/purine steps.

separated by 7 base pairs, and this is the only strictly conserved motif in the different known binding sites for FIS. The bending of double stranded DNA in complex with FIS has been computer modelled using automatic docking procedures and energy minimization Fig. 4.

Tet repressor was isolated from *Escherichia coli*. It regulates the resistance against the antibiotic tetracycline. In absence of the antibiotic, Tet repressor binds to two operators, thereby inhibiting the expression of two genes. One of the genes codes for the membrane intrinsic protein Tet A that exports tetracycline out of the cell and causes resistance, and the other codes for Tet repressor itself. If tetracycline enters the cell, it binds to Tet repressor and induces a conformational change so that DNA is released and expression of Tet A and Tet R can occur.

The crystal structure shows that the 207 amino acids long polypeptide chain of Tet repressor is folded into ten α-helices, and two monomers interact to form a homodimer with twofold rotation symmetry [14]. α-Helices 2, 3 and 2', 3' form helix-turn-helix motifs, and helices 5 to 10 and 5' to 10' are involved in stabilization of the homodimer. They provide two pockets for the tetracycline, which binds to Tet repressor in presence of a divalent cation. The X-ray structure has given a view of the interaction between the antibiotic and the protein (Fig. 5), and shows that the DNA binding helices in the helix-turn-helix motifs are separated by 39 Å if tetracycline is bound to Tet repressor. This is 5 Å too wide to insert into adjacent major grooves of DNA, which explains why this complex is not able to bind to the operator with palindromic DNA sequence. Consequently, Tet repressor whithout tetracycline must have the correct separation of the helix-turn-helix motifs to be able to bind to DNA, and the interaction with tetracycline forces the helix-turn-helix motifs apart so that binding can not occur.

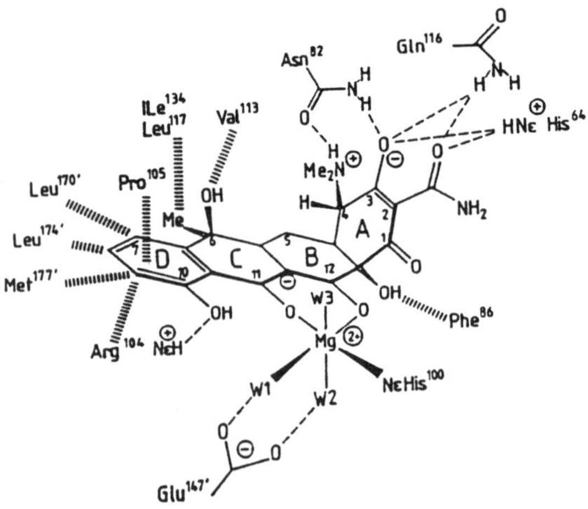

Fig. 5. Schematic drawing of the very specific recognition between Tet repressor and tetracycline; hydrogen bonding and hydrophobic contacts are indicated. Tetracycline is coordinated to magnesium ion Mg^{2+} which interacts with water molecules W1, W2, W3. Tet repressor binds to operator DNA only in absence of tetracycline. Taken from [14].

The zinc finger motif

The classical zinc finger motif consists of a short stretch of protein that forms a small domain which is stabilized by coordination of two histidine imidazoles and two cysteine S groups to a central zinc ion [8]. The N-terminal part of the peptide forms an antiparallel pleated sheet and the C-terminal part forms an α-helix. Meanwhile ten different types of zinc-fingers have been observed, which all have a central zinc ion. Most of the proteins contain three or more fingers in direct succession. The sequences of the zinc-fingers show the following consensus:

$$\mathrm{o\ x\ C\ x_{2\text{-}4}\ C\ x_3\ o\ x_5\ o\ x_2\ H\ x_{3\text{-}4}\ H\ x_{2\text{-}6}}$$

where o is hydrophobic amino acid and x is variable in character. The invariant Cys and His residues define the structures as zinc fingers. A good example of the interaction between a zinc-finger and DNA is given in Fig. 6.

where o is hydrophobic amino acid and x is variable in character. The invariant Cys and His residues define the structures as zinc fingers. A good example of the interaction between a zinc-finger and DNA is given in Fig. 6.

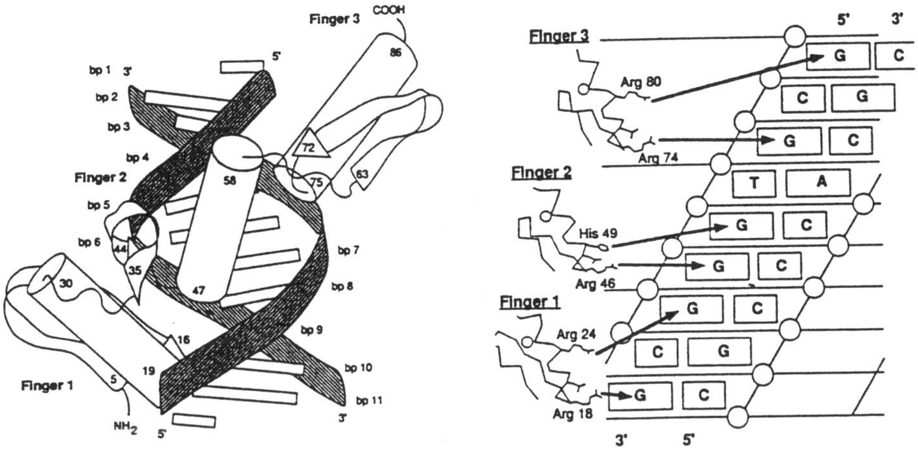

Fig. 6. Interaction between three zinc fingers in Ziff268 and DNA. Left, schematic description of the complex (Zn ions omitted), right recognition of DNA occurs via guanine bases; DNA represented in cylindrical projection. Taken from [15].

The leucine zipper

The leucine zipper is named according to the interaction between the two α-helices in the coiled coil. The distribution of amino acids along the sequence of the coiled coil segments is such that every seventh amino acid is hydrophobic (mostly leucine). Since the α-helix has 3.6 amino acids per turn, this means that every second turn a hydrophobic amino acid is located on the same side of the α-helix. The two α-helices interact such that the hydrophobic side chains interdigitate and give rise to a zipper-like appearance which stabilizes the interaction between the two α-helices in the coiled coil. In this coiled coil, the two α-helices are parallel to each other and they carry at their C-terminal ends two short α-helices with a fork-like structure. Since they contain a number of basic amino acids, they are suited to interact with the phosphate groups in the sugar-phosphate backbone of the DNA double helix, and these and other amino acids can interact with the major grooves of the DNA. A picture of the recently determined X-ray structure of the jun-fos hetrodimer structure is shown in Fig. 7

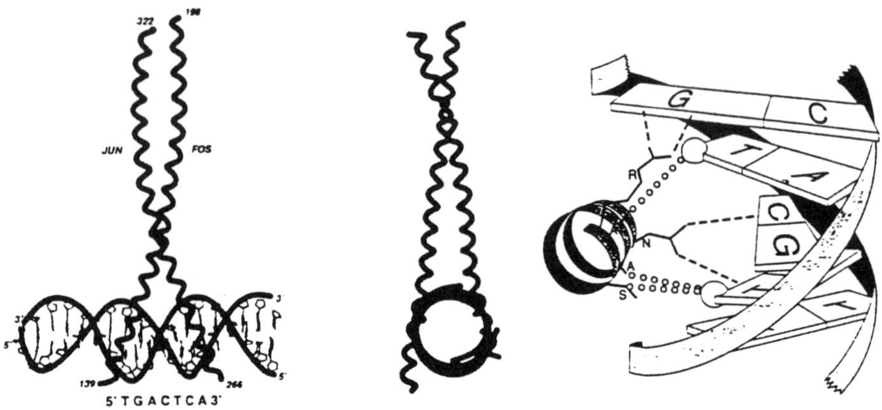

Fig. 7. Structure of the FosJun-DNA complex. Left, Interaction of the heterodimer protein with major grooves of DNA in two views, right, detail of protein-DNA contacts. Taken from [9].

Interaction of an antiparallel β-pleated sheet with a DNA major groove

The MetJ repressor occurs as a homodimer and its binding to DNA is regulated by the corepressor S-adenosylmethionine (AdoMet). In presence of AdoMet, MetJ binds to a specific DNA sequence. It inserts an antiparallel β-pleated sheet, which is formed by the interaction of the two monomers in the homodimer, into the major groove of a DNA segment with palindromic nucleotide sequence. This means that the twofold axis of the MetJ homodimer coincides with the two fold axis of the palindromic sequence DNA double helix (Fig. 8).

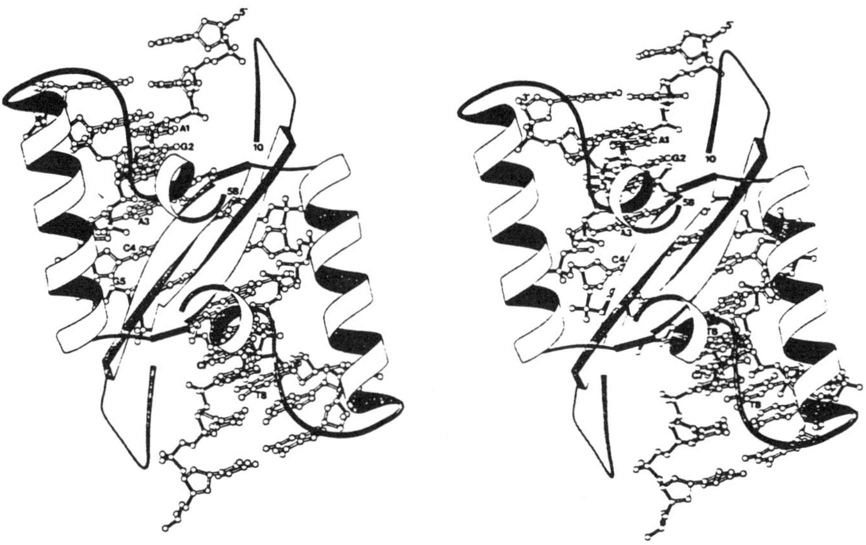

Fig. 8. Partial structure (amino acids 10-58) of the MetJ Repressor-operator complex. The structure has twofold symmetry, the β-pleated sheet inserts into the DNA major groove (Taken from [10]).

Conclusions:

The specific interactions which have thus far been observed between the DNA double helix and proteins involved in transcriptional regulation can be grouped into four main classes. With other proteins, the interaction with DNA can be different as shown for restriction endonucleases, DNase I, and methyltransferases. These proteins occur predominantly as monomers and recognize non-palindromic DNA sequences. In the RNA field, the situation is even more complex because RNA frequently forms non-double helical structures with loops and single strands which can interact with proteins in a variety of different schemes.

are arranged in certain protein motifs. In addition, water molecules can mediate between hydrogen bonding donors and acceptors, and hydrophobic contacts come into play. In principle, the number of interaction schemes can be virtually "indefinite", but there are preferred binding and recognition modes so that approaches towards predictions of protein-nucleic acid interactions have been put forward [16, 17].

Acknowledgements

This work was supported by Deutsche Forschungsgemeinschaft (Sonderforschungsbereich 344), Bundesministerium für Forschung und Technologie (Project 41.13), Fonds der Chemischen Industrie.

References

1. Struhl, K. (1989), Helix-turn-helix, zinc-finger, and leucine-zipper motifs for eukaryotic transcriptional regulatory proteins. *Trends in Biochem. Sci.* **14**, 137-140.

1a. Travers, A. (1993), DNA-Protein Interactions. *Chapman & Hall, London.*

2. Harrison, S. C. (1991), A structural taxonomy of DNA-binding domains. *Nature* **353**, 715-719.

3. Pabo, C. O. and Sauer, R. T. (1992), Transcription factors: structural families and principles of DNA recognition. *Ann. Rev. Biochem.* **61**, 1053-1095.

4. Saenger, W. (1983), Principles of Nucleic Acid Structure. *Springer-Verlag* New York.

5. Seemann, N. C., Rosenberg, J. M. and Rich, A. (1976) Sequence-specific

recognition of double helical nucleic acids by proteins. *Proc. Natl. Acad. Sci. USA* **73**, 804-808.

6. Brennan, R. G. and Matthews, B. W. (1989), The helix-turn-helix DNA binding motif. *J. Biol. Chem.* **264**, 1903-1906.
7. Gehring, W. J., Affolter, M. und Bürglin, T. (1994), Homeodomain proteins. *Ann. Rev. Biochem.* **63**, 487-526.
8. Schwabe, J. W. R. and Klug, A. (1994), Zinc mining for protein domains. *Nature Struct. Biol.* **1**, 345-349.
9. Glover, J. N. M. and Harrison, S. C. (1995), Crystal structure of the heterodimeric bZIP transcription factor C-Fos-c-Jun bound to DNA. *Nature* **373**, 257-261.
10. Somers, W. S. and Phillips, S. E. U. (1992), Crystal structure of the met repressor-operator complex at 2.8 Å resolution reveals DNA recognition by β-strands. *Nature* **359**, 387-393.
11. Otwinowski, Z., Schevitz, R. W., Zhang, R.-G., Lawson, C. L., Joachimiak, A., Marmorstein, R. C., Luisi, B. F. and Sigler, P. B. (1988), Crystal structure of trp repressor/operator complex at atomic resolution. *Nature* **335**, 321-329.
12. Jordan, S. R. and Pabo, C. O. (1988), Structure of the lambda complex at 2.5 Å resolution: details of the repressor-operator interactions. *Science* **242**, 893-899.
13. Kostrewa, D., Granzin, J., Stock, D., Choe, H.-W., Labahn, J. and Saenger, W. (1992), Crystal structure of the factor for inversion stimulation FIS at 2.0 Å resolution. *J. Mol. Biol.* **226**, 209-226.
14. Hinrichs, W., Kisker, C., Düvel, M., Müller, A., Tovar, K., Hillen, W. and Saenger, W. (1994), Structure of Tet repressor-tetracycline complex and regulation of antibiotic resistance. *Science* **264**, 418-420.
15. Pavletich, N. P. and Pabo, C. O. (1991), Zinc finger-DNA recognition: Crystal structure of a Zif268-DNA complex at 2.1 Å. *Science* **252**, 809-817.

16. Suzuki, M. and Yagi, N. (1994), DNA recognition code of transcription factors in the helix-turn-helix, probe helix, hormone receptor, and zinc finger families. *Proc. Natl. Acad. Sci. USA* **91**, 12357-12361.

17- Suzuki, M., Brenner, S. E., Gerstein, M. and Yagi, N. (1995), DNA recognition code of transcription factors. *Protein Engin.* **8**, 319-328.

CATALYTIC ANTIBODIES
An Intriguing Host-Guest System

BERNARD S. GREEN
The Hebrew University of Jerusalem
Faculty of Medicine - School of Pharmacy
Department of Pharmaceutical Chemistry
P.O. Box 12065; Jerusalem 91120, Israel

Discussions of host-guest chemistry generally involve simple, but sometimes more complex, naturally occuring or synthesized molecules as models or mimics of biological host systems, such as enzymes. In general, the host molecule mimics a very narrow, restricted aspect of the biological binder, catalyst, or transporter. In this presentation we are discussing a relatively complex biological system, the antibody molecule, as a host molecule. One can profitably discuss antibodies as host molecules which reversibly bind guest molecules with high specificity (one of the traditional, long-studied aspects of "immunochemistry"). In addition, one can discuss the newer, exciting, enzyme-like catalytic aspects which may also be found in selected antibodies. It will probably be useful for the audience to be exposed to both of these features. It will be seen that antibody-based systems are thus far superior to any other as tailor-made, enzyme-like catalysts.

Antibodies

Typically, antibody proteins (IgG) have a Y-shaped structure where the two identical tips of the "arms" contain the binding sites. These combining sites (or complementarity determining regions, CDRs) of the antibody are formed by six peptide loops whose varying amino acid sequences provide the incredibly large number of possible antibody combining sites and their ability to tightly bind virtually any chemical structure (association constants are typically in the range 10^4 - 10^8 M^{-1} but values up to $10^{10} M^{-1}$ and higher are also possible) . Antibodies have evolved to

G. Tsoucaris et al. (eds.), Crystallography of Supramolecular Compounds, 395–399.

bind to invading pathogens as well as biopolymers such as proteins. However, they can also be generated to bind to small molecules (haptens) and much of the knowledge of antibody specificity was obtained by detailed studies of antibodies with organic substances [1,2].

Binding Antibodies

The host properties of antibodies are exploited in the wide use of immunoassay techniques for the accurate and sensitive detection of chemicals, especially for biomedical applications. Thus, it is routine to be able to detect the presence of a particular steroid or alkaloid in the presence of very closely similar structures, even in very low concentrations in complex biological fluids.

Much host-guest chemistry has been devoted to chiral discrimination, a topic of particular interest in many areas of chemistry and biology. High chiral discrimination by antibodies was reported early on (1928!) by immunochemists such as Landsteiner [3]. Antibodies have also been used, similar to more conventional organic hosts, to bind metal ion complexes with high selectivity [4]. And antibodies can be raised which bind a single ion, such as mercuric ion [5]. The notion of using antibodies as templates for bringing together two species has also been demonstrated, for example, for photocycloaddition reactions [6].

Catalytic Antibodies

About tweny years ago Jack Dunitz predicted that,"the day is not too far off when comprehensive examinations in organic chemistry will include questions such as the following: Outline the synthesis of an 'enzyme' that will effect the conversion of acetophenone to R(+)-2-phenylethanol" and he further mused that there would be many correct answers including transition metal catalysts, NADH analogs, functionalized crown ethers, diimide precursors, activated crystal surfaces, etc. At the time, he didn't include the possibility that antibodies would provide one of the best approaches for answering this question. (The philosophical question regarding the experimenter's *synthesis* of an antibody vs his or her choosing the conditions to ensure that a specific antibody is generated may be raised but today "man-made antibodies" [7] are common as genetic engineering provides many antibodies that would not naturally be produced. In this connection, it is interesting to note that the idea of *synthesizing* an enzyme was proposed as early as 1904 by Emil Fischer.)

Recently, Schultz and coworkers carried out remarkable antibody-catalyzed reductions of the type that Dunitz had posed: one antibody selectively reduced a ketone to an alcohol with high enantiomeric excess, even though the substrate contained two very closely similar substitutuents. In a second example, this group used an antibody-mediated reaction for the reduction of only one keto group in a diketone, even though the chemical environments of the two ketone functional groups were nearly identical. [8] This kind of selective chemistry not only rivals what can be obtained with enzymes but it illustrates the capacity for experimenter-defined design, a feature that is, thus far, not possible with enzymes or modified proteins.

Another example of the extraordinary selectivity of catalytic antibodies is represented by antibodies which catalyze the hydrolysis of an ester substrate but not of the corresponding carbonate (the sole change in structure being replacement of O-C(C=O)CH_2- by O-C(C=O)O-) [9]. All known enzymes which catalyze the hydrolysis of an ester invariably also catalyze the hydrolysis of the corresponding carbonate.

The approach to obtaining catalytic antibodies is conceptually very simple: to select the antibodies that bind to a structure that mimics the transition state (TS) of the reaction which one wants to catalyze. The idea that enzymes act by stabilizing transition states was proposed by L. Pauling and he also first formulated the idea of inducing catalytic antibodies by using a transition state analog. The idea was first succinctly stated by Jencks [10].

Although an unambiguous antibody-promoted ester hydrolysis reaction was reported in 1980 [11], the catalytic antibody field was launched with the reports of R.A. Lerner and his colleagues [12] and P.G. Schultz and his group [13] at the end of 1986. Over 80 different chemical reactions have been catalyzed by antibodies (see review by Thomas [14] and many recent publications.) The most numerous are acyl transfer reactions (ester and amide hydrolysis; ester exchange reactions; amidation of esters; etc.) in which a tetrahedral, negatively charged phosphorus species is the hapten used to elicit the antibodies. Some reactions have mimicked actual enzyme catalyzed reactions (chorismate mutase) using the TSA which is the most powerful inhibitor of the enzyme as the hapten [15, 16]. The 'dial-a-property' character of the catalytic antibody approach is perhaps best illustrated by the production of a number of "Diels-Alderase" antibodies, which catalyze one of the most important and versatile synthetic reactions used by organic chemists but one which is not known in Nature. Here, the possibility of selective reaction pathway was demonstrated by producing separate antibodies that catalyze only the *endo* or only the *exo* product whereas normally mixtures of *endo* and *exo* adducts are formed [17].

Many applications have been alluded to in earlier studies, but recent reports describe explicitly designed catalytic antibody uses: as biosensors [18], prodrugs [19,20], therapeutic agents [21], etc. As expected, high enantiomeric excess asymmetric synthesis can be achieved using catalytic antibodies (e.g., [22]). In addition, multigram-scale organic synthesis has been performed using catalytic antibodies [23] and a catalytic antibody played a key role in a natural product synthesis [24].

Recent and continuing advances have involved mechanistic, x-ray structural [25, 26,27] and site-directed mutagenesis (selective changes of amino acids in the combining site) studies. A full discussion of catalytic antibodies must include recent, highly effective methods for selecting catalytic antibodies (a method termed catELISA) [28] and improved immunization protocols which afford greatly increased numbers of catalytic antibodies [29]. Finally, applications of catalytic antibodies hinge on their efficiency, turnover, and our ability to catalyze reactions thus-far not reported using antibodies. The rapid growth of this field and the active current work gives promise that many practical applications as well as a deeper understanding of the nature of catalysis will surely result.

References

[1] K. Landsteiner The Specificity of Serological Reactions (C.C. Thomas, Springfield, Ill, 1936).

[2] D. Pressman and A.L. Grossberg, "The Structural basis of Antibody Specificity", New York, W.A. Benjamin, 1968.

[3] K. Landsteiner and J. van der Scheer *J. Exp. Med.* **48**, 315-320 (1928); K. Landsteiner, *Naturwissenschaften*, **18**, 653-659 (1930).

[4] D. T. Reardan, C. F. Meares, D. A. Goodwin, M. McTigue, G. S. David, M. R. Stone, J. P. Leung, R. M. Bartholomew, and J. M. Frincke, *Nature* **316**, 265-268 (1985).

[5] D.E. Wylie, D. Lu, L.D. Carlson, R. Carlson, K.F. Babacan, S.M. Schuster, F.W. Wagner. *Proc. Nat. Acad. Sci. USA*, **89**, 4104-4108 (1992).

[6] Balan A, Doctor BP, Green BS, Torten M, Ziffer H: *JCS Chem Comm* 106-108, 1988.

[7] G. Winter and C. Milstein, *Nature* **349**, 293 (1991).

[8] Hsieh, L.C., Yonkovich, S., Kochersperger, L. and Schultz, P.G. (1993) *Science* **260**, 337-339.

[9] D.S. Tawfik, R.R. Zemel, R. Arad-Yellin, B.S. Green, and Z. Eshhar *Biochemistry*. **29**, 9916-9921 (1990).

[10] Jencks, W.P. (1969) *Catalysis in Chemistry and Enzymology*, McGraw-Hill, New York, p. 288.

[11] Kohen F, Kim JB, Lindner HR, Eshhar Z, Green BS:*FEBS Lett.*, **111**, 427, 1980.

[12] A.Tramontano, ,K. D. Janda, and R. A. Lerner, *Science* **234**, 1566-1570 (1986)

[13] S.J. Pollack, J. W. Jacobs, and P. G. Schultz, *Science* **234**, 1570-1573 (1986)

[14] N. R. Thomas *Appl. Biochem. Biotech.* **47**, 345-372 (1994).

[15] D. Hilvert, and K. Nared J. Am. Chem. Soc. 110, 5593 (1988).

[16] D.Y. Jackson, et al. J. Am. Chem. Soc. 110, 4841 (1988).

[17] V.E. Gouverneur, K.N. Houk, B. de Pascual-Teresa, B. Beno, K.D. Janda, and R.A. Lerner *Science* **262**, 204-208 (1993).

[18] G.F. Blackburn, D. B. Talley, P. M. Booth, C. N. Durfor, M. T. Martin, A. D. Napper, and A. R. Rees, *Anal. Chem.* **62**, 2211-2216 (1990).

[19] D.A. Campbell, B. Gong, L.M. Kochersperger, S. Yonkovich, M.A. Gallop, and P.G. Schultz, *J. Am. Chem. Soc.* **116**, 2165-2166 (1994).

[20] H. Miyashita, Y. Karaki, K. Masakazu, I. Fujii, *Proc. Nat. Acad. Sci., U.S.A.* **90**, 5337-5340 (1993).

[21] Landry, D. W., K. Zhao, X. Q. G. Yang, M. Glickman, and T. M. Georgiadis *Science* **259**, 1899-1901 (1993).

[22] K.D. Janda, S. J. Benkovic, and R. A. Lerner, *Science* **244**, 437-440 (1989).

[23] J.-L. Reymond, J.-L. Reber and R.A. Lerner *Angew. Chem. Int. Ed. Engl.* **33**, 475-477 (1994).

[24] E. Keinan, et al. In press.

[25] M.R. Haynes, E.A. Stura, D. Hilvert, and I.A. Wilson science, 263, 646 (1994).

[26] B. Golinelli-Pimpaneau, B. Gigant, T. Bizebard, J. Navaza, P. Saludjian, R. Zemel, D.S. Tawfik, Z. Eshhar, B.S. Green and M. Knossow *Stucture* **2**, 175-183 (1994).

[27] W.G. Zhou, J. Guo, W. Huang, R.J. Fletterick, and T.S. Scanlan. Science, 265, 1059 (1994)

[28] D.S. Tawfik, B.S. Green, R. Chap, M. Sela, and Z. Eshhar, *Proc. Nat. Acad. Sci., U.S.* **90**, 373-377 (1993).

[29] D.S. Tawfik, R. Chap, B.S. Green, M. Sela, and Z. Eshhar, *Proc. Nat. Acad. Sci., U.S.* **92**, 2145-2149 (1995).

Supramolecular Organization of Keratinized Tissue

Jean-Luc LEVEQUE
Laboratoires de Recherche de L'OREAL
AULNAY-SOUS-BOIS - FRANCE

1 - Introduction -

Keratins are a family of proteins produced by certain epidermal cells in higher vertebrates. Their function is to act as an interface between the living medium and the external environment. This function is crucial, as life could not exist in the various conditions encountered on earth without a fine physical and chemical equilibrium. By way of an example, hair clearly protects against extreme temperatures but also shields the against UV radiation. Similarly, the stratum corneum (the first 10 to 20 first microns of the epidermis) limits exchanges between the outer and inner environments. Feathers both act as a thermal shield and permit flight, while hooves, nails and horns are necessary for the travel, food-seeking and fighting behaviours necessary to maintain given species.

This interface between the living medium and the outside environment can only be fulfilled by these proteins thanks to their remarkable physical and chemical properties which have been the subject of intensive research for the last 50 years or more. Although they all belong to the same family, keratins differ in their chemical nature and, above all, in their macromolecular organisation.

All keratins possess cystine residues in their amino acid sequence, and this is an essential feature. This particular amino acid (R - CH - CH2 - S-S-CH2 -CH-R) is responsible for very strong intra- and inter-chain bonds that confer not only a high degree of chemical inertia but also highly unusual physical properties.

The first motor behind keratin research was the international wool trade. Chemists in universities and private companies of the countries concerned (mainly Australia and the United Kingdom) studied the properties of wool in order to develop ways of changing its texture and other aspects. Keratins later became a model protein investigated by biologists, clinicians, biophysicists and physicists. This research partially elucidated the structure and amino acid sequence of keratins, but also covered the biological aspects of this protein in

G. Tsoucaris et al. (eds.), Crystallography of Supramolecular Compounds, 401–413.

settings as varied as pathological dysfunction, synthesis, and evolution of the species (1).

As stated above, keratins act as an interface between the biological medium and the environment, a function that depends on the physical and chemicophysical properties confered on this remarkable molecule by a particular tissue organisation and structure, which were first studied in the 1930s(2). In the 1960s, Fraser and Baden made a major advance by showing the existence of two or three α helices in the protofilaments of wool fibers (1).

This chapiter will briefly recall the morphometric organisation of two keratinized tissues -- hair and the horny layer or stratum corneum (SC) -- before looking at molecular structure. The latest structural data, obtained by means of synchrotron (X ray) investigations, which have also revealed supramolecular organisation, will be reported and discussed. The chapter will end with the influence of water, which can modulate all the properties of keratins. These studies have thrown light on the part played by lipids in the supramolecular structure of keratins, together with the biological functions of the SC.

2 - Morphohistology of Hair and Stratum Corneum

The hair shaft results from intense biochemical activity in different cells of the hair follicle, under the control of the papillary dermis. This complex cell division and differentiation activity leads specialized cells to form the main elements of hair, i.e. the cuticle, cortex and medulla.

Figure 1 illustrates the complex manner in which these main elements are organized. The cuticle is composed of about 10 layers of cuticular cells with a thickness of 0.5 μm. Each of these cells is composed of an internal part (the endocuticle) and an external part (the sulfur-rich exocuticle). The intercuticular spaces could contain organized lipids bound to sulfur-containing proteins (3).

The cortex is the mainstay of the fiber. It is composed of fusiform cortical cells of about 3 to 5 microns in diameter and a few tens of microns long. These cells only contain nuclear remnants. They are limited by membrane zones which sometimes have a layered aspect. Grains of melanin (about 1 μm in diameter) show up as dark spheres embedded in the cortex. Cortical cells contain subassemblies known as macrofibrils (diameter 0.3 microns) about 80 Å in diameter that contain microfibrils of about 20 Å in diameter. Although conventional electron microscopy is of little value in determining the ultrastructure of the hair shaft, mathematical processing methods applied to ultra-high resolution images have provided a model compatible with X-ray diffraction data. It is thought that each microfibril is composed of 9 + 2 protobrils in turn composed of 2 or 3 left-handed α helices (1).

Protofibrils, microfibrils and macrofibrils are embedded in a sulfur-rich amorphous protein matrix. Decomposition of hair protein into its constituent amino acids yields a majority of cystine residues, which are thought to stabilize the structure and to confer remarkable mechanical properties. Lipids are present in very small quantities (a few percent), and mainly consist of fatty acids, wax and triglycerides. The most polar lipids present in hair are cholesterol (about 5%) and ceramides (4).

The SC is a membrane about 10 to 20 microns thick composed of flattened corneocytes with an individual thickness of about 0.3 μm. These flattened cells, which do not possess a nucleus, have a fairly regular geometric shape (pentagonal or hexagonal) and are embedded in a lipid matrix.

Studies of corneocyte structure have revealed a lipid membrane bound covalently to the protein composing the cystine-rich envelope, which is about 150 Å thick. The keratin filaments (about 80 Å in diameter) are randomly dispersed in the cellular plane. These cells are connected by desmosomal plates about 1 μm in length, which have a glycoproteic composition. The cells are released during desquamation by enzymatic digestion of the plates. Landmann's electron microscopy studies revealed the bilayer organisation of intercellular lipids (5). Figure 2 shows the two types of organisation present in the SC. These lipids represent about 10% of the SC by weight. They are composed fatty acids, cholesterol and cholesterol diesters. The most polar fraction (about 30% of the total lipid pool) appears to be composed of the 7 types of ceramides described by Downing (6).

3 - Oriented structure in keratinized tissue -

Figure 3 shows one example of the various shapes and intensities of the different diffraction patterns. It is particularly remarkable to observe very fine diffraction rings and arcs indicating the presence of large crystallites. The aspect can vary somewhat according to the sample, even though the main motifs remain the same. These differences between samples may be explained by the fact that the proliferation and differentiation of the cells synthesising the SC are modulated by both the environment and the biological medium.

3.1. TYPES OF KERATIN

The main diffraction patterns obtained with the SC and hair are shown in tables 1 and 2, based on work by Garson (7) and Franbourg (8).
In the case of hair the keratin spectrum is perfectly typical of keratin α as described by Astbury in 1930. The mean distance between two helices is 9.8 Å, while the step of the helix is 5.15 Å.

TABLE 1 : Main Stratum Corneum Scattering features -

1- Protéic Components -

d (Å)	Aspect	Intensity	Interpretation
9.4	Equatorial sharp arc	vs	
4.7	" "	s	β pleated sheet of
3.15	" "	w	an non identified protein
9.5	Non isotropic broad band	vs	β-pleated sheet of keratin

2 - Lipidic Components -

d (Å)	Aspect	Intensity	Interpretation
62.5	Equatorial broad arc	s	Ceramide bilayers type 1
45	" "	s	Ceramide bilayers type 2
22.6	" "	s	" "
14	" "	w	" "
42	Equatorial sharp arc	w	Bilayers composed of neutral lipids
33.5	Equatorial sharp arc	vs	Cholesterol crystals
16.8	" "	w	" "

The relatively distinct patterns in the equatorial position at 30 and 88 Å appear to reflect the mean distances between keratin protofibrils and microfibrils, respectively. The meridional arc at 67.7 Å may be a harmonic of super-coiled α keratin. As regards the SC, it is astonishing to observe no meridional arc at 5.15 Å representing the step of the α helix. Indeed, according to most data obtained by X-ray diffraction of the SC, this tissue is composed of keratin α. In fact, close examination of the plates shows a degree of anisotropy in this diffraction zone situated around 9.5 Å, which would be more characteristic of keratin β as described by Fraser and Mc Rae (9). Garson et al stressed an important point: the X-ray diffraction pattern of the SC generated by the callus (the thick horny layer on the hands and feet) clearly shows a ring at 5.15 Å characteristic of keratin α. This might mean that the macromolecular structure of keratin depends on the conditions in which it is synthesised.

3.2. OTHER PROTEINS

There are at least two diffraction features probably attributable to protein constituents of the horny layer, but which remain to be interpreted. In parallel geometry, i.e. during irradiation of the sample parallel to its plane, the equatorial axis contains an extremely fine arc corresponding to a reticular distance of 9.4 Å (harmonics 2 and 3 too are often clearly visible). This structural element of the SC, previously identified in the mouse SC by White (10), appears to be organised perpendicular to the plane and could only be situated in the cells or the most dilated areas of the intercellular spaces. This points to the corneocyte envelope protein or desmosome glycoprotein.

The second uninterpreted diffraction spot is that located at about 125 Å, close to the center of the diagramme, on the equator. It appears to have a proteic or lipoproteic composition, as this diffraction motif does not disappear after delipidation.

Further studies of the structural elements generating these diffraction patterns will focus on membranes of isolated corneocytes or desmosomes.

TABLE 2 : Main Hair Scattering features -

d (Å)	Aspect	Intensity	Interpretation
1- Proteic Components			
9.8 e	band	vs	α-Keratin
5.15 m	arc	vs	α-Keratin
30 e	band	w	mean protofibril distance
67.7	arc	s	pitch of the supercoil (Harmonic 7)
88 e	band	s	mean microfibril distance
2 - Lipidic Components			
40	ring	s	Lipid bilayer (?)

3.3. LIPIDS

About 20 diffraction patterns attributable to lipids can be recorded after narrow and wide-angle diffraction in the SC. Delipidation with neutral (hexane) or polar solvents (chloroform / methanol, 2/1) gives rise to a preliminary distribution of these lipids according to their nature and, thus, their origin (sebaceous or epidermal, for example).

Our results, obtained at small angles in parallel geometry, point to the existence of lipid bilayers oriented in the plane of the SC. There appear to be two types of organisation corresponding to reticular distances of 65 and 45 Å. This type of organisation seems to be due to the most polar lipids (ceramides), that are removed by delipidation with chloroform / methanol. It is these lipids and this

particular organisation (see below) that ensure the barrier function of the SC against water vapor. Other arcs in the equatorial position are due to neutral lipids. Finally, it appears that cholesterol is often present in crystalline form in the SC, as revealed by a very fine arc situated at 33.5 Å (11). Numerous rings appear at wide angles, corresponding to the mean distances between the aliphatic chain of lipids engaged in the bilayers.
Hair yields a single, relatively distinct ring which varies from one sample to the next and is located at 40 Å; Franbourg attributed this ring to lipids (8).

4 - Influence of water on the structure of keratinised tissues -

The hydration state has a marked influence on the properties of keratins. Water molecules interact with protein endings (amino acid side chains for example) with binding energies that vary according to the nature of these endings. Water molecules can also interact with each other via hydrogen bonds. According to the energy of these bonds, keratin-associated water is described as highly bound, bound and free, but there is clearly a continum between these types of water, and "free water" does not have the properties of liquid water (12) (13) (14).
When hair fibres are gradually hydrated the reticular distances of the diffraction spots and arcs characterising the α helix are displaced.
These displacements, which remain small (about 2%) and are always smaller than the diameter of a water molecule (13). Water cannot thus penetrate the helices. In contrast, the displacements noted at reticular distances corresponding to mean distances between protofilaments and microfilaments (22 Å and 88 Å) are compatible, according to Franbourg et al, with swelling of amorphous interfibrillar keratin (8). As regards the ring at 40 Å, which seems to be characteristic of supramolecular lipid organisation in hair, its intensity is markedly increased when the water content is very high. This result is difficult to interpret, as our understanding of lipid-protein interactions in hair, and even the location of lipids in the structure, are poorly characterised.

As regards the SC and the organisation of keratin, the distance between the helices increases from 9.46 to 9.92 Å when the conditioning RH increases from 55% to 100% (15). In the case of lipids, the arcs situated at about 65 and 45 Å are modified in terms of their intensity but not their position. This latter experimental observation signifies that the lipid bilayers do not swell, unlike phospholipid bilayers.

The lipid diffraction pattern becomes increasingly distinct as hydration increases (see figure 4), reflecting better structural organisation. According to Ribaud (15), this structural organisation, previously reported by Bouwstra (16), results from an interaction with the polar heads of ceramides, without intercalation in the space between the bilayers. This hypothesis is difficult to check by means of wide-angle diffraction, in which no change in the spaces between aliphatic chains is observed. It is also conceivable that cell swelling creates pressure on the intercellular spaces which, in turn, leads to better lipid bilayer organisation.

5 - Relation between lipid organisation and SC function

As pointed out in the introduction, the most important function of the SC is to protect the living medium from the environment. The SC barrier function maintains epidermal hydration and inhibits the penetration of xenobiotics. Studies over the last 10 years have clearly demonstrated the role of intercellular lipids and their nature in the efficacy of this double barrier formed by the SC. It was, however, interesting to investigate the part played by the supramolecular organisation of these lipids. To do so it was necessary to determine the experimental conditions in which changes in barrier function were not due to lipid extraction. There are at least two such sets of conditions. Heating the SC to above 90°C makes it far more permeable to water vapor (and other molecules). The other example is the effect of sodium laurylsulfate (SLS), which increases SC permeability without removing significant amounts of ceramides. It was shown that these two experimental situations disorganised the bilayers characterized by the two equatorial diffraction spots at 65 and 45 Å(17). At the same time, the DSC peaks corresponding to the change in the state of these lipids was displaced towards lower temperatures, reflecting a degree of fluidification. These results are in keeping with other studies showing that certain molecules used to increase penetration, such as azone ® and oleic acid, had the same disorganising effect on the supramolecular scaffold of lipids (15).

6 - Conclusion -

Keratins have always been considered as highly complex tissues with littler biological interest. As regards the SC, its function has been undergoing a reassessment over the last 10 years. It is not a dead tissue but rather a tissue whose properties both result from and generate a biological mechanism the end purpose of which is to preserve life. This regulatory function depends of the exchange of water vapor across its structure (18). These phenomena have not been fully elucidated, even if recent biophysical and biochemical data have greatly improved our understanding of the SC. The possibility of working on cultured epidermis, in which the conditions of proliferation and differentiation can be modulated, should be of great help in this direction. Structural studies, which have been rendered possible by the use of synchrotron analysis, will no doubt be the experimental approach of choice.

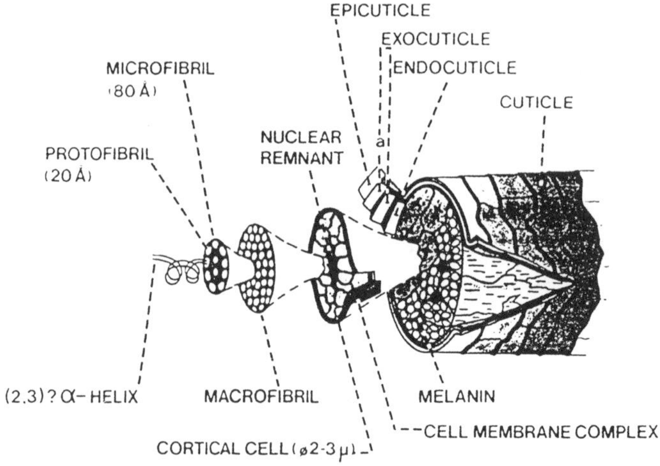

Fig. 1 : Schematized structure of Hair

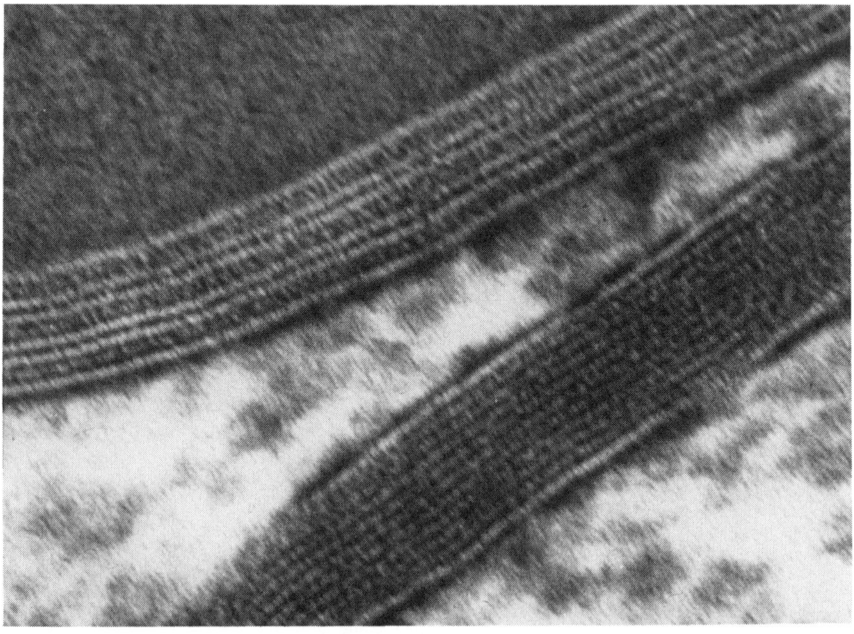

Fig. 2 : Supramolecular organization of lipids in the intercellular cements of Stratum Corneum (Courtesy F. FIAT - L'OREAL)

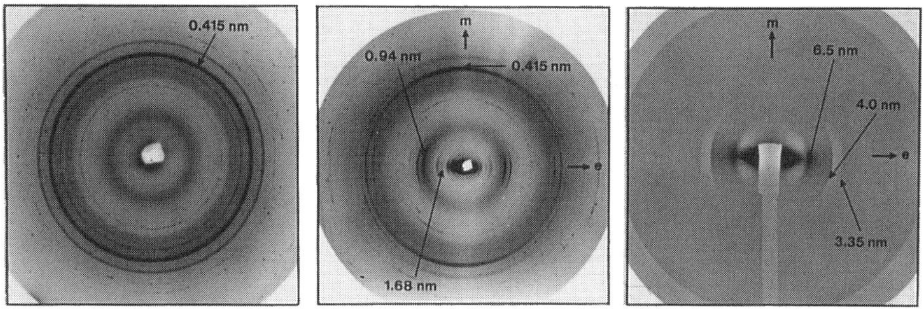

Fig. 3 : X-ray patterns of normal Stratum Corneum sample (7).

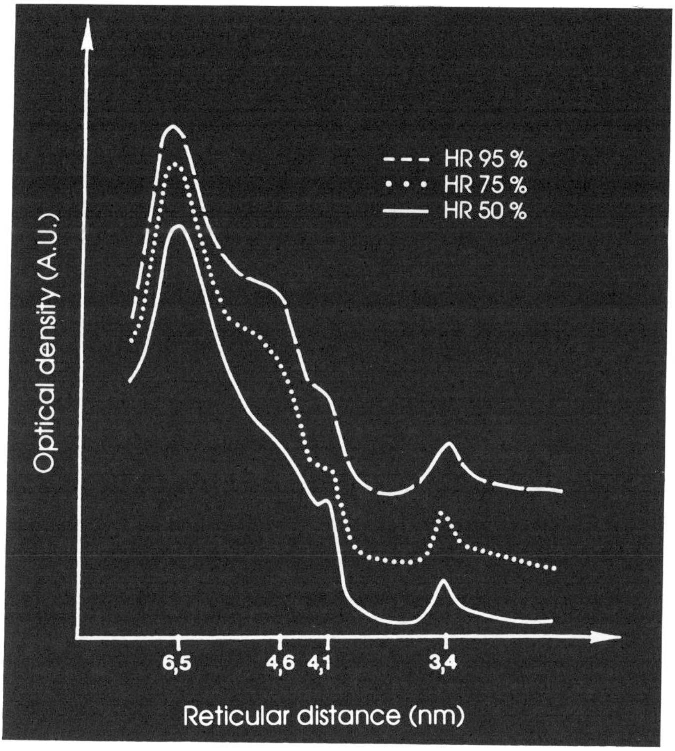

Fig. 4 : Modification of the X-ray pattern (at small angle) of Stratum Corneum under water vapor influence (17).

References

1. Fraser, R., Mac Rae.T.P., Rogers. A. (1972) Keratins : their composition, structure and biosynthesis. New York City Ed. C. Thomas.

2. Astbury, W.T. and Street, A. (1931) X-ray studies of the structure of wool and related fibres,Philosophical Transations of the Royal Society of London Series A, **230**, 75-101

3. Naito, S., Takahashi, T., Hattori, M. and Arai, K. (1992) Histochemical observation of the cell membrane complex of hair *Sen'i Gakkaishi* **48 N° 8**, 420-426.

4. Hussler, G., Kaba, G., François, A.M. and Saint-Léger, D. (1995) Isolation and Identification of human hair free ceramides. *Int. J. Cosmet.* **17**, (to be published)

5. Landmann, L. (1986) Epidermal permeability barrier : transformation of lamellar granule-disks into intercellular sheets by a membrane-fusion process, a freeze-fracture study. *J. Invest. Dermatol.* **87**, 202-209.

6. Wertz, P.W., Miethke, M.C., Long S.A., Strauss, J.S., Downing, D.T. (1985) The composition of the ceramides from human Stratum Corneum and from comedones. *J. Invest. Dermatol.* **84**, 410-412.

7. Garson, J.C., Doucet, J. Lévêque, J.L., and Tsoucaris, G. (1991) Oriented structure in human Stratum Corneum revealed by X-ray diffraction. *J. Invest. Dermatol.* **96**, 43-47.

8. Franbourg, A., Leroy, F., Lévêque, J.L. and Doucet, J. (1995) Influence of water on the structure of α keratin fibres. Biophysics and Synchrotron Radiation. Grenoble, August 21-25.

9. Fraser, R.D.B. and Mac Rae, T.P. (1962) An investigation of the structure of β keratin. *J. Mol. Biol.* **5**, 457-466.

10. White, S.H., Mitejovsky, D. and King, G.I. (1988) Structure of lamellar lipid domain and corneocyte envelopes of mice Stratum Corneum. An X-ray diffraction study. *Biochemistry* **27**, 3725-3732.

11. Garson, J.C., Doucet, J., Tsoucaris, G. and Lévêque, J.L. (1990) Study of lipid and non-lipid structure in human Stratum Corneum by X-ray diffraction. *J. Soc. Cosmet. Chem.* **41**, 347-358.

12. Walkley, K. (1972) Bound water in Stratum Corneum measured by differential scanning calorimetry. *J. Invest. Dermatol.* **59**, 225-227.

13. Hey, M.J., Taylor, D.J. and Derbyshire, W. (1978) Water sorption by human callies. *Biochem. Biophys. Acta.* **540,** 518-533.

14. Lévêque, J.L., Escoubez, M. and Rassneur, L. (1987) Water-keratin interaction in human Stratum Corneum. *Bioeng. Skin.* **3**, 227-242.

15. Ribaud Ch. (1994). Relations entre les propriétés structurales des lipides et les fonctions barrière du Stratum Corneum. Thesis. Université de Paris Sud.

16. Bouwstra, J., Gooris, G. Salomons de Vries, M., Van der Spek, J. and Braes, W. (1992) Structure of human Stratum Corneum as a function of temperature and hydratation : a wide-angle X-ray diffraction Study. Int. *J. Pharm.* **84**, 205-216.

17. Garson J.C., Lévêque, J.L., Ribaud, Ch. and Doucet, J. (1995) Contribution of Synchrotron radiation and DSC in comprehension of lipid Stratum Corneum organisation in relation to permeability. Biophysics and Synchrotron radiation. Grenoble, August 21-25.

18. Grubauer, G. Elias, P.M. and Feingold, K.R. (1989) Transepidermal water loss: the signal for recovery of barrier structure and function. *J. Lipid. Res.* **30**, 323-333.

MOLECULAR RECOGNITION: THE LIPOCALINS

E. ELIOPOULOS
Dept. of Agr. Biology and Biotechnology,
Agricultural University of Athens,
Athens 118 55,
GREECE.

1. Introduction

The importance of non-covalent interactions in biological systems motivates much of the current interest in supramolecular assemblies. The recently obtained inclusion complex of alpha-cyclodextrin with polyethylene glycol bisamine [1] provides a good example of protection of a long molecule by the more robust cyclodextrins and their intermolecular interactions.

The inclusion of hydrophobic molecules for protection, transport as well as to provide the environment for enzymatic and photo activity has been the common functional identity of a family of proteins called the lipocalins (from the Greek words **lipos** (meaning hydrophobic) and **calyx**) [2].

There are approximately 60 non-redundant lipocalin sequences that can be divided into at least 12 non-orthologous sequence families [3].They are composed of eight or ten antiparallel beta strands arranged as two stacked orthogonal sheets that form a compact, protease resistant calyx like framework (Fig.1).

Lipocalins, whose primary structure homology between subfamilies exist at the N-terminal but is not retained beyond the first strand of the structure, have a wide range of functions and include odour and tastant binding proteins, immunoglobulin associated glycoproteins, enzymes and photoactive proteins.
Members of this family possess dual molecular recognition properties for the ligand bound by the protein and for a receptor cell to which the ligand is transported.

G. Tsoucaris et al. (eds.), Crystallography of Supramolecular Compounds, 415–427.

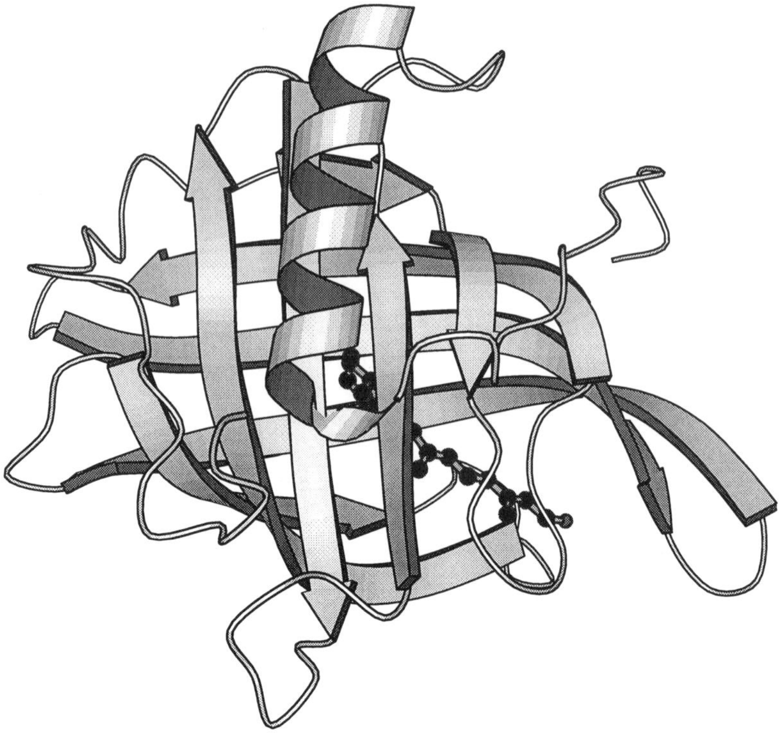

Figure 1. The lipocalin framework. (The ribbon diagram was made using the program MOLSCRIPT [29])

2. The Lipocalin family and the Sensory Systems.

Sensory transduction in taste and olfaction, the principal chemical senses, seems to be mediated by membrane associated proteins on the apical surfaces of the respective receptor cells.

Olfactory transduction is initiated when odorants enter the nasal cavity, where they interact with the cilia protruding from the dendritic tips of the chemosensory neurons. Interaction between odorants and putative ciliary receptors enhances adelylyl cyclase activity through G-protein mediated activity. Animals can detect nanomolar concentrations of odorants in ambient air despite the low sensitivity of olfactory receptors to direct stimulation by odorants [4] and the existence of the hydrophilic mucus barrier between air and receptors that the highly hydrophobic odorants have to traverse. Odorant binding proteins participate in olfaction by binding to hydrophobic odorants and transporting them accross the mucus layer to the receptor sites where recognition occurs. The differential odorant specificity may contribute to odorant recognition prior to interaction with the ciliary receptors.

The primary reception of taste stimuli occurs in specialized neuroepithelial receptor cells clustered in taste buds sheltered in epithelial folds of the foliate and circumvalate papillae [5]. The perireceptor milieu consists of taste pore mucus and secretions from salivary glants. Several different G-proteins have been identified from taste tissue. The sequence homology between the odorant binding proteins and proteins highly expressed in the salivary gland (VEG) indicate that prereceptor events could also be necessary for the concentration and delivery of sapid molecules in the gustatory system. Like odorants many sapid chemicals including most bitter-tasting substances, are highly hydrophobic and the structural resemblance of the taste lipocalins with the odour ones emphasize the close relationship of taste and olfaction [6].

The role of retinol binding protein (RBP), another lipocalin, in visual perception is well characterized and studied [7]. It involves the strong binding of retinol by RBP and its transport in plasma in association with prealbumin (transthyretin), formation of complexes with membrane associated receptors [8], and intracellular transport to photoreceptors by the cellular RBP.

Beyond the involvement of lipocalins in sensual mechanisms, members of the family act as chromophore (Crustacyanin [9], Photoactive Yellow Protein [15]) and lipid containers (Fatty acid binding proteins) or enzymes (Postaglandin D synthetase) [11].

The monomeric chain of these proteins are 130-190 aminoacid long. Comparisons of aminoacid sequences shows two regions of above average homology. One is located near the N-terminus (around residue 17 in most members of the family) where the tripeptide Glycine-X-Tryptophane (G-X-W) is heavily conserved. The other is located towards the middle of the aminoacid sequence with the hexapeptide Theonine-Aspartate-Tyrosine-X-X-Aromatic. In addition a disulphide bridge is formed between a cysteine near the C-terminus of the molecule and either a cysteine around position 36 or 66 both alternatives being compatible with the overall chain fold. Topologically is interesting to mention that the conserved tryptophane is found at the bottom of the calyx, while the hexapeptide is a surface feature beyond the tryptophane. Its conservation is more apparent in the transport members of this family and it could be involved in protein-receptor recognition. Since very little is known about the structure of the membrane associated receptors and their interaction more emphasis will be given in the binding site features in association with the different functionalities.

```
                                                10        20        30        40        50        60        70        80        90       100
Secondary Structure of RBP RBP-SEC      -------------------------------------EEEEEE--------------EEEEEEEE--------EEEEEEEEEEEEEE-----EEEEEEEE
Cow Odorant Binding Protein PBP-C- 1- 74--------------------AQEEAEQN-----LSELSGPWRTVYIGSTNPEKIQ-ENGPFRTYFRELV-FDDEKGTVDFYFSVKRDGKWKNVHVKATKQ
Rat Odorant Binding Protein OBP-R- 1- 90----MVKFLLIVLALGVSCAHHENLDIS-----PSEVNGDWRTLYVIVADNVEKVAEGGSLRAYFQHME-CGDECQELKIIFNVKLDSECQTHTVVGQKH
Histocomp. Complex Protein  PR-HC- 1- 77-----------GPVPTPPDNIQVQENFN-----ISRIYGKWYNLAIGSTCPWLKKIMDRMTVSTLVL------GEGATEAEISMTS-TRWRKGVCEETSG
Insecticyanin               INSCT- 1- 72-------GDIFYPGYCP--DVKPVNDFD-----LSAFAGAWHEIAKLPL-ENENQGKCTIAEYKYD-------GKKASVYNSFVSN------GVKEYMEG
Bilin-binding Protein       BBP-I- 1- 71--------NVYHDGACP--EVKPVDNFD-----WSNYHGKWWEVAKYPN-SVEKYGKCGWAEYTPE-------GKSVKVSNYHVIH------GKEYFIEG
Apolipoprotein-D            ALP.D- 1- 71---------AFHLGKCP--NPPVQENFD-----VNKYLGRWYEIEKIPTT-F-ENGRCIQANYSLMEN-----GKIKVLNQELRAD------GTVNQIEG
Frog Bowman Gland Protein   BGP-F- 1- 73-----------QCQADLPPV---MKGLE-----ENKVTGVWYGIAAASNCK-QFLQMKSDNMPAPVNIYSLNNGHMKSSTSFQTEK-------GCQQMDV
Purpurin                    PURPU- 1- 84---IFSAVEVSLAQTCAVDSFSVKDNFD-----PKRYAGKWYALAKKDP-EGLFLQDNISAEVTVEED-----GTMTASSKGRVKLFGFW--VICADMAA
Human plasma Ret. Binding   BPRBP- 1- 75------------ERDCRVSSFRVKENFD-----KARFSGTWYAMAKKDP-EGLFLQDNIVAEFSVDET-----GQMSATAKGRVRLLNNW--DVCADMVG
Alpha-1-Ac.Glycoprotein     A1GLP- 1- 78------------QNPEPANI--TLGIPITNET-LKWLSDKWFYMGAAFRDP-VF-KQAVQTIQTEYFYLT---PNLINDTIELREF--QTTDDQCVYNFT
Androgen-dependent Protein  AND-P- 1- 65---------------------AVVKDFD-----ISKFLGFWYEIAFASKMGTPGLAHKEEKMGAMVV-------ELKENLLALTTT--YYSEDHCVLEKV
Major urinary Protein       URP-M- 1- 73-----------CVHAEEASS--TGRNFN-----VEKINGEWHTIILASDKR-EKIEDNGNFRLFLEQI--------HVLENSLVLKFHTVRDEECSELSM
Alpha-2-microglobulin       A2-MG- 1- 73-----------CGHAEEASS--TRGNLD-----VAKLNGDWFSIVVASNKR-EKIEENGSMRVFMQHI--------DVLENSLGFKFRIKENGECRELYL
Crustacyanin A              CRSTA- 1- 78------DGIPSFVTAGKCASVANQDNFD-----LRRYAGRWYQTHIIEN-AYQPVTRCIHSNYEYSTNDY---GFKVTTAGFNPN-------DEYLKIDF
Crustacyanin C              CRSTC- 1- 81------DKIPDFVVPGKCASVDRNKLWAEQTPNRNSYAGVWYQFALTNN-PYQLIEKCVRNEYSFDGKQF---VIKSTGIAYDGN---------LLKRNG
Cow Beta Lactoglobulin B    BLG-C- 1- 71----------------LIVTQ-TMKGLD-----IQKVAGTWYSLAMAASDISLLDAQSAPLRVYVEELK-------PTPEGDLEILLQKWENGECAQKKI
Aphrodicin                  APHRO- 1- 62--------------------------QD-----FAELQGKWYTIVIAADNL-EKIEEGGPLRFYFRHID------CYKNCSEMEITFYVITNNQCSKTTV
Horse Beta Lactoglobulin -I BLGH1- 1- 71----------------TNIPQ-TMQDLD-----LQEVAGKWHSVAMAASDISLLDSEEAPLRVYIEKLR-------PTPEDNLEIILREGENKGCAEKKI
Horse Beta Lactoglobulin-II BLGH2- 1- 71----------------TDIPQ-TMQDLD-----LQEVAGRWHSVAMVASDISLLDSEEAPLRVYVEELR-------PTPEGNLEIILREGANHACVERNI
Pig Beta Lactoglobulin   -I BLG-P- 1- 71----------------VEVTP-IMTELD-----TQKVAGTWHTVAMAVSDVSLLDAKSSPLKAYVEGLK-------PTPEGDLEILLCKRENDKCAQEVL
Dolphin Beta LactoglobulinI BLG-D- 1- 61----------------VSVIR-TMEDLD-----IQRVAGTWHSVAMAASDISLLDTEEAPLRVNVEELR-------PTPQGDLEIFLQK----------I
Secondary Structure OF BLG BLG-SEC --------------------------------------EEEEEEEEEEEE-----------EEEEEEE---------EEEEEEEEEEE---EEEEEEEEEE

                    110       120       130       140       150       160       170       180       190       200       210       220
RBP-SEC -    EEEEEE--------EEEEEEEEEEE--EEEEEEEEEEEEEE--EEEEEEEEEE-------EEEEEEEEEE------HHHHHHHHHHHHHH-------EEEEEEEEE
PBP-C 75-160 D------------DGTYVAIYEG-------QNVFKIVSLSRTHLVAHNINVDK----HGQTTELTELFVKLNV-EDEDLEKFWKLTEDK---GIDKKNVVNFLENEDHPHPE
OBP-R 91-174 E------------DGRYTTDYSG-------RNYFHVLKKTDDIIFFHNVNVDE-----SGRRQCDLVAGK-----REDLNKAQKQELRK-----LAEEYNIPNENTQHLVPTDTCNQ
PR-HC 78-184 AYEKTDT------DGKFLYHKSKW-----NITMESYVVHTNYDEYAI-FLTKKF-SRHHGPTITAKLYGRAPQ-LRETLLQDFRVVAQGVGIPEDSIFTM--ADRGE-CVPGEQEPEPILIPR
INSCT 73-190 DLEIAPDAKYT-KQGKYVMTFKFGQ---RVVNLVPWVLATDYKNYAINYNCDYH-PDKKAHSIHAWILSKSKV-LEGNTKEVVDNVLKTFSHLIDASKFISNDFSEAACQYSTTYSLTGPDRH
BBP-I 72-174 TAYPVGDS----KIGKIYHKLTYGG---VTKENVFNVLSTDNKNYIIGYYCKYD-EDKKGHQDFVWVLSRSKV-LTGEAKTAVENYLIGSP-VVDSQKLVYSDFSEAACKVN
ALP.D 72-169 EATPVNLT----EPAKLEVKFS-W----FMPSAPYWILATDYENYALVYSCTCI--IQLFHVDFAWILARNPN-LPPETVDSLKNILTS---NNIDVKKMTVTDQVN-C-PKLS
BGP-F 74-165 EMTTVE-------KGHYKWKM-------QQGDSETIIVATDYDAFLMEFTKIQM---GAEVCVTVKLFGRKDT-LPEDKIKHFEDHIEKVGLKKE--QYIRFHTKAT-CVPK
PURPU 85-185 QYTVPDPTT----PAKMYMTYQGLASYLSSGGDNYWVIDTDYDNYAITYACRSLK-DGSCDDGYSLIFSRNPRGLPPAIQRIVRQKQEEICMSG---QFQPVLQSGA-C
RBP-H 76-183 TFTDTED------PAKFKMKYWGVASFLQKGNDDHWIVDTDYDTYAVQYSCRLLNLDGTCADSYSFVFSRDPNGLPPQAQKIVRQRQEELCLAR---QYRLIVHNGY-CDGRSERNL
A1GLP 79-188 HLGVQRENGTLSKCAGAVKIFAHLIVLKKHGTFMLAFNLTDENRGLSFYAKKPDL--SPELRKIFQQAVKDVGM-DESEIVFVDWTKDK-CSEQQKQQLELEKETKKETKKDP
AND-P 66-167 TATEGDG------PAKFQVTRLS-------GKKEVVVEATDYLTYAIIDITSLVA---GAVHRTMKLYSRSLD-DNGEALYNFRKITSDHGFSE-TDLYILKHDLT--CV-KVLQSA-AESRP
URP-M 74-167 VADKTEK------AGEYSVTY--------DGFNTFTIPKTDYDNFLMAHLINE---KDGETFQLMGLYGREPD-LSSDIKERFAQLCEE--------HGILRENIIDLSNANRCLQARE
A2-MG 74-167 VAYKTPE------DGEYFVEY--------DGGNTFTILKTDYDRYVMFHLINF---KNGETFQLMVLYGRTKD-LSSDIKEKFAKLCEA--------HGITRDNIIDLTKTDRCLQARG
CRSTA 79-173 KVYPTKEF-----PAAHMLIDAP-----SVFAAPYEVIETDYETYSCVYSCITTD---NYKSEFAFVFSRTPQTSGPAVEKTAAVFNKNG---VEFSKFVPVSHTAE-CVY
CRSTC 82-182 KLYP-NPF-----GEPHLSIDYE-----NSFAAPLVILEYDYSNYACLYSCIDYN--FGYHSDFSFIFSRSANLADQYVKKCEAAFKNIN---VDTTRFVKTVQGSS-CPYDTQKTL
BLG-C 72-163 IAEKTKI------PAVFKIDA--------LNENKVLVLDTDYKKYLLFCMENSAEP---EQSLACQCLVRTPEV-DDEALEKFDKALKA-----LPMHIRLSFNPTQLEEQCHI
APHRO 63-152 IGYLKG-------NGTYQTQF--------EGNNIFQPLYITSDKIFFTNKNMDRAG---QETNMIVVAGKGNA-LTPEENEILVQFAHE--------KKIPVENILNILATDTCPE
BLGH1 72-163 FAEKTSQ------PAQFKINA--------LDEDTVFYLDTDYKNYLFLCMKNAATPG---QSLVCQYLARTQ-MVDEEIMEKFARALQP-----LPGRVQIVPDLTRMAERCRI
BLGH2 72-164 VAQKTES------PAEFKTED--------SAVFTVNYQPGERKNYLFLCMKDVGPCL--PASMVCQYLARTQ-KVDEEVMEKFSRALQP-----LPGRVQIVQDPSGGQERCGF
BLG-P 72-160 LAKKTDI------PAVFKINA--------LDENQLFLLDTDYDSHLLLCMENASQEH----SLVTQCLARTL-EVDDQIREKFEDALKT-----LSVPMRIL--PAQLEEQCRV
BLG-D 62-132 IAEKTEI------PAVF--NF----------------LNSDYTNYLLFCME---------VSLTCQYLARTLQ-VDDGVMEKFNKAIKP-----LPMHIRL-FSPTQLEE
BLG-SEC      EEEEE----------EEEEE-----------EEEEEEEEE----EEEEEEEE--------EEEEEEEEEE------HHHHHHHHHHH---------EEEEEE
```

Table 1. Sequence alignment of members of the lipocalin family. The observed secondary structures of Retinol Binding Protein and Beta-Lactoglobulin are given in bold at the top and bottom lines of the alignment (**E**= beta strand, **H**= alpha helix).

3. Plasma Retinol Binding Protein (RBP) , β-Lactoglobulin (BLG) and Crustacyanin (CRT).

β-lactoglobulin is a very stable dimeric protein found in milk. Although it shows limited homology to RBP their three dimensional structures are closely similar, with 129 out of 162 residues of BLG having an rms difference in position of less than 3A [12]. Both proteins bind retinol with RBP showing a 100-fold stronger binding.

The aminoacid side chains lining the cavity of RBP are hydrophobic or uncharged (Leu 35, Phe 36, Ala 43, Phe 45, Ala57, Val 61, Met 73, Gly 75, Met 88, Leu 97, Gln 98, His 104, Gln 117, Tyr 133 and Phe 135) [7]. The β-ionone ring of the retinol lies deepest in the pocket, with the isoprene tail stretching out almost to the surface of the protein. It is worth noting the packing of the aromatic residues against the polyene tail of retinol.

Figure 2. The Plasma Retinol Binding Protein Binding Pocket

In BLG the hydrophobic character of the cavity is retained but the replacement of certain residues to smaller hydrophobic makes the pocket deeper. In particular three phenylalanine residues in RBP (Phe 15, 20 and 45) are replaced by smaller ones in BLG (Leu 10, Val15 and Val 41). Other changes that increase the volume of the cavity include RBP Met 53 to BLG Thr49, RBP His104 to BLG Val92 and RBP Ile106 to BLG Val94. No specific interactions other than hydrophobic are involved in binding except the hydrogen bond formed by the alcohol hydroxyl of retinol to a side chain of RBP.

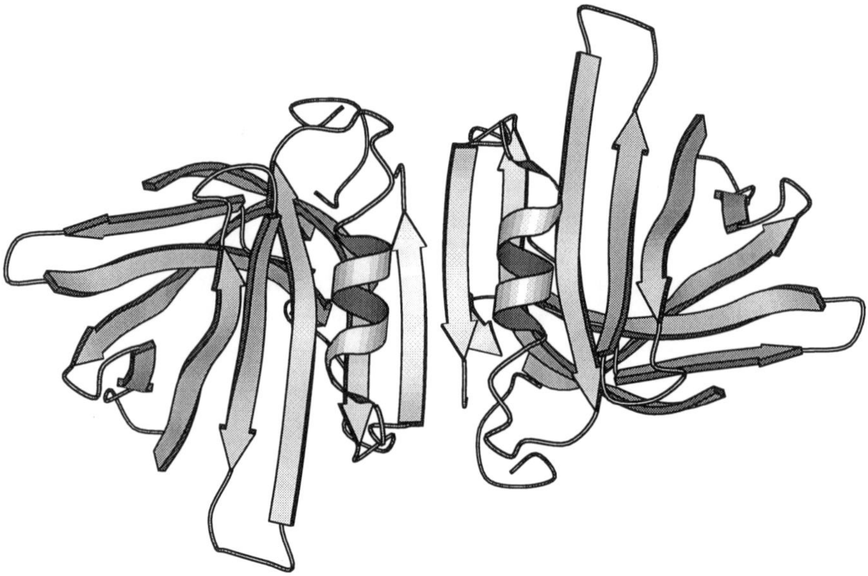

Figure 3. The structure of the β-lactoglobulin dimer.

A variation is provided by Crustacyanin, one of a family of carotenoid-binding proteins found in the carapace of crustacea, where it is responsible for the animal's coloration through its pigment astaxanthin. The absorption maximum characteristic of free astaxanthin is shifted by more than 100nm as a result of interaction with the protein moeity [9]. Crustacyanin shows limited sequence homology with RBP with 33 identities.

The crystallography of the molecule is still in the early stages but a model consistent with accepted protein structure principles and biochemical and spectroscopic observations has been built [10].

The binding pocket is quite deep but not sufficiently long to accomodate entirely a molecule of astaxanthin. While the major part of the pigment molecule is bound within the pocket of one subunit of the dimer the remaining part is positioned in a depression in the surface of the opposite monomer in agreement with the observation that the affinity of the monomer for astaxanthin is lower than the affinity of the dimer.

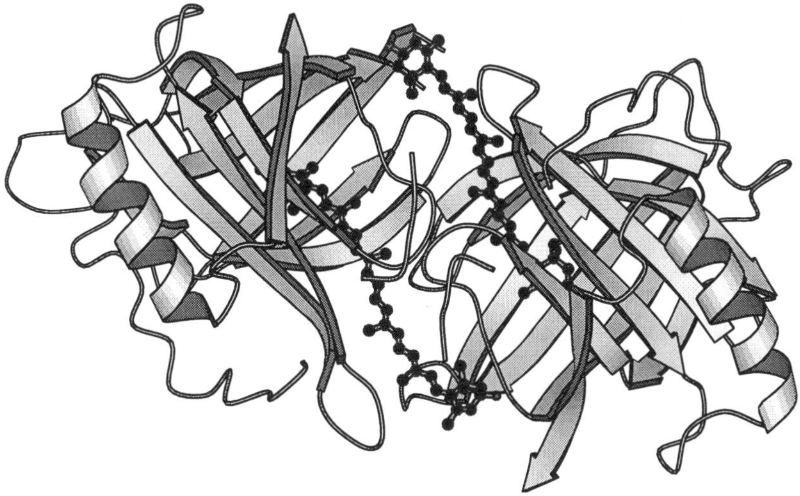

Figure 4. The model of β-crustacyanin dimer

The side chains lining the binding pocket are mainly hydrophobic and close to the polyene chain of the pigment deforming it from its normal all-trans conformation. Deep within the binding cavity lie a threonine and a tyrosine side chains which could form hydrogen bonds to the keto and hydroxy groups of astaxanthin .

4. Fatty Acid Binding Protein (FABP) and Photoactive Yellow Protein (PYP).

Fatty acids are being carried through intra and extracellular aqueous phases by binding to proteins that also protect cell membranes and enzymes from their detergent effect. Intestinal fatty acid binding protein (FABP-I) is a representative of this group that uses the same beta-calyx motif to enclose the fatty acid [13]. With P2-myelin [14], cellular RBP and other eight members form a family of cytoplasmic hydrophobic ligand-binding proteins.

Figure 5. Stereodiagram of the Fatty Acid Binding Protein

The interior of the binding pocket is composed primarily of the side chains of hydrophobic aminoacids. Two interesting clusters of aromatic side chains are present. One consists of Phe 47, 62, 68 and Trp 82 with most aromatic rings stacked one upon the other and are located near the bottom of the binding cavity near the carboxylate group of the bound fatty acid. The other involves Tyr 14,117,119 and Phe 17 that are located near the methylene chain of the bound fatty acid. The lipid molecule is forced by the position of hydrophobic side chains to take a bent conformation with its polar carboxylate group buried near the center of the cavity. Of particular interest is the presence of Arg106 and Gln115 and two bound water molecules in the cavity and their charge interaction with the carboxylate group of the fatty acid. It has been shown that the guanidium group of Arg106 is in the same plane to the carboxylate group, forming hydrogen bonds, while the nitrogen of the arginine side chain is hydrogen bonded to the carbonyl of glutamine. With the additional hydrogen bonds formed by two solvent molecules towards the glutamine and the carboxylate of the fatty acid a charge neutralising quintet is formed.

A similar example of a protein from the same family but with a completely different function is given by photoactive yellow protein (PYP) [15]. Its unique ability to capture and use energy from a photon of light, through a photoisomerisation that causes a structural rearrangement that is completely reversible in the absence of other substrates or redox activity, depends on the chromophore, 4hydroxycinnamyl [16], its covalent linkage to the protein (a thioester to Cys69) [17] and the surrounding protein environment. There is an asymmetric distribution of charged and hydrophobic side chains surrounding the chromophore. Fifteen aromatic (Phe, Tyr and Trp) and 20 negatively charged (Glu and Asp) aminoacids lie in an ellipsoidal band around it. Separate from any aromatic or negatively charged side chains lie six positively charged side chains (Lys, Arg). The electron rich enviromnent around the chromophore and the asymmetric charge distribution maybe functionally important for the proteins photocycle. The interaction of the buried Lys111, Glu56 and Tyr 96 may contribute to the observed red shift of the absorption maximum of the chromophore relative to that in solution. It is the first example where the free sulphydryl of Cys69 covalently links to the substrate which can only be removed by bleaching.

5. Prostaglandin D2 synthetase. The first lipocalin enzyme.

Glutathione (GSH)-independent PGD syntetase [prostaglandin-H2 D-isomerase] (PGDS) catalyses the conversion of prostaglandin H2 to prostaglandin D2 in the presence of various sulphydryl compounds [18]. This enzyme is responsible for the biosynthesis of PGD2 in the brain. It also binds bilirubin, indocyanine green and other hydrophobic molecules in line with other lipocalins. Together with alpha1-microglobulin [19] and the γ chain of the complement component C8 [20], form a separate branch of the superfamily with two interesting properties. They are glycosylated and use one free sulphydryl (in position 34 for Protein HC [21,22] and at position 65 for PGD-S) placed in the binding cavity to covalently bond to a substrate. No three dimensional information is available but from their aminoacid sequences, in conjunction with their homology with other known lipocalins, models have been built. In PGD-S, which is a membrane associated enzyme both the characteristic motifs of lipocalins are conserved. These maybe responsible for membrane recognition through mechanisms similar to those found for other members of the family involved in transport of hydrophobic molecules (RBP, OBP, etc.).

6. Urinary, Odorant and Tastant Proteins. A Biological signalling system.

Chemical communication by means of urine appears to be the basis of animal sexual behaviour [4]. The fact that the primary structure of urinary proteins is highly homologous with those of odorant and tastant binding proteins [23] provides clues for the transmition and reception mechanisms of pheromone signaling. This biological signalling system includes other sexual behaviour proteins (aphrodicin [24], epididymal secretory protein [25]) that structurally belong to the lipocalins. Although the threshold of odorant and tastant detection can reach 0.002 ppb of water for geosmin [28] ligands as diverse as geosmin, decahydronaphtalene, thymol, 2-isobutyl-3-methoxy-pyrazine, β-ionone and 2-nonenal bind in the hydrphobic cavity. Their common characteristic being a polar atom or group in close proximity to 5-12 carbon atoms in a chain or ring arrangement.

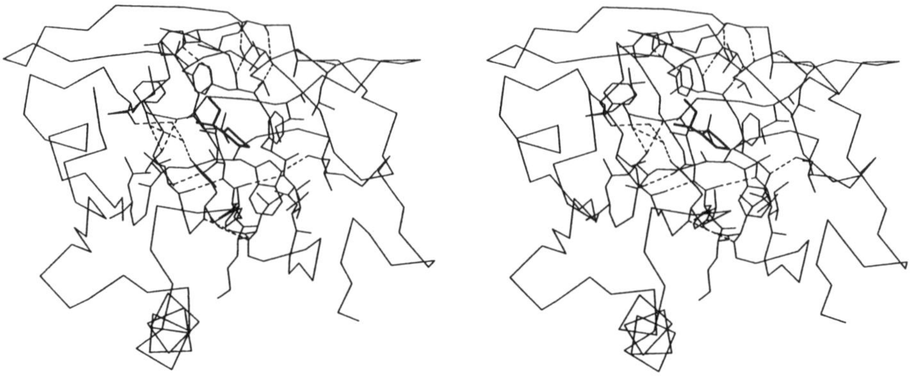

Figure 6. Stereodiagram of the major urinary protein.

The crystal structures of two urinary proteins have been determined [26] and a model of the highly homologous nasal pyrazine binding protein has been constructed [27]. The 8-stranded β-calyx is conserved together with the highly apolar binding pocket. There are differences in the type and position of the side chains lining the pocket. Val 58 and Ala107 of the Major Urinary Protein (MUP), located at the base of the binding pocket are substituted by phenylalanines in Urinary α2-globulin (A2U) introducing volume restrictions and change of shape in the binding pockets. These are reflected in the different binding affinities of ligands. A polar tyrosine (124) in A2U points directly to the bound ligand forming a hydrogen bond, while in MUP two solvent molecules form hydrogen bonds with the equivalent tyrosine. It is evident that by retaining the hydrophobic character of the cavity and relocating charge, polarity and reshaping the volume of the pocket, ligand selectivity can be achieved.

Molecular recognition occurs as a result of favorable interactions between the molecules concerned. Steric and electrostatic complementarity of the combining surfaces, contribute in such favourable interactions together with the hydrophobic effect on the interacting molecular surfaces, dispersion forces and hydrogen bonding. The compact structure of the lipocalin β-calyx provides the framework for such processes. With variation in shape, volume, charge distribution and reactive side chains variation in function occurs with only constant the hydrophobic character of the ligand/substrate involved.

7. References

1. A. Harada, J. Li & M. Kamachi (1992) The molecular necklace: a rotaxene containing many threaded α-cyclodextrins Nature 365:325-327
2. J. Godovac-Zimmermann (1988) The structural motif of beta-lactoglobulin and retinol-binding protein: a basic framework for binding and transport of small hydrophobic molecules? Trends Biochem Sci 13: 64-66
3. M. C. Peitsch & M. S. Boguski (1991) The first lipocalin with enzymatic activity. Trends Biochem Sci 16: 363
4. J. Pevsner, R. R. Reed, P. G. Feinstein & S. H. Snyder (1988) Molecular cloning of odorant-binding protein: member of a ligand carrier family. Science 241: 336-339
5. K. Kock, C. Ahlers & H. Schmale (1994) Structural organization of the genes for rat von Ebner's gland proteins 1 and 2 reveals their close relationship to lipocalins. Eur J Biochem 221: 905-916
6. H. Schmale, H. Holtgreve-Grez & H. Christiansen (1990) Possible role for salivary gland protein in taste reception indicated by homology to lipophilic-ligand carrier proteins. Nature 343:366-369
7. M. E. Newcomer, T. A. Jones, J. Aqvist, J. Sundelin, U. Eriksson, L. Rask & P. A. Peterson (1984) The three-dimensional structure of retinol-binding protein. EMBO J 3: 1451-1454
8. A. Sivaprasadarao and J.B.C. Findlay (1988) The interaction of retinol binding protein with its plasma-membrane receptor. Biochem. J. 255: 561-569.
9. J. N. Keen, I. Caceres, E. E. Eliopoulos, P. F. Zagalsky & J. B. Findlay (1991) Complete sequence and model for the A2 subunit of the carotenoid pigment complex, crustacyanin. Eur J Biochem 197: 407-417
10. J. N. Keen, I. Caceres, E. E. Eliopoulos, P. F. Zagalsky & J. B. Findlay (1991) Complete sequence and model for the C1 subunit of the carotenoprotein, crustacyanin, and model for the dimer, beta-crustacyanin, formed from the C1 and A2 subunits with astaxanthin. Eur J Biochem 202: 31-40
11. M. Igarashi, A. Nagata, H. Toh, Y. Urade & O. Hayaishi (1992) Structural organization of the gene for prostaglandin D synthase in the rat brain. Proc Natl Acad Sci U S A 89: 5376-5380
12. M. Z. Papiz, L. Sawyer, E. E. Eliopoulos, A. C. North, J. B. Findlay, R. Sivaprasadarao, T. A. Jones, M. E. Newcomer & P. J. Kraulis (1986) The structure of beta-lactoglobulin and its similarity to plasma retinol-binding protein. Nature 324: 383-385

13. J. C. Sacchettini, J. I. Gordon & L. J. Banaszak (1989) Crystal structure of rat intestinal fatty-acid-binding protein. Refinement and analysis of the Escherichia coli-derived protein with bound palmitate. J Mol Biol 208: 327-339
14. S. W. Cowan, M. E. Newcomer & T. A. Jones (1993) Crystallographic studies on a family of cellular lipophilic transport proteins. Refinement of P2 myelin protein and the structure determination and refinement of cellular retinol-binding protein in complex with all-trans-retinol. J Mol Biol 230: 1225-1246
15. D. E. McRee, J. A. Tainer, T. E. Meyer, J. Van Beeumen, M. A. Cusanovich & E.D. Getzoff (1989) Crystallographic structure of a photoreceptor protein at 2.4 A resolution. Proc Natl Acad Sci U S A 86: 6533-6537
16. M. Baca, G. E. Borgstahl, M. Boissinot, P. M. Burke, D. R. Williams, K. A., Slater & E. D. Getzoff (1994) Complete chemical structure of photoactive yellow protein: novel thioester-linked 4-hydroxycinnamyl chromophore and photocycle chemistry. Biochemistry 33: 14369-14377
17. J. J. Van Beeumen, B. V. Devreese, S. M. Van Bun, W. D. Hoff, K. J. Hellingwerf, T. E. Meyer, D. E. McRee & M. A. Cusanovich (1993) Primary structure of a photoactive yellow protein from the phototrophic bacterium Ectothiorhodospira halophila, with evidence for the mass and the binding site of the chromophore. Protein Sci 2: 1114-1125
18. A. Nagata, Y. Suzuki, M. Igarashi, N. Eguchi, H. Toh, Y. Urade & O. Hayaishi (1991) Human brain prostaglandin D synthase has been evolutionarily differentiated from lipophilic-ligand carrier proteins. Proc Natl Acad Sci U S A 88: 4020-4024
19. H. Vetr & W. Gebhard (1990) Structure of the human alpha 1-microglobulin-bikunin gene. Biol Chem Hoppe Seyler 371: 1185-1196
20. J. A. Haefliger, M. C. Peitsch, D. E. Jenne & J. Tschopp Structural and functional characterization of complement C8 gamma, a member of the lipocalin protein family.
21. J. Escribano, A. Grubb, M. Calero & E. Mendez (1991) The protein HC chromophore is linked to the cysteine residue at position 34 of the polypeptide chain by a reduction-resistant bond and causes the charge heterogeneity of protein HC. J Biol Chem 266: 15758-15763
22. M. Calero, J. Escribano, A. Grubb & E. Mendez (1994) Location of a novel type of interpolypeptide chain linkage in the human protein HC-IgA complex (HC-IgA) and identification of a heterogeneous chromophore associated with the complex. J Biol Chem 269: 384-389
23. A. Cavaggioni, R. T. Sorbi, J. N. Keen, D. J. Pappin & J. B. Findlay (1987) Homology between the pyrazine-binding protein from nasal mucosa and major urinary proteins. FEBS Lett 212: 225-228
24. W. J. Henzel, H. Rodriguez, A. G. Singer, J. T. Stults, F. Macrides, W. C. Agosta & H. Niall (1988) The primary structure of aphrodisin. J Biol Chem 263: 16682-16687
25. L. Morel, J. P. Dufaure & A. Depeiges (1993) LESP, an androgen-regulated lizard epididymal secretory protein family identified as a new member of the lipocalin superfamily. J Biol Chem 268: 10274-10281

26. Z. Bocskei, C. R. Groom, D. R. Flower, C. E. Wright, S. E. Phillips, A. Cavaggioni, J. B. Findlay & A. C. North (1992) Pheromone binding to two rodent urinary proteins revealed by X-ray crystallography. Nature 360: 186-188
27. R. Tirindelli, J. N. Keen, A. Cavaggioni, E. E. Eliopoulos & J. B. Findlay (1989) Complete amino acid sequence of pyrazine-binding protein from cow nasal mucosa. Eur J Biochem 185: 569-572
28. P. Pelosi (1994) Odorant-binding proteins. Crit Rev Biochem Mol Biol 29: 199-228
29. Kraulis Per J. (1991) MOLSCRIPT: A Program to Produce Both Detailed and Schematic Plots of Protein Structures, J. App. Cryst. 24 : 946-950.

SUPRAMOLECULAR PHOTOCHEMICAL REACTIONS OF ORGANIC MOLECULES ADSORBED ON POROUS CRYSTALLINE ZEOLITES

NICHOLAS J. TURRO
Chemistry Department, Columbia University
New York, New York 10027 (USA)

Abstract

Supramolecular chemistry refers to the chemical structures, dynamics and transformations of the chemistry of guest/host (substrate/receptor) systems (supramolecular assemblies) which are "connected" by intermolecular interactions and non-covalent intermolecular "bonds". By considering zeolites as rigid counterparts of classical supramolecular host (receptor) systems such as biological enzymes, an understanding of the photochemistry of supramolecular assemblies consisting of organic molecules adsorbed in porous crystalline zeolites may be acquired.

1. The Supramolecular Paradigm

Structure is at the heart of systematic scientific thinking. The paradigm of **molecular structure** is at the heart of virtually all chemical thinking. Any molecular structure may be analyzed in terms of the following four forms or levels [1]: *composition*; *constitution*; *configuration*; and *conformation*. At the level of composition only the number and kinds of "elements" or atoms of the structure are of chemical interest. At the level of constitution, not only is the composition of chemical interest, but also, the number of possible ways in which the atoms of a given composition are connected to one another. A unique molecular structure at this level is identified as one for which all of the

G. Tsoucaris et al. (eds.), Crystallography of Supramolecular Compounds, 429–450.

atoms of the composition possess a unique connectivity relationship. At the level of configuration, in addition to the composition (numbers and kinds) and constitution (connectivity relationships), the number of possible different orientations in three dimensional space of "neighboring atoms" about a central atom is of chemical interest. A unique structure at this level may be identified as one for which all of the neighboring atoms surrounding a central atom possess a distribution in three dimensional space that is unique. Finally, at the level of conformation, the possible shapes of a given composition, constitution and specific set of configurations are of chemical interest. A unique structure at this level may be identified as one which differs from others in terms of shapes resulting from rotations about single bonds.

Chemists have grown so familiar with the paradigm of molecular structure that its validity and centrality to all of chemistry are no longer seriously debated. However, chemists are now seeking the development of a paradigm that attempts to go beyond that of molecular structure, namely the **supramolecular level**. Supramolecular structures are related to molecules in the same way that molecular structures are related to atoms [2]. Although atoms play a critical role in the paradigm of molecular structure, it is the molecular structure which is generally at the center of the chemistry of molecules. The "intellectual glue" of molecular structure is the concept of the covalent bond and the stereochemistry associated with the consequences of the covalent bonds. The paradigm of the covalent chemical bond provides chemists with the rules that govern the structures, dynamics, characteristics, and transformations of molecules. *We therefore expect that a paradigm of non-covalent bonds will provide the rules that govern the structures, dynamics, characteristics, and transformations of supramolecular assemblies.*

Just as the level of simple atomic structure is inadequate for understanding aspects of chemistry in which molecular aspects dominate, the level of molecular structure is inadequate for understanding aspects of chemistry in which supramolecular aspects dominate. The emerging paradigm of supramolecular chemistry emphasizes the **intermolecular bond which holds molecules together, rather than the covalent bond which holds atoms together.** Non-covalent intermolecular bonds are more varied and complex than covalent intramolecular bonds. In general, intermolecular, non-covalent bonds are held together by weaker forces than intramolecular, covalent bonds. For example, the important forces holding molecules together consist of dispersion forces, hydrogen bonds, hydrophobic bonds, etc. Although these forces, **per atom contact**, are relatively small, in many supramolecular systems many such weak bonds are formed in a cooperative manner, thereby causing the supramolecular structure to be energetically stable; however, because each "bonding contact" is relatively weak, the bonds may be broken relatively easily so that the supramolecular structure often possesses a certain amount of flexibility.

1.1 SUPRAMOLECULAR CHEMISTRY. TOWARD SOME WORKING DEFINITIONS

According to Lehn [2], "Supramolecular chemistry may be defined as *chemistry beyond the molecule*, bearing on the organized entities of higher complexity that result from the association of two or more chemical species held together by intermolecular forces." This definition is in contrast with molecular chemistry, "the chemistry of the covalent bond." Lehn has emphazied aspects of supramolecular chemistry which are apparent in enzyme catalysis, chemical systems for which chemists have developed an excellent paradigm for the structure and dynamics of exceedingly complex chemical systems. Thus, appealing to the enzyme paradigm, an operating enzyme may be considered as a bimolecular supramolecular system whose composition is a receptor (or host) and a substrate (or guest). Although binding is a characteristic of the entire supramolecular assembly, usually the receptor is considered to be a large molecule compared to the substrate and to have numerous binding sites to which the substrate may become noncovalently attached in various degrees.

1.1.1 *How Can a Non-Covalent Bond Be Defined?*

The definition of a non-covalent bond is necessarily somewhat arbitrary, so that the term supramolecular will mean different things to chemists working in different areas. However, even a covalent bond is sometimes difficult to define and may be described in one of several terms: (1) distance relationships between atoms; (2) energetic relationships between atoms; and (3) kinetic or temporal relationships between atoms. Distance or energy relationships are probably the most common means of determining whether or not a covalent bond exists. However, the notion of the **temporal persistence** of a distance or energetic relationship between atoms may be a useful device for defining certain supramolecular assemblies.

For example, a solvent cage (as host) containing two radicals (as substrates) may be considered as a supramolecular system during the time period for which the radical pair is geminate, i.e., possesses a compositional relationship to a precursor molecule which has dissociated to produce the radical pair. During this time scale, because the two partners of the pair are much closer to each other than they are to radicals produced from other dissociations, the reactivity of the geminate partners in the pair toward each other is much higher than the reactivity of either of the partners with other radicals in the system. However, for non-viscous solvents, the lifetime is of the order of picoseconds after which the radicals of the pair become random free radicals. Because of the short time scale of geminate composition a radical pair in a solvent cage is not usually considered as a supramolecular system.

Consider now the same radical pair (substrate) in a micelle (receptor) in an aqueous solution. The environment of the radical pair is liquid-like and quite analogous to the environment that the pair experiences in a homogeneous solvent, but now the **temporal persistence** of the geminate character of the pair may extend over many orders of magnitude, and a radical pair in a micelle may exhibit some of the characteristics expected of a supramolecular assembly. The "walls" of the micelles, although soft and flexible, present an energy barrier for hydrophobic organic radicals, which are more soluble in the hydrophobic micellar environment than in the aqueous phase; therefore, the radicals' exit from the micellar environment is inhibited. Experimentally, the lifetime of such supramolecular assemblies may last many orders of magnitude longer than the lifetime of the pair in a solvent cage. Finally, consider the same radical pair adsorbed on a zeolite crystal. Adsorption may occur on the external surface of the crystal or on the internal surface of the crystal, or on both. If the radical pair is adsorbed on the external surface of the crystal, it will be able to execute a certain diffusional motion along the external surface. If the radical pair is adsorbed on the internal surface, the pair may find itself in a cavity that causes the geminate character to persist for many orders of magnitude beyond the time scale in solution. In each case the adsorbed radical pair and the zeolite constitute a guest (radical pair) host (zeolite) supramolecular assembly.

1.1.2 *Catalysts as Reaction Filters Rather than Reaction Accelerators*

A typical textbook definition [3] of a catalyst is "a substance that increases the rate of a reaction without being consumed in the reaction." Perhaps the more fundamental idea behind the definition is the concept that "a catalyst is a substance which influences the course of reactions without being consumed". Sometimes the amounts of the substances (in a stoichiometric sense) are much smaller than the amounts of reactants that are converted to products, hence a common term that "a catalytic amount" was added. When this is the case, the concept of "turnover" or the ratio of moles of product to moles of catalyst is commonly employed. Enzymes are textbook examples [4] of catalysts that capture the features of accelerating rates and have a high turnover. However, perhaps the most spectacular feature of an enzyme is its ability to cause acceleration with high turnover and still be selective, i.e., in addition to being a reaction accelerator, an enzyme is a reaction "filter", employing the charateristics of the supramolecular host guest assembly to filter out all but one of a number of possible reactions.

Fast reaction rates are often associated with poor selectivity. The rationale behind this paradigm is that a fast rate implies a low activation barrier for reaction and a low barrier implies numerous possibilities that can be explored. For example, the rate of reaction of chlorine atoms with saturated hydrocarbons in solution (eq. 1) is close to diffusion (or collision) controlled [5]. The abstraction of a hydrogen atom by a chlorine atom is essentially nonselective, i,e., on a statistical basis, a primary hydrogen is abstracted with a selectivity no different from a tertiary hydrogen. In this case, the reaction rate is of not of great concern, **because the rate of reaction is already**

diffusion controlled! Thus, the catalytic influence desired is not kinetic acceleration, but enhancement of **chemical selectivity.**

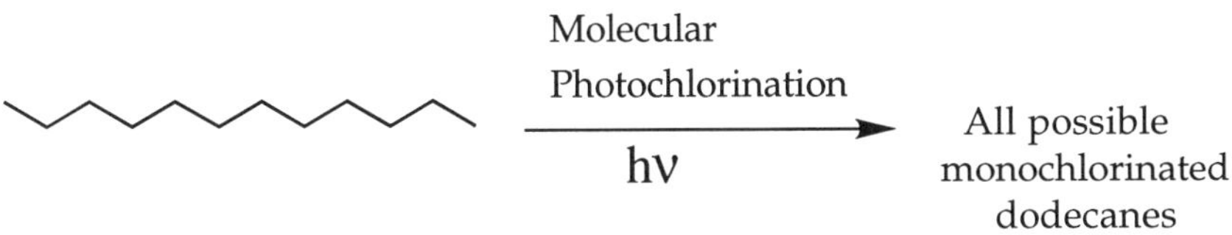

(1)

As another example, the possible combination reactions of two reactive carbon radicals in solution generally occur at or close to the diffusion controlled rate [6]. In addition, the cage effect for reactions of a geminate radical pair in solution are generally small and the pair separates and becomes a pair of free radicals, which then react non-selectively with one another (eq. 2).

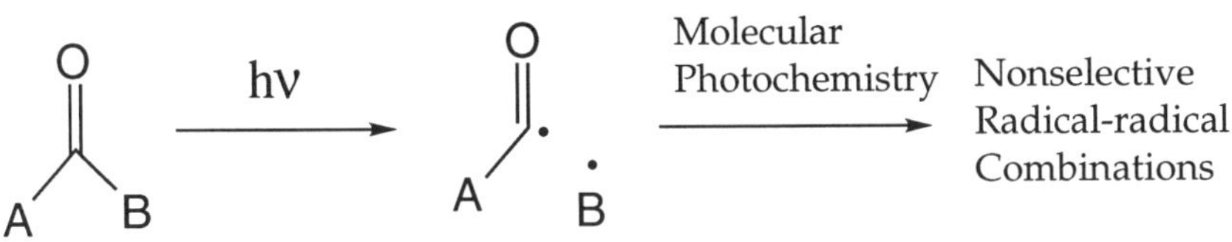

(2)

We now seek to determine how we can transform the non-selective molecular photochemistry of eqs. 1 and 2 into selective supramolecular photochemistry implied in eqs. 3 and 4.

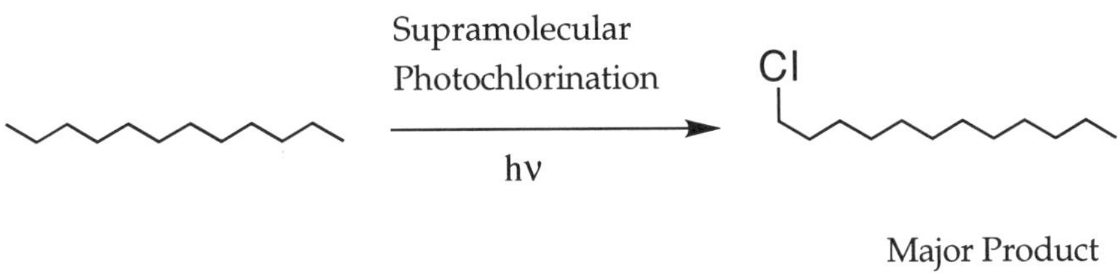

(3)

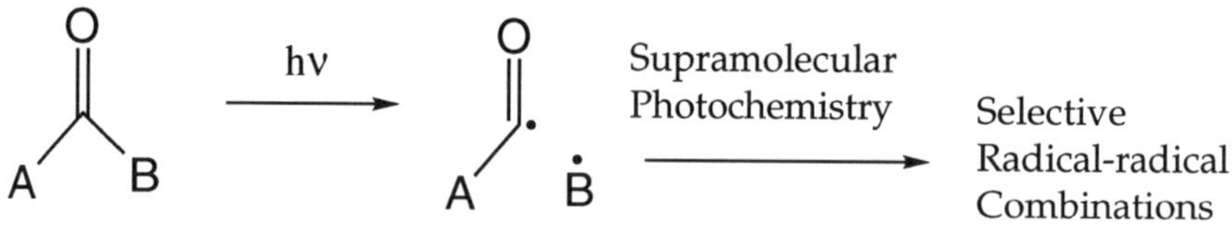

(4)

1.1.3 *Merging of the Supramolecular Paradigm with Photochemical Paradigms*

From the above considerations, the simplest supramolecular assembly is a complex between two molecules. In general, chemists do not consider solvated molecules as supramolecular species, but there are many examples of solvent effects that appear to be interpreted in terms of features expected of well defined supramolecular assemblies. For example, the reactions of chlorine atoms are very solvent dependent [5,7]. In a saturated hydrocarbon solvent, chlorine atoms are nonselective toward hydrogen abstraction, whereas in benzene the selectivity is quite significant. One interpretation of the solvent effect is that a complex is formed between a chlorine atom and benzene and that this complex is the chlorinating agent, not a "free" chlorine atom. Indeed, spectroscopic evidence [8] for such a complex is available from laser flash photolysis and thereby gives a definite structural validity to this complex.

Photochemists have become quite accustomed to dealing with "complexes" between electronically excited states and ground state molecules of substrates or solvent [9]. A supramolecular assembly consisting of an electronically excited molecule and a ground state molecule is termed an "exciplex" and can be fully characterized with a unique set of spectroscopic properties in addition to unique photochemistry. The ubiquitousness of such species in photochemistry makes a photochemist much more receptive to supramolecular concepts even at the level of small molecules. We shall return to this theme.

2. Zeolites as Enzyme Models

An analolgy may be made [10] between thea rigid framework structure (the "walls" of the internal surface) of a zeolite possessing internal void space of molecular dimensions (ca. 10 Å) and the cavity that forms the active site resulting from the tertiary protein structure of an enzyme which operates on a molecular substate of size ca. 10 Å. To some extent, enzymes at the active sites incorporate metal reagents that are relatively non-selective in their reactions with substrates in homogeneous solutions. The phenomenal selectivity of the metal when it is incorporated into the enzyme is due to the environment, which may both discipline a metal ion's tendency toward non-selectivity by modifying the redox parameters and more importantly, by providing an environment that encourages the transport of substrates to the active site and which selectively recognizes only the "correct" substrate on which to operate at the active site. The latter molecular discrimination or molecular recognition is achieved through a stereochemically void space in the environment of the active site that is exquisitely designed to select substrate via size/shape characteristics.

The overall enzymatic catalysis can be viewed [4] as the process of (1) relatively indiscriminate adsorption of the substrate on the surface of the protein; (2) diffusion of the substrate to the region in the vicinity of the active site; (3) size/shape "molecular sieving" of the substrate into the active site; (4) selective transformation of the substrate into a new chemical species; (5) release of the transformed substrate from the active site; and (6) accumulation of the transformed substrate in the bulk system. Effectively, the enzyme structure has molecular features which serve as (1) "fly paper" to which the substrate in the bulk solution becomes attached; (2) a "policeman" to exert "molecuar traffic control" on the diffusion of substrate to the active site; (3) "molecular sieves" which allow access to the active site only to those substrates possessing the appropriate size and shape characteristics; (4) discipline for the reactivity of reagents which cause the transformations that occur in the active site and orient the substrate for transformation of high selectivity; (5) release of the transformed substrate.

We shall now propose that in spite of the rigidity of the framework of a zeolite, it does possess many of the structural and dynamic characteristics of an enzyme as expressed in the previous paragraph, i.e., (1) The external surface of a zeolite serves as "fly paper" collecting substrate molecules that diffuse through a bulk solvent and then adsorb on the external surface. (2) The adsorbed molecule may then diffuse on the external surface that serves as the "policeman" and directs the substrate to a pore that allows access to the internal surface. (3) The size and shape of the pore determines which substrate molecules can sieve into the internal surface. Thus, zeolites are clearly capable of mimicing the fly paper, molecular traffic control, and molecular sieving characteristics of zeolites. It has been proposed that analogies in the transformations of substrate also exist. In this report, however, we depart from the analogy of the thermal action of zeolites in substrate transformations and concentrate on *photochemical production of non-selective reagents and non-selective reaction partners, and show how zeolites can serve to enhance selectivity as an enzyme mimic.*

2.1 POROUS ZSM-5 ZEOLITE CRYSTALS AS HOSTS IN SUPRAMOLECULAR STRUCTURES

What is a zeolite? In broad terms, a zeolite [11] is a crystalline solid possessing an enormous internal surface or void space consisting of precisely identical and repeating pores, channels, and cavities that allow access to small molecules (ca. 10Å). More precisely, zeolites are cystalline inorganic polymers whose constitution consists of corner sharing AlO_4 and SiO_4 tetrahedra which form a 3D framework [1]. The rigid framework structure encapsulates an intracrystalline void space of channels, pores, and cages that are accessible to guest molecules of appropriate size and shape which are capable of diffusing from the external surface into the intracrystalline surface. The enormous internal surface area (typically ca. 500-1000 m^2/g) of zeolites is capable of high selectivity of adsorption of molecules of appropriate size and shape, and the high thermal stability and outstanding catalytic ability of zeolites place them as one of the most important classes of materials employed for industrial catalysis, e.g., the catalysis of petroleum formation and cracking processes.

The pentasils [12] are a family of high silica zeolites possessing an internal framework of intersecting channels whose 10 membered rings created by Si,Al centered tetrahedra control the size of the pores and the internal void space of the crystal, and therefore control the size and shape of molecules capable of penetrating and diffusing through the internal surface. LZ-105 and silicalite are representative members of the pentasil family of zeolites and are related to the important industrial family of ZSM-5 zeolites, which are well known for their outstanding size/shape sensitive catalytic characteristics [13]. Although LZ-105 and silicalite possess identical internal void space morphologies, LZ-105 has a relative high Al content for a pentasil (Si/Al ca. 20) and silicalite contains the lowest Al content of any pentasil zeolite (Si/Al > 1000).

Their practical importance aside, zeolites possess fascinating structural features as "microscopic restricted spaces" or nanosized reaction vessels for chemical reactions and as such are excellent hosts for the investigation of the role of size-shape features in determining the effects of reactions in restricted spaces [14]. The remarkable catalytic properties and outstanding size/shape adsorptive selectivity characteristics of pentasil zeolites [13] provide a tremendous incentive and motivation for basic research that will further the understanding of the connection between the pentasil void space structure, the sitings of substrates in the accessible void space, and catalytic properties. Information concerning the siting of adsorbed molecules, the dynamics of exchange of adsorbed molecules between sites, and the number and location of active sites are critical issues to be addressed by fundamental investigations. Another important issue is the role of the Si/Al ratio in determining the catalytic and adsorptive properties of pentasils, since this ratio can be varied over several orders of magnitude and the number of Al atoms is generally agreed to be equal to the number of active sites. These issues have all been addressed theoretically [15], but have not been definitively established experimentally.

In addition to these structural issues, an understanding of the dynamics of diffusion of reagents and substrates throughout the void space presents the possibility of the fascinating phenomenon of "molecular traffic control" of catalytic reactions [16]. The traffic metaphor elicits microscopic analogies of "parking problems" as a function of sites occupied and of "traffic jams along the diffusional highway". We shall keep these analogues in mind since the idea of mass transport is very effective as a basis of interpreting the results.

Finally, since photochemical excitation is being employed in our investigations, complications due to the high temperatures commonly employed to examine the catalytic properties of zeolites are avoided and the size-shape and diffusional aspects of catalytic action are emphasized.

2.1.1 Supramolecular Structures and Dynamics

What do the structures of micelles, cyclodextrins, liquid crystals, polymers, DNA, zeolites, porous silica, and other microheterogeneous systems have in common? From the standpoint of Euclidean geometry the answer would appear to be "very little." However, from the standpoint of topological geometry [1] these apparently structurally distinct systems have much in common. To the extent that the topological features of these systems are correctly identified, their sizes, shapes, and chemical aspects are irrelevant and each system can be considered to be *locally or globally homeomorphic*. A feature that is common to all these systems is that they all have the ability of restricting the rotational and translational molecular freedom of the substrate by virtue of relatively long-lasting nonbonded associations. This feature may include highly specific binding in preorganized systems such as enzymes, relatively static media such as molecular crystals, as well as non specific and often less organized environments such as glassy matrices and amorphous surfaces. Micelles, liquid crystals, inclusion complexes, porous solids, and other systems of current interest may lie somewhere in between.

The lifetimes of photochemically excited states (singlets and triplets) and reactive intermediates (radicals pairs, free radicals, biradicals, etc.) produced by primary photochemical primary processes serve as a "reaction clock" which determines the time scale during which these species can explore the accessible void space of a host such as a porous crystalline solid. The ability of the guest to globally explore the host free volume depends on the relative time scales of the reaction clock. Three common cases are: (1)

guest diffusional motion is rapid compared to the time scale of reaction; (2) guest diffusional motion is slow relative to the time scale of reaction; and (3) guest diffusional motion is of the order of the time scale of reaction. Local rotational guest motion may be rapid relative to the time scale of reaction, although global motion may be slow relative to the time scale of reaction.

We shall present examples demonstrating how the interacting supramolecular structure and supramolecular dynamics can lead to conditions controlling the selectivity of supramolecular photochemical reactions.

3. Supramolecular Organic Photochemistry

For the purposes of this report, the organic photochemistry involved is simply that of producing a pair of reactive radicals (e.g., two chlorine atoms or a pair of reactive carbon radicals) by the absorption of light. In homogeneous solutions, these radicals are known to be very reactive reagents and to be involved in non-selective reactions. However, these radicals are not at all reactive with the aluminosilicate framework of the external or internal framework of zeolites. Indeed, the framework of zeolites is similar in structure to glass and is therefore expected to be inert to radicals of all types.

3.1 TRANSFORMING NON-SELECTIVE MOLECULAR PHOTOCHEMISTRY INTO SELECTIVE SUPRAMOLECULAR PHOTOCHEMISTRY OF ORGANIC MOLECULES ADSORBED IN ZEOLITES

The issue that we now wish to address is how to exploit the paradigms of supramolecular chemistry, enzyme chemistry, and molecular photochemistry to transform the reactions of photochlorination and radical pairs, which are non-selective when they are conducted in the non-restrictive space of non-viscous homogeneous solutions, into reactions which are selective when they are conducted in a restricted space of a supramolecular assembly. In particular, we are concerned with porous, crystalline zeolites as the host in the supramolecular assembly.

From the standpoint of a photochemist concerned with photoreactions in restricted spaces, such as porous zeolite crystals, the numerous alternatives of binding can be broadly classified in terms depending on the freedom enjoyed by the substrate during the time scales of interest in photochemical phenomena. Probes that are tightly bound at the molecular level or that experience high local translational or rotational barriers, for instance, can be considered to be in a *closed or restricted environment* or *reaction cavity* [14]. We seek to model these reaction cavities for two families [11,12] of porous crystalline zeolites. We shall examine the photochlorination of n-dodecane in pentasil zeolites [17] and the photochemistry of dibenzyl ketones in both the pentasil [18]. Other examples, not mentioned here involve the faujasite zeolites [19].

3.1.1 *Photolysis of n-Dodecane Adsorbed on ZSM-5 Zeolites*

The product distributions [17] produced by the photoinduced dichlorination of n-dodecane (nD) and photoinduced monochlorination of 1-chlorododecane (1CD) adsorbed on two pentasil zeolites (silicalite and LZ-105) provide an informative example of supramolecular photochemistry of an organic molecule adsorbed in a porous crystal. The photochlorination of the adsorbed guest molecule is very selective toward terminal chlorination compared to the totally non-selective photochlorination of the same molecule in fluid solution (eq. 1).

The results are explained in terms of a supramolecular model for which the mobile and diffusing chlorination reagents ($Cl\cdot/Cl_2$) enter the zeolite particle from the external surface and diffuse preferentially along the linear channels of the zeolite internal surface containing immobile adsorbed nD (or 1CD) molecules in one of two possible topological sites of the internal surface (an outer layer and an inner core) and in terms of six plausible supramolecular isomeric structures which may exist in either topological site. The model assumes that the adsorbed guests in the outer layer are attacked preferentially, and furthermore, that the attack occurs at the end of adsorbed nD molecules closest to the external surface. After the first chlorination the substrate molecules in an inner layer are protected from chlorination by "blocking" molecules parked in the outer layer via a supramolecular "steric" effect. The model describes each substrate molecule adsorbed on the internal surface in terms of supramolecular isomeric structures that are capable of characterizing the specific void space sites (linear channels, zig-zag channels, intersections) occupied by the substrate. Examination of the details of the local supramolecular structures of nD/Z also reveals both a **regiochemical and a stereochemical** aspect. Specifically, four of the six supramolecular isomers described above (Figure 1) possess one end of the nD molecule within the structure that is closest or **proximal** to the external surface of the crystal and a second end that is further away or **distal** to the external surface of the crystal. The two exceptions are the IZI and ZIZ isomers for which the two ends of the nD molecule are equivalently positioned with respect to the distance from the external surface. Thus, for these two isomers the proximal/distal relationship does not apply (compare the term meso for diastereomers). These considerations reveal a regiochemical aspect of the local nD/Z systems resulting from the formation of the guest host complex that does not exist for the guest when considered only in molecular terms in the absence of the host. A detailed analysis of the results allows the conclusion that when compensating cations are present they tend to be preferentially located in the zig-zag channels rather than in the linear channels or

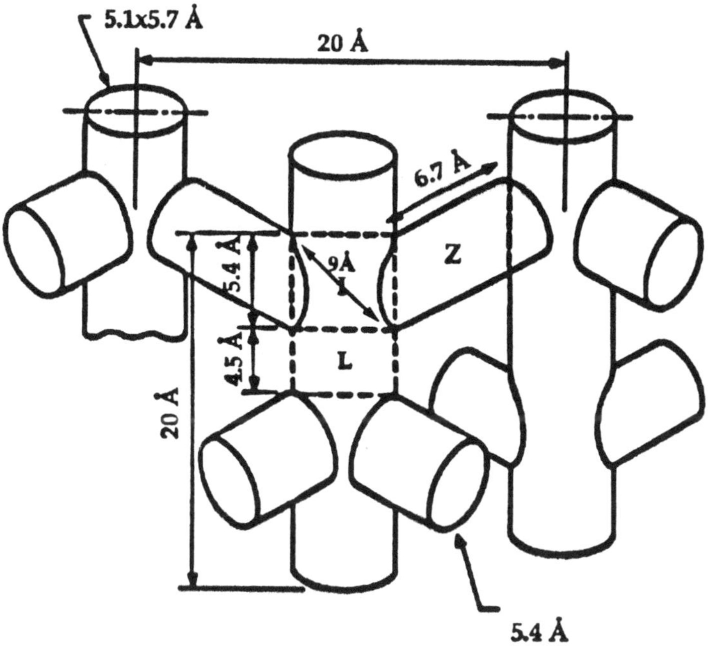

Figure 1. Schematic representation of the void space of a pentasil zeolite. There are three distinct forms of void space in a unit cell: the Z (for zig-zag) space, the I (for intersection) space, and the L (for linear space. A n-dodecane molecule can span about 3 of these units if they are contiguous. For example, the ILI and LIL combinations will hold a n-dodecane molecule intersections and that the variation of selectivity of chlorination with experimental conditions results from redistribution of the isomeric supramolecular structures.

intersections and that the variation of selectivity of chlorination with experimental conditions results from redistribution of the isomeric supramolecular structures.

3.1.2 *Photochemistry of Dibenzyl ketones (DBKs) Adsorbed on Pentasail Zeolites*

The photochemical excitation of ketones of appropriate structure has proven to be an excellent method of preparing geminate triplet radical pairs. Because the radical pair is born as a triplet, it is inert to radical pair combination reactions and cannot react until it finds a mechanism to undergo intersystem crossing to a reactive singlet geminate radical pair. In non-viscous solutions, diffusional separation is so fast that geminate pair reaction (cage effect) is very low (typically less than a few percent). However, when a triplet geminate radical pair is produced in the constrained reaction cavity of a porous crystalline zeolite, diffusional separation and local rotational may be inhibited and thereby enhance the cage effect.

The use of dibenzyl ketone (**DBK**) and its derivatives as probes for studying the effect of organized media and restricted environments follows from their ability to report on events that cover relatively large spatial and dynamical ranges. The correlation times in the photochemical processes of dibenzyl ketones are closely related to the rates commonly found in the topological isomerization processes of a number of organized media and restricted environments. Thus, the photobehavior of **DBK** reflects the topological and geometrical features of the **void space** of the porous crystalline solid in which the photochemistry occurs.

Irradiation of **DBK** in unrestricted environments such as solution media is characterized by the formation of benzyl radicals following a-cleavage and decarbonylation processes [20]. The termination process of these radicals is generally the diffusional controlled radical-radical coupling to yield 1,2-diphenylethane. Photolysis of unsymmetrically substituted derivatives (eq.5) results in the totally non-selective formation of three radical coupling products, **AA**, **AB**, and **BB**, in a ratio of 1:2:1. In

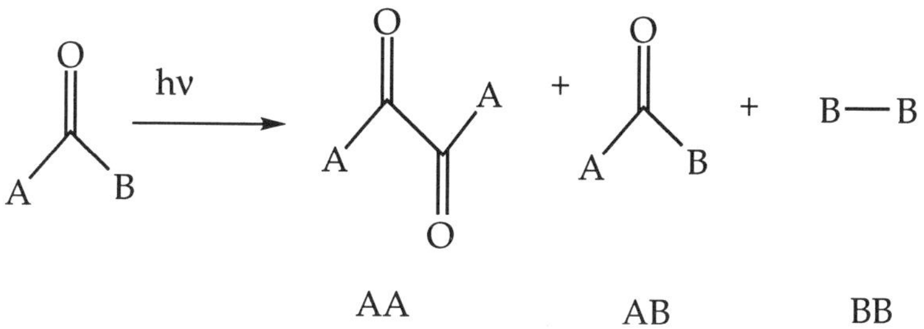

(5)

contrast, when escape of the radicals is impeded by confining environments, the radical recombination probability is modified strongly and can become selective. We shall show how through simple inferences derived from product structures, the initially occupied site of the DBK, the diffusional history and the rotational history of radicals produced in the photoreaction of DBK adsorbed in zeolites can be determined.

3.1.3 *Product Distributions from DBK Adsorbed on Pentasil Zeolites*

The photochemistry of substituted DBKs adsorbed on pentasil zeolites is found to depend on whether the ketone is present on the external surface or on the internal surface of a zeolite [18]. For example, ortho-methyl DBK, which cannot enter the internal surface of pentasil zeolites, is adsorbed on the external surface exclusively and therefore absorbs light while adsorbed on the external surface. The radicals' products are of two types: one possessing an ortho-methyl moiety which cannot enter the internal surface and one possessing a benzyl moiety which can enter the internal surface. Under the proper conditions, the smaller benzyl radicals are sieved into the internal surface faster than they react with ortho-methyl radicals on the external surface. Once the benzyl radicals have entered the internal surface, they react with each other faster than they exit from the internal surface. The net result is that the random coupling of benzyl radicals in solution is replaced with a more selective reaction producing coupling of only self coupling (**AA** and **BB**) of ortho-methyl radicals and self coupling of benzyl radicals.

A totally different situation occurs when the DBK adsorbed on the internal surface is photolyzed. In this case, the close contact of the radicals with the walls leads to a large cage effect in which radicals produced from the same molecule recouple rather than react with radicals produced from dissociations of other molecules. The new result is that the random coupling of benzyl type radicals in solution is replaced with a very selective reaction producing only one geminate coupling product (**AA**).

Conclusion

Supramolecular chemistry involves at one extreme receptor structures which are extremely "soft" and binding of substrate may be sufficiently weak such that the supramolecular characteristics may be subtle or marginal because of the mobility of the substrate and the total averaging of its environment. Examples are the environment of a solvent cage and the environment of a micellar aggregate. At another extreme, the receptor structure may be "hard" and the binding sufficiently strong such that the substrate is completely immobile and reaction of any type may be difficult. Examples are tightly packed molecular crystals. Intermediate cases, such as those provided by enzymes, contrive to exploit features of supramolecular systems that allow for high reaction rates and selectivities. Zeolites may be viewed as enzyme models where the substrate has extremely "hard walls". Photochemical reactions allow the convenient production of extremely reactive species (chlorine atoms, reactive carbon radicals) adsorbed on the external or internal zeolite surface. The surface or the void space of the zeolite serves to control the products of reaction such that species which are very reactive and non-selective in their solution chemistry, become very selective when they are substrates bound to zeolite receptors.

References

1. Turro, N.J. (1986) Geometric and topological thinking in organic chemistry, *Angew. Chem. Int. Ed., Engl.* **25,** 882 .
2. Lehn, J.-M. (1988) *Angew. Chem. Int. Ed. Engl.* **27**, 90; Lehn, J.-M. (1978) *Pure Appl. Chem.* **50**, 871 .
3. McMurry, J. and Fay, R.C. (1995) *Chemistry*, Prentice Hall, Englewood Cliffs, NJ, p. 483.
4. Carey, F. A. (1992) *Organic Chemistry*, 2nd Ed., McGraw-Hill, New York.
5. (a) For a general review of photochlorination see Huyser, E. (1965) *Adv. Free Radical Chem.,* vol.1, William, G. H.(ed.) Logos, London, p. 79; Russell, G. A. (1960) *Tetrahedron* **8**, 101; (b) For attempts to achieve selectivity by adsorption on substrates see: Perry, E. (1970) *J. Org. Chem.* **35**, 2053; Eden, C. and Shaked, Z. (1975) *Israel J. Chem.*, **13,** 1.
6. Ingold, K. (1973) Rate Constants for Free Radical Reactions in Solution, *Free Radicals*, vol. 1, J.K. Kochi (ed.), John Wiley & Sons, New York.
7. Russell, J. A. (1955) *J. Am. Chem. Soc.* **80**, 4987.
8. Bunce, N. J., Ingold, K. U., Landersk, J. P., Lusztyk, J. and Scaiano, J. C. (1985) *J. Am . Chem. Soc.* **107**, 5464.
9. Turro, N. J. (1992) *Modern Molecular Photochemistry*, University Science Books, Menlo Park, CA.
10. Herron, N. (1990) *Biomimetic Polymers,* C. G. Gebelein (ed.), Plenum Press, New York, p. 81.
11. Breck, D. W. (1974) *Zeolite Molecular Sieves*, John Wiley & Sons, New York. Barrer, R. M. (1978) *Zeolites and Clay Minerals as Sorbents and Molecular Sieves*, Academic Press, London; Barrer, R. M. (1986) *Pure & Appl. Chem*, **58**, 1317.
12. Flanigen, E. M., Bennett, J. M., Grose, R. W., Cohen, J. P., Patton, R. L., and Kirchner, R. M. (1978) *Nature* **271**, 512; Kokotailo, G. T., Lawton, S. L. and Olson, D. H. (1978) *Nature* **272,** 437; Kototailo, G. T., Lawton, S. L. and Meier, W. M. (1981) *J. Phys. Chem.* **85**, 2238.
13. Csicsery, S. M. (1976) *Zeolite Chemistry & Catalysis*, J.A. Rabo (ed.), ACS Monograph 171, Washington, D.C., chapter 12; Derouane, E. G. (1984) *Zeolites: Science & Technology*, F.R. Ribeiro (ed.) , Nijhoff, Boston, 347; Derouane, E. G. and Gabelica, Z. (1980) *J. of Catalysis* **65**, 486; Weisz, P. B. (1980) *Pure Appl. Chem.*, **52**, 2091; Weisz, P. B. (1986) *Ind. Eng. Chem. Fundam.*, **25**, 53 ; Derouane, E. G. (Ed.) (1982) *Diffusion and Shape Selective Catalysis in Zeolites.*, Academic Presss, New York.
14. Turro, N. J. (1986) *Pure & Appl. Chem.*, **58**, 1219; Weiss, R. G., Ramamurthy, V. and Hammond, G. S. (1993) *Acc. Chem. Res.* **26,** 530; Ramamurthy, V., Weiss, R. G. and Hammond, G. S. (1993) *Adv. Photochem.* **18,** 69.
15. Titiloye, J. O., Parker, S. C., Stone, F. S. and Catlow, C. R. A. (1991) *J. Phys. Chem.*, **95**, 4038.
16. (a) Derouane, E. G. and Gabelica, Z. (1980) *J. of Catalysis*, **65**, 486; (b) Derouane, E. G. (1980) *Catalysis by Zeolites*, Imelik, B. et al. (eds), Elsevier, Amsterdam, p. 5.

17. (a) Turro, N. J., Fehlner, J., Hessler, D. P., Welsh, K. M., Ruderman, W, Firnberg, D. and Braun, A. M. (1988) *J. Org. Chem.*, **53,** 3731.

18. Turro, N.J., Lei, X., Cheng, C.C., Corbin, D.R. and Abrams, L. (1985) *J. Am. Chem. Soc.*, **107**, 5824.

19. (a) Turro, N.J. and Zhang, Z. (1987) *Tetrahedron Letts.* **28**, 5637; (b) Turro, N.J. and Zhang, Z. (1989) Photochemistry of Dibenzyl Ketone Adsorbed on Size/Shape Selective Faujasite Zeolites. Steric Effects on Product Distributions, *Photochemistry on Solid Surfaces*, Anpo, M. and Matsuura, T. (ed.) Elsevier, Amsterdam, pp. 197-215.

20. Engle, P.S. (1970) *J. Am. Chem. Soc.*, **92**, 6074; Robbins, W.K. and Eastman, R.H. (1970) *ibid.*, **92**, 6076.

TOPICS IN SUPRAMOLECULAR CHEMISTRY

J.-M. LEHN
Université Louis Pasteur, Strasbourg, and Collège de France, Paris, France

1. Towards Molecular and Supramolecular Devices

Synopsis. *Molecular* and *supramolecular* devices may be defined as structurally organized and functionally integrated chemical systems built on supramolecular architectures. The development of such devices requires the design of molecular components performing a given function (*e.g*. photoactive, electroactive, ionoactive, thermoactive or chemoactive) and suitable for assembly into an organized array.

Photonic devices may undergo energy transfer or photoinduced electron transfer processes. Light-conversion devices and charge separation centers have been realized with photoactive cryptates formed by receptors containing photosensitive groups such as bipyridines and porphyrins.

Electronic and *ionic devices* are based on electroactive and ion binding components. Redox-active polyolefinic chains, such as the "caroviologens", represent *molecular wires* for electron transfer through membranes. Push-pull polyolefins possess marked non-linear optical properties.

Tubular mesophases, formed by organized stacking of suitable macrocyclic components, and "chundle" type structures, based on bundles of chains grafted on a macrocyclic support, represent approaches to *ion*

G. Tsoucaris et al. (eds.), Crystallography of Supramolecular Compounds, 451–458.

channels. Lipophilic macrocyclic units form LB-films that may perform molecular recognition at the air-water interface.

Light- and redox-triggered *switches* may be introduced into electronic and ionic devices leading for instance to switchable molecular wires and to materials for information storage. These processes also represent entries into *semiochemistry*, the chemistry of molecular signalization.

Of special interest is the possibility to design devices that may form by *molecular self-assembling*, i.e. by the spontaneous generation of the desired molecular architecture from its components.

Developments in molecular and supramolecular design and engineering open perspectives towards the realization of molecular photonic, electronic and ionic devices based on energy transfer, electron transfer and ion transfer processes, and capable of undergoing self-organization into a desired architecture. Such *programmed* molecular and supramolecular systems may be able to perform highly selective operations of recognition, transfer and structure generation for signal and information processing at the molecular and supramolecular levels.

For some recent references see:

- Lehn, J.-M. (1990) *Angew. Chem. Int. Ed.* **29**, 1304; (1995) *Supramolecular Chemistry*, VCH, Weinheim, Ch. 8.
- Pregel, M.J., Jullien, L., and Lehn, J.-M. (1992) *Angew. Chem. Int. Ed.* **31**, 1637.
- Pregel, M.J., Jullien, J., Canceill, J., Lacombe, L., and Lehn, J.-M. (1995) *J. Chem. Soc. Perkin Trans. 2* 417.
- Goulle, V., Harriman, A., and Lehn, J.-M. (1993) *J. Chem. Soc., Chem. Commun.* 1034.
- Gilat, S.L., Kawai, S.H., and Lehn, J.-M. (1993) *J. Chem. Soc., Chem. Commun.* 1439.
- Kawai, S.H., Gilat, S.L., and Lehn, J.-M. (1994) *J. Chem. Soc., Chem. Commun.* 1011.
- Tsivgoulis, G. and Lehn, J.-M. (1995) *Angew. Chem. Int. Ed.* **34**, 1119.

2. From Molecular Recognition Towards Self-Organization

Synopsis. Supramolecular chemistry has relied on more or less preorganised molecular receptors for effecting molecular recognition, catalysis and transport processes. A step beyond consists in the design of systems undergoing *molecular self-organization*, i.e. systems capable of spontaneously generating a well-defined supramolecular architecture by *self-assembling* from their components in a given set of conditions.

The *molecular information* necessary for the process to take place must be stored in the components and acts through selective molecular interactions. Thus, these *programmed supramolecular systems* operate via molecular recognition.

Several approaches to self-assembling systems have been pursued:

1) the generation of *mesophases* and *liquid crystalline polymers* of supramolecular nature from complementary components, amounting to macroscopic expression of molecular recognition;

2) the molecular recognition directed formation of *ordered solid state structures*;

3) the *self-assembly of inorganic species* based on ligand design and on the use of suitable coordination algorithms as expressed in

a) the formation of helical metal complexes, the *double-stranded and triple-stranded helicates*, that result from the spontaneous organization of two or three linear oligobipyridine ligands respectively into a double or a triple helix by binding of specific metal ions;

b) the spontaneous generation of closed cage structures from two different types of ligands from metal ions in a process involving the self-assembling of up to 19 particles !

c) the self-assembling of inorganic grids of metal ions based on suitably designed ligands.

The design of molecular information controlled, "*programmed*" and functional self-organizing systems represents new horizons in supramolecular chemistry towards "*intelligent*", functional supramolecular materials, network engineering and polymolecular patterning.

For some recent references, see:

- Lehn, J.-M. (1990) *Angew. Chem. Int. Ed. Engl.* **29**, 1304; (1995) *Supramolecular Chemistry*, VCH, Weinheim, Ch. 9.
- Fouquey, C., Lehn J.-M., and Levelut, A.-M. (1990) *Adv. Mater.* **21**, 254.
- Gulik-Krzywicki, T., Fouquey, C., and Lehn, J.-M. (1993) *Proc. Natl. Acad. Sci. USA* **90**, 163.
- Drain, C.M., Fischer, R., Nolen E., and Lehn, J.-M. (1993) *J. Chem. Soc., Chem. Commun.* 243 .
- Lehn, J.-M., Mascal, M., DeCian, A., and Fischer, J. (1990) *J. Chem. Soc., Chem. Commun.* 479.
- Lehn, J.-M., Mascal, M., DeCian, A., and Fischer, J. (1992) *J. Chem. Soc. Perkin Trans* **2** 461.
- Lehn, J.-M. (1993) *Makromol. Chem., Makromol. Symp.* **69**, 1.
- Lehn, J.-M., Rigault, A., Siegel, J., Harrowfield, J., Chevrier, B., and Moras, D. (1987) *Proc. Natl. Acad. Sci. USA* **84**, 2565.
- Lehn, J.-M. and Rigault, A. (1988) *Angew. Chem. Int. Ed. Engl.* **27**, 1095.
- Krämer, R., Lehn, J.-M., DeCian, A., and Fischer, J. (1993) *Angew. Chem. Int. Ed. Engl.* **32**, 703.
- Krämer, R., Lehn, J.-M., and Marquis-Rigault, A. (1993) *Proc. Natl. Acad. Sci. USA* **90**, 5394.
- Baxter, P., Lehn, J.-M., DeCian, A., and Fischer, J. (1993) *Angew. Chem. Int. Ed. Engl.* **32**, 69.
- Baxter, P., Lehn, J.-M., Fischer, J., Youinou, M.-T. (1994) *Angew. Chem. Int. Ed. Engl.* **33**, 2284.

3. Organization in Fluids, a Frontier in Supramolecular Chemistry - Some Remarks

Since species making up solids are held together by a variety of non-covalent interactions, the solid may be considered to represent an extended supermolecule or supramolecular assembly and solid state chemistry as part of a general supramolecular chemistry [1a-4].

Solids presenting long range order (LORO) may be studied by X Rays crystallography that establishes the molecular structure of the components and their supramolecular arrangement. The crystallographic data also provide very important information about the non-covalent bonding features (distances, angles) [3, 5]. This has been the main topic of the present meeting.

One may then ask the question of the operation of supramolecular effects in the fluid phase, of the way in which they determine the properties of fluids and of the possible existence of limited range order (LIRO) leading to more or less organized fluids and to the formation of aggregates and assemblies in liquids, and even in gases, through non-covalent interaction. This also raises the problem of the collective properties present in condensed phase and of the conditions leading to the appearance of such collective properties.

Is there LIRO, local order in fluids? When does it appear? What are the molecular structural features and the supramolecular interactions involved? Which are the local architectures generated? What are their dynamic properties, their time dependence?

Considering the three basic states of matter, the Solid, the Liquid and the Gas, the most highly ordered entity is the crystalline periodic solid and the least ordered (if at all) the gas. Various other entities bridge these two extremes such as amorphous solids, liquid crystals, etc.

Order may appear at the *interface* between two states, as is the case in the generation of molecular layers at the solid/liquid interface by self-assembly of molecules on solid surfaces [6, 7] or at the gas/liquid interface as in the formation of Langmuir-Blodgett monolayers or multilayers at the air/water interface [8, 9]. Incidentally, one may wonder whether the nature of the gas or of the molecules in the liquid have an effect on the features of the molecular layer. This may be the case if the molecules of the gas or in the subphase interact with the molecules forming the layer(s). Then, sensor type devices based on substrate selective membranes may be envisaged.

A most interesting question is that of the presence of LIRO in *bulk fluids*, liquids, solutions, as well as in entities such as sols and gels. Local organization appears to be present in pure liquid quinoline as indicated by NMR investigations, a rather unexpected observation [10]. Some degree of order can be induced in liquids by intense magnetic fields as also found in NMR experiments [11].

The generation of local order through designed self-organization in liquids, in solutions, and even in (dense) gases, represents an exciting line of research in supramolecular chemistry. One may envisage the derivation of molecular components presenting the required structural and interactional features for generating well-defined short or medium range organized entities in the fluid state. Sols and gels are obvious candidates for the exploration of local organization. Pure liquids that, like liquid quinoline [10], would present some kind of local order are of special interest for the study of the physics of liquids.

Aggregates and clusters found in mass spectrometry [12] or in molecular beam experiments represent supramolecular species with local organization occuring in the gas phase.

Solutions of specially designed components may contain specific architectures conferring LIRO. Various spectroscopic means (e. g. NMR, relaxation and correlation times, etc.), neutron diffraction, vapour phase osmometry, may be used to obtain information about the nature of the species formed. In view of the lability of the latter, such studies require collecting a set of data from different physical methods and giving special attention to the dynamic properties of these supramolecular assemblies [1b]. Recent results obtained by mass spectrometry indicate that direct injection of solutions may reveal at least some of the different solute species formed by self-assembly [12].

The temporal behavior of the system concerns namely the *persistence time* of a given local structure and its *dynamic features* (in particular the rate of component exchange, of dissociation with respect to the rate of the overall motions of the local supramolecular entity).

Thus, organization and order extending to variable ranges may be present in bulk fluids as well as in solids or at interfaces (gas/liquid, gas/solid, liquid/liquid, liquid/solid), induced by multiple interactions between components leading to the generation of supramolecular arrays and architectures. In particular, (locally) *organized fluids* represent a new frontier in the chemistry as well as in the physics of the structure and dynamics of fluids and of the supramolecular features of matter.

Supramolecular science as the *science of organized matter* paves the way *from condensed matter to organized matter.* The generation and control of order in fluids also leads to the inception of collective behaviour and represents a step up the ladder of complexity, a basic goal for supramolecular research [1].

References

1. Lehn, J.-M. (1995)*Supramolecular Chemistry* VCH, Weinheim, a) p. 174; b) p. 142.
2. Dunitz, J.D. (1995) G. Desiraju (ed.) in *Perspectives in Supramolecular Chemistry* Wiley, Chichester, **2**, in press.
3. Pascard, C., this book, ch. 8, p. 127.
4. See poster presented at this ASI: E.V. Boldyreva, "Solid state chemistry taught as supramolecular chemistry: a comprehensive university course".
5. Allen, F. H., this book, ch. 5, p. 55.
6. Ulman, A.(1991)*Ultrathin Organic Films: From Langmuir-Blodgett to Self-Assembly* Academic Press, Boston; Ulman, A. (1990) *Adv. Mater.*, **2**, 573; Sellers, H., Ulman, A., Shnidman, Y., and Eilers, J.E. (1993) *J. Am. Chem. Soc.* **115**, 9389.
7. Bain, C.D. and Whitesides, G.M. (1989) *Angew. Chem. Int. Ed. Engl.* **28**, 506; Kumar, A., Abbott, N.L., Kim, E., Biebuyck, H.A., and Whitesides, G.M. (1995) *Acc. Chem. Res.*, **28**, 219.
8. Lahav, M., this book, ch. 18, p. 331.
9. Moebius, D., this book, ch. , p. ; (1987) *Acc. Chem. Res.* **14**, 63.
10. Jalabert, D., Robert, J.B., Roux-Buisson, H., Kintzinger, J.-P., Lehn, J.-M., Zinzius, R., Canet, D., and Tekely, P.(1991) *Europhys. Lett.* **15**, 435.
11. Bastiaan, E.W. and Maclean, C. (1990) *NMR Basic Principles and Progress* **25**, 17.
12. Leize, E., Van Dorsselaer, A., Krämer, R., and Lehn, J.-M. (1993) *J. Chem. Soc., Chem. Commun* 990; Russell, K.C., Leize, E., Van Dorsselaer, A., and Lehn, J.-M. (1995) *Angew. Chem. Int. Ed. Engl.* **34**, 209.

APPENDIX

TUTORIALS ON MOLECULAR MODELLING

TUTORIAL: *Introduction to Conformational Analysis with the MacroModel Software*

Alexandre Varnek, Georges Wipff *

Laboratoire de Modélisation et Simulations Moléculaires
Université Louis Pasteur, Strasbourg

This tutorial introduces molecular modeling of simple systems, such as hydrocarbons and oligopeptides, using the MACROMODEL software. The students were provided with documents concerning Force Fields, Molecular Mechanics, Molecular Dynamics, and structural features of proteins. In the following, we only summarize the questions which were addressed to the students.

PART 1 Conformational analysis of hydrocarbons

Topic 1. Relative stability of ***ee, ea*** and ***aa*** conformers of *1,3*-dimethylcyclohexane

1. Molecular Building of *1,3*-dimethylcyclohexane conformers
2. Energy minimization.
3. Analysis of the results (total energies and their components, Me...Me distances).

Topic 2. Calculations of the torsional barrier in *n*-butane and *2,3*-dimethylbutane.

1. Molecular Building of *n*-butane and *2,3*-dimethylbutane
2. Energy minimization keeping one torsional angle (C-C-C-C) constrained.
3. Analysis of the results (making plot E_{tot} *vs* Θ and calculations of the rotational barrier).

Topic 3. Ramachandran plots of *n*-pentane and *1*-phenylbutane

1. Molecular Building of *n*-pentane and *1*-phenylbutane
2. Variation of energy with respect to two torsional angles in a structure.
3. Analysis of the results (search for the global minimum of E_{tot})

G. Tsoucaris et al. (eds.), Crystallography of Supramolecular Compounds, 461–463.

PART 2 Molecular Modelling of Oligopeptides.

Method: Molecular Mechanics (MM) and Molecular Dynamics (MD) simulations using the AMBER* force field.

Topic 1. *Conformational analysis of N-Acetyl-Aanyl-N'-methylamide ((Ala)$_2$).*

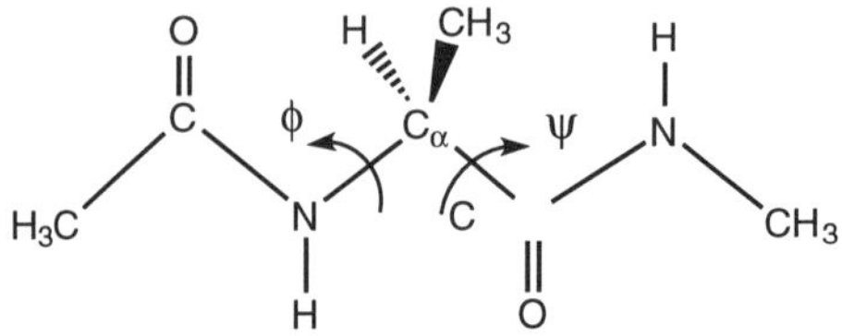

1. Modelling of C5, C7$_{eq}$, C7$_{ax}$ and α_R conformers
2. Energy minimization.
3. MD simulations in the gas phase at 300K and at 500K.
4. Analysis of the results (energies and geometry parameters).

Conformer	ϕ / ψ starting	ϕ / ψ final	N–H...O=C, Å final	E_{opt}, kJ/mol
C5	-150 / 150			
C7$_{eq}$	-75 / 65			
C7$_{ax}$	70 / -65			
α_R	-50 / -70			

Questions:
- Which conformer is the most stable in the gas phase ?
- Do the relative conformer stabilities correlate with the N–H...O=C distances?
- Which energy components are responsible for larger stability (or instability) of a given conformer?
- Do you observe conformational transformations at 300 K within 5 ps? 50 ps?
- How do the distances of 1...5 or 1...7 N–H...O=C "hydrogen bonds" and the torsional angles ϕ and ψ change as a function of simulation time?
- What differences in the time evolution of the studied conformers do you observe at 300 K and at 500K ?
- What is the relative population of the C7 forms found in the MD simulations starting with C7$_{ax}$ conformer at 300 K ? at 500K ?

Topic 2. *Simulations of the oligomers (Ala)$_{11}$ in the gas phase.*

Energy Minimization and Molecular Dynamics of three structures: α-helix, β-sheet and collagen helix.

Conformer	Number of atoms in the H-bonded ring	N–H...O=C, Å	ϕ / ψ	E_{opt}, kJ/mol
α-helix				
β-sheet				
collagen helix.				

Questions:

- Which conformer is the most stable in the gas phase ? Why ?
- Compare the distances of N–H...O=C hydrogen bonds in $(Ala)_2$ versus $(Ala)_{11}$ conformers
- Do you observe conformational transformations in the three forms of the $(Ala)_{11}$ at 300 K ? at 500 K ?
- Which conformer may form H-bonds with other oligopeptides ?

TUTORIAL: *Display of Dynamic Structures from Molecular Dynamics Simulations in Aqueous / Non - Aqueous Solutions. Comparison with X-Ray Structures.*

A. Varnek, E. Engler, M. Lauterbach, L. Troxler, G. Wipff *

Laboratoire de Modélisation et Simulations Moléculaires
URA 422 CNRS
Université Louis Pasteur, Strasbourg

Introduction. Goals. Representation of molecular motions in traditional publications is quite difficult. They are generally shown as cumulated views, of time-dependent selected parameters (distances, energies, ..). Two years ago, we made a video movie representing structures and dynamics of typical molecules, in supramolecular chemistry (crown ethers, cryptands, calixarenes, helicates, catenates, knots ..) [1]. Although this represents a clear improvement, compared to static color pictures, it does not allow to analyze interactively complex systems, especially in the solution state. The purpose of this tutorial is to examine interactively structures and motions of ionophores and salts *in solution*, previously simulated at MSM, and to compare these structures with the ones in the *solid state*, when available.

Please focus on the following general questions as guidelines:

- how does the X-ray structure (when available) compare with the structure(s) simulated in solution ?

- how does the structure(s) simulated in solution differ from water to a non-aqueous solvent (here, acetonitrile) ?

- how "dynamic" is the structure ? Some molecules simply oscillate around an average equilibrium structure, other become metastable or unstable in the conditions simulated. For instance, conformational changes of the ionophore, or separation of ion pairs make take place.

Simulation of the trajectories requires quite long simulations (typically, one day to simulate 50 ps of dynamics one a workstation). Some of the systems shown were simulated for 50 ps, others up to 2 ns (1 ns = 1000 ps = 10^{-9} s).

G. Tsoucaris et al. (eds.), Crystallography of Supramolecular Compounds, 465–476.

A general discussion of these questions is given in references [2- 4]. A copy of the files displayed can be obtained from G. Wipff on request.

<u>**Procedure:**</u>

1 - first call interactively (with the “mouse”), sequentially, the system to be displayed.

e.g. 18-crown-6 -> K^+ Cl^- -> in water

This will automatically call the display program *MD-DRAW* [5] and the files containing the trajectory.

2 - then use *MD-DRAW* to manipulate the system. *E.g.* animate the dynamics forwards or backwards, select the solvent shells, display the solvent box, measure distances, energies, etc ... . See enclosed instructions.

1) 18-crown-6 free and complexed.

Three typical forms are found in the solid state: D_{3d} (from the K^+ complex), C_1 (from the Na^+ complex), C_i (crown uncomplexed).

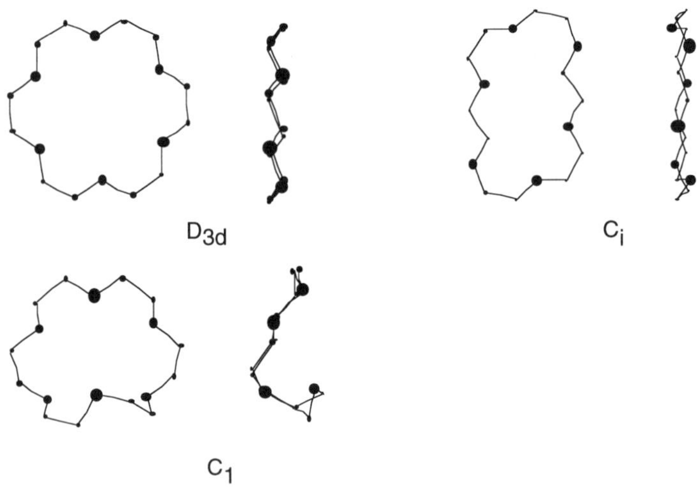

D_{3d}, C_i and C_1 conformers of 18-crown-6 (orthogonal views).

18-crown-6 uncomplexed. [6]

a) Compare the solvation of these three forms in water / in acetonitrile. For this purpose, the MD they have been simulated in solution, with constraints on the starting coordinates of the solute, in order to avoid conformational changes.

Look in particular at the first solvation shell. How many solvent molecules are coordinated to the solute ? What can you say about their lifetime ?

Select a sphere of 10 Å radius around the center of 18-crown-6, and determine the solvent / solute interaction energies (min / max, and average).

b) Examine 18-crown-6 uncomplexed free of constraints, simulated for 2 ns in acetonitrile. Notice the overall translation, rotation motions and conformational changes of the solute.

Can you recognize, among the many conformations generated by MD: the C_i form ? the D_{3d} form ? the C_1 form ?

The K^+ and $CaCl_2$ 18-crown-6 complexes in the water / in acetonitrile. *[6]*

- Does the cation sit at the center of the crown ?
- Is it coordinated to solvent molecules ? How many ?
- What about the ion pair ? Does it remain intimate ?
  Does it dissociate ? How is it solvated ?

2) The bicyclic 222 cryptand free and complexed.

The 222 cryptand uncomplexed. [6, 7]

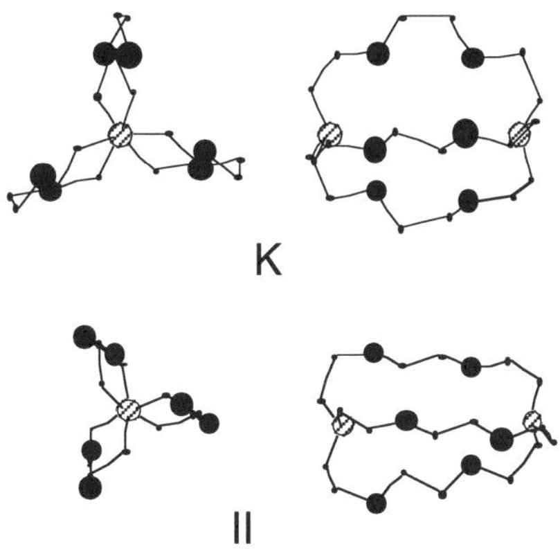

K and II conformers of 222 cryptand (orthogonal views).

Compare two conformers 222-II (from the solid state of 222 uncomplexed) and 222-K (extracted from the K^+ complex) in solution.

- Is thre solvent inside the cavity ?

- The 222-K form has larger interactions with water than 222-II . What specific hydration pattern do you notice ? Look at the hydration of 222-II: do you observe bridging hydrogen bonds ?

- Compare the interaction energies of 222-II and 222-K with acetonitrile and water (*e.g.* in a 10 Å radius sphere).

The 222 K^+ Cl^- and 222 $Ca^{2+}2Cl^-$ cryptates in water / acetonitrile solution. [6, 7]

- Does the ion pair remains intimate ? Is this behaviour expected ?

3) The t-butyl-calix[4]arene-tetraamide ("Calix-N") and its cation complexes. [8, 9]

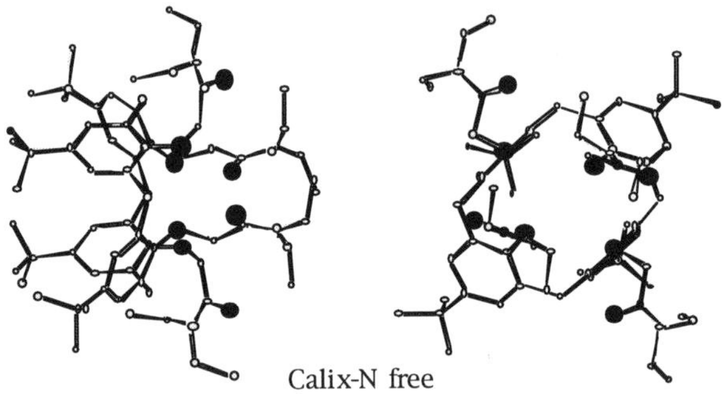

Calix-N free

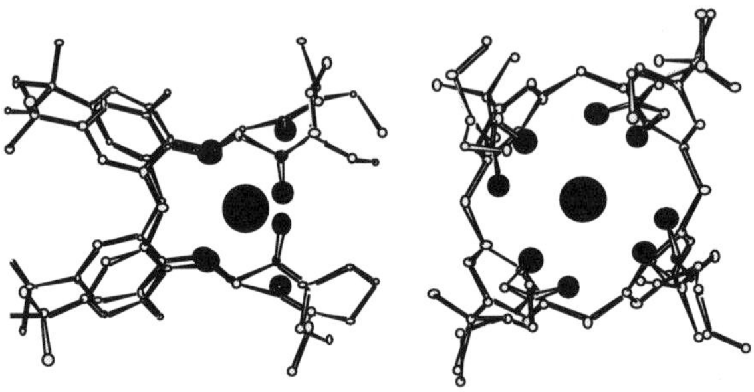

Calix-N K^+

Calix-N free and its complex with K^+ in the solid state (orthogonal views).

The X-ray structure of the K^+ complex. Notice the "cone" shape of the phenolic groups and that K^+ is symmetrically surrounded by 4 carbonyls and 4 ether oxygens with delineate a "pseudo-cavity".

The solid state structure of Calix-N: do you observe such a pseudo-cavity, preorganized to complex a cation? Compare the dimension of its cone, with the cone of the K^+ complex.

The Calix-N K^+ complex in solution. First in water, than in acetonitrile.

- Does it retain the cone shape ? What are the dimensions of the cone (min / max / average distances between opposed aromatic carbons) ?
- Is K^+ shielded from the solvent as in the solid state structure ?
- Do solvent molecules fill the cone ? How do they interact with the complexed cation ? with the ligand ?
- Explain why the complex "opens" markedly in water, but much less in acetonitrile.

The Li^+, Cs^+ and Eu^{3+} complexes of Calix-N in a given solvent:

- To what extent is the "structure" of the K^+ complex representative of the structure of these other complexes ? Discuss the effect of size and charge of the cation.
- Notice the dramatic solvent effect on the structures, for a given cation.

4) Comparison of ion pairs simulated in Water / in Acetonitrile

The Picrate- K+ ion pair [10]: measure the *ion... ion* distance as a function of time in the two solvents. Comments ?

Do like-ion always "repulse" each other ? [4,11,12]

The *Cl^- ... Cl^-* pair and *Guanidinium$^+$... Guanidinium$^+$* pair remains "intimate" in water, but dissociate in acetonitrile. Provide some simple arguments, based on the first shell solvation pattern.

5) Harmonic dynamics: Normal Modes of Vibration ("NMV") of cryptand *222*. [13]

Display the harmonic motions, starting with the lowest frequencies (<100 cm^{-1}) . Notice that the 6 first correspond to 3 translations and 3 rotations of the molecule. What are the corresponding “frequencies” ?

- Notice the "opening" / "closing" motions of the cage in some of the Normal Modes. Why is this important in view the the complexation process?
- What trend follows the amplitude of the motions, as the frequency increases?
- Look again at a normal (Newtonian) molecular Dynamics trajectory.

Why is it difficult to recognise any of the motions of the normal modes ?

References.

1. J.-M. Lehn, J.-P. Sauvage and G. Wipff, *"Molecular Modeling in Supramolecular Chemistry", Image- Media CNRS* (1992).

2. G. Wipff, "Molecular Modeling Studies on Molecular Recognition: crown ethers, cryptands and cryptates. From static models in vacuo to dynamical models in solution.", *J. Coord. Chem.*, **27**, 7-37 (1992).

3. G. Wipff, *Computational Approaches in Supramolecular Chemistry*, Eds. Kluwer, (1994).

4. G. Wipff and L. Troxler, "MD Simulations on Synthetic Ionophores and on their Cation Complexes: Comparison of Aqueous / non-Aqueous Solvents" in ref. 3, pp. 319-348.

5. E. Engler and G. Wipff, "MD-DRAW, a program to display MD results.",(1992).

6. L. Troxler and G. Wipff, "Conformation and dynamics of 18-crown-6 , cryptand 222, and their cation complexes studied by MD simulations", *J. Amer. Chem. Soc.*, **116**, 1468-1480 (1994).

7. P. Auffinger and G. Wipff, "Hydratation of the 222 cryptand and 222 cryptates studied by molecular dynamics simulations.", *J. Amer. Chem. Soc.*, **113**, 5976-5988 (1991).

8. P. Guilbaud, A. Varnek and G. Wipff,"MD study of Calix[4]arene tetraamide with neutral and cationic guests. influence of solvation on structures and stabilities", *J. Amer. Chem. Soc.*, **115**, 8298-8213 (1993).

9. A. Varnek and G. Wipff, "Dramatic Solvent Effect on the Ligand Wrapping around a Complexed Cation: a MD Study of *p-tert*-butylcalix[4]arene Tetramide and its Complexes with Alkali Cations and Eu^{3+} in Acetonitrile.", *J. Phys. Chem.*, **97**, 10840-10848 (1993).

10. G. Wipff and L. Troxler, *"The Picrate ion in water and acetonitrile solutions""*, American Institute of Physics, Eds., 1st European Conference on Computational Chemistry, Nancy (1994).

11. J. Gao, S. Boudon and G. Wipff, "Ab initio and Crystal Structure Analysis of Like-Charged Ion Pairs.", *J. Amer. Chem. Soc.*, **113**, 9610-9614 (1991).

12. S. Boudon, G. Wipff and B. Maigret, "Monte Carlo Simulations on the Like-Charged Guanidinium-Guanidinium Ion Pair in Water.", *J. Phys. Chem.*, **94**, 6056-6061 (1990).

13. G. Wipff and J. M. Wurtz, "Dynamics Views of Macrocyclic Receptors: Molecular Dynamics Simulations and Normal Modes Analysis." in *"Transport through Membranes, Carriers, Chanels and Pumps."*, Reidel, Eds. Pullman, R., Dorbrecht (1988), pp. 1.-26

ANNEX.1: **MD-DRAW** Software
Display of Dynamic Structures from M D Simulations.

Etienne Engler, Georges Wipff *

Laboratoire de Modélisation et Simulations Moléculaires
URA 422 CNRS, Université Louis Pasteur, Strasbourg

The MD-DRAW software was developed at MSM to display and analyze interactively molecular trajectories calculated with AMBER, in the gas phase, or in solution.
It can also display the Normal Modes of Vibration ("Harmonic Dynamics").

Files used. MD-DRAW reads:
- the cartesian coordinates of all atoms: x, y, z = f(time). This file has no other information.

- a PDB file containing the list of atoms of the solute molecule(s): ATOMS, RESIDUES*, ATOM LABELS*, ATOM TYPES*, X, Y, Z, Charge.
It is appended by the List of connectivities, calculated from the PDB coordinates and a Table of van der Waals radii.

- a "draw.in" input file. To specify the characteristics of the display.
E.g. : files to be read; title; mode of representation; step for reading the MD coordinates; colouring features; velocities*; MD at constant Volume / Pressure;
Some of the options present in this file can be also selected interactively via the menu (see Annex).

Operating tools
The Mouse has 3 buttons, "Left", "Middle" and "Right", referred to as
Mouse-L, Mouse-M; or *Mouse-R*.
The Keyboard: see enclosed scheme, with numbers.

Here are now some specific operations to perform in the tutorial:

Display the Dynamics Forward / Backward
Press the *Bar* at the *Keyboard* to move to the next set

Display interactively shells of solvent (within a sphere):
- select from the menu *(Mouse-R)*: *Solvent -> Clipping*
(A menu All/ *End* appears with *Mouse-R*)

- select the center of the shell,
either as a single atom or the center-of-mass of one group of atoms:
* pick one or more atoms with *Mouse-L*, then select *End* with *Mouse-R*

or the center-of-mass of the whole solute

* select *End* without picking

- increase / decrease the radius of the sphere of solvent displayed:
* hold *Mouse-M* and move horizontally or use the *Keyboard.*

<u>Calculate the solute / solvent interaction energy E (within a sphere)</u>

Select at the menu *(Mouse-R)*: *Energy*

(When the radius of the sphere is zero, E is the energy of the solute alone)

<u>Calculate geometrical features: distances, angles, dihedral angles.</u>

Select at the menu (*Mouse-R*): *GEO-MEASURES*

Pick the atoms (*Mouse-L*):

pick1 , pick2 , END	for the distance
pick1 , pick2 , pick3 , END	for a dihedral angle
pick1 , pick2 , pick3 , pick4 , END	for a dihedral angle

then press *Mouse-R* and select *End.*

<u>Display Normal Modes of Vibration</u>

Select at the menu (*Mouse-R*): *Representation -> Norm. Modes*

To move from one Normal Mode to the next, press the *Keyboard* bar (forwards) or the *Backspace Key* (backwards).

<u>Quit the Program</u>

On the*Keyboard*: press *ESCAPE* ,
then select *Yes* from the *Menu* with *Mouse-R*

ANNEX.2: **MD-DRAW** MENUS

Etienne Engler, Georges Wipff*

Laboratoire de Modélisation et Simulations Moléculaires
URA 422 CNRS, Université Louis Pasteur, Strasbourg

Main menu

VIEW MODE	Drawing mode of molecular system
COLOR	Color mode
POSITION	Display of Solute atom Characteristics
SOLVENT	Optional Display for the Solvent
REPRESENTATION	Optional Representation
SUPERPOSE	Structure Superposition
GEO. MEASURES	Geometrical measurements
ENERGY	AMBER Force Fields Energies
SURFACE	Solvent Accessible Surface
OPTION	Additional informations
MODIFY	Modify the molecular structure

Drawing mode of molecular system

STICK	
BALL&STICK	
CPK	
MIXTE	Ball representation only for isolated atoms
ORTHO	Two orthogonal views are simultaneously displayed
STEREO	Stereo mode (for stereo glasses)
PHOTO(STEREO)	Left and right stereo pictures

Color mode

Atoms are colored according to their :

ATOM	atomic types	
RESIDU	residu types or numbers	
MOLECULE	molecule numbers	
VELOCITY	atomic velocities	
ENERGY	energy ("energy" after selection of interactions)	
CLASS	class numbers	
SHELLS		ATOM type
		SHELL number
		VELOCITY
GRAY LEVEL	Gray predefined colors according to atomic types	
B&W	Black and white	
BACKG. COLOR	Background color in RGB components	

Display of Solute atom Characteristics

The atomic position is represented by :

DOTS	a dot
ATOMS	the atom name/number
RESIDUES	the residue name/number
PARAMS	the AMBER van der Waals parameters/charge
ENERGY	the interaction energy (previously selected in main menu)

Optional Display for the Solvent

CLIPPING	Interactive display of spherical shell (center is chosen by user)		
SHELLS	Hide/show entire box or previous shell(s)		
NUMBER	// // the residu number of each solvent molecule		
BONDS	// // bonds		
BOX	// // edges of box		
HBONDS		solut-solv	Calculate and show
		solv-solv	H-bonds interactions
COLOR		ATOM type	
		SHELL number	
		VELOCITY	

Optional Representation

PDB	Display the PDB structure
AVERAGE	// // the average structure (of each atomic position)
CUMUL	Superpose selected sets
NORM. MODES	Normal modes of vibration (arrows/moving structures)
BONDS	Hide/show bonds
VELOCITY	// // velocity vectors
TRAJECTORIES	// // trajectories
HCatoms	// // hydrogen atoms linked to Carbon atoms

Structure Superposition

AVERAGE	These structures may be superposed to (or removed from) the current structure
PDB	
OPT	

Geometrical measurements

DIST.	Print the distance between two selected atoms
ANGL.	// the angle // // three // //
DIH.	// the dihedral angle between four atoms
HIDE/SHOW	Hide or Show printed values of measurements
REMOVE	Remove all previous measurements

AMBER Force Fields Energies

INIT	Initialisation (Energy=0)
VDW	Van der Waals interactions
EEL	Electrostatic interactions
NBD	VDW and EEL interactions
BAD	Bond + Angle + Dihedral interactions
1 - 4	Non-Bonded Dihedral interactions
BND	1-4 + BAD interactions
TOT	All interactions are taken into account

Solvent Accessible Surface

Previous	Draw of a previously calculated surface
EEL Pot.	Color by electrostatic potentials
VDW Pot.	// // VDW potentials

Additional informations

SUBJECT		Print(or remove) the Number of current visualized set
COORD. SYSTEM		Show two coordinates references (absolute and rotating)
TITLE		Print a title
CURVE		Draw a curve (read in a file)
BITMAP		Draw a 2D-image (read in a file)
SNAPSHOT		Create a 2D-image from a selected area of the screen
ANTIALIASING		
DEPTHCUEING		
ANIME	BACKWARD	Rotation coupled
	FORWARD	with set of
	END	dynamics

Modify the molecular structure

UNDRAW	ATOM	Hide/show selected atom(s)
	BOND	Hide/show selected bond(s)
INIT(UNDRAW)		
DRAW BOND		
MOVE ATOM	CARTESIAN	Move selected atom(s) in cartesian
	INTERNAL	or in internal coordinates (BAD)
CHANGE COLOR	ATOM type	Change
	RESIDUE number	predefined
	MOLECULE number	colors
WRITE	PDB	Write the visualized structure in PDB format
	OPT	// // // // in AMBER opt. format

HARMONY: THE SUPRAMOLECULAR DATABASE MANAGEMENT SYSTEM

M. BAILLY #, E. ELIOPOULOS * and G. TSOUCARIS #

CNRS-UPR 180,
Universite Paris-Sud,
Centre Pharmaceutique,
92290 Chatenay-Malabry,
FRANCE.

* Dept. of Agr. Biology and Biotechnology,
Agricultural University of Athens,
Athens 118 55,
GREECE.

1. Introduction.

1.1 THE USE OF A SUPRAMOLECULAR DATABASE.

The number of publications refering to supramolecular compounds exceeds ten thousand and rises continuously. These cover several scientific fields (chemistry, physics, biology, agricultural sciences) and deal with various compounds of ionic, molecular or macromolecular type. Supramolecular compounds are involved both in fundamental studies and in applications in areas such as pharmaceuticals, plant protection, industrial processes, electronics, optics etc.

Communication between scientists in these diverse fields can be difficult due to different terminology and focus of interest. A database that covers the research areas and applications of supramolecular compounds is under development in order to help efficient communication, with the use of keywords and synonyms that cover a broad spectrum. Physicochemical properties and three dimensional structure (when available) give the potential of predicting the interaction between host and guest based on stereochemical, conformational and potential energy calculations with the use of computer graphics.

The general notion of assembly of two or more molecules in mutually defined positions is the concept of our bibliographic selection. Since this

G. Tsoucaris et al. (eds.), Crystallography of Supramolecular Compounds, 477–489.

concept can be general there are many necessary exclusions to be considered. Protein complexes and assemblies are not included as well as zeolites, liquid crystalls or colloids. The rules for selection include names of compounds, phenomena and processes concerned with molecular recognition and inclusion compounds.

1.2 THE PARTS.

The Database Management System (DBMS) for a personal is being developed as independent modules with the following functions:

1.2.1 The Bibliographic Database Management System.

The bibliographic DBMS supports at the present stage 8500 bibliographic entries, keywords , synonyms and properties for potential linking with other custom databases (e.g. physicochemical, crystallographic etc.).

1.2.2 The Crystallographic Database Management System.

The crystallographic DBMS handles entries from the Cambridge structural (mainly crystallographic) database that relate to the supramolecular field.

1.2.3 Molecular Graphics.

In order to visualise the structural information extracted from the databases and for the comparison and three dimensional manipulation of supramolecular structures molecular graphics interfaces are available. Software for the calculation and visualisation of accessible surface area of molecular compounds has also been written and incorporated.

1.2.4 The Modular User Interfaces.

In order to facilitate independent development of software utilities, a DOS and a Windows based modular user interfaces have been developed to control DBMS, graphics and utilities (Fig.1).

1.2.5 Utilities.

The variety of different sources for both bibliographic and structural data as well as the every day maintenance of the databases lead to the development of a series of management programs for efficient input and output of data, cross checking of entries for the databases, analysis of entries (overlap, duplicates etc.). Utilities for handling the structural data in different input and output formats, coordinate transformation, symmetry operations and definition of structural elements (hydrogen bonding, mean plane calculation etc.) have also been included.

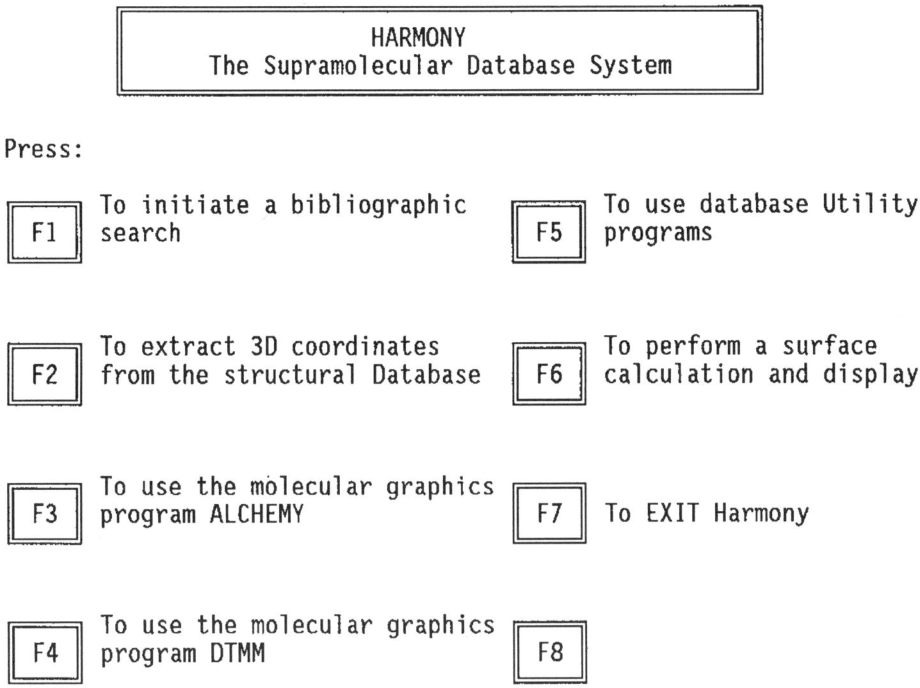

Figure 1The DOS Harmony user interface

2. The Bibliographic Database.

2.1 SOURCES.

The information comes from the INIST bibliographic database PASCAL, the Cambridge Structural Database and a selection of papers/publications. Information concerning solubility and stability constants is also derived from selected publications.

2.1.1 <u>INIST Bibliographic Database.</u>

The Institute of Scientific and Technical Information (INIST) has a developed a general bibliographic database PASCAL from where we extracted the entries for the supramolecular field (Fig. 2). The extraction criteria were based on keywords describing families or properties of supramolecular compounds.

-1- PASCAL

NO : 88-0067829
TI : Two high-performance liquid chromatographic columns for analyses of malto-oligosaccharides
AU : KOIZUMI (K.);UTAMURA (T.);KUBOTA (Y.);HIZUKURI (S.)
AF : N1 Mukogawa women's univ., fac. pharmaceutical sci.;V1 Nishinomiya;P1 JPN
DT : Publication en serie
SO : JN J. Chromatogr.;TS Journal of Chromatography;SN 0021-9673;OR NLD;DA 1987;VO 409;PA 396-403;BI 9 ref.;SH 4 fig.;PR 3 tabl.
LA : ENG
FA : Comparaison des performances de deux nouvelles colonnes (YMC-Pack PA gel et TSK gel Amide-80) appliquees au dosage des cyclodextrines, de l'amylose chaine courte et des malto-oligosaccharides. La colonne TSK apparait la plus performante (bonne stabilite et possibilite d'application des temperatures elevees)
FD : Amylose;Dextrine;Dosage;Cyclodextrine
ED : Amylose;Dextrin;Assay

Figure 2. A typical INIST database entry.

The search was performed for years 1987 to 1994 inclusive. This database is biannually updated. Since the PASCAL keywords were not always necessary oriented to the supramolecular field most of the entries were reassesed manualy and their keywords modified to describe more accurately the entry's contents.

Family of compounds	Property	Authors
cyclodextrin	supramolecular	
calixarene	assembly	
crown ether	macrocyclic + inclusion	
cryptate	macrocycle	
clathrate	molecular recognition	
werner		
hofman		
trivera		
oxycholin		
thymotide		

Table 1. **Examples of k**eywords used for searching the PASCAL database.

2.1.2 Cambridge Structural Database.

The Cambridge Structural Database (CSD) contains bibliographic references for the years 1964 to 1995. Based on keyword extraction similar to those for PASCAL one thousand two hundred entries related to supramolecular compounds were located and extracted and the bibliographical information was imported to the Bibliographical DBMS. The non availability of keywords is being handled semiautomaticaly and so far for nearly 200 entries additional keywords have been included to relate the entry more accurately to the subject described. There is limited (about 7 per cent) overlap between PASCAL and CSD and cross referencing utilities have been developed. The Cambridge structural codes have also been included in relevant PASCAL entries necessary to develop the link bibliography-structure-visualisation.

2.1.3 Selection of Manual Entries (Personal Archives).

In addition to the PASCAL and CSD entries a large selection of recent publications relevant to the field were processed manually and classified and keyworded with the same criteria (Table 2.) used for the modified keywords in PASCAL and CSD.

2.1.4 Nucleus of factual data.

Solubility and stability constants play a particularly important role in the assembly. In most practical applications the inclusion molecule is being carried by the receptor (host) molecule and then released. This action is heavily dependent on the physicochemical properties which again are dependent on the structural aspects of the supramolecules. So far 70 entries that concern mainly cyclodextrins are included and contain information on host/guest solubilities and concentrations (Fig.3).

```
HOST   : beta cyclodextrin
GUEST  : ipriflavone  (7-isopropyloxy-isoflavone)
SOLVENT: water
SOLUBHO:    1.85     gr/100ml
SOLGUPU:   0.0012   mg/ml
SOLGUHO:    0.021    mg/ml
CONC_HO:   10.0      mg/ml
AUTHOR : Weiszfeiler (V.);Stadler-Szoke (A.);Stejtli (J.)
SOURCE : Proc. of the 4th Int.Symp. on Cyclodextrins;PA 329-335;
--
```

Figure 3. Entry for solubility constants

2.2 FORMAT.

Originaly the database was developed in the form of DOS binary indexed files[1]. With the expansion of Windowing facilities the database is now operational under Microsoft ACCESS 2 for Windows as well with its own Windows Utility[2]. Various input formats exist and software utilities to convert to the adopted format have been developed.

---

[1] *Pressing F1 on the menu initiates the DOS supramolecular bibliographic database handling program. Searches can be initiated using keywords etc. as described in the program's operating instructions. Highlighted letters are for operations such as display, print, view, next, example etc.*

[2] *The Windows version of the bibliographical database is operational by entering Windows and double clicking the ACCESS2 application or the* ESBDLI application icon.

MOLECULAR DESCRIPTION

crown ethers
cryptands
spherands
cavitands
calixarene
saccharides
cyclodextrines
peptide
steroid
liquid clathrate
cyclophane
gas clathrate
lipid
macrocyclic
porphyrins
molecular clefts
molecular surfaces
proteins
lectins
carbohydrates
ionophores
charged

CHEMISTRY

host
guest
synthesis
new
improved
modification
selectivity
molecular
recognition
conformation
enzyme mimics
organic coordination
metal
chain
non-aqueous
chiral
catalysis
stereoselective
inclusion
complexation
complex
dimer
racemic
diastereome

PHYSICAL CHEMISTRY

solubility
viscosity
chromatography
chemiluminescence
thermogravimeric
analysis
thermodynamics
binding constant
association constant
stability constant
kinetics
calorimetry
photochemical
phase transfer
structure
dye
aggregation
mode

PHYSICS

spectroscopy
fluorescence
NMR
NMR1H
conductivity
crystallography
structure
IR
ESR
spectroscopy
Mossbauer
electronic
spectroscopy
circular dichroism
[alpha]d
irradiation

INTERACTIONS

Hydrogen bond
charge transfer
hydrophilic
hydrophobic
electrostatic
polarity
solvent
shape
Van der Waals

STATE

solid
gas
liquid
aggregated liquid
solution
polymers
monomer
oligomers
gels
membranes
layer
liquid crystal
micelles
lamellar

FRAGMENTS

O,N,P,S,As,Sn,C,
H,others
phenol
amine
ether
hydroxyl
imidazol
olefinic
acetylenic
acidic
aldehyde

APPLICATIONS

reactivity
electronics
pharmaceutical
catalysis
alimentary
synthesis
biotechnology
materials
separation
cosmetic
inclusion
photochemical
molecular films
extraction
molecular epitaxy

Table 2. Related Keywords

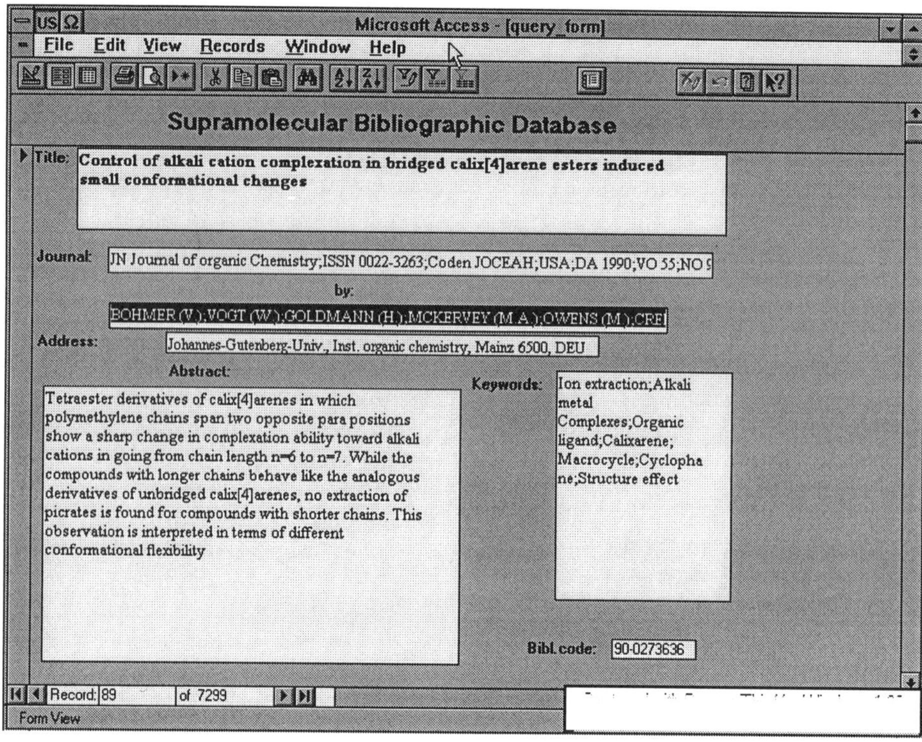

Figure 4. The query form of the Access 2 HARMONY bibliographic database

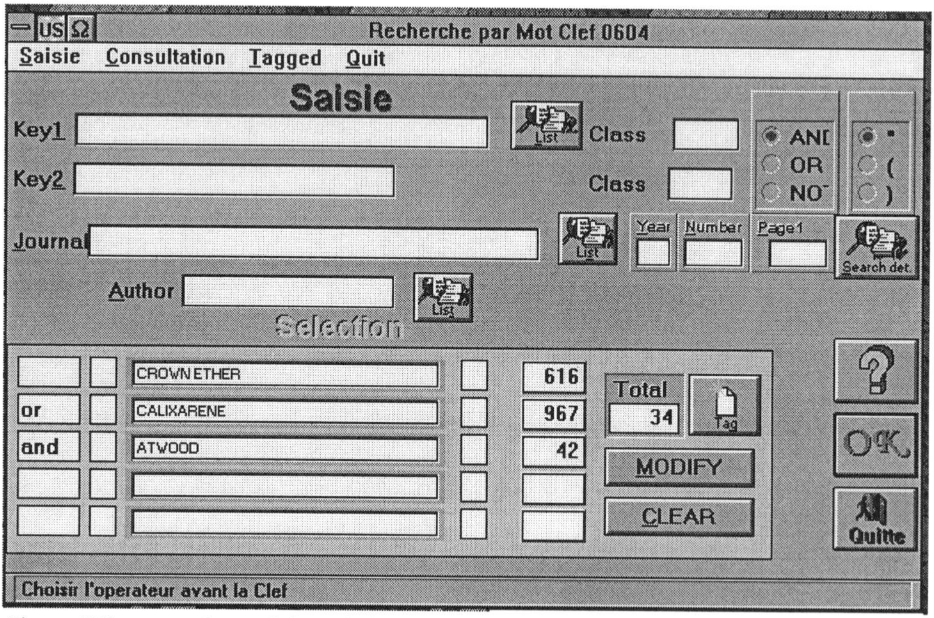

Figure 5 The query form of the Windows HARMONY bibliographic database utility

3. The Crystallographic Database.

3.1 SOURCES.

3.1.1 Cambridge Structural Database.

The Cambridge Structural Database (CSD) contains structural information for the years 1964 to 1995. Using the selected entries as described in section 2.1.2, 1200 entries with structural information can be imported in the crystallographic database. [3]

3.2 FORMAT.

3.2.1 Input format.

In constructing the supramolecular database the Cambridge FDAT format in ASCII is adopted so additional entries can be checked or manually integrated in the database. A utility to create the indexed structural database has been developed were the administrator has the control of what information is integrated in the database in its binary form. This DBMS does not include the Cambridge Database, which is commercially available, but the necessary software and entry codes to construct the supramolecular indexed database.

---

[3] *Pressing F2 initiates the supramolecular crystallographic database handling program which accesses the 3-d cartesian coordinates of all supramolecular compounds that the crystal structure exists and has been deposited in the Cambridge Crystallographic Database. The link between the crystallographic and the bibliographic databases is via the Cambridge alphanumeric identification code (e.g. FOWZOG, CIDMOR10) which is also the only available key so far to enquire the crystallographic database. The database enquiry program can handle the whole database that has been created from an FDAT Cambridge file using the utility program FDATBIN.*

The enquiry program prompts for the filename only of the database (e.g. HARMONY) while it expects the extension to be .BIN). If the database file is found, the program gives the available number of entries and prompts for the Cambridge code of the entry required. If the code exists some information on the entry is displayed and files compatible with common molecular graphics programs (e.g. ALCHEMY, DeskTop Molecular Modeller, SYBYL etc) are written, containing orthogonalised 3D coordinates. The filename of the relevant files is the Cambridge code while the file extension reflects the relevant molecular graphics program (e.g. .ALC, .MOL, .PDB).

When storage space is at a premium specific subject databases (e.g. calixarenes and cryptates only) can be constructed for personal use.

The structural database entries are linked to the bibliographic database via the unique "cambridge code".

3.2.2 Output format.

The enquiry and extraction program for the structural database automaticaly generates orthogonal atomic coordinates in formats adopted by the most popular PC or workstation available formats (Brookhaven, Alchemy, DTMM, MAGE etc.).

3.3 UTILITIES.

Several utilities are available with the structural database and are accessible from the top menu by pressing F5. The Utility menu contains:

a. A program to make a database file from an FDAT file.
b. A program to extract orthogonal coordinates in Alchemy or DTMM format from an FDAT (Cambridge Data File) file.
c. A program to apply coordinate transformations (fractional to orthogonal, orthogonal to fractional).

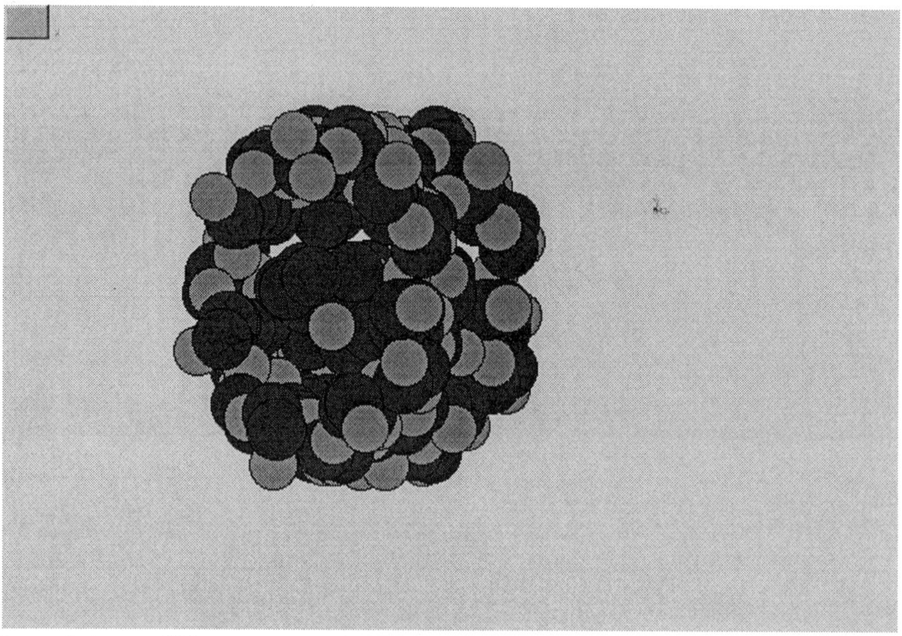

Figure 6. A spacefilling representation of the a beta cyclodextrin complex using Harmony graphics

4. Molecular Graphics.

4.1 PURPOSE.

It is in the philosophy of the Harmony package to be able to visualise at the molecular level the structural information available and to correlate properties, function and structure (Fig. 5). Several commercial packages are available to ease visualisation of structural information, compare or even built structures. For the PCs chemical structure packages will perform standard manipulation and visualisation operations (rotations, translations, space filling models, partial rotations, calculations, superimpositions etc.) but usually lack the operations necessary to the supramolecular structural chemist (host-guest fitt, accessible surface area, crystallographic operation, complex building etc.).

In the first instance we have incorporated interfaces to the most popular molecular graphics programs available for the PC while in parallel we develop and test a menu driven molecular graphics program oriented more towards the supramolecular chemist's needs.

4.2 FORMAT.

The output from the structural database retrieval program can be directly used by Alchemy, DTMM, MAGE, VACO and Harmony Graphics[4].

4.3 PACKAGES/ USE.

Apart from the commercial and public domain molecular graphics packages, which are briefly described in appendix 1 a program for the display and manipulation of supramolecular complexes has been developed.

4.3.1 Harmony Graphics.

The program MOLWIN is a windows application directed to the needs of the supramolecular chemist. The input is one or two separate molecules (host and guest) which can be displayed as spheres or lines and rotated globally or separately. Sections or slices of the host guest complex can be displayed to concentrate on specific volume constraints and interactions. The molecules can also be translated separately to obtain the best fit. Symmetry operators can be applied to easily display complexes formed during crystal packing.

---

[4] *Pressing F8 initiates Harmony Graphics. Modules include Host-Guest fit, superposition, Accessible surface calculation.*

For those with no graphics facilities the program VACO has been developed. VACO does not require elaborate graphics since it uses line printer characters to display sections of the Van der Waals spheres of the molecules. Equivalent manipulations to the windows graphics exist although they cannot be concidered “real time”.

5. References

1. INIST-CNRS, Direction Technique des Bases des Donnees, 2 Allees du Parc de Bravois, 54 514 Vandoeuvre les Nancy.
2. Allen, F.H., Davies, J.E., Galloy, J.J., Johnson, O., Kennard, O., Macrae, C.F., Mitchel, E.M Mitchel, G.F., Smith, J.M., and Watson, D.G. (1991) *J. Chem. Inf. Comp. Sci.,* **31**, 187-195.
3. ALCHEMY (1988), A program to display molecules on a PC.
4. Desktop Molecular Modeller (1988). A program to display molecules on a PC.
5. RASMOL (1995), TIBS. *20,* 374-376.
6. Kinemage by D.C. Richardson and J.S. Richardson (1992) Protein Science 1, 3-9.

APPENDIX

PC based Molecular Graphics Packages that readily interface to HARMONY

A.1 Alchemy.

Pressing F4 initiates ALCHEMY, a popular PC based commercial program available from Tripos Inc. Separate operating instructions for this mouse operated program are available.

A.2 DeskTop Molecular Modeller.

Pressing F3 initiates DTMM, a commercially available program from Oxford Publishing. Separate Instructions are available for this program.

A.3 Rasmol.

RasMol2 is a public domain molecular visualisation program developed by Roger Sayle of Glaxo Research and Development and is intended for the visualisation of proteins, nucleic acids and small molecules. The program is aimed at display, teaching and generation of publication quality images. RasMol runs on Microsoft Windows, Apple Macintosh, UNIX and VMS systems. The program reads in a molecule co-ordinate file and interactively displays the molecule on the screen in a variety of colour schemes and molecule representations. Currently available representations include depth-cued wireframes, 'Dreiding' sticks, spacefilling (CPK) spheres, ball and stick, solid and strand biomolecular ribbons, atom labels and dot surfaces.

To use RasMol initiate Windows and double click the RasWin image.

A.4 MAGE

The MAGE software (D.C. Richardson and J.S. Richardson (1992) Protein Science 1, 3-9) which allows molecules of known 3-dimensional structure to be displayed in a variety of ways and a limited range of measurements to be made. Operations on the displayed kinemage respond within a fraction of a second: the entire image can be rotated in real time, parts of the display can be turned on or off, any point can be identified by picking it, and the change between different forms can be animated. The image can be recentered, zoomed, put in stereo, or the front and back clipped away; distances, angles, and dihedrals can be measured.

MOLECULAR GRAPHICS APPROACHES IN STRUCTURE PREDICTION AND DETERMINATION

E. ELIOPOULOS * and I. M. MAVRIDIS #
* Dept. of Agr. Biology and Biotechnology,
Agricultural University of Athens,
Athens 118 55,
GREECE.

Institute of Physical Chemistry,
NRCPS Demokritos,
Aghia Paraskevi 153 10,
Athens,
GREECE.

1. Introduction

The ability to easily display graphically a molecule or a molecular assembly is very helpful in gaining quick comprehension of its detailed geometry and the interactions among the components in three-dimensions. Interactive molecular graphics provide the above, plus the opportunity to manipulate molecules or parts of them i.e. move specific atoms, fragments of molecules or whole molecules in space [4].

Moreover, it is possible to display a three-dimensional grid representing a property by contouring isoparametric levels. Such a property can be the electron density of a molecule or part of it as determined by X-ray crystallography, displayed at isoelectronic levels. This kind of representation is very useful in finding the positions of a disordered component of a molecular assembly by fitting a model of that component at various positions into the electron density corresponding to it . Similarly, given the coordinates of a molecule, one can display a surface calculated by surrounding each atomic center by a sphere of specific radius i.e. the van der Waals radius. Thus, in inclusion compounds one can visualise inside the host cavity the space available for the guest molecule and determine its position.

G. Tsoucaris et al. (eds.), Crystallography of Supramolecular Compounds, 491–498.

2. The three dimensional exclusion volume renderer.

Even with the modern methods of dynamic stereoscopy using electronic and liquid crystal devices and real time rotation, the depth perception remains an important issue in three dimensional manipulation on a two dimensional display. Volumes represented by points, lines or a wire mesh are still difficult to comprehend. Comparison of two volume projections on a flat screen is entirely dependent on the relative orientation of the projections.

In order to eliminate much of the spatial element in closely interacting molecules, as in the cases of host-guest or drug-receptor supramolecular systems, the approach of the exclusion volume is used. If two interacting atoms A_{host} and B_{guest} have Van der Waals radii $\mathbf{R_A}$ and $\mathbf{R_B}$, then the nearest distance the centre of atom B_{guest} can approach the centre of atom A_{host} is $\mathbf{R_A} + \mathbf{R_B}$. If we now represent a surface formed by the surfaces of spheres centered on the atoms of the host but with radii $\mathbf{R_{A_{host}}} + \mathbf{R_{B_{guest}}}$ the surface will include all the volume **excluded** to the centers of atoms of the guest, satisfying the stereochemical restrictions posed by the Van der Waals interactions between atoms of the host and guest. The remaining restricted space, outside the enveloping surface, will be the volume where the centers ofthe atoms of the guest molecule will be contained in a wire model representation since the intramolecular bond length and angle constrains of the guest remain (Fig. 1).

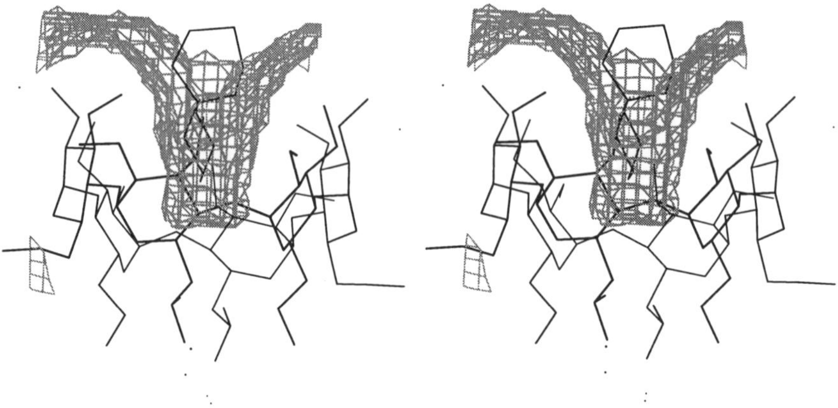

Figure 1. Stereo diagram of the 2,3,6- (Tri - O - methyl) - α - cyclodextrin (TMαCD) / 1,7 - Dioxaspiro [5,5] undecane (spiroacetal) inclusion complex. The grid represents the exclusion volume surface map.

This way the containing volume is more restricted and easier to perceive. It also gives an idea as to the shape limitations of the interacting molecule since we can now readily compare the restricted accessible volume determined by the host and a wire model of the guest.

3. The surface calculation algorithm

For our approach a parallel computational algorithm has been developed to calculate the exclusion and Van der Waals surface maps. Each atomic volume of radius $\mathbf{R_i}$ is represented by a cubic grid (a three dimensional array) of predetermined spacing (e.g. 0.2 Å), of which the nodes that are at a distance $0.1\text{Å} + \mathbf{R_i} < \mathbf{R} > \mathbf{R_i} - 0.1\text{Å}$ take the value of 1 and the nodes that are at a distance $\mathbf{R} < \mathbf{R_i} - 0.1$ Å take the value of 2. These precalculated and stored cubic grids (arrays) are as many as the types of atoms (carbon, oxygen, nitrogen etc.) of the molecules involved.

During the exclusive volume map calculation the cubic grids are centered on the atomic cartesian positions and the array values are added on a cell grid (array) of the same spacing containig all host atoms. In this way points of the cell grid that are less than a distance $\mathbf{R_i}$ from any atom have an array value of >1, points at a distance $\mathbf{R_i}$ from one atom only retain the value of 1, while points at a greater distance than $\mathbf{R_i}$ from all atoms retain the value of 0.

Because the calculation requires only addition of small integers and can be performed in parallel over all atoms of the host/receptor model on the cell grid, it takes advantage of the RISC and parallel processing computer technology. This novel algorithm is fast enough to be used in real time rendering of Van der Waals, exclusion volume and other property surfaces.

The following two examples show clearly the advantages in the use of surfaces for structure prediction and determination.

4. 2,3,6- (Tri - O - methyl) - α - cyclodextrin (TMαCD) / 1,7 - Dioxaspiro [5,5] undecane (spiroacetal) inclusion complex. Fitting spiroacetal model into a difference electron density.

The 1:1 complex mentioned above was formed from TMαCD and racemic spiroacetal. Half of the cyclodextrin molecule is in the asymmetric unit. As the difference electron density (F_{obs}-$F_{TM\alpha CD}$) in Fig. 2 shows, the guest must be located at the entrance of the cavity from the secondary hydroxyl side and it must be sitting on a 2-fold crystallographic axis. Since the spiroacetal molecule does not possess a 2-fold axis along its length, it is apparently disordered over two positions related by the 2-fold axis.

Figure 2. Stereo diagram of the 2,3,6- (Tri - O - methyl) - α - cyclodextrin (TMαCD) / 1,7 - Dioxaspiro [5,5] undecane (spiroacetal) inclusion complex. The grid represents the difference electron density map of the guest.

It has already been determined that the crystals obtained from TMαCD and racemic spiroacetal contained only the R enantiomer(1). In order to find the position of the guest, a model of the R enantiomer of spiroacetal was used. The model was created from the coordinates of the (S)-spiroacetal/ overmethylated β-cyclodextrin (2) and it is fitted into the electron density . Note the non-symmetrical positioning of the model inside the electron density which exhibits 2-fold symmetry along the axis of the cavity (Fig. 2).

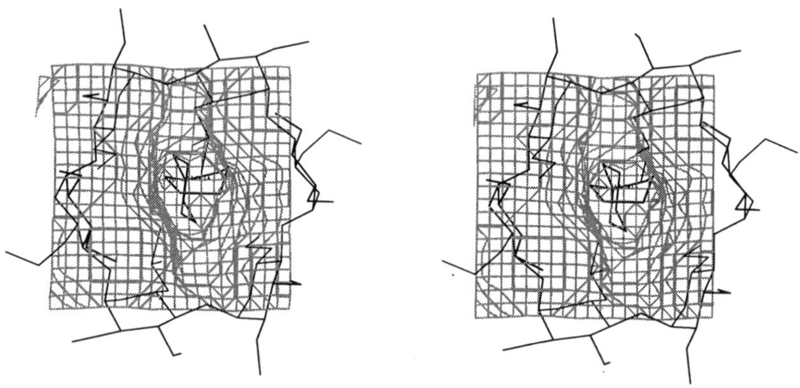

Figure 3. Stereo diagram of the 2,3,6- (Tri - O - methyl) - α - cyclodextrin (TMαCD) / 1,7 - Dioxaspiro [5,5] undecane (spiroacetal) inclusion complex. The grid represents a different view from Fig. 1 of the exclusion volume surface map.

The fit can be checked by calculating the exclusion volume map (Figs. 1 and 3) which provides the restricted volume where the guest molecule should be located. Indeed the difference electron density falls whithin the exclusion volume in agreement with the stereochemical restrictions.

5. Inclusion complexes of 2,3,6-(tri-O-methyl)-α-cyclodextrin and (S) and (R)-Mandelic Acid (MA). Fitting of the guests into the exclusive surface area of the host.

The crystal structures of the inclusion complexes of TMαCD with the (S) and (R) enantiomers of MA have been determined separately (3). In the (S)-MA complex the cyclodextrin molecule has a pseudo 2-fold axis and includes loosely the phenyl group of the guest while the hydroxyl and carboxyl groups protrude from the the secondary side of the cavity (Fig. 4).

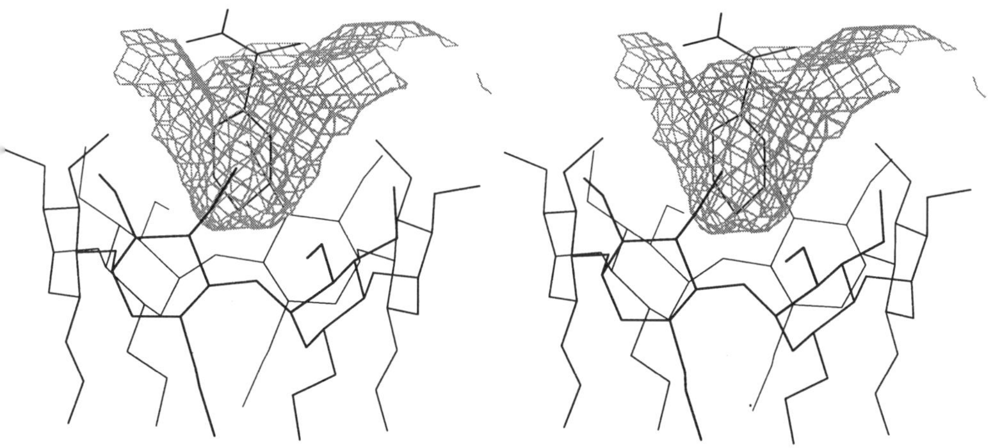

Figure 4. Stereo diagram of the observed structure of 2,3,6-(tri-O-methyl)-α-cyclodextrin and (S)-Mandelic Acid complex. The grid represents the exclusion volume surface map.

In the complex of (R)-MA the cyclodextrin ring is less symmetrical than that of the (S)-MA complex and the (R)-MA molecule is deeply inserted into the host cavity (Fig. 5). The hydroxyl group is H-bonded to a secondary hydroxyl group of the host, while the carboxyl group, situated outside the host cavity, is H-bonded to a primaty hydroxyl of an adjacent host molecule.

By calculating the exclusive volume surface for the hosts of the above complexes, it is evident that the volume available for the guest inside each host is different. It is more shallow for the macrocycle of (S)-MA (Fig. 4) than for that of (R)-MA enantiomer (Fig. 5).

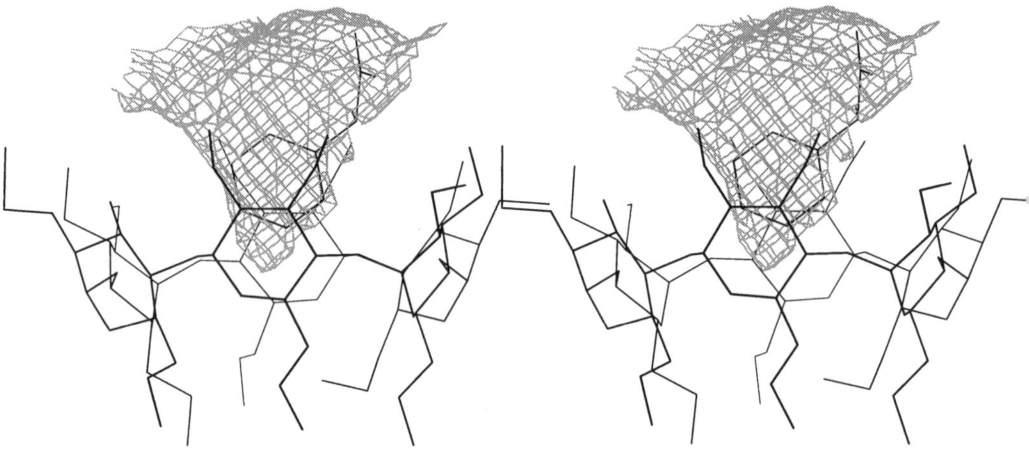

Figure 5. Stereo diagram of the observed structure of 2,3,6-(tri-O-methyl)-α-cyclodextrin and (R)-Mandelic Acid complex. The grid represents the exclusion volume surface map.

The (S)-MA enantiomer fits perfectly, as it can be seen in two different views (Figures 4 and 6), while attempting to fit the (R)-MA model in the exclusion volume of the TMαCD complexed to (S)-MA shows that the end carboxyl group violates the stereochemical restriction posed by the boundary of the exclusion volume map (Fig. 7).

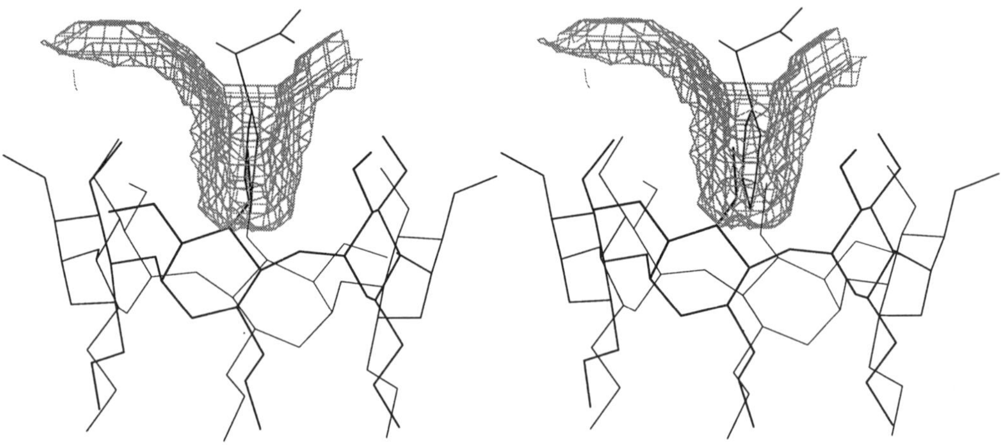

Figure 6. Stereo diagram of the observed structure of 2,3,6-(tri-O-methyl)-α-cyclodextrin and (S)-Mandelic Acid complex. The grid represents the exclusion volume surface map.

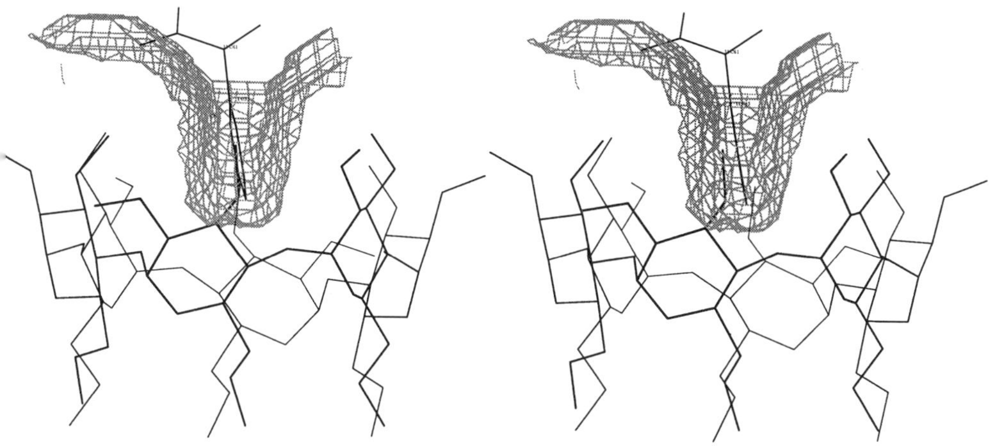

Figure 7. Stereo diagram of the observed structure and the exclusion volume map of 2,3,6-(tri-O-methyl)-α-cyclodextrin complexed to (S)-Mandelic Acid together with a model of (R)-Mandelic Acid .

In the case of the TMαCD-(R)-MA complex, it is impossible to fit the guest, sitting parallel to the macrocycle axis, into the available space left in the exclusion volume map (Fig.8).

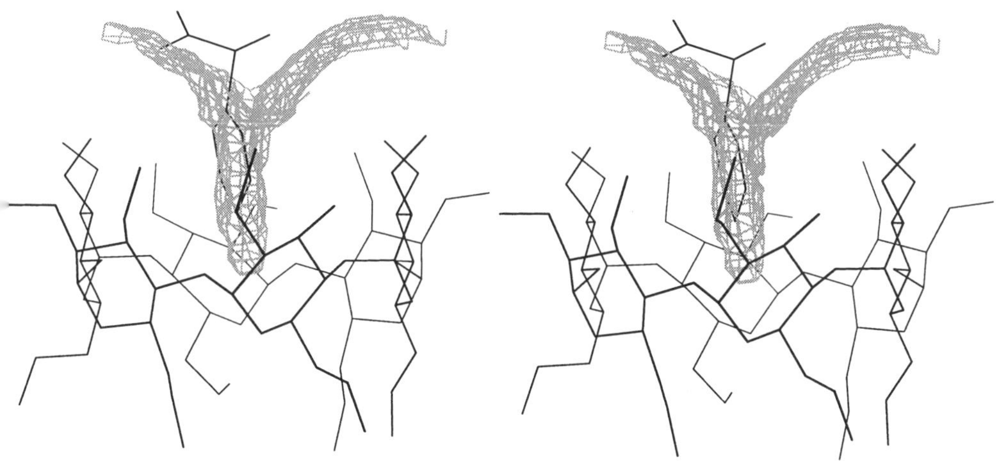

Figure 8. Stereo diagram of the observed structure and the exclusion volumemap of 2,3,6-(tri-O-methyl)-α-cyclodextrin complexed to (R)-Mandelic Acid together with a model of (R)-Mandelic Acid placed perpendicular to the plane of the cyclodextrin.

The same is true of course for the model of the (S)-MA enantiomer (Fig. 9). However, by tilting the model of the guest, (R)-MA, as in Fig. 5, it can be seen that a perfect fitting is observed. This is not possible for the (S)-MA enantiomer.

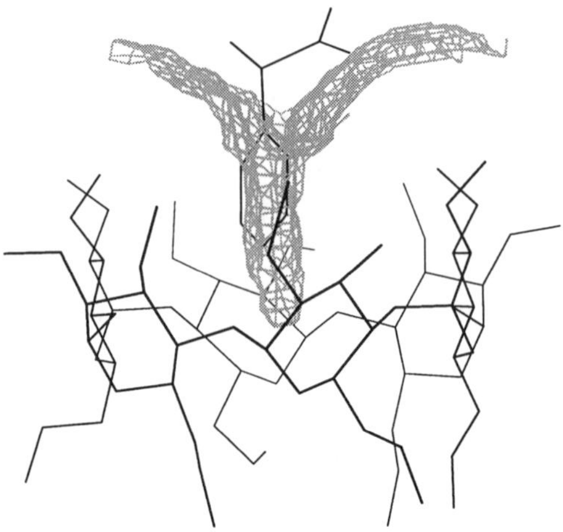

Figure 9. Stereo diagram of the observed structure of 2,3,6-(tri-O-methyl)-α-cyclodextrin and the exclusion volume of the TMαCD complexed to (R)-MA. together with a model of (S)-Mandelic Acid placed perpendicular to the plane of the cyclodextrin.

Thus, the exclusion volume surface represents very clearly the induced fit i.e. the conformational change of the host induced by (S) and (R) mandelic acids, especially (R)-MA, upon complexation in order to accomadate them more suitably in the cavity.

Moreover, the exclusion volume map can be used as a tool for molecular modelling studies since it provides an outline for the design of the wire model of the guest.

6. References

1. D. Mentzafos, K. Yannakopoulou and I. M. Mavridis Poster presented in this meeting.
2. N. Rysanec, G.Le Bas, F. Villain and G. Tsoucaris (1992) Acta Cryst. C48, 1466-14471.
3. K. Harata, K. Uekama, M. Otagiri and F. Hirayama, (1987) Bull. Chem. Soc. Jpn. 60, 497-502.
4. T.A. Jones, J-Y.Zou, S.W. Cowan and M. Kjeldgaard (1991) Acta Cryst. A47, 110-119.

AN INTERACTIVE-KINETIC APPROACH TO THE STUDY OF SUPRAMOLECULAR STRUCTURES.

E. ELIOPOULOS * and I. M. MAVRIDIS #
* Dept. of Agr. Biology and Biotechnology,
Agricultural University of Athens,
Athens 118 55,
GREECE.

Institute of Physical Chemistry,
NRCPS Demokritos,
Aghia Paraskevi 153 10,
Athens,
GREECE.

1. Introduction to the Practical

This practical makes use of the MAGE software [1] which allows molecules of known 3-dimensional structure to be displayed in a variety of ways and a limited range of measurements to be made. The entire image can be rotated in real time, parts of the display can be turned on or off, any point can be identified by picking it, and the change between different forms can be animated.

Cyclodextrins (CDs) are cyclic oligosaccharides produced by partial hydrolysis of starch. Best characterized are those built from 6,7,8 glucose residues (α,β,γCD) linked by α(1-4)glucosidic bonds. They have the shape of a truncated cone with a hydrophobic cavity at its center. The narrow rim is occupied by the primary i.e. O(6)H hydroxyl groups, while the wide rim by the secondary i.e. O(2)H and O(3)H ones [2,3] (Fig. 1) *{**Kinemage 1, View 1**}* . ♦

♦ Operations on the displayed kinemage respond within a fraction of a second: The image can be recentered, zoomed, put in stereo, or the front and back clipped away; distances, angles, and dihedrals can be measured. In the text it is indicated to what image (Kinemage) each figure corresponds. Although the views that have been chosen are appropriate, the pictures can be moved around.

G. Tsoucaris et al. (eds.), Crystallography of Supramolecular Compounds, 499–511.

They are important systems for the study of weak interactions because they serve as host molecules for a wide variety of compounds. Recently, because of their low cost, water solubility and natural origin they have come to prominence as carriers of drugs, essential oils and various key compounds in the pharmaceutical, food, cosmetic and other industries [4]. The packing and hydrogen bonding in the cyclodextrin family has been extensively used as prototypes in tutorials in previous supramolecular schools [5].

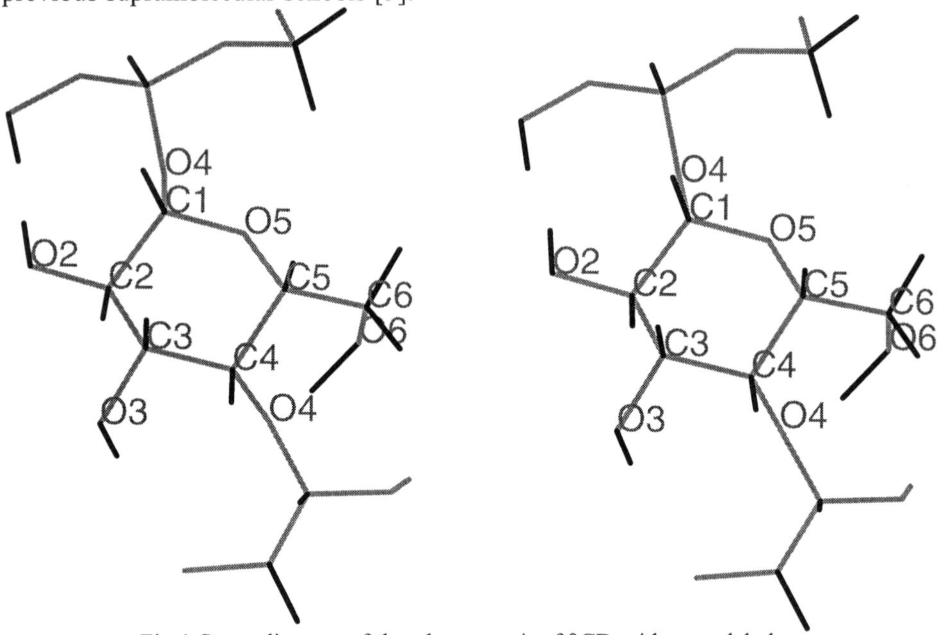

Fig.1 Stereodiagram of the glucose unit of βCD with atom labels.

β-Cyclodextrin is the most commonly used. Most complexes of βCD crystallize as dimers *{**Kinemage 2, View 1**}* in four space groups P1, C2, $C222_1$ and $P2_1$ (Q1, Q2) • [5,6]. The dimers in all known structures are formed by hydrogen bonds between the secondary hydroxyls O3-H (distances between the O3 atoms about 2.8Å), Q3. The βCD monomers are aligned in the dimer with their seven fold axes parallel. The dimers are linked together into infinite 2D layers by a dense network of hydrogen bonds involving primary and secondary hydroxyl groups as well as molecules of solvent, usually water (Fig. 2) *{**Kinemage3, View1**}*. These dimeric layers are perpendicular to the dimer axis; they have a C-centered or pseudo C-centered geometry, depending on the space group, with two orthogonal axes of dimensions close to 19Å and 24Å respectively (Fig. 3) and width that of one dimer (about 16 Å). A remarkable fact is the presence of a quasi-invariant two dimensional hydrogen bond network. The 3D crystal structure is built up by linking together the layers by hydrogen bonds involving primary hydroxyl groups and water molecules [6,7,8].

• At points indicated by Q_n the question listed in appendix is inserted.

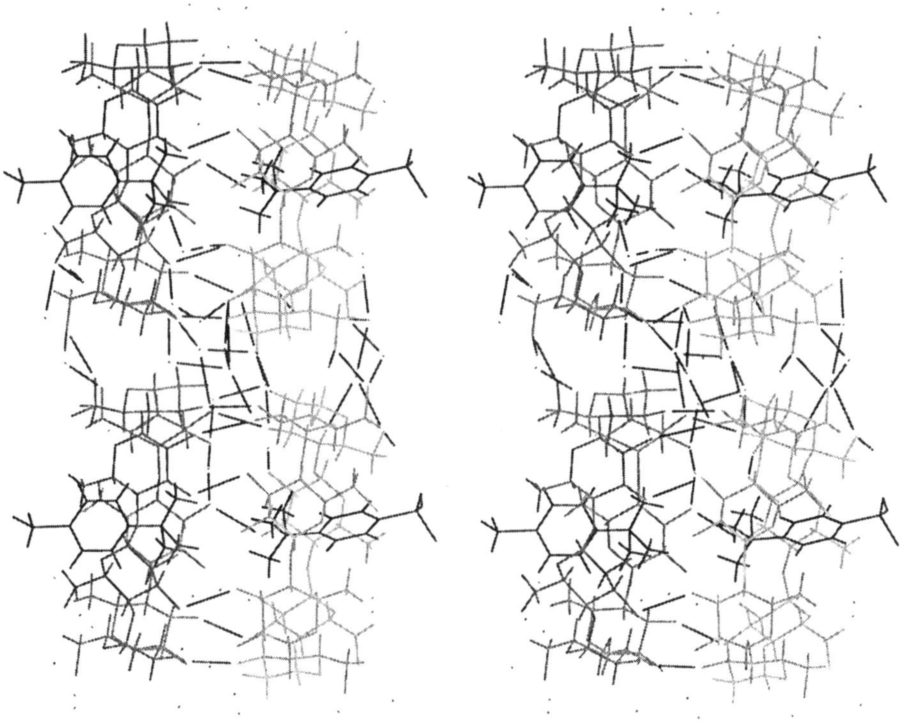

Figure 2. Stereo diagram showing the hydrogen bonding within the layer in the 4-tert-butyl-benzyl-alcohol βCD complex.

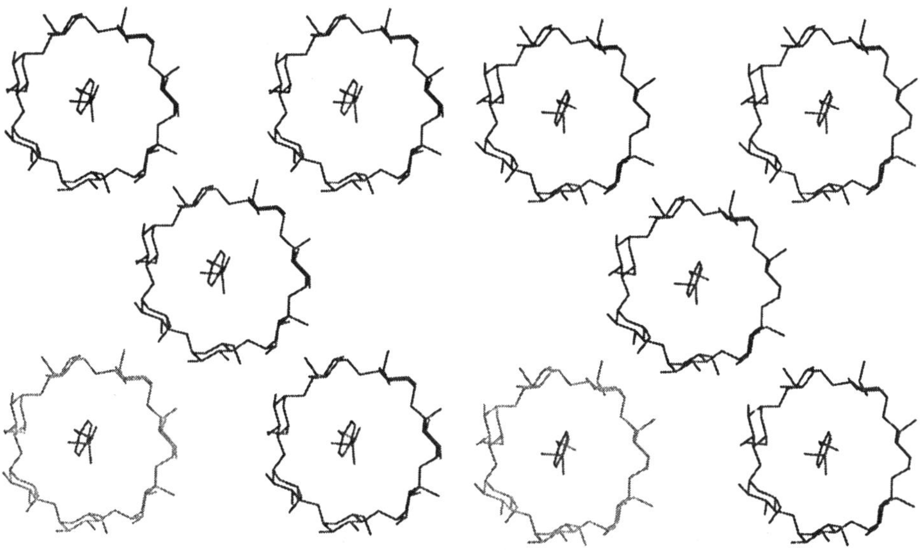

Figure 3. Stereo diagram showing the C-centering in the layer.

2. Packing Modes of β-Cyclodextrins

All known crystal structures of βCD dimeric complexes are generated from a superposition of the above mentioned practically invariant layers. They fall into four main classes which differ in the relative position of the layers.

2.1 CHANNEL PACKING (CH)

The channel forming dimers crystallize in space groups C2 or P1. The dimeric plane is almost parallel to the ab crystal plane. The dimer's 7-fold axis form an angle close to 10 degrees with the **c** axis along which they stack forming columns (Fig. 4) *{**Kinemage 4, View 1**}*. Thus the dimeric layers are translated along the **c** axis. The relative average shifting of two consecutive dimers, and thus layers, with respect to each other is 3.1 Å for P1 and 2.7Å for C2 structures. The channel is only slightly deformed at the interdimeric interface.

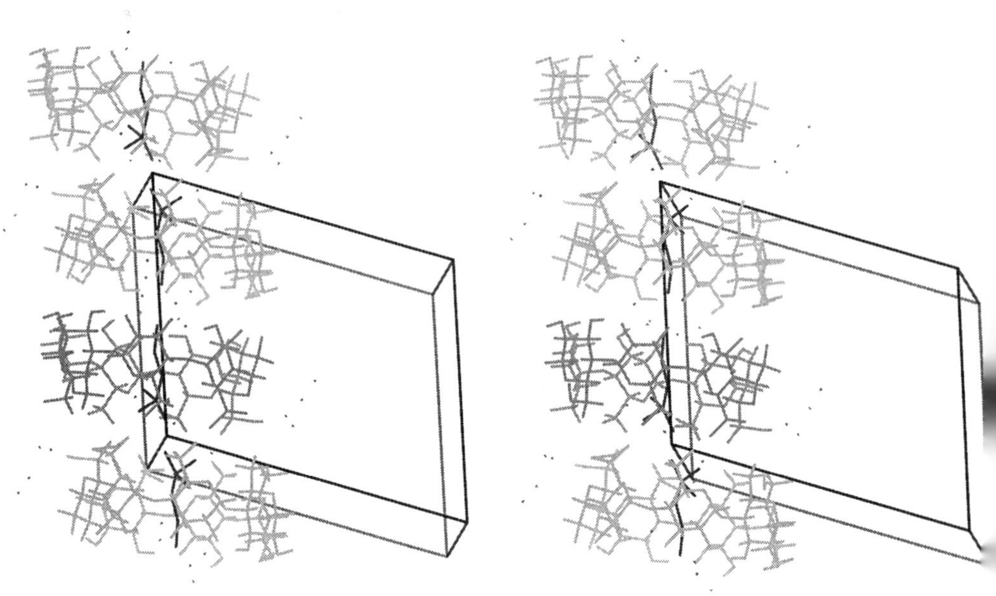

Figure 4. Channel packing. Stereo diagram of βCD- 3,3-dimethylbutylamine

2.2 CHESSBORD PACKING (CB)

In the **chessbord** packing the dimeric complexes crystallize in space group **$C222_1$**. The dimer's 7-fold axis form an angle of about 10 degrees with the **c** axis, thus the dimeric layers are almost parallel to the **ab** crystal plane. Adjacent layers are related by a screw 2-fold axis perpenticular to the **ab** plane. Each dimer is thus located above the space filled by solvent. This arrangement leads to columns with alternating CD dimer and solvent regions and confers on this class a chessbord appearance. Although the average planes of the O(4) heptagons of the same dimeric layer are still parallel (Fig. 5) *{**Kinemage 5, View 1**}* they form an angle of about 20 degrees with the O(4) planes of consecutive layers. The mean distance of the centers of two successive dimers related by the 2-fold screw axis projected on the **ab** plane is 8.7Å. Therefore, given that the βCD cavity radius is of the order of 4Å, there is almost no superposition of the cavity projections of two adjacent layers.

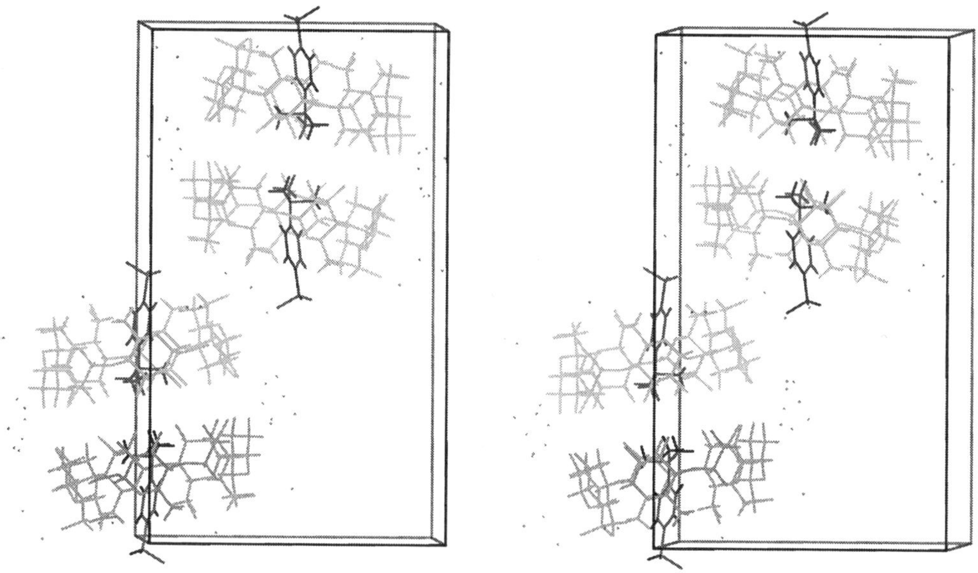

Figure 5. Chessboard packing. Stereo diagram of βCD-4-tert-butylbenzyl alcohol

2.3 INTERMEDIATE PACKING (IM)

In the **intermediate** class dimeric βCD complexes crystallize in space group P1. The dimeric layers are almost parallel to the **bc** plane. The layers are interrelated by a lattice translation along the **a** axis, as in the CH class but the 7-fold dimer axis forms an angle of about 20 degrees with the **a** axis and consequently adjacent layers are far from exactly aligned (Fig. 6) *{**Kinemage 6, View 1**}*. The relative average shifting of two consecutive dimers is 6 A, almost equal to the inner diameter of the cavity at the primary edge. The dimer cavity entrance is thus partially closed by the CD atoms of adjacent layers. Such a large shift results in discontinuity of the channel and justifies the term intermediate: between the CH and CB packing modes.

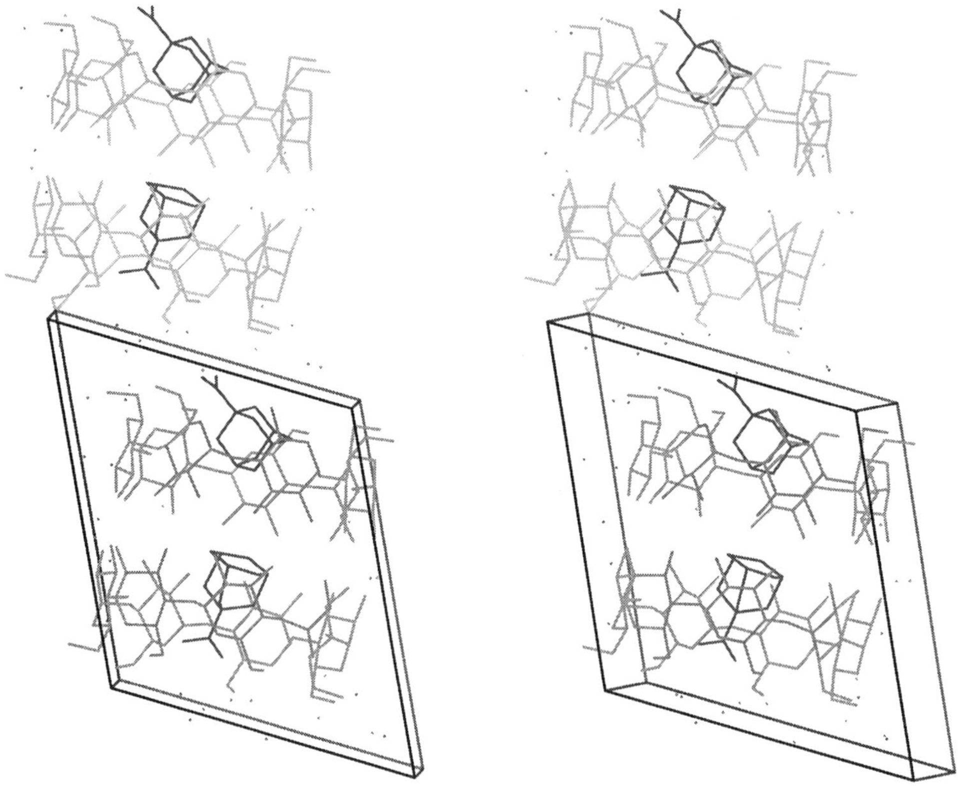

Figure 6. Intermediate packing. Stereo diagram of βCD-adamantene-carboxylic acid. [9]

2.4 SCREW CHANNEL PACKING (SC)

In the forth packing mode, SC, dimeric βCD complexes crystallize in space group **P2$_1$**. Adjacent layers, almost parallel to the **ac** plane and related by a 2-fold screw axis, form another variant of a deformed channel. In SC, each dimer packs above another dimer as in the CH mode but the channel has a 'screw' appearance because the concecutive dimers are not parallel but they form an angle of 20 degrees (Fig. 7) *{**Kinemage 7, View 1**}*. The mean distance of the projections of the centers of two such dimers on the **ac** plane is 2.6Å, similar to that of the CH packing mode.

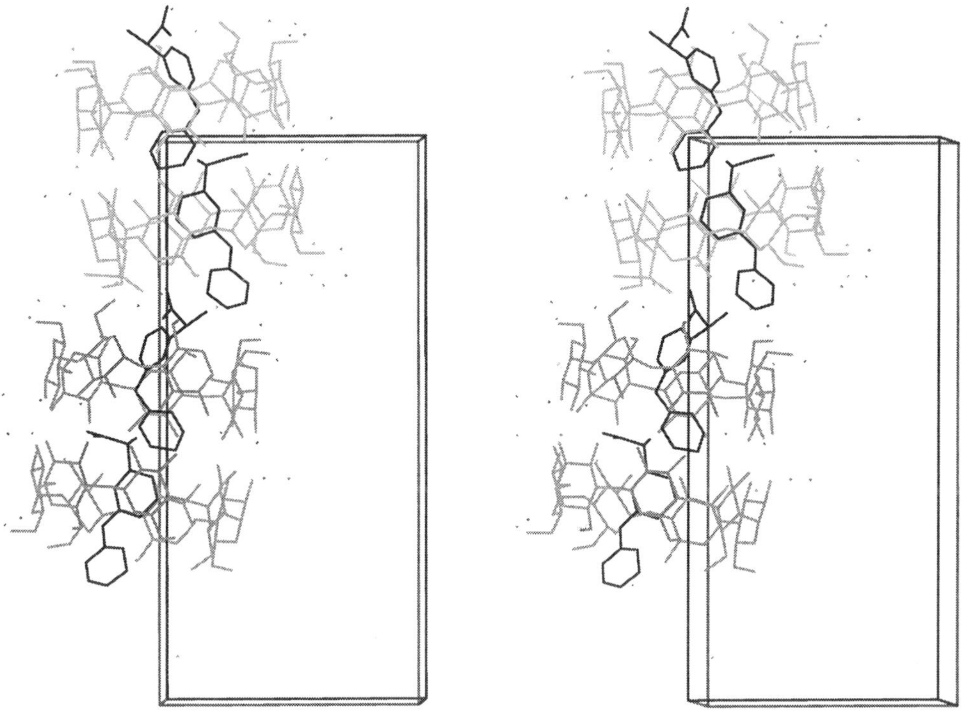

Figure 7. Screw Channel packing. Stereo diagram of βCD-fenoprofen.

3. Hydrogen Bond Networks

An intramolecular ring of hydrogen bonds between the secondary hydroxyl atoms O(2) and O(3) of adjacent glucose residues is found in cyclic CDs. However, βCD complexes forming dimers have in addition seven O(3)...O(3') H-bonds linking the two molecules of the dimer. There are overall 21 intramolecular and intradimer H-bonds (the acceptance of H-bond between two atoms is based both on the O...O distance of <3.1Å and on the value of the involved angles between 90-135 degrees).

In addition, βCD hydroxyl groups are involved in H-bonding with the surrounding solvent molecules, usually water. Despite the intrinsic disorder of the water of crystallization a quasi invariant water network exists [6, 7,8, 10] consisting of two separate sub-networks of H-bonds one involving the primary and another the secondary hydroxyl groups of the βCD dimer (Fig.2) *{**Kinemage 3, View 1**}*. Irrespective of the different packing arrangements, the H-bond network within a layer is very similar.

The secondary sub-network links the βCD dimers in the 2D dimeric layers. In most of the cases there are nine water molecules of the first hydration sphere per βCD dimer and two direct interdimer hydrogen bonds (Q4). One to three water molecules belonging to the second hydration sphere, link water molecules together and extend the H-bonding in the sub-network. The primary network involves hydroxyl-solvent and direct hydroxyl-hydroxyl interactions within the layer (Fig. 8) as well as hydroxyl-solvent interactions between two layers. However, in packing modes where the dimers come close together, as in the CH packing, there are also direct H-bonds between primary hydroxyl groups of adjacent dimeric layers (Q5).

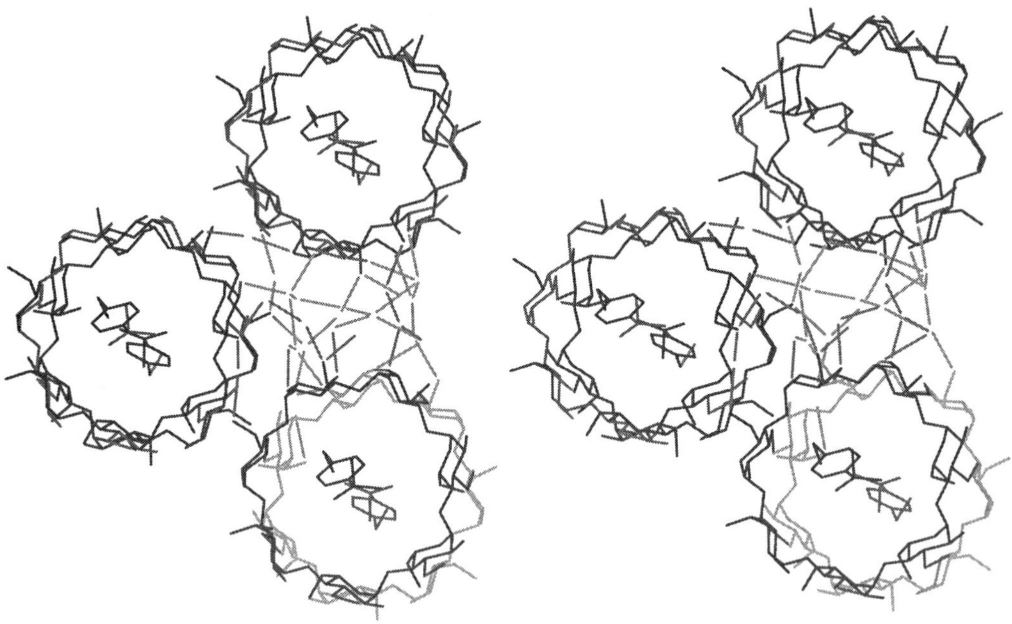

Figure 8. Stereo diagram of the hydrogen bonding networks between dimers.

The existence of the packing independent H-bonding network around the βCD dimers in the layers described above and, in particular, the quasi-invariant primary sphere of water molecules around each dimer, point towards the hypothesis that the latter might exist in solution as a stable entity after complex formation and prior to nucleation and crystallization. The nature of the guest and the mutual geometric arrangement of the host(s) and guest(s) in the complex is possible to affect the equillibrium:

$$\text{monomeric complex} \Leftrightarrow \text{dimeric complex.}$$

If we accept this hypothesis, it is easy to explain the invariability of the dimeric layers formed subsequently *{**Kinemage 8**}* ♦. The step which relates to the packing mode, will result from the layers according to the geometric and energetic requirement of the dimeric complex. The guest should play a crucial role in that step i.e. a guest with a hydrophilic, H-bond forming group will favor extensive hydration at the primary faces of the dimer and therefore will induce CB packing. On the contrary, a hydrophobic group at the primary faces will favor channel formation in order to be isolated from the aqueous environment. Guests having polar groups with intermediate tendency to be hydrated or with the tendency to associate with themselves, such as the carboxylic groups, pack in one of the other modes. The above hypothesis is also supported by the existence of complexes where only one guest is included inside the cavity of the dimer [11] or a guest molecule is found in the interdimer interface [12].

4. Chiral Recognition

βCD is composed of α-d-glucose and thus is an optically active compound that might be expected to show enantioselectivity towards enantiomeric guest molecules. Such a case has been observed in the inclusion complex of β-cyclodextrin with a racemic mixture of fenoprofen(FP) [13]. βCD forms dimers with one FP molecule per βCD monomer. Both enantiomers of FP exist in the crystal of the complex; each dimer contains only an R/R or an S/S pair of crystallographically independent FP monomers. However, the molecular ratio of S/R is 3:1 in the crystal which is confirmed by the change in the specific rotation of racemic FP-βCD complex after complexation.

♦ The stages of the assembly monomer ⟶ dimer ⟶ layer ⟶ crystal can be visualized at kinemage 8 by stepping through an animation procedure.

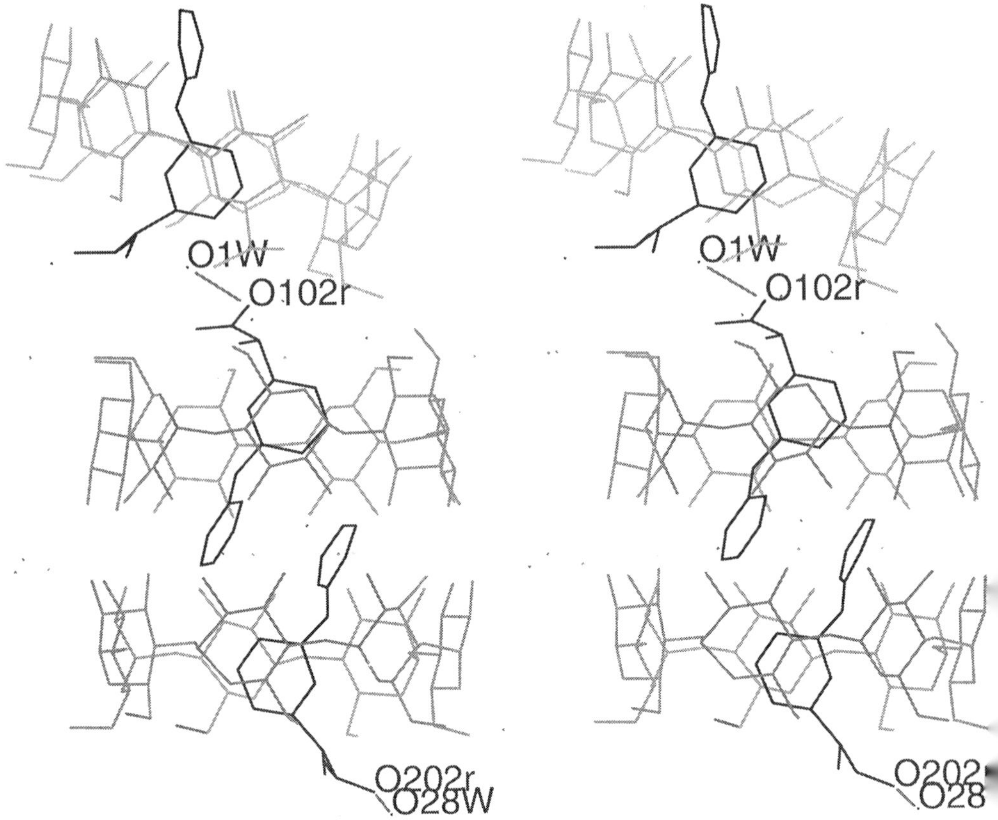

Figure 9. Enantioselectivity. Stereo diagram of βCD with R fenoprofen.

The structures of the individual R and S isomers with βCD have been solved also [14] (Fig. 9,10 respectively) *{**Kinemage 9, View 1**}*. The host-guest interactions are completely different in the R and S complexes, although the βCD is isomorphous in both complexes. The two (S)-FP enantiomers are found in a head to tail arrangement inside the dimer cavity. One of the (S)-FP carboxyl group found at the interdimer region forms a H-bond with a secondary hydroxyl group of βCD while the other at the primary region is H-bonded to a primary hydroxyl group of a neighbouring βCD molecule. On the contrary, in the dimeric structures of βCD with (R)-FP the two (R)-FP enantiomers of each dimer pack head to head with the two carboxyl groups at the primary faces of the dimer H-bonded to water molecules.

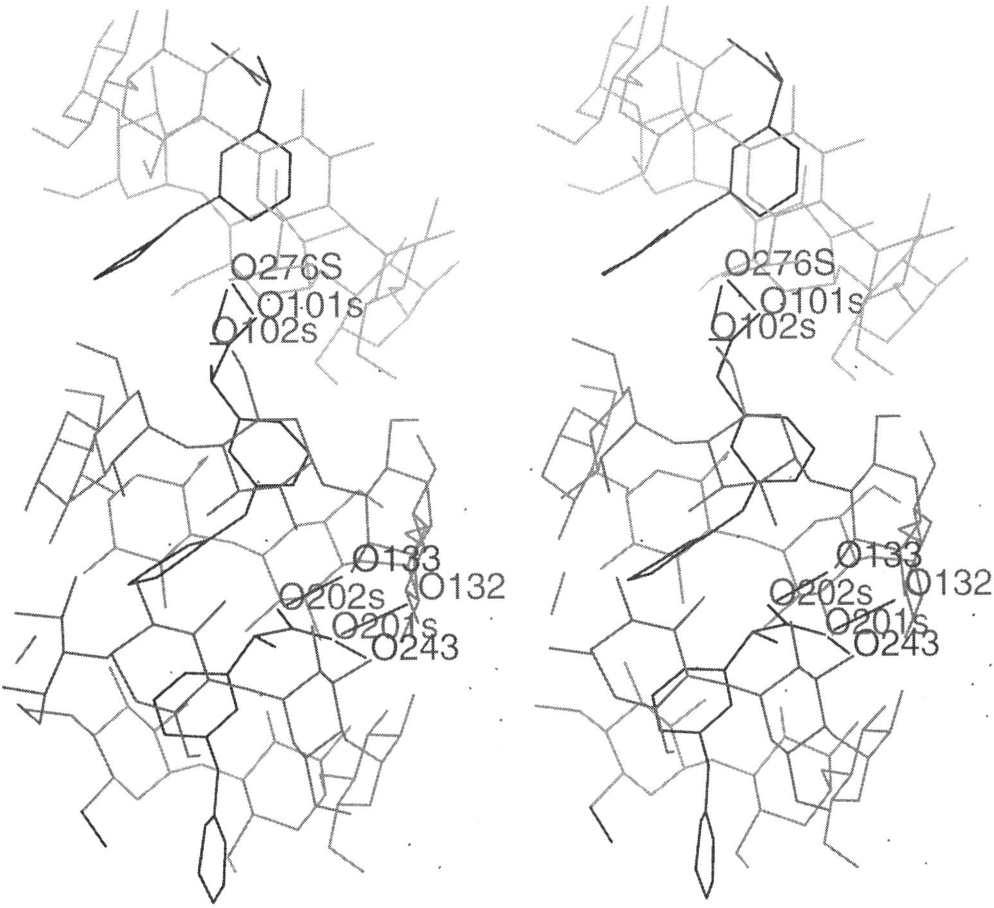

Figure 10. Enantioselectivity. Stereo diagram of βCD with S fenoprofen.

In the crystal of the racemic FP-βCD complex the conformations of (R) or (S)-FP and their position in the dimer cavity are the same as those found in the individual R and S complexes. One monomer of each βCD dimer holds the R or S enantiomers with their carboxyl groups at the primary side. The latter show the expected stable conformation (-CH3 and -COOH groups staggered with respect to the phenyl group and the hydrogen atom lying in its plane). They also occupy similar positions inside the βCD cavity. However, the carboxyl group of R is H-bonded to a water molecule because it is not able to form H-bonds with any hydroxyl group of the host, in contrast to the S enantiomer in which the carboxyl group is H-bonded to a primary hydroxyl group of an adjacent dimer.

The second βCD monomer includes the (R) and (S)-FP enantiomers quite differently. The latter finds it energetically more favorable to invert itself in the cavity and have its carboxyl end stabilized by H-bonds to three secondary βCD hydroxyl oxygen atoms of the interdimer region. This stabilization would not have been possible for the corresponding R enantiomer (the carboxyl group would point down the cavity rather than towards the side). Therefore, the latter orients its carboxyl group towards the exterior of the cavity in the primary hydroxyl region as the first R enantiomer (Q6).

Hamilton et al [10] claim that the differences in positioning of (R) and (S)-FP inside βCD are due to the hydrophobicity of the βCD cavity and the steric restrictions imposed by the tight fit of the guest inside the the latter, steric restrictions on the propionic acid substituent resulting in different orientations of the carboxylic group and subsequently different ways of forming H-bonds. In addition, the high binding affinity of (S)-FP to βCD which is approximately 3 times stronger than that of the (R)-FP is due to the strong H-bonds between the carboxyl groups and the βCD-hydroxyl groups, which cannot be formed with (R)-FP. Moreover, the methyl-phenyl interaction in the interface between the dimers is more favorable for the (S)-FP complex than in the propionic acid / propionic acid interaction (no H-bonds between the carboxylic groups) for the (R)-FP. Those differences are due to the steric restrictions imposed by the the tight fit of the guest inside the βCD cavity.

5. References

1. D.C. Richardson and J.S. Richardson (1992) Protein Science 1, 3-9
2. W. Saenger (1980), Angew. Chem. Int. Ed. Engl. *19*, 344-362.
3. G. Wenz (1994), Angew. Chem. Int. Ed. Engl. *33*, 803-822.
4. J. Szejtli (1988), "Cyclodextrin Technology", Kluwer Academic, Dordrecht.
5. G. Le Bas, Int.School on Supramol. Chem, Strasbourg, 1990.

6a. G. Le Bas, Doctorat d'Etat, Univ. Paris VI, 1985.

6b. Le Bas, G. and Rysanek, N. (1987) Structural Aspects of Cyclodextrins, in *Cyclodextrins and their Industrial Uses*, Chap.III, p.105 ; Ed. Duchкne, D., Editions de Santı, Paris.

7. D. Mentzafos, I. M. Mavridis, G. Le Bas and G. Tsoucaris (1991), Acta Cryst., B47, 746-757.
8. G. Le Bas and G. Tsoucaris (1994), Supramolecular Chem. *4*, 13-16.
9. J.A. Hamilton and M.N. Sabesan (1982), Acta Cryst. B38, 3063-3069.
10. G. Le Bas, N. Rysanek and G. Tsoucaris (1990), "Minutes International Symposium on Cyclodextrins, Paris, p. 114, Paris: Editions de Sante; G. Le Bas, G. Tsoucaris (1986), Mol. Cryst. Liq. Cryst. *137*, 287-301.
11. I. M Mavridis, K. Yannacopoulou, J. Papaioannou, Poster in this School.

12. A Rontoyianni and I. M. Mavridis (1994), J. of Inclusion Phen. and Mol. Recognition 18, 211-227.
13. J. A. Hamilton and L. Chen (1988), J. Am. Chem. Soc., 110, 5833-5841.
14. J. A. Hamilton and L. Chen. (1988), J. Am. Chem. Soc., 110, 4379-4391.

6. Appendix

The questions inserted at points indicated by Q_n .

Q1: *\{Q: Identify the origin and the three axes of the cell. To which axis does the 7-fold axis of the dimer run parallel? (The X , Y, Z, other)\}*.
Q2: *\{Q: Identify the origin and the three axes of the cell. To which axis does the 2-fold axis of the dimer run parallel? (The X , Y, Z, other)\}*.
Q3: *\{Q: The intradimer H-bonds are between O3n...O3'(8-n). Can you identify those H-bonds and give an average length of them?\}*.
Q4: *\{Q: Identify direct H-bonding between dimers in the same layer using Kinemage 3. Click on two atoms forming such H-bonds and indicate their distance.\}*.
Q5: *\{Q: Identify direct H-bonding between layers using Kinemage 4. Click on two atoms forming such H-bonds and indicate their distance.\}*.
Q6: *\{Q: Identify direct H-bonding interactions of FP with βCD and find the length of H-bonds using Kinemage 9.\}*.

Subject Index